Power Electronics and Power Systems

Series editors

The Power Electronics and Power Systems Series encompasses power electronics, electric power restructuring, and holistic coverage of power systems. The Series comprises advanced textbooks, state-of-the-art titles, research monographs, professional books, and reference works related to the areas of electric power transmission and distribution, energy markets and regulation, electronic devices, electric machines and drives, computational techniques, and power converters and inverters. The Series features leading international scholars and researchers within authored books and edited compilations. All titles are peer reviewed prior to publication to ensure the highest quality content. To inquire about contributing to the Power Electronics and Power Systems Series, please contact Dr. Joe Chow, Administrative Dean of the College of Engineering and Professor of Electrical, Computer and Systems Engineering, Rensselaer Polytechnic Institute, Jonsson Engineering Center, Office 7012, 110 8th Street, Troy, NY USA, 518-276-6374, chowj@rpi.edu.

More information about this series at http://www.springer.com/series/6403

Sarma (NDR) Nuthalapati
Editor

Power System Grid Operation Using Synchrophasor Technology

Springer

Editor
Sarma (NDR) Nuthalapati
PEAK Reliability
Vancouver, WA
USA

ISSN 2196-3185 ISSN 2196-3193 (electronic)
Power Electronics and Power Systems
ISBN 978-3-030-07754-9 ISBN 978-3-319-89378-5 (eBook)
https://doi.org/10.1007/978-3-319-89378-5

Printed on acid-free paper

This Springer imprint is published by the registered company Springer International Publishing AG part of Springer Nature
The registered company address is: Gewerbestrasse 11, 6330 Cham, Switzerland

This book is dedicated to the memory of
Dr. James S. Thorp
....who is one of the pioneers in the
development of PMUs and
Synchrophasor Technology....

(February 7, 1937—May 2, 2018)

Foreword

I am pleased to write this foreword to the volume "Power System Grid Operation Using Synchrophasor Technology", edited by Dr. Sarma (NDR) Nuthalapati. The contributors to the volume are outstanding researchers in the field of phasor measurement units (PMUs) and wide-area measurement systems (WAMSs). Their papers clearly show that this technology has reached a high degree of maturity, and that the use of PMUs as modern instruments of precise measurements of power system voltages and currents is well accepted by power systems throughout the world. As the person involved with PMU development from mid-1970s to the present time, I am particularly happy that this technology has received universal acclaim as the measurement system of choice for power systems.

Now that the PMU and WAMS technology is mature, the researchers' attention naturally shifts to applications of these measurements. State estimation using PMU measurements has always been the most promising application—one that can be implemented as soon as enough measurements become available. In this present volume, Chaps. 13 and 14 discuss this subject in detail. It is gratifying to see papers from India. I participated in Indian WAMS project in an advisory capacity to a small extent. Chapters 8, 16, and 17 deal with developments in India, one of the major efforts to launch WAMS on a large power grid. China has also seen major developments in WAMS technology. Chapter 5 provides a comprehensive treatment of these developments.

A natural extension of the state estimation effort is to involve WAMS measurements in power grid operations. Chapter 1 provides a good introduction on the applications of PMU and WAMS to managing grid operations using synchrophasor technology. Chapters 2 and 3 consider the impact of data quality from these measurements on grid operations.

The effort to use PMU measurements for control purposes has been a dream for many researchers, but control applications using WAMS have been somewhat slow in evolving. Chapter 4 discusses BPA view of wide-area monitoring and wide-area control. It should be noted that BPA was one of the first power systems to start using PMUs on their system when Macrodyne PMUs became commercially available.

Sustained oscillations on the power grid were discovered to be widespread when PMUs were placed on the power grids. It should be noted that before the advent of PMUs, the existence of such sustained oscillations was not commonly known. Chapters 6–11 consider the evaluation of stability and sustained oscillations taking place on the power grid due to various triggering events. Characterizing the modes of these oscillations and the strategies for damping or eliminating these oscillations has turned out to be an important topic.

Another topic which has received considerable attention is the use of WAMS data for improving the protection functions—particularly those protection tasks which are slow and have time to access PMU measurements. It is noted that only one chapter in this collection deals peripherally with this topic: Chap. 16 which deals with remedial action schemes in the presence of PMU measurements.

I hope this volume will be read carefully by power system engineers throughout the world and will lead to more applications of this technology in the coming years.

Wilsonville, Oregon
January 2018

Arun G. Phadke

Preface

When I started my career in Power Systems in the early 1990s in India, we just started to hear of PMUs (Phasor Measurement Units). I was intrigued, but I did not get a chance to work closely with that synchrophasor technology until a few years ago while working at the Electric Reliability Council of Texas (ERCOT). I was involved in a Synchrophasor project funded by US Department of Energy (DOE) at ERCOT. It was exciting to finally see the scope of PMUs data, especially as it applied to real operational use. Visualizing PMU data allowed us to detect oscillations and in turn mitigate them through corrective actions. Best of all was being able to share our experiences through North American SynchroPhaor Initiative (NASPI) working group meetings and at the IEEE Power and Energy Society (PES) General Meetings. These platforms provided a forum to discuss how synchrophasor data is used in managing the grid in different regions.

In 2014, I proposed to have an IEEE PES Magazine issue focusing on the use of synchrophasor technology in managing the grid. The editorial board readily agreed and Dr. Arun Phadke, a Board Member himself, agreed to help me on the issue. Our work resulted in the September/October 2015 PES Magazine Issue which included articles from across the world with experiences of using synchrophasor technology. Dr. Phadke, along with Dr. James Thorp, invented PMUs and thus, working with him was a wonderful experience. It was that experience that later gave me the idea of this book.

It is sad that Dr. Thorp passed away on May 2, 2018. I would like to dedicate this book to his memory.

As technology developed and funding for deploying synchrophasor technology increased, new there are large scale implementations of Synchrophasor Technology in managing the grid across the world (China 4000 PMUs, India 1800 PMUs, North America 2000 PMUs, etc.). Efforts are in progress to take this technology into control center operations and develop operational procedures to better manage the grid with wide-area visualization tools using PMU data.

The key benefits of this technology include the following:

- Wide-area visualization
- Oscillation detection and monitoring
- Phase-angle monitoring
- Generator model and parameter validation
- Island detection
- Event detection
- Post-event analysis
- Voltage instability monitoring
- Identification potential malfunction of devices in the grid

This book discusses the use of synchrophasor technology in managing the grid. It is organized into 20 chapters written by professionals from different organizations who are using synchrophasor technology. These chapters cover PMU data quality and applications of synchrophasor technology for oscillation detection, linear state estimation, phase angle alarming, event analysis, operation of HVDC systems, validation of remedial action schemes, etc.

Chapter 1 provides the context for the other chapters and explains the importance of using synchrophasor technology in managing the grid.

Chapter 2 discusses the impact of PMU data quality and suggests methods for validating data in real-time. It also presents effects of various data impairments and provides examples for different kinds of data errors. Recommendations for error control and detection are also included through all sections of this chapter.

Chapter 3 presents procedures for testing synchrophasor devices and applications before field deployment. Specifications of the testing equipment required for PMU testing Lab and architecture of the test bed are presented. Test bed environment is needed before field deployment of any synchrophasor applications and this chapter provides a good overview of testing and validation of synchrophasor device and applications.

Chapter 4 discusses the details of deployment of synchrophasor technology at Bonneville Power Administration (BPA). BPA was among the first adopters of the synchrophasor technology in the early 1990s. This chapter describes in detail how synchrophasor technology is deployed and used in managing the grid at BPA. It provides the details of control room applications of synchrophasors at BPA. The value realized from BPA Synchrophasor Project are also discussed in this chapter.

Chapter 5 discusses the use of synchrophasor technology in China. It discusses the communication and synchrophasor network and presents the details of some of the major advanced applications deployed in China utilizing synchrophasor measurement technology.

Chapter 6 presents a generalized approach to identify signature oscillations in any system. It provides the details of the data mining for oscillations on the Electric Reliability Council of Texas (ERCOT).

Chapter 7 discusses use of synchrophasor technology for Oscillation Detection at Electric Reliability Council of Texas (ERCOT). It discusses some of the oscillation events that happened in their system and demonstrates on how synchrophasor technology can help to detect and mitigate oscillations in the system.

Chapter 8 discusses use of synchrophasor technology for oscillation detection and mitigation in the India Power Grid. It discusses different cases of oscillations and provides the details of measures taken to improve the damping of oscillations.

Chapter 9 presents the experiences of oscillation detection and mitigation in grid operations at PEAK Reliability which is the reliability coordinator of the Western Interconnection. It provides the details of implementation of tools for oscillation detection and source location.

Chapter 10 discusses online oscillations management using synchrophasor technology at ISO New England. It presents observed oscillations in their grid, forced oscillation mitigation, and the source locating application which is an online tool for robust estimation of the source of sustained oscillations.

Chapter 11 discusses the use of synchrophasor technology for oscillation monitoring at California ISO (CAISO). It provides the architecture for gathering synchrophasor data for the purpose of oscillation monitoring in CAISO.

Chapter 12 discusses the use of synchrophasor technology for wide-area power system phase angle monitoring and provides the details of implementing the use of phase angle monitoring at CAISO.

Chapter 13 discusses synchrophasor-based Linear State Estimation (LSE) Techniques. It provides details of both a linear weighted least-squares (WLS) and a robust linear least-absolute-value (LAV) power system state estimator in both a positive sequence and a three-phase formulation. It also presents several applications that employ estimation techniques and/or estimation results were presented including dynamic load modeling, exciter failure detection, and symmetrical component calculation.

Chapter 14 discusses some of the implementations of synchrophasor based linear state estimator. It provides some of the practical LSE use cases of real-time operations. It presents some of the use cases at Bonneville Power Administration (BPA), Duke Energy and Southern California Edison (SCE). These use cases demonstrate that LSE can improve synchrophasor data quality and provide reliable and accurate state estimation solution for synchrophasor applications. It also illustrates the effectiveness of using LSE to expand PMU observability which would benefit the downstream applications.

Chapter 15 presents experiences by ERCOT in performing post-event analysis based on PMU data. It also presents a case involving a compound event in the ERCOT system involving a fault, which, due to relay misoperation, induced a loss of generation event.

Chapter 16 describes case studies to explain the usage of Synchrophasors for evaluation and review of Special Protection Schemes (SPS) in Indian Grid. It also demonstrates how grid operator has utilized the Synchrophasor data in improving the system performance under various contingencies and laid down the foundation for effective utilization of PMU data for evaluating SPS operation, modifying SPS design and designing as new SPS.

Chapter 17 discusses the utilization of synchrophasors in control room application for both real-time dispatch and post-dispatch analysis. It presents different case scenarios which illustrate the benefit of synchrophasors in understanding behavior of HVDC Systems. Apart from easily identifying HVDC restarts, Reduced Voltage Operation (RVO) mode, HVDC islanding, synchrophasors helped in understanding voltage changes associated with filter bank switching.

Chapter 18 explains the use of synchrophasor technology to aid model validation and improve accuracy of models. It describes how to use synchrophasor measurements for validation of power system models used for power systems planning and operations.

Chapter 19 presents the details of a software suite named Grid Stability Awareness System (GSAS) that has been developed to monitor and analyze power grid stability in real-time using wide-area synchrophasor measurements. GSAS consists of five analytical and monitoring tools including an Oscillation Monitoring Tool, a Voltage Stability Monitoring Tool, a Transient Instability Monitoring Tool, an Angle Difference Monitoring Tool and an Event Detection Tool. System architecture and functionalities of the tools are described. The alarming mechanisms implemented in GSAS are also presented. In addition, a series of off-line simulations were developed to test, evaluate and validate the immediate suitability of the tools for use in an operational environment of a North American utility, and to identify and support potential improvements in the tools. The testing methodology and the procedures to create the simulation cases are also presented.

Chapter 20 discusses a new model for PMU data sharing, based upon today's cloud computing technology base. Cloud-based data capture, archiving, analysis and sharing represents a new paradigm, and provides access to flexible resources well-suited to intermittent bursts of heavy computational work. In addition, collaboration among entities could be greatly facilitated by hosting common applications in the cloud, with the further assurance when different operators examine the same data, they will see consistent information. The chapter presents GridCloud, a new cloud-hosted synchrophasor data sharing platform and demonstrates security, scalability, low latency and cost effectiveness of the model.

All these chapters provide a complete overview of synchrophasor technology and its role in managing the grid across different parts of the world. It is evident that this technology is now mature and is surely improving reliability of the grid. I am hoping that this book will also serve as good reference material to different utilities who are looking forward to using this technology in managing their grid. It also serves as a textbook to teachers and students in understanding the benefits of synchrophasor technology in managing the grid.

I would like to thank Dr. Joe Chow, the Editor of Springer Power Electronics and Power System Series and Springer Publishers for agreeing to my proposal on this book and for publishing it. I hope this will benefit professionals interested in synchrophasor technology.

I would like to thank all the authors for their time and efforts in preparing the chapters for this book. I am grateful to Dr. Arun G. Phadke for kindly writing the foreword. I would also like to thank my former colleagues Bill Blevins, Sidharth Rajagopalan, Patrick Gravois, Tom Mortensen at ERCOT and my colleague Hongming Zhang and other team members at PEAK Reliability for their support in my learning of synchrophasor technology. I am enjoying not only learning, but sharing and contributing to the community through IEEE and NASPI activities. Organizing panel sessions/tutorials, preparing focus area documents, etc., in promoting this technology for managing the grid has been an honor. I would like to thank all the professionals who actively participate at these meetings; I learn a great deal from them.

Today, I am what I am because of my teachers at the Railway High School (Kazipet), Regional Engineering College (Warangal) and the Indian Institute of Technology (Delhi) in India and my dear parents Sri N. Hanumantha Rao and Smt. Kamala Devi. I am indebted to all of them for their guidance. They have made me a good engineer and a good human being. I also appreciate my wife Vasudha and our daughter Sruti for their understanding and constant patience regarding my passion to my professional life.

Vancouver, USA
9th May 2018

Sarma (NDR) Nuthalapati

Contents

Chapter 1
Importance of Synchrophasor Technology in Managing the Grid

Jeff Dagle

1.1 The Value of High-Speed Time-Stamped Data

Historically, wide-area electricity infrastructure telemetry utilized a technology called supervisory control and data acquisition (SCADA). Developed over the span of several decades, SCADA initially pushed the technological envelope as the state-of-the-art means for providing remote monitoring and control. SCADA provides the telemetry between a utility's control center and their substations and power plants, spanning vast geographical distances. Supervisory control refers to communicating set-points (such as voltages at substations and power plants) or switching (e.g., controlling circuit breakers on transmission lines). The data acquisition portion of SCADA is referring to the collection of measurement information from substations or power plants. This could include parameters such as voltages and line flows. The data rate of SCADA was driven primarily by technological limitations of the time, associated with available communications bandwidth. Most SCADA systems collect data once every few seconds, typically 2 or 4 s between measurements. The data is gathered through a polling architecture, whereby each substation reports its information to the master front-end processors sequentially. Time stamps are then applied at the point of assembly at the control center, which may not necessarily coincide exactly with the events being measured, particularly with timing delays inherent in wide-area communications. Therefore, it is common for SCADA measurements to have errors on the order of up to a few seconds between when the power system event occurs and the corresponding time stamp of the data record as it is received at the control center.

Synchrophasors address both of these issues—the relatively slow data rate of SCADA and the inherent timing errors associated with the measurements.

J. Dagle (✉)
Pacific Northwest National Laboratory, Richland, USA
e-mail: jeff.dagle@pnnl.gov

S. Nuthalapati (ed.), *Power System Grid Operation Using Synchrophasor Technology*, Power Electronics and Power Systems,
https://doi.org/10.1007/978-3-319-89378-5_1

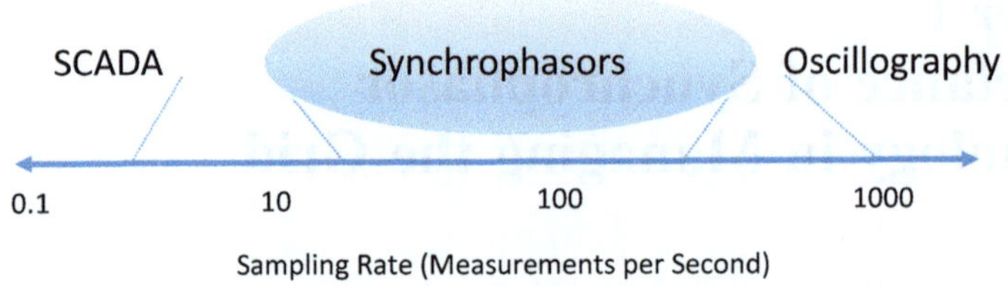

Fig. 1.1 Synchrophasors fill a gap between traditional SCADA measurements and high-speed oscillography (point-on-wave measurements). Figure credit: author

Beginning as a research topic in the 1980s, and accelerating with utility-scale deployments through the 1990s and beyond, utilities were able to leverage advances in precise wide-area time synchronization enabled by systems such as global navigation satellite systems, and higher-bandwidth communication technologies enabled by Internet-era advances.[1] Similar to SCADA technology in the sense that the data collection equipment resides in the substation, and the parameters that are measured include power system quantities such as voltages at the bus and current flows on the lines, the Phasor Measurement Unit (PMU), enabled by wide-area time synchronization, also provides a measurement of the relative phase angle of these parameters (in addition to the magnitude). Thus, a synchrophasor measurement is like a vector; it contains both magnitude and angular information.

One of the earliest benefits of this technology was more granular measurement compared to traditional SCADA measurements. Data measurement rates of 30 samples per second are necessary to provide visibility for dynamic phenomena that might occur up to about 15 Hz (or less depending on the sampling techniques and measurement filters). More modern systems can have even higher sampling rates, which enable measurement of higher-frequency phenomena. While 30 samples per second are still quite common today, some utilities have deployed synchrophasor systems with 60 or 120 samples per second. Even higher speed systems are being investigated.

Figure 1.1 depicts the typical measurement range for synchrophasor technology. It measures more data than slower traditional SCADA measurements, and is more amenable to wide-area applications than higher speed "point-on-wave" oscillography that is used for applications such as fault recorders and power quality meters that measure multiple times per cycle on the 60 Hz waveform.

The value of dynamic measurements is depicted in Fig. 1.2, which shows a power system parameter (such as a voltage or line flow) that contains dynamic information. These dynamics are clearly evident at 30 measurements per second, but are not measurable at a 4 s scan rate. The dynamic signal shown in Fig. 1.2 is

[1]Phadke, A.G. (Virginia Polytechnic Institute and State University) (2002-10-10). "Synchronized phasor measurements-a historical overview" (pdf). Transmission and Distribution Conference and Exhibition 2002: Asia Pacific. IEEE/PES. Institute of Electrical and Electronics Engineers (IEEE). 1: 476. https://doi.org/10.1109/tdc.2002.1178427.

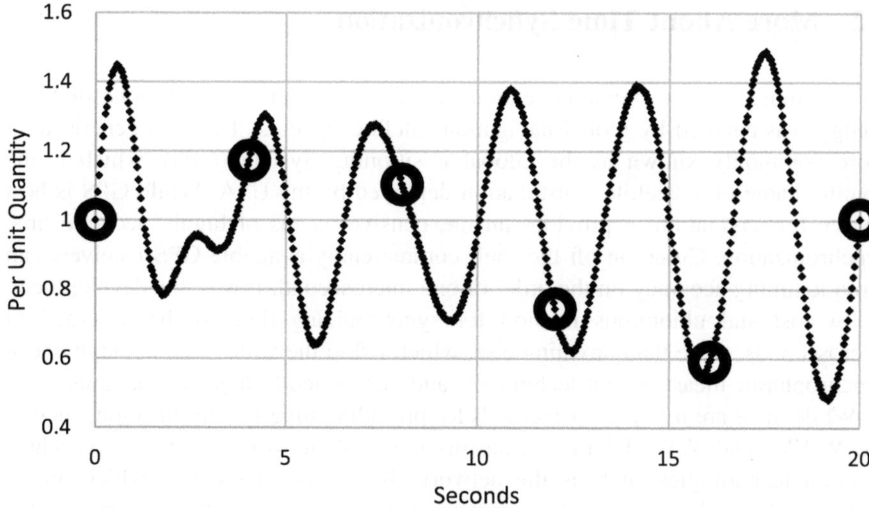

Fig. 1.2 Notional representation of the difference between synchrophasor and SCADA measurement. Figure credit: author

fictitious, but is representative of the electromechanical modes of oscillation present in all large interconnected systems. Particularly in the Western Interconnection of the North American power system, low-frequency oscillatory dynamics have been a well-known and troublesome aspect of system reliability for many years. Interarea electromechanical modes of oscillation around 0.3 Hz were observed in this power system dating back to the 1970s.[2] Over the years, other power systems around the world have also experienced operational challenges associated with these oscillations. They tend to manifest in large interconnected power systems that transfer power over vast distances. These oscillations become critically important if they are poorly damped. Having a means to directly observe these oscillations is a valuable benefit of high-speed measurements, described in greater detail later in the chapter.

Furthermore, synchronizing the measurements at the substation device more accurate time-stamping of the data. The time stamps are passed along with the measurements to the data collection point. This enables more accurate data representing events as they are occurring on the system, which is critical for time-aligning sequences of events, or characterizing the system's dynamic characteristics.

[2]Cresap, RL and JF Hauer. "Emergence of a New Swing Mode in the Western Power System", IEEE Transactions on Power Apparatus and Systems, vol. PAS-100, no. 4, pp 2037–2043. April 1981.

1.2 More About Time Synchronization

As mentioned in the previous section, the development of synchrophasor technology was enabled by global navigation satellite systems. This is a generic term more commonly known as the Global Positioning System (GPS), which is the specific name of a satellite constellation deployed by the USA. While GPS is best known for navigation, it provides an inexpensive means of highly accurate time synchronization. Common off-the-shelf commercially available GPS receivers can provide timing accuracy on the order of one microsecond. It was this development, a low-cost and ubiquitous method for synchronizing time to the accuracy of microseconds, more than anything else, which led to the wide-scale deployment of synchrophasor measurement techniques, and consequently their applications.

While there are many other methods for providing time synchronization, such as the WWV[3] and WWVB[4] radio transmission, and network-based time synchronization technologies such as the network time protocol (NTP),[5] which are in widespread use, these approaches can provide time synchronization accuracy in the millisecond range. This is insufficient to provide the phase-angle calculation that is a crucial element of the synchrophasor technology. Newer technologies, such as the Precision Time Protocol,[6] are capable of providing network-based time synchronization in the microsecond accuracy range, and hence are suitable for use synchronizing the measurements. Some PMU deployments are exploring the use of this technology. Other methods of accurate timing have been proposed, such as other satellite-based systems, eLORAN[7] (a terrestrial-based radio system), and atomic clocks (that provide adequate accuracy and holdover time stability). These other methods may eventually be deployed when their cost and adoption challenges have been addressed.

One of the key value propositions for PMUs using these previously discussed methods of highly accurate time synchronization is the direct measurement of relative phase angle between different buses in the power system. With microsecond timing accuracy, it became possible to measure phase angle to less than a degree of measurement uncertainty given that power system parameters (e.g., voltage, current) vary 360° each 1/60th of a second for a 60 Hz system. Therefore, theoretically, a one microsecond error corresponds to 0.0216° of uncertainty in the

[3]The radio station with call sign WWV, operated by the National Institute of Standards and Technology (NIST), broadcasts time and frequency information from Fort Collins, Colorado.

[4]Located at the same location as WWV, the WWVB broadcasts are used to synchronize consumer timekeeping applications.

[5]NTP has been widely used for synchronizing computer networks for many years.

[6]IEEE Standard 1588 provides a specification for the precision time protocol, which can provide network time synchronization with microsecond accuracy.

[7]The enhanced LOng RAnge Navigation (eLORAN) system has been proposed as a follow-on radio-based navigation technology initially developed in the 1940s and retired by the United States in 2010 (reference: https://www.gps.gov/policy/legislation/loran-c/).

phase-angle measurement. This provides sufficient accuracy for any phase-angle measurement application that is being contemplated today. Prior to this, phase angles were computationally estimated.

More details on these aspects of the technology will be revealed as we discuss the various synchrophasor applications below. They will also be covered in greater detail within subsequent chapters of this book.

1.3 The Principal Applications and Benefits of Synchrophasor Technology

Situational Awareness and Wide-Area Monitoring: Synchrophasor data enables grid operators to visualize the power system across an entire interconnection, understand grid conditions in real time, and diagnose and react to emerging problems. Synchrophasor-enabled visibility, and the deployment of modern detection technologies to alert system operators to impending system instability, can reduce the likelihood of blackouts.

For example, while still a new technology, having WAMS data available to support the investigation of the August 10, 1996 system separation event in the Western Interconnection of the North American grid was fortuitous, and provided a significant impetus for the further deployment of advanced measurement technology.[8]

As synchrophasor data coverage continues to improve, more data are being integrated into existing control applications, gaining broader acceptance for synchrophasor-enhanced wide-area monitoring.

Real-Time Operations: Synchrophasor data is being used today to improve state estimator models for better understanding of real-time grid conditions. It is being used to detect and address grid oscillations and voltage instability, and integrated with SCADA data, to drive real-time alarms and alerts. Analysts are looking at PMU data to expedite resolution of operating events such as fault location, and to quickly diagnose equipment problems such as failing instrument transformers and other system imbalances.

More advanced applications are considering synchrophasor data as an input to special protection systems (SPS) or remedial action schemes (RAS), and can trigger automated equipment controls. Phase-angle data has already been used to monitor and manage system islanding and black-start restoration.[9] More of these applications are being deployed.

---

[8]Kosterev, DN, CW Taylor and WA Mittelstadt. "Model Validation for the August 10, 1996 WSCC System Outage" IEEE Transactions on Power Systems (Volume: 14, Issue: 3, Aug 1999).

[9]Galvan, F., M. Sujit, and M. Thomas. "Phasor Measurement Units (PMU) instrumental in detecting and managing the electrical island created in the aftermath of Hurricane Gustav", Power

Power System Planning: Good dynamic models allow a better understanding of how power systems respond to grid disturbances; better prediction enables better system planning with improved asset utilization. Synchrophasor data has been particularly useful for validating and calibrating models of power plants and other grid equipment.[10] Some grid operators and generator owners are leveraging synchrophasor data to comply with North American Electric Reliability Corporation (NERC) modeling and calibration standards, performing online testing with better results at lower cost than what was previously available through off-line testing methods. Ultimately, these data are also being used to improve system models and calibrating state estimators and dynamic system models. For multiple decades, the Western Interconnection of North America has been a leader in using synchrophasor data for planning applications.[11]

Forensic Event Analysis: Synchrophasor data is invaluable for post-event analysis of disturbances and blackouts. Because synchrophasor data is time-stamped with microsecond accuracy, it can be used to accurately determine the sequence of events in a grid disturbance and facilitate better model analysis and reconstruction of the disturbance.[12] These enable a faster and deeper understanding of the disturbance causes and inform development of ways to avert such events in the future.

1.4 The Role of the US Department of Energy Promoting the Deployment of Synchrophasors in the North American Power System

In 1995, DOE partnered with the Bonneville Power Administration (BPA) and the Western Area Power Administration (WAPA) to launch the Wide Area Measurement Systems (WAMS) project.[13] This project drew on many years of industry effort and collaboration among engaged stakeholders in the western North American power system. The primary objective of this effort was demonstrating the value

Systems Conference and Exposition, 2009. PSCE '09. IEEE/PES. Institute of Electrical and Electronics Engineers, Piscataway, New Jersey.

[10]NASPI Model Validation Workshop, October 22, 2013. https://www.smartgrid.gov/files/NASPI_model_validation_workshop.pdf.

[11]Hauer, J.F., et al. "Use of the WECC WAMS in Wide-Area Probing Tests for Validation of System Performance and Modeling", IEEE Transactions on Power Systems (Volume: 24, Issue: 1, Feb. 2009). Institute of Electrical and Electronics Engineers, Piscataway, New Jersey.

[12]Hauer, J.F., N.B. Bhatt, K. Shah, and S. Kolluri, "Performance of "WAMS East" in providing dynamic information for the North East blackout of August 14, 2003", Power Engineering Society General Meeting, 2004. Institute of Electrical and Electronics Engineers, Piscataway, New Jersey.

[13]Mittelstadt, WA, PE Krause, PN Overholt, DJ Sobajic, JF Hauer, RE Wilson, and DT Rizy, "The DOE Wide Area Measurement System (WAMS) Project—Demonstration of Dynamic Information Technology for the Future Power System", EPRI Conference on the Future of Power Delivery, Washington DC. 1996 (https://www.osti.gov/scitech/servlets/purl/254951).

of time-synchronized wide-area measurement for the purposes of characterizing the dynamic behavior of the interconnected power system. An important element of this research was showing the value of information sharing among utility organizations so that wide-area dynamic behavior could be gathered and accurately assessed. This collaboration was primarily driven through committees and working groups of the Western Electricity Coordinating Council (WECC) and its predecessor organizations.

The sequence of events that occurred on the afternoon of August 10, 1996 that led to an unstable oscillation that precipitated a major blackout in the western power system was captured by early elements of the existing WAMS information network, and provided highly valuable information for the subsequent investigation. Furthermore, when it was revealed that the contemporary dynamic models were unable to accurately reproduce these oscillations, major model validation efforts were undertaken over the ensuing years. Through Congressional support during this time frame, DOE was able to enhance its focus on transmission reliability research, leading to new research initiatives such as the Consortium for Electric Reliability Technology Solutions (CERTS).[14]

In parallel with these efforts in the western North American power system, progress was being made deploying PMUs in the Eastern Interconnection. Several utilities were experimenting with the applicability of synchrophasors, and while there was less interest in dynamic behavior than in the Western Interconnection, there was nevertheless a strong interest in exploring the value of measuring phase angles to support a variety of applications. However, there was a lack of widespread information sharing at the time, which resulted in less than full value being extracted from the technology by limiting the utilities from obtaining a wide-area view of system conditions throughout the interconnection. In part to provide an impetus for greater information sharing, but also to provide a greater awareness of bringing this advanced technology to operating entities, DOE launched the Eastern Interconnection Phasor Project (EIPP) beginning in 2002.[15]

When the lack of wide-area situational awareness was identified as one of the key root causes of the August 14, 2003 blackout,[16] there was a broader industry recognition of the importance of synchrophasors to support transmission reliability objectives. Therefore, industry interest in, and participation with DOE's EIPP project soared in the years following this blackout.

By 2006, it was apparent that there was a need to harmonize the efforts underway in the Western Interconnection with WAMS, and the progress being

---

[14]J.F. Hauer, and J.E. Dagle, "Review of recent reliability issues and system events," PNNL-13150, Pacific Northwest National Laboratory, Richland Washington, 1999.

[15]M.K. Donnelly, M. Ingram, and J.R. Carroll, "Eastern Interconnection Phasor Project," Hawaii International Conference on System Sciences, HICSS-39, January 2006. Published by the IEEE Computer Society, Piscataway, New Jersey.

[16]U.S.-Canada Power System Outage Task Force. "Final Report on the August 14, 2003 Blackout in the United States and Canada: Causes and Recommendations." April 2004.

made in the Eastern Interconnection with the EIPP. It was decided to bring these efforts together into the North American SynchroPhasor Initiative (NASPI).[17]

During the formation of NASPI, DOE reached out to NERC to help provide a formalized structure for "mainstreaming" the technology. By linking the various synchrophasor-related efforts to elements of the NERC committee structure, including various subcommittees and working groups, the electric power industry could attain the maximum benefit through the development of best practice documents that are codified as NERC activities. This gave rise to the eventual formation of the NERC Synchronized Measurement Subcommittee (SMS) and also the WECC Joint Synchronized Information Subcommittee (JSIS). Many of the guidance documents initiated in NASPI are now supported by these NERC and WECC activities.

The Electric Power Research Institute (EPRI) was added as a key partner with DOE to provide greater industry connectivity and logistical support for hosting biannual work group meetings.

Another major development that facilitated additional deployment of synchrophasor technology was the American Recovery and Reinvestment Act (ARRA) of 2009, which was used to provide "stimulus" funding to restart the US economy following the economic downturn of 2008. Included in this funding package were smart grid investment grants, administered by DOE to cost-share recipients, to provide funding for infrastructure projects including but not limited to additional synchrophasor deployments.[18] Given the scope and scale of these investments, a dramatic increase in phasor measurement units in North America occurred in the years immediately following this development.

A map depicting networked PMUs is shown in Fig. 1.3. During the implementation phase of the ARRA grants, NASPI provided a key role facilitating the coordination of best practices and lessons learned associated with synchrophasor development that greatly expanded the value added to the government and other industry stakeholders.

NASPI has become recognized as the premier organization in North America (and internationally) for assembling the world's leading researchers, vendors, and deployers of synchrophasor technology. The work group meetings are highly regarded forums for exchanging information that is vital to the successful application of this technology. In addition, various subject matter experts volunteer their time through five technical task teams, where much of the cutting-edge work developing documents and other products takes place. These task teams are:

Control Room Solutions: The Control Room Solution task team's mission is to work collectively with other NASPI task teams to advance the use of real-time synchrophasor applications for the purpose of improving control room operations

[17]Dagle, JE. "The North American SynchroPhasor Initiative (NASPI)", IEEE Power & Energy Society General Meeting, 2009.

[18]U.S. Department of Energy. "Advancement of Synchrophasor Technology in Projects Funded by the American Recovery and Reinvestment Act of 2009", March 2016 (https://www.smartgrid.gov/document/synchrophasor_Report_201603.html).

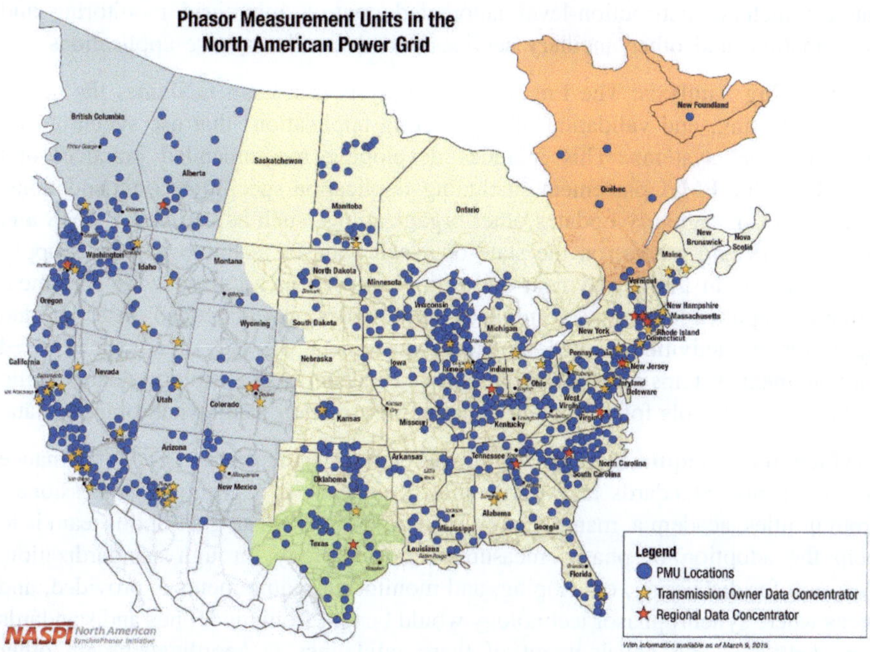

Fig. 1.3 Networked PMUs in North America. Figure: Courtesy of NASPI

and grid reliability. In order to successfully deploy many of the reliability-enhancing applications previously described, issues such as application robustness and operator training must be addressed. This team utilizes its experience and regional diversity to provide advice, direction, support, and guidance to NASPI stakeholders and other organizations involved in the development and implementation of real-time synchrophasor applications.

Data and Network Management: The Data and Network Management task team is a forum focused on tactical and strategic elements of deploying and maintaining synchrophasor technology within current networking and data management disciplines. These are largely data quality, data management, and networking issues. Many of the supporting details, such as naming conventions, interoperability, and standardization, occur within the jurisdiction of this technical task team.

Distribution: The mission of the Distribution task team is fostering the use and capabilities of networked PMUs at the medium-voltage distribution level, beyond the substation. This group shares information about effective research, development and deployment of distribution PMUs, and their applications. By creating a community to solve technical and other challenges specific to distribution PMU technology and their applications, this task team is enabling greater deployment of synchrophasors among distribution utilities. Topics of interest include distribution-level automated control, microgrid monitoring and coordination, and other ancillary services and demand response applications.

Engineering Analysis: The Engineering Analysis task team facilitates the development, testing, and validation of engineering applications that use synchronized measurements systems. This includes developing recommended practices and guidelines for PMU placement, including application-specific recommendations. The task team regularly updates other organizations, such as the NERC SMS and WECC JSIS and others, on the state of synchrophasor technology and its application, assists in the deployment and utilization of synchronized wide-area measurement applications, and formulates and guides recommended research and development activities related to the advancement of wide-area synchronized measurement systems and their applications. The team is also exploring data mining techniques and tools for extracting valuable information from synchrophasor data.

Performance Requirements, Standards, and Verification: The Performance Requirements, Standards and Verification task team is a group of professionals from utilities, academia, manufacturers, and government. The goal of this team is to help the adoption of phasor measurement technology through standardization. A forum for discussing, developing, and monitoring requirements is provided, and areas where synchrophasor technology would benefit from guidelines and standards are identified. The development of these guidelines is coordinated with other NASPI task teams and, as appropriate, those guidelines are migrated to Institute of Electrical and Electronics Engineers (IEEE) Power & Energy Society Working Groups and the IEEE Power System Relaying Committee.

Specifically, standardization of measurement requirements and communication protocols are codified by IEEE Standards C37.118.1 and C37.118.2, respectively. Standardization is also occurring with the International Electrotechnical Commission (IEC) through efforts associated with the 90-5 profile of the IEC 61850 standard and other activities. NASPI provides liaison with these standard-setting bodies and often provides coordination among the subject matter experts who are developing guidance documents that are ultimately leveraged in the development of these standards. A more comprehensive description of how these standards enable and support various power system applications will be covered in greater detail throughout subsequent chapters of this book.

In addition, NASPI has convened task forces to focus on a specific technical area. For example, there was a PMU Application Requirements Task Force (PARTF) recently convened to create a document characterizing the quantitative requirements of PMU applications. The necessary measurement precision, accuracy, data quality, and timeliness are driven by the applications requirements. These application requirements can be represented by various PMU data classes. This report is used by the synchrophasor community as a basic reference for synchrophasor data specifications and applications design. It has led to defining the need for an ongoing activity to test application requirements. This activity is now leveraging an open-source and automated standardized platform for the testing of applications that receive power system sensor data and act upon it.

During the past several years, NASPI has also contributed important documentation to the synchrophasor community, including a number of guidelines, best practice documents, and success stories. NASPI has produced a business case, a document inventorying value propositions, and a starter kit to assist utilities that are now embarking on a synchrophasor deployment journey to be able to leverage the knowledge and experience of other utilities that are further along in their deployment.

More recently, NASPI has been providing important contributions in the areas of alternative methods for robust wide-area time synchronization, cyber security of synchrophasors, and leveraging the advances with next-generation networking technologies for ultra large scale deployment of synchrophasors. Up-to-date information about the various NASPI activities is available on the Web site: https://www.naspi.org/.

The remaining chapters of this book will provide additional details about various technology advancements and applications associated with synchrophasor deployments to enhance the reliability of the electric power system.

Chapter 2
Impact of Phasor Measurement Data Quality in Grid Operations

Kenneth Martin and Kevin Chen

2.1 Introduction

Data quality is a term that is often used but seldom defined. Consequently, it is used in many different ways and applied in inconsistent ways. For this discussion, the following definition will be used:

Data quality: "Any aspect of data that bears on its ability to satisfy a given purpose.[1]"

It is worth noting that this definition:

- Includes any aspect of data, not just accuracy, corruption, or data loss,
- Relates the purpose—intended use—of the data, so disregards problems that will not have an effect on the intended use.

This definition is broad enough that it can be applied to practically any situation where the data used in an application can adversely affect performance. It is also important to note that data may have an aspect that impairs it for the use in one application but not another.

This chapter focuses on the use of synchrophasor data in applications used for power system operations. Operations are carried out in real time, so the data is used soon after it is created, and there is little time to check and validate measurements. Methods for validating data in real time will be described. Effects of various data impairments will be explored, and examples will be given for different kinds of data errors.

[1]Wikipedia, definition of "Data quality", October 2015.

K. Martin (✉) · K. Chen
Electric Power Group, Pasadena, CA, USA
e-mail: martin@electricpowergroup.com

K. Chen
e-mail: chen@electricpowergroup.com

S. Nuthalapati (ed.), *Power System Grid Operation Using Synchrophasor Technology*, Power Electronics and Power Systems,
https://doi.org/10.1007/978-3-319-89378-5_2

The first section of this chapter breaks down data quality into six categories based on the type of impairment and the way the user will observe it. The next section discusses overall data error control and aspects of the measurement system to reduce and manage errors. The last section explores the impacts on operations applications from each type of error category. Recommendations for error control and detection are included in all sections.

2.2 Categories of Data Impairment

For ease of discussion, data impairments are categorized here based on source and type. This is only one such categorization; others could be used and will be found in the literature [1]. The following six categories align well with what users will experience and are easy to account in practice. These categories of data impairments are:

1. Data loss,
2. Data corruption,
3. Inaccurate representation of engineering quantities,
4. Lack of precision,
5. Incorrect measurement identification,
6. Excessive or inconsistent latency.

Each of these categories is explained and analyzed in the following sections.

2.2.1 *Data Loss*

Data loss is defined as data that is planned for and expected by a particular application but not received. This includes losses in data creation (PMU measurements), data communication, and data storage devices. Phasor measurement systems typically involve many communication steps including handoffs between measurement equipment, communication systems, and processing applications (Fig. 2.1). With so many steps, there are many things that can go wrong, and data is lost in the process. Data loss is the most common problem reported by most users of synchrophasor systems.

A phasor measurement unit (PMU) samples voltage and current waveforms and calculates phasor equivalents using a precise time reference. The values, called "synchrophasors," are sent in real time to a phasor data concentrator (PDC) or other applications for processing. The measurements are usually sent continuously at a constant rate, called the "reporting rate," that is usually between 25 and 60 measurements per second. Blocks of measurements corresponding to the same point in time are sent as a frame of data with the timestamp. Thus, the reporting rate is called "frames per second" (fps). The reason for sending data at a high rate is to capture the dynamic activity of the power system. It is sent as a frame since the

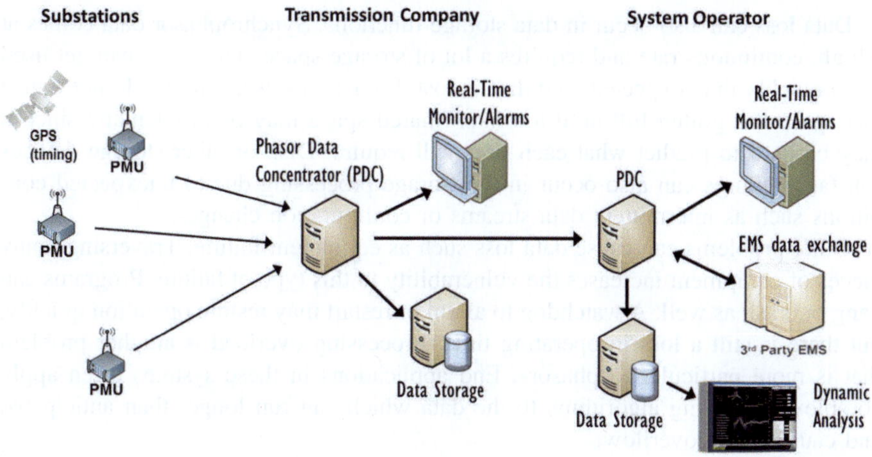

Fig. 2.1 Typical phasor measurement system showing the processing and communication elements

measurements must be matched precisely by timetag to determine system phase angles; having reports from across the system all taken at the same time provides a snapshot of the system. Successive snapshots show dynamic system response in system phase angles, voltage and current magnitudes, frequency, and calculated information such as power.

With data sent continuously at a high rate, any short communication interruption can easily cause data loss. Communication methods, such as TCP/IP, that detect and replace lost data can be used. But lost data replacement requires restoration time and more communication bandwidth. This can cause problems with constrained communications and time-critical applications, so it is not always used. Other communication methods like UDP/IP have lower bandwidth use and no restoration delays, but can lose frames of data. High rate data is also more prone to collisions and overrun buffering, particularly with shared communication channels. With so many ways that the data stream can be interrupted, a small amount of data loss is widely experienced.

A bigger problem than dropouts is lost connections. Once a connection breaks, all the data is lost pending restoration. Synchrophasor systems are inherently wide area, which requires traversing several subsystems. The typical system entails a PMU in a substation environment, a wide-area communication system that connects the substation to a control center, and a control center system. A switch or router connects the subsystems. There may be more than one subnet in each location connected by switches. Security layers are often imposed at the transition points. The system requires careful configuration to complete all the connections from source to destination. Additionally, security will be periodically updated. If the complete synchrophasor system is not thoroughly documented, it is likely that some portion of it will be missed during updates leading to a broken connection.

Data loss can also occur in data storage functions. Synchrophasor data comes at a high, continuous rate and requires a lot of storage space. The space can get used more quickly than expected, and data is lost. Sometimes users simply do not notice that space has gotten full until too late. Shared space may be problematic since it may be hard to predict what each user will require. Disk or other storage devices can fail. Failures can also occur in the storage processing due to unexpected conditions such as intermittent data streams or configuration changes.

Other problems can cause data loss such as equipment failure. Traversing many pieces of equipment increases the vulnerability to this type of failure. Programs can hang or crash as well. A watchdog to alarm or restart may restore operation quickly, but there is still a loss in operating time. Processing overload is another problem that is more particular to phasors. End applications in these systems often apply complex processing algorithms to the data which can run longer than anticipated and cause buffer overflow.

2.2.2 Data Corruption

Data corruption refers to data that has been received by the end application but does not represent the measurement in a meaningful way. Typically data is corrupted in the communication subsystem, but the corruption could occur at the measurement point or the end-use application as well.

Data corruption in communication can be caused by interference that changes bits in the transmission stream. Another cause is a mismatch between transmitter and receiver modulation speeds and formats. These problems are fairly common in serial communications but have been largely eliminated by the extensive use of Ethernet. Ethernet techniques for synchronization and error detection are "baked in" and very effective. In addition, most Ethernet interfaces discard errored packets, so the user never even receives corrupted data. However, this means "corrupted data" will become "lost data" that was discarded by the communication interface. Differentiating between corrupted and lost data can be helpful in determining the cause and location of the problem; in some cases, the user may want to use equipment that can read flags from the communication interface to find if data was discarded.

Data can also be corrupted by the PMU. A typical PMU basically consists of an A/D front end that interfaces to analog signals, a time input that synchronizes the measurements, and a processing section makes the phasor, frequency, and rate of change of frequency (ROCOF) estimates (Fig. 2.2). The processor also manages the output formatting and data transmission. A PMU could be a stand-alone device or a function in a multi-function device such as a relay or a fault recorder (DFR). The PMU function is assumed to handle the calculations correctly and manage the resources to prevent processing and buffer overruns. However, the PMU may not be tested under worst-case communication or input change. If it is a multi-function device, the other functions could also contribute to interruption of the expected PMU operation. Overruns or other interruption of the normal processing can cause

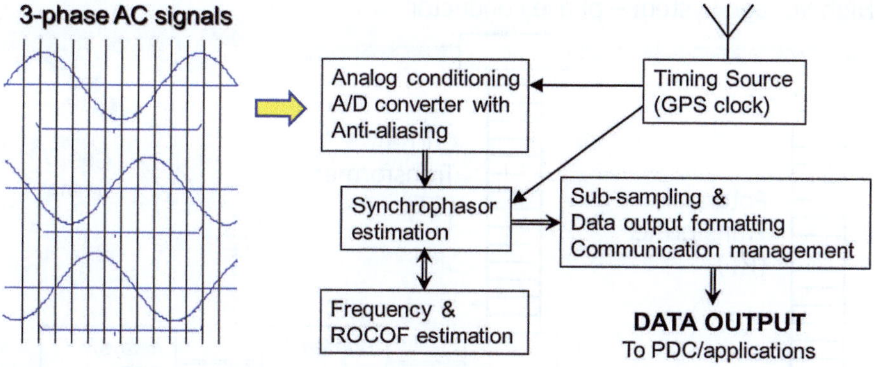

Fig. 2.2 PMU function block diagram

data corruption. If not detected by the PMU before transmission, corrupted data will be sent to the destination where the user must detect it.

At the receiving end, the data may be corrupted in a variety of ways. Processing overruns due to insufficient processing speed or memory can occur. Data fragments can be incorrectly distributed. Control center programs are typically assigned to virtual machines that serve multiple applications. They run in virtual time which may be fast, but does not respond directly to real time. The buffering for the real-time data stream may be insufficient to prevent overruns that corrupt data. Data can also be corrupted at the receiver through misapplication of data types. Data may have been sent as floating point, but taken from the receiver as integer and then incorrectly converted to floating point. Phasors may be sent in rectangular coordinates but received and handled as polar. An application may expect the angle in degrees, but the data is scaled in radians. These and other data handling problems can corrupt the data in the end-use processing.

2.2.3 Inaccurate Representation

Inaccurate representation includes all accuracy-related errors that cause the engineering value to incorrectly represent the engineering quantity. The source of the inaccuracy can range from the original transducer to the display/recording device. Scaling, linearity, estimation algorithms, timing, interference, noise, and correction factor errors can all create inaccuracy. The measurement chain starts with the primary transducer that brings high voltage and currents into the measurement range.

The primary transducer is usually a PT or CT device (Fig. 2.3). These come in a variety of types and configurations. They are specified by ratio, accuracy, range of operation, type of construction, and other environmental and performance factors. This basic device accuracy limits the measurement accuracy. If the PT accuracy is

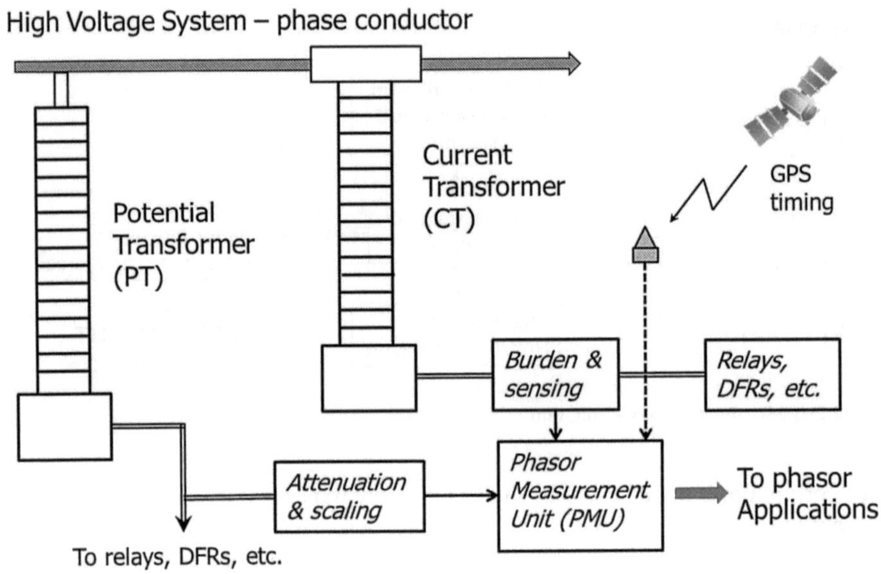

Fig. 2.3 Typical substation measurements using PT and CT sensing devices

1%, the PMU measurement accuracy cannot be expected to be better than 1% unless special calibration is used. Special calibration of primary measurement devices is rarely done because it requires stable, high accuracy test instruments which are expensive and not generally available. So PMUs are used with primary transducers that provide signals with accuracies ranging from 0.1 to 2% and possibly larger. This sets a base accuracy that cannot be improved without corrections provided by other means, such as state estimation.

Phase angle calculation is based on a time signal. There is a direct correlation between time and phase angle errors; the larger the time error, the larger the angle error. Timing errors include errors in PMU internal process synchronization as well as errors in synchronizing to the input signal and its synchronization with the reference. Phase angle measurements can only be compared together if they are based on the same time reference. Grid-wide angle comparison requires synchronization to a grid-wide time source. Angle measurement must be synchronized to the local time source which in turn must be synchronized to the grid-wide time source in order to make an accurate angle measurement.[2]

Noise is present in all signals and must be considered when assessing accuracy. A single measurement will deviate according to the value of the noise signal at the

[2]Note that phase angle measurements which are locally synchronized to the same time source can be used together. This may occur in a single device, a group of devices, or perhaps a whole substation. These signals can be used for power calculation within that sphere, but not with other signals.

moment. If the noise is zero mean and random, averaging several measurements will reduce the error due to noise and give a better-measured value. However, some applications that use measurements cannot wait to gather several measurements for an averaged value. A high-speed control will use each value as soon as it is obtained for making a decision; it does not wait for a number of values to averaging for taking action. If phasor measurements are to be used for control, averaging does not make a realistic assessment. Consequently, testing under the C37.118.1 standard [2] specifies using the worst case in a group of measurements to be sure all measurements are considered. This approach emphasizes the effect of noise on accuracy, so it requires both accurate processing and control of noise.

Noise is generated by the many processes on the power system as well as in transducer and connecting devices. It is usually very low amplitude, appearing as a little fuzz on the waveform. Synchrophasor estimation tends to filter this very well, so the measurement has very little noise. However, interfering signals and harmonics can create significant errors. These signals may be a low frequency that does not average out or may fall inside the phasor filter characteristic. Phasor estimation and filtering algorithms can have an impact on accuracy in unusual noise and interference conditions, but in most cases will provide similar results. The user needs to assess whether their system has excessive interference or harmonics, and if so, what special measurement characteristics are needed to deal with it.

Measurement processing errors by the receiving PDCs, data storage, and application programs can also impact accuracy. Application of incorrect scale and correction factors distorts the data. Misinterpretations of units, such as amps for kilo-amps or degrees for radians, misrepresent the data. Incorrect time alignment, such as caused by communication latency or timetag issues, mismatches phase angles causing error in angle processing. These and other processing errors misrepresent the measurement or reduce accuracy.

2.2.4 *Lack of Precision*

Precision refers to how finely the measurement can be resolved. It directly relates to the number of numerical digits that represent it. If there are more digits, the representation is more precise, even if they only represent noise.

Precision is often mistaken for accuracy. A measurement can be precise without being accurate and can be accurate without being precise. Accuracy specifies how closely a measurement compares to a standard reference. An accuracy specification is limited by the reference accuracy and how close the comparison to that reference can be made. However, if the precision is lower than the reference and test accuracy, it will limit the measurement accuracy. (Precision limits accuracy, but accuracy does not limit precision.) Conversely, a measurement whose precision is higher than the accuracy can resolve small differences in the input signal but does not represent a more accurate measurement. For example, 1% measurement accuracy means that if a measurement is 10.1 V, the input is between 10.0 and 10.2 V

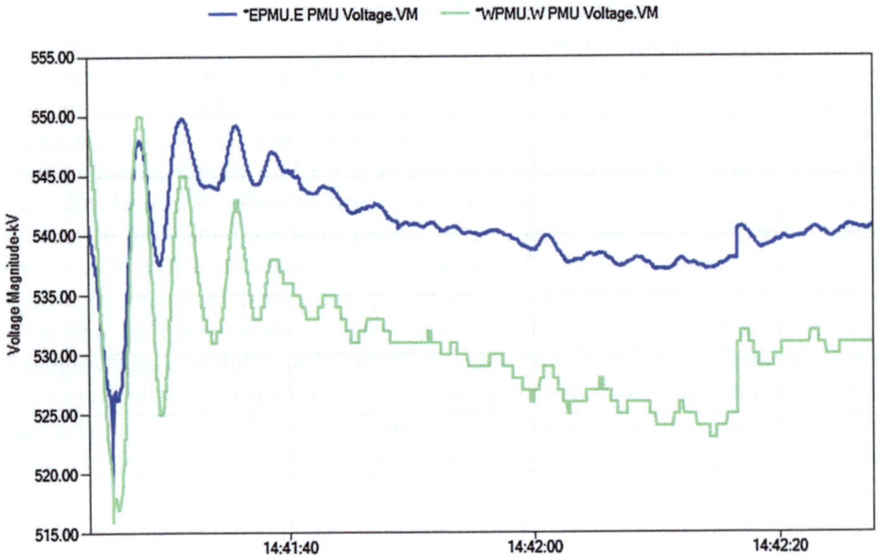

Fig. 2.4 Lower resolution in the W PMU signal as compared with the E PMU

which is ±1% of the reading. If the measurement precision is 5 digits, a reading of 10.095 V still only says the voltage is between 9.994 and 10.196 V (±1%).

Low precision can make waveforms look "steppy" or "jagged" (Fig. 2.4). The actual signal value may change slowly over several sample points, but the representation stays the same until the change is large enough to make a digit change causing a step in value. Note that every digital representation has a limited number of discrete values, so if the presentation scale is expanded enough these steps will be seen. Precision must be considered in relation to the signal range and processing needs.

Low precision also limits noise and noise-like signals. Small signals in a power system often contain significant information about processes across the system. This information is largely hidden in background and measurement noise but can be detected and analyzed with advanced techniques. The synchrophasor estimation process adds very little measurement noise, so retention of the maximum precision has been valuable for small signal analysis.

Differences in output values with no change in the input value represent noise within the measurement process and limit useful precision. Greater precision is meaningful when changes in the output value reflect changes in the input value. The amount of precision that can produce meaningful results depends on many factors in the PMU including the A/D conversion, filtering, and phasor estimation. Most PMUs will produce a measurement with much more meaningful precision than accuracy. Consequently, it is good practice to retain phasor values in their highest precision.

Precision may be reduced if a PMU calculates phasors using floating point or 32-bit integer values and then converts the output to 16-bit integers. This can be made worse by using scale factors that do not maximize the number of usable bits. For example, if an internal measurement value of 456,000 (32-bit integer) is scaled by a scale factor of 10,000 to 45.6 (456,000/10,000 = 45.6) which rounds to 46 as a 16-bit integer, 3 bits of resolution have been lost. By contrast, scaling by 1000 gives an output value of 456 which loses no resolution in the given value. The use of floating point representation eliminates the problem of loss of precision from scale factor application. Floating point representation (32-bit) has enough precision to fully represent phasor values computed from 16 to 18 bit A/D values.

Precision can also be lost by compression. Data compression is used to reduce data for transmission and storage to save bandwidth and disk space. Compression may throw away low-order bits under the assumption that the smallest change does not represent important changes. Lossless compression techniques eliminate this problem, but may require some added programming complexity.

2.2.5 Incorrect Identification of Data

Signals are measured in a substation and the values are sent to a control center where they are processed and recorded. In many cases, values are further sent to other control or monitoring centers. The data is only meaningful when it is identified with the engineering quantity it represents. Correct identification has to be applied in many places, such as displays on a system diagram, time series data plots, and data storage devices that will be used for later analysis. Misidentified data leads to errors in each of these processes.

The first place misapplication occurs in the primary wiring of the signal into the measurement device. This can be detected and corrected with proper installation procedures and usually will not change accidentally in time. However, with the emerging use of sample values for connection between the PT/CT rather than direct wire, changes in software or networking associated with other equipment are more prone to causing inadvertent changes to existing circuits, leading to misidentified signals.

A voltage measured on a bus is essentially the same over the whole bus and on lines directly connected to it. If there is no bus PT, the measurement can be taken from a line PT that is connected to the bus. The application may use this measurement as the bus voltage, which is correct as long as the breaker is closed. But when it opens, the voltage may go to zero or some other value (due to induction) creating a false indication of the bus voltage. Similarly, a bus could be split into several sections where a bus PT does not represent the whole bus. If every separable element is measured, the node voltage can be correctly identified; otherwise, the application needs a more detailed identification system.

Measurement data is transmitted as a series of bits in a long string of binary values. These values have to be separated out of the string as numerical or

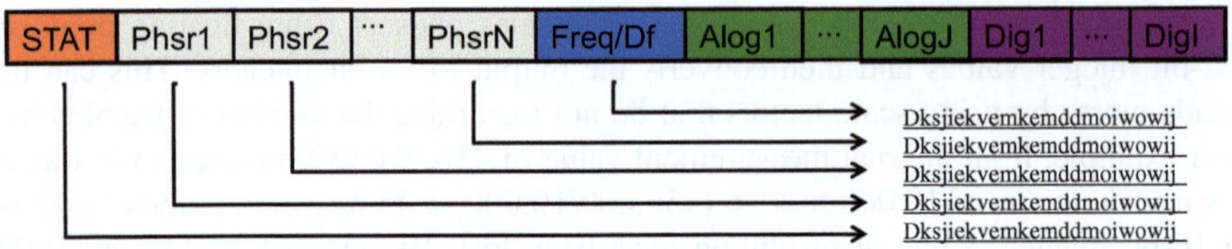

Fig. 2.5 Data is sent as a string of values and has to be parsed using a matching guide

indication data according to the meta-data that identifies each measurement with a name and data type (Fig. 2.5). Any mismatch in these will give incorrect identification. Synchrophasor values are often sent between entities that use different naming conventions when relating the measurement to the power system. Translation between naming can be in error or cause confusion. Synchrophasor data seems to be particularly prone to this since it is streamed a raw data message with meta-data sent separately.

2.2.6 *Excessive or Inconsistent Latency*

Synchrophasor data is measured and reported without delay to enable its use in real-time applications. Some applications, like map board displays, may not require updates very quickly, but controls based on measurements usually require very little delay to operate effectively. Excessive or variable delay, called latency, can significantly degrade the value of the measurement.

Phasors provide system phase angles. Angle measurement is made relative to UTC time and rotates by the difference between the nominal and the actual system frequency. Since the angles constantly change, calculating system angles as the differences between the voltage angles requires using measurements taken at exactly the same time. This requires aligning the measurements by the timetag. Time alignment also yields a synchronized system "snapshot" which improves all analyses with the data.

A phasor data concentrator (PDC) is usually used to combine data into a time aligned frame, but it can be done by any application. Data is sent from multiple locations over a variety of communication links. Though data is sent at the same rate from all PMUs, the delay in delivery time can be substantially different. If all the delays were the same and did not change, this would be simple. But they do not, so the PDC function requires the capability to accommodate varying latency. If the latency becomes excessive, the PDC may not wait long enough and lose the data (Fig. 2.6). If the PDC wait is too long, the applications it serves may not be as effective.

Most applications that receive data in discrete packages are not sensitive to some variation in delays; they simply process the data when it arrives. Large delay

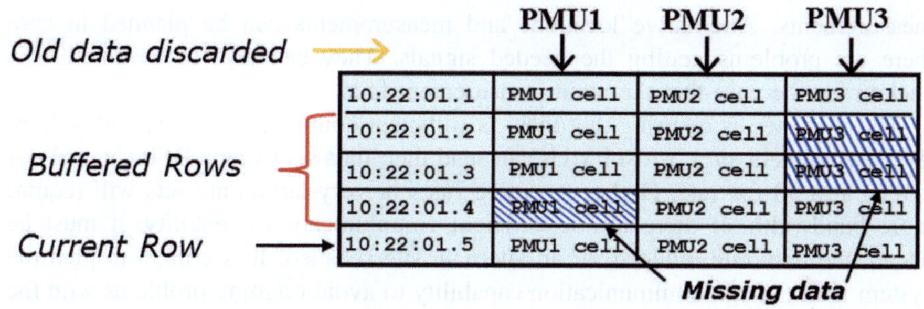

Fig. 2.6 Data table created by a PDC function showing data correlation, dropouts, and high latency loss

variation can be seen in some displays, such as a strip chart that stops momentarily and then jumps ahead. But if the surges are big enough, they can overrun buffers causing data loss. Applications need to provide enough buffering to accommodate the largest block of data that can result from the worst-case variability in latency. However, some applications may be sensitive to surges, such as real-time controls that expect to see data in intervals that match real time. Latency should be monitored so problems can be detected and addressed before they cause application failures.

2.3 Data Error Control and Detection

Factors that degrade the data have been described. The next question might be "how are these data problems prevented?" Certainly, it is not possible to eliminate all problems for all time, but it is usually the case that problems can be reduced through careful consideration of the factors that cause them. The following sections detail some considerations.

2.3.1 Measurement System Planning and Design

Planning for a measurement system has to start with the intended applications. This determines which signals will be measured and where. It also determines the measurement characteristics including measurement class and reporting rate which will, in turn, affect the system performance.

Most applications require certain signals such as voltages from certain buses and currents from specific feeders. This relates the measurements to the overall power system, giving the "view" that the application requires. Multiple applications usually share many of the same measurements, which simplifies selection of measurements. Alternative locations and measurements can be planned in case there are problems getting the needed signals. They can also be installed as a backup for the case that the main measurement fails.

The next step is assuring that there is sufficient communication capability from the measurement sites. Most PMUs can send their data set using a DS0 channel (64 kBPS) at a 30 fps rate. Higher reporting rates or very large data sets will require more bandwidth. If there is not sufficient communication capability, it must be added, the data rate reduced, or an alternate site selected. It is critical to plan the system with enough communication capability to avoid ongoing problems with the system.

PMU timing is usually provided by a GPS receiver. This device has an internal clock which is synchronized to UTC time from the GPS satellite system. It sends a timing signal to devices within the substation using an IRIG-B, IEEE 1588, or other timing signal. High accuracy is required by the PMU, preferably at least 1 μs. A GPS receiver can assess the accuracy of its signal based on the received GPS signal and its own capability. This information should be included in the timing signal, so the PMU can also assess its timing accuracy. In some cases, a GPS receiver is built into the PMU which simplifies the timing communication, but requires an antenna and cabling. It is important that the time synchronization and its accuracy is monitored and reported with the synchrophasor data, as time is the basis for the phase angle measurement. Time quality reporting also helps location and resolution of problems.

Measurement classes are described in the standard IEEE C37.118.1-2011 [2]. M-class requires filtering to remove out-of-band signals to prevent aliasing. The filtering will create longer latency than the P-class which does not require filtering. Applications that are sensitive to aliased signals will do better with M-class; those requiring minimal latency should use P-class. If the user is assured there are no higher frequency signals, aliasing is not needed and either class is fine. Higher reporting rates reduce filtering requirements so that at a reporting rate of $2 \times f_0$ the filtering of both classes is essentially the same.

Reporting rate affects the required filtering as noted above, and also the bandwidth and latency. The bandwidth of the measurement should be effectively about 1/4 of the reporting rate. (i.e., a 30 fps reporting rate should have an effective bandwidth of about 7.5 Hz.) The actual bandwidth will depend on the particular estimation algorithm. C37.118.1 only requires a 5 Hz bandwidth for all but the lowest reporting rates, but most PMUs do better than that minimum. The user needs to assess the measurement bandwidth required for the application. In addition, the reporting rate sets the minimum latency. At a 30/s rate, there are 33 ms between reports. Each estimate is calculated over a window of data, so this adds to the latency. For example, an M-class PMU reporting 30 fps using the filtering described in C37.118.1 would have an added 160 ms giving about 200 ms of maximum total latency. By contrast, a P-class PMU would have a maximum total latency of about 50 ms. While well-designed filtering can substantially improve M-class performance, there is added delay that could affect some applications.

The communication system in the substation and the control center also need to be designed to handle the data loads. Synchrophasor data is sent at a constant rate, so bandwidth and data handling requirements are easy to calculate. Sharing resources like routers and switches that are subject to bursts of traffic should be avoided. Synchrophasor data is sent constantly, and delays can result in buffer overflow and lost data. As described above, inconsistent latency caused by shared resources can also cause problems for some applications.

Once the planning group has analyzed the application requirements to determine the required signals and measurement characteristics, designers will select the equipment and design the system. In some cases, existing equipment, like relays or DFRs with PMU capability, can be used instead of procuring and installing new equipment. This option has been found desirable by some utilities. However, the advantage of multi-function equipment has to be balanced with the difficulty of coordinating maintenance and upgrades. In some cases, mixed-use equipment complicates meeting security requirements, but it does reduce purchase and installation delay and expense.

Similar considerations apply to communication facilities. Shared channels reduce the bandwidth and equipment expenses, but also can degrade performance and create maintenance issues. It is important to consider the longer term and possible added maintenance costs. Saving a little up front can be more costly in the long term.

The control center often uses racks of servers and allocates virtual machines to run applications. This simplifies wiring and allocation of computing resources. It is also much easier to add CPU time to applications that need it. However, virtual servers run in virtual time, expecting to process when they are ready. Synchrophasor systems run continuously in real time. The virtual machine has to be set up to act in real time and have enough processor allocation so that synchrophasor applications can run in the time frame they require and there are no data overruns.

A few of the significant considerations for planning and design have been discussed here. These are some that occur with synchrophasor systems that may not occur with other installations. A utility should follow its own processes to be sure all requirements are met and details are fully completed.

2.3.2 Installation and Validation

Equipment and system installation is a critical step that is often given insufficient attention. Synchrophasor systems start with the PMU measurement in the substation and continue through the control center, often with more than one layer. They are similar to SCADA in that way, but more demanding due to the precise timing and higher data rate characteristics. Local and end-to-end validation is essential to assure the system is installed and operating correctly and accurately.

A PMU that is certified as conforming to a standard will comply with all the basic performance criteria. PMUs can be spot checked in a test laboratory before installation or sampled for quality. A utility may have some special requirements that require additional testing that also needs to be done in a laboratory. It is generally not necessary to fully test each unit. When installed in a substation, the phasor measurements should be checked carefully against the input signals. Measurements with portable test instruments made while the power system is quiet (that is, at a time when there are no events happening and the system is not in an hourly ramp) should compare reasonably closely with the PMU. Since the PMU measures phase angle against UTC time, phase meter comparisons can only be made on the angle between signals, not the absolute angle measurement. Phase meter comparisons assure that the phasing of the PMU inputs is correct and can indicate if there are wiring errors. Local measurement comparisons should also be made with installed substation instrumentation. PMU voltage and current measurements can be combined for power flow comparisons.

The next stage of comparison is with SCADA measurements. These can be compared by capturing readings from the phasor system and SCADA at the same time. Direct comparison between the voltage and current magnitudes can be made where they are both available. SCADA usually reports power rather than current, so the PMU equivalent values need to be computed from V and I. Periodic EMS state estimation values including voltage angles will often be available along with the measured SCADA data. By timing when the estimation is done, a reasonable comparison between the phasor and the estimation angles can be made. Comparisons with SCADA measurements should be within 2–3% though usually not much better due to differences in measurement timing, reporting, and the averaging interval. Differences larger than this warrant investigation to find a reasonable cause for the greater difference. A little experience with this validation should make problems easy to find and resolve.

If the phasor data is forwarded to a higher level control center such as an independent system operator (ISO), the same process of validation with SCADA data should be followed. In this case, the main problems to check for are data identification and regional phase angle differences. Different utilities may label their phase rotation differently and may choose a different phase as the reference ("A phase"). A single utility system will not see these differences, but the ISO will. Also, the ISO may have a different naming convention which requires reapplying a different name to each signal. An independent reference such as SCADA will help to spot those errors.

2.3.3 Error Detection and Mitigation

Even the best designed and validated system can experience failures. Failures range from hardware and firmware in individual devices, to wiring and fiber connections, and to end-user applications. System failures can also result from accidental intrusion where a change is made in a configuration or communication link which inadvertently alters the phasor communication path. And intentional intrusion is a growing concern. Continuous monitoring with appropriate alarms helps assure rapid problem detection, so action can be taken to promptly remedy the problem.

The simplest kind of monitoring scheme simply counts incoming data from each expected source and compares that with the actual received data. This function should also keep track of whether data is lost as individual samples or groups of samples, how frequently it occurs, and the time of occurrence. A random loss of about 0.1% of data where data is being reported at 25 fps or faster is generally not very harmful and considered acceptable. (At 25 fps, this represents the loss of 1.5 samples per minute.) Higher data losses indicate a problem in the path or equipment and should be investigated. The pattern of loss can help to locate the problem.

Another step in monitoring is looking at the message details. Corrupted data may have formatting, addressing, check word (such as cyclic redundancy check), or fragmentation errors. The message time stamps should conform to the given data rate and be received in monotonic order. A status included in the C37.118 message indicates errors detected by the PMU regarding time synchronization, message calculation, and processing errors. It also indicates if the data is invalid (such as data received with a bad CRC or inserted as a placeholder for missing data). All of these conditions are well described and should be used to validate the data or indicate the kind of error if needed. These message examinations also are good tools for intrusion detection. Spoofing, meaconing, and other kinds of intrusion often will corrupt the normal message format and time sequence, which can be detected by these examinations.

A further step in monitoring is looking at the measurements themselves. They should have reasonable values, and that allows setting limits for comparison and validation. For example, a frequency measurement of 69 Hz on a 60 Hz system is very likely an error and the data should be discarded. Other examples are extremes in voltage and current magnitude, though error detection needs a delay of a few samples for fault conditions. Other detectable conditions include "flat-line" (also called "stale value") where constant readings have been caused by a stuck value, and "noisy" where noise that is higher frequency and amplitude than real signal noise is present, usually caused by hardware or processing failure. More advanced measurement monitoring can include the comparison of measurements based on the topology of the system they are derived from. For example, the sum of all currents out of a bus should add to zero; if not, one or more of the measurements is in error. This complete data examination can be combined into a quality checking application that provides indications of problems and removes bad data to prevent use in applications (Fig. 2.7).

More complete examination of the measurements can be made with a linear state estimator (LSE). The LSE creates a set of consistent measurements based on the model and reported measurements. A mismatch between the estimated and measured values indicates either the measurement or the model is in error. Unfortunately, this calls the model into question as well, but over time, it is possible to correct the model to the extent that differences reliably indicate measurement errors.

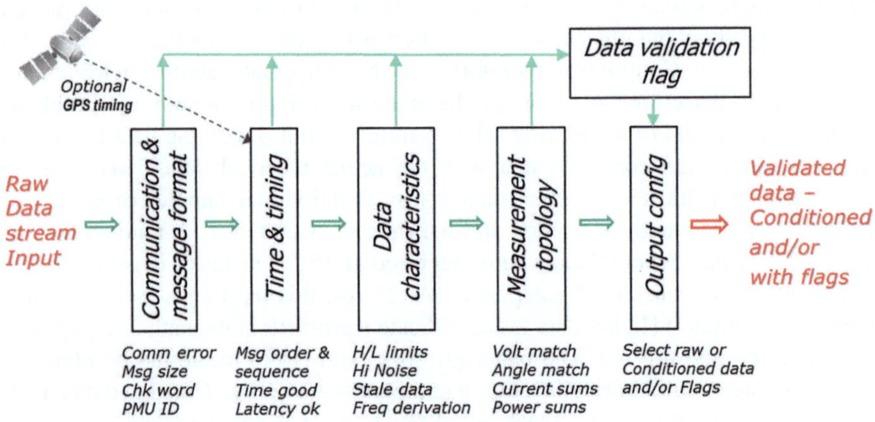

Fig. 2.7 Bad data detection processing—without the system model

Once a failure is detected, an action is needed to remedy it. The two important elements are locating the problem and notifying the responsible party to correct it. Equipment, methodology, and procedures to accomplish these objectives need to be built into the system, so these elements can be achieved effectively and quickly. A regular report that reports identified the problems, the frequency and time of occurrence, and likely source location is an important element of a data quality system (Fig. 2.8).

Location of a point of failure can be difficult. The typical measurement system traverses many communication links and types of equipment. A failure at one point can result in failures or errors at other points. Fortunately, different types of errors can only occur in certain ways, so some types of errors clearly indicate the point of

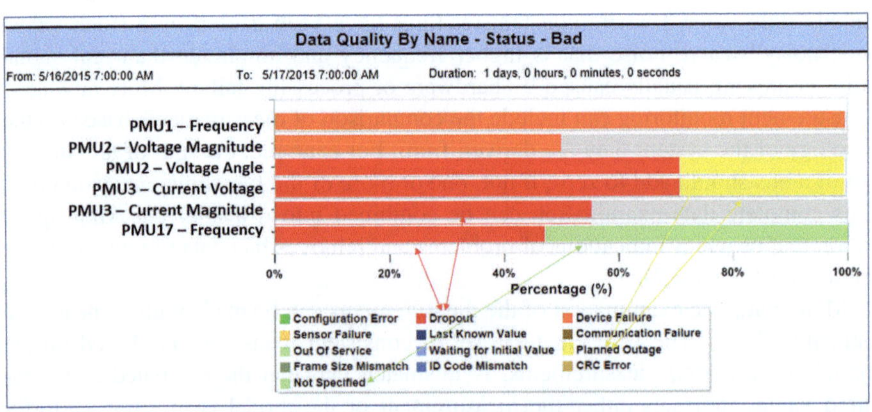

Fig. 2.8 Example of a data quality report showing errors identified and frequency of occurrence (DataNXT^{TM} application report, courtesy Electric Power Group)

failure. For example, an oscillation in a single measurement does not result from communication or post-processing, so must emanate from the original source transducer or PMU. However, most errors are not so clear, the worst being data loss which can occur at nearly any step in the system. This chapter does not attempt to analyze data loss and the potential ways to differentiate causes. Refer to the DOE report "Best Practice Recommendations for Synchrophasor Systems" [3] and the IEEE Guide C37.242 [4] for further detail on problem determination. Product vendors are also a good source of white papers and other recommendations for maintaining synchrophasor systems.

Establishment of remediation procedures is just as important as locating the problem source. Most utilities have several organizations that are responsible for maintenance of the equipment and systems. Established notification processes are essential for accomplishing prompt repairs. Often there are separate organizations for resolving control center, communications, and field equipment problems. Links are needed to assure the proper organization can be contacted effectively. Knowing the likely source of a problem can be extremely helpful: If the wrong group is contacted first, they will look for a problem they cannot find wasting time, resources, and goodwill. Alerting every organization at every problem may save time, but is costly and demotivating. So it is important to both have procedures for locating the most probable location of the problem and the organization that can correct the problem.

2.4 Impacts of Data Impairments

2.4.1 Introduction

Phasors are good for observing dynamics in power systems, so can be valuable in both the control room and for engineering analysis. This chapter focuses on grid operation which is impacted by all aspects of data quality including delays. However, in this context grid operations do not include high-speed automatic controls like relaying and special protection systems (SIPS and remedial actions), so the time frame for delays that are of concern is in the ½ to two second range, not milliseconds. Data analysis is off-line, so the time of delivery is not a concern as long as it meets the requirements of data recording devices. It is also assumed that lost data is detected and indicated so that "filler" data is not interpreted as usable. The following sections discuss the impacts of data impairments as described in the first section of this chapter.

First, it may be important to describe data handling using C37.118. Data is sent in frames that include a number of measurements corresponding to a single time sample. These measurements are identified by their position in the frame of data. If a measurement is omitted, all the other measurement would move forward in the frame and be identified as the incorrect data item. Consequently when a measurement is missing, filler data must be inserted, so the data items following will be correctly identified. The PMU always inserts the full data set. If any measurement is identified as bad, the PMU inserts a flag in the status word to indicate a measurement error. The next step is when data is concentrated in a PDC or application where data from an entire PMU data set may be missing. Using the standard PMU protocol defined in C37.118, the missing data will again be filled in and the data block listed as bad data in the status word. In all cases, the filler data is set to an invalid value, 8000 (hex) for integer values and not-a-number (NaN) for floating point. Applications should be programmed to check both the status flag and the numeric value so that a bad numbers are not used.

Data handling by individual applications can be simplified by using a preprocessing application that examines all incoming data for errors and creates a data quality "validation" flag (Fig. 2.7). An application like this that looks at the whole data set together can make comparisons across data that may not be available to all applications. It also simplifies the data examination required by individual applications. It can supply separate data streams to different applications with individualized patching and corrections. However, it will introduce some delay to all data streams that processes and requires coordination with the applications it serves.

An important aspect of data quality monitoring is cyber infrastructure protection (CIP). CIP is concerned with intrusion into an overall system which alters, corrupts, or otherwise changes the normal flow of information. In a synchrophasor measurement system as addressed here, this would consist of altering, corrupting, or delaying the normal data flow. Data errors can cause operation or control errors if used in these functions. As data systems have become more and more integrated, CIP has become a major concern and is specifically being addressed by power system regulators. For example, in North America, the North American Electric Reliability Corporation (NERC) has created guidelines to determine when a data system could impact operation of the power system and should be considered for CIP [5]. NERC has created regulations for how CIP systems need to be operated and managed. Whether it is the result of intrusion or some kind of equipment or system failure, this section discusses data impairment and its impacts on power system operation. Detection of data errors is mostly the same regardless of the cause, though some kinds of intentional data alteration may be difficult to detect.

Basic preprocessing looks at the data itself, including communication and format errors, data flags, and measurement comparisons. While these examinations can be complex, they can be handled using only the data itself. More comprehensive preprocessing can include data corrections based on the system model. The incoming data can be fed into a linear state estimator (LSE) that applies the data to the system model to estimate better values (Fig. 2.9). This introduces the complexity of maintaining an accurate model and keeping it in sync with the operating power system. However, this technique can provide accuracy improvement for the input devices (like the PT and CT transducers) which cannot be done with data examination only. These accuracy improvements depend on measurement redundancy and model accuracy, so careful assessment is required in choosing the best value. LSE values can be used when the measured value has an error indication or

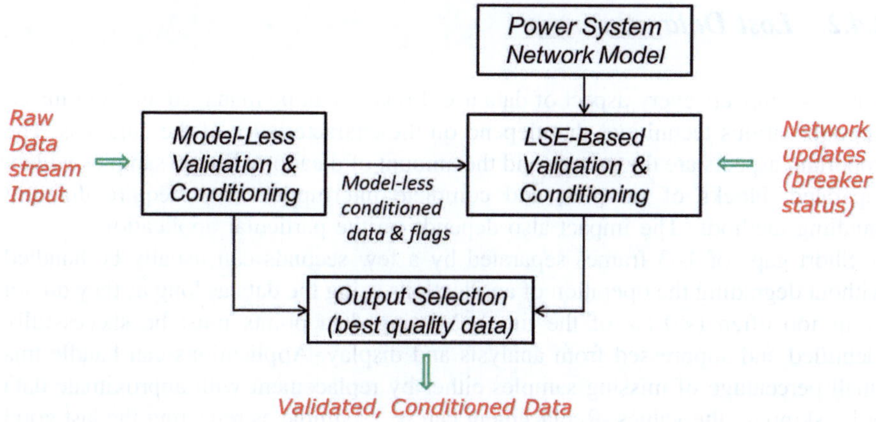

Fig. 2.9 Data stream with error preprocessing including an optional LSE

exceeds an error estimate; reported measurements may be preferred when the LSE estimate exceeds an allowed threshold. An additional benefit of LSE preprocessing is the extension the measurement set with pseudo-measurements, improving synchrophasor coverage of the system (Fig. 2.10). This is a low-cost way to extend coverage and a good way to provide redundancy for the case of failures.

Fig. 2.10 Extending measurement coverage with an LSE using voltage and current measurements

2.4.2 Lost Data

Data loss impacts every aspect of data use. Impacts can be managed and minimized through various techniques that depend on the characteristics of the data loss. The important aspects are the pattern and the amount of data loss. Single samples widely separated, blocks of samples, and complete interruption each require different handling methods. The impact also depends on the particular application.

Short gaps of 1–3 frames separated by a few seconds can usually be handled without degrading the operation of applications using the data as long as they do not occur too often (<0.1% of the time). Missing data points must be successfully identified and suppressed from analysis and display. Applications can handle this small percentage of missing samples either by replacement with approximate data or by skipping the values. Replacement can be as simple as repeating the last good sample or estimating values by approximation. Repeating the last good sample is fast and can easily be done for real-time applications as long as it is curtailed in a few samples. Approximations such as a linear or second-order fitting require holding the data flow for the end point of the gap to make the fitting. This introduces delay, so is better suited to off-line analysis.

Missing data is a problem for auto-scaling displays. Missing values are usually interpreted as 0 or an extreme value, so they cause an auto-scaling to expand to a point where the normal data is not readable. Replacing data with an approximate value or suppressing it from the display can alleviate this problem (Fig. 2.11).

Analysis applications, like oscillation detection, can handle short gaps with approximated filler data. It may distort the result slightly, but the effect is small

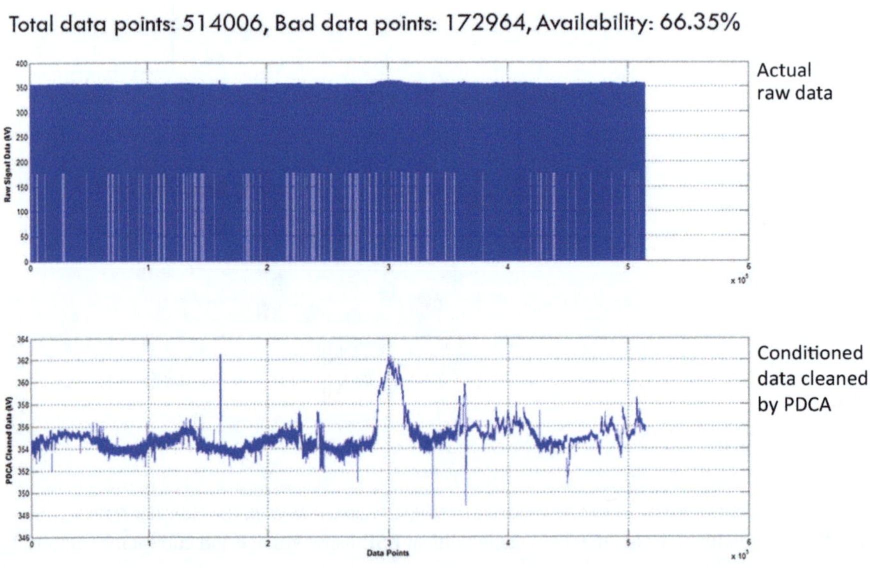

Fig. 2.11 Plot with dropouts that make the scale unreadable; conditioning restores usefulness

enough that it usually can be ignored. Short-term analysis in operations programs usually works on a block of data, so a small delay caused by filling a short gap is not a problem. Functions that need to operate in a sample or two would be impacted by short dropouts but are not generally used in control center applications.

Dropouts longer than a few frames require different treatments. These gaps cannot simply be bridged over with a few approximated samples. They will affect analysis as well as display. When data disappears, it is not known when it will start again and how long it will continue after it restarts. If a data cleaner inserts last good sample as filler until data resumes, it may continue indefinitely since it does not know where the data will restore. Such an application should insert approximation (such as last good sample) for only a few samples and then switch to NaN (not a number) until good data restores. The preprocessing program should flag missing data as bad regardless of what is inserted as a placeholder. Applications downstream must be compatible with the data preprocessing so that inserted values will not be used as valid measurements.

Analytical applications have to be able to start and stop their functions to accommodate missing data. They have to be able to discard data with gaps that are too large or have too much inserted filler data and approximated values. The user must be sure the application will work with the approximation method such as repeated values or linear interpolation.

A failed data stream creates a long-term dropout. In this case, the data feed has to be restored by some kind of system repair. An alarm is needed to alert users to the loss, so they can initiate repair. To prevent excessive alarms, notification should be based on outage duration that is long enough that self-restoration is not likely to occur. Applications have to deal with no data (or filler data) until the data stream restores. Displays need to indicate outage, so users know that data is no longer coming. Analysis applications must suspend operation until data is restored.

2.4.3 Data Corruption

Corruption can cause data to change to any value randomly and so can produce unpredictable errors. Applications need extreme value checking to avoid presenting wild results from corrupted data. A preprocessor is the best approach to protect applications from the varied effects of corrupt data.

As previously described, data corrupted in the communication system will be discarded by the Ethernet interface, so will be seen as lost data. However, data corrupted in the PMU or application may find its way to the application.

Data corrupted in the PMU can take the form of noise on a signal or values misplaced in the data stream. Random noise superimposed on a measurement does not undergo the same processing that the signal has gone through. Consequently, it can have much higher frequency components than are found in the measured signal, so this type of corruption may be detected by looking for excessive high frequency in the signal. Misplaced values in the data stream show up as jumps, extreme, and inconsistent values. These errors can be detected and flagged with a preprocessor.

Data can be corrupted by the control center distribution system and the applications themselves. Data distribution may parse data incorrectly or insert the wrong set of values. Changes intended for one subsystem can affect many others. This kind of corruption can only be seen in the receiving application, so each application needs to detect and call attention to data values that make no sense. Applications can make processing errors, particularly when an event causes large system changes. If the application itself is making errors, the user will receive corrupted results and will have to make the decision as to what is reasonable. In this case, having more than one source of information that can be compared is essential to making good decisions. All control center applications should include logging that allows users to go back to examine indications and alarms. Problems caused by corruption are unpredictable, so a good system design and constant vigilance are the best approaches.

2.4.4 Inaccurate Representation of Engineering Quantity

Inaccurate data can be difficult to detect, since it can result from transforming devices, bad scale factors, and processing errors that are small enough to pass as good data but large enough to be misinterpreted. Timing problems are particularly difficult to detect since a very small time error can result in a large phase angle deviation. Inaccurate data can effect practically any application, though not necessarily with the same impact.

Magnitude accuracy errors will particularly affect applications that compare or display voltage, current, or power (computed from voltage and current) and make decisions on it. These include displays, flow gates, voltage profiles, and line loading limits. Things like sensitivity and oscillation detection are not particularly affected since they measure changes relative to the signal or other signals. Magnitude errors usually occur in configuration or by a failed component rather than during continuous operation. Consequently, they are usually corrected in the validation process. However, they can occur at any time, so critical functions need a cross-check from a secondary measurement before making critical operational decisions. This is, of course, true for any control system, not just synchrophasor systems.

Time is used directly with the cosine function to determine the phase angle ($x(t) = A\cos(2\pi f_0 t + \varphi)$). With a 60 Hz system, a timing error of 46 μs creates a 1° measurement error ($\Delta\varphi = 2\pi f_0\,\Delta t \times 360°/2\pi = 60 \times 0.000046 \times 360° \approx 1°$). Thus, a small time error creates a big phase angle error (Fig. 2.12). Typical substation monitoring used for fault and system analysis requires 1 ms accuracy. Synchrophasor data is reported at rates comparable to power system cycles, typically at 16–50 ms intervals. So the timing required for accurate phase angle measurement is 1–2 orders of magnitude more accurate than other timing systems which could be used for comparison. Consequently, timing errors must be detected and flagged at the measurement source since they cannot be reliably assessed by the

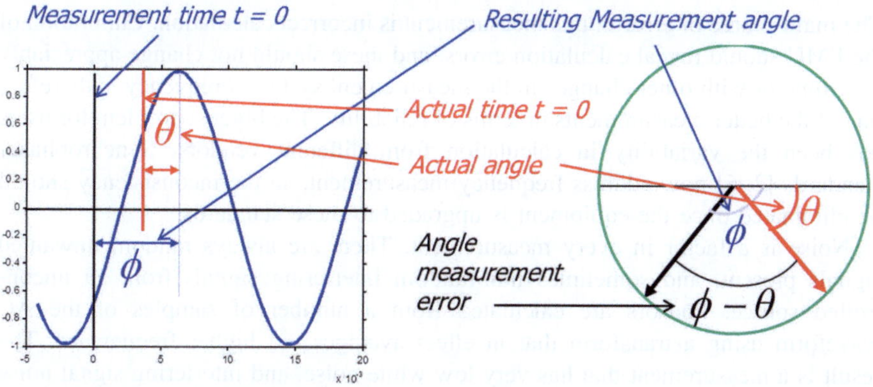

Fig. 2.12 Timing offsets directly translate to angle displacement

end application. All applications need to check this flag to prevent trusting unreliable phase angles.

Timing errors also occur with data alignment. Data from different measurements and locations is aligned by timestamp to allow direct computation using both magnitude and angle. Data may arrive with different latencies, so the alignment process has to hold data for correct alignment. This can create data loss if the wait time is not longer than the delay. Data sent at different rates requires sorting or discarding data to keep the time alignment. A function performing data alignment must perform best guess alignment and flag data that has faulty timetags or other timing errors encountered in the alignment process. The status of every measurement indicates if there is a timing error or approximated alignment. Any application using data from a set must check the timing flag of each measurement to be sure the angle from a PMU with bad timing is not used with data from another for phase angle calculations.

Phase angle errors do not interact directly with magnitude errors, but can throw off calculations where they are combined. For example, power calculated from a voltage and a current uses both magnitude and phase angle. However, as with magnitude errors, relative measurements like sensitivity and oscillation detection are not affected by angle errors created by time offsets.

Applications use system voltage phase angles to indicate system stress, power flow, and to enhance state estimation. Undetected angle errors directly impact these applications. Unlike magnitudes, angle errors can pop up any time there is a problem with the timing signal and alignment processing. So the user always needs to cross-check signals before making critical decisions based on angle measurements.

Frequency offset is defined as the derivative of phase angle relative to the nominal system frequency. It is usually estimated as numerical differencing of angle measurements. Sometimes it is filtered or calculated with many points to reduce noise. Time offset errors do not have much effect on this measurement. Since frequency is computed from the angle, amplitude errors have very little effect.

The main source of error in this measurement is incorrect calculation. Calibration of the PMU should reveal calculation errors, and these should not change appreciably over time or with other changes in the measurement system. Frequency is therefore one of the better measurements in terms of reliability. The biggest problem for users has been the variability in calculation from different vendors. Synchrophasor standards [2, 6] now address frequency measurement, so the inconsistency should be eliminated once the equipment is upgraded to these standards.

Noise is a factor in every measurement. There are always random unwanted signals present, and sometimes non-random interfering signals from an uncontrolled source. Phasors are calculated from a number of samples of the AC waveform using a transform that in effect averages out higher frequencies. The result is a measurement that has very low white noise, and interfering signal noise that is limited by the Nyquist sampling frequency band. Out-of-band filtering used in some PMUs (e.g., M-class) will reduce the noise even further as well as reduce any contribution from higher frequency signals.

Noise has a direct impact on nearly every monitoring and analytical process. The noise makes it difficult to determine when the measurement achieves a certain value. Limits on trigger points or alarms need to be set wide enough to allow for noise. Graphs of signals with high noise look like "woolly worms" and do not clearly show the true signal value at a given point (Fig. 2.13). Noise is present in

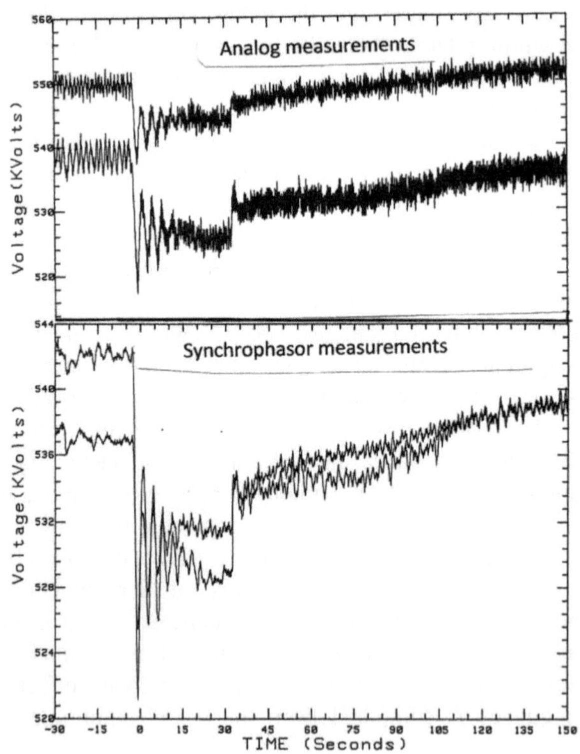

Fig. 2.13 Comparison of synchrophasor and analog measurements at a similar scale

both phase angle and magnitude. Noise becomes worse in frequency and worse still in ROCOF measurements since they are derived with successive differentiation. Consequently, equipment and techniques that provide the lowest possible measurement noise are critical in supporting ever more precise applications. Synchrophasors generally provide power system measurement with lower noise than previous techniques and continue showing improvement.

2.4.5 Lack of Precision

The most obvious characteristic of low precision measurement is a "steppy" waveform appearance. This happens because the discrete steps in the measurement values are visible within the range of the plot. The range coupled with the resolution can show or hide the discrete measurement steps. For example, a plot of a waveform of a 500 kV signal may look steppy when plotted with a range of 2 kV but look smooth when plotted with a range of 200 kV. While this can occur with any digital signal representation that is magnified enough, it only becomes a problem when the steps occur in within the range of analysis. It is important that measurement is scaled to produce the precision that applications require and communications are allocated to preserve the precision. Though the measurement may only start with a 12- or 16-bit sample representation, the estimate improves the resolution by using a number of samples which effectively reduces the step size. Coupled with low noise, synchrophasor estimates can produce a very high-resolution estimate approaching the limit of 32-bit floating point representation.

The biggest problem with the lack of precision is in small-signal analysis. This analysis looks at system oscillation and parameter sensitivity changes that are riding on the high system voltages and currents. A measurement has to be scaled to include the highest expected full value for voltage and current, but it also has to include enough resolution (bits of representation) to resolve the small changes in that value. There is a trade-off between the two requirements that the user has to consider when scaling measurements. The use of floating point (FP) representation greatly helps resolving this conflict.

Analysis that looks at small changes in voltage or frequency to assess sensitivity to other changes is also degraded by low precision. An example is looking at voltage changes related to power flow to determine the proximity to voltage collapse. In extreme cases, low precision can make it difficult to set alarm limits. For example, if a voltage was scaled to 5 kV/bit, an alarm could not be set to operate reliably closer than 10 kV of a given value (± 5 kV $\sim \pm 1$ bit). Integer representation for most PMUs is scaled to 10–100 V/bit and 1–10 A/bit, which is enough resolution for most applications. FP gives even better resolution. Integer phase angle is specified in the C37.118 standard to 0.0345 °/bit and gives resolution close to the level of measurement noise. Integer frequency is specified to 0.001 Hz/bit which is better than most analog measurement, but can be limiting in some cases.

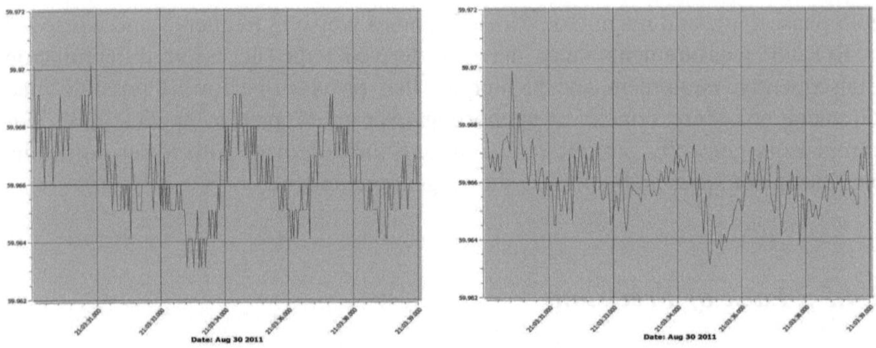

Fig. 2.14 Comparison of frequency plots using integer values on left and floating point on the right

However, frequency in FP greatly improves the resolution, as can be seen in the comparisons in Fig. 2.14.

2.4.6 Incorrect Measurement Identification

Incorrect identification is usually resolved during the installation and validation process. However in a typical utility, changes in other systems, software upgrades, and other system maintenance can result in inadvertent changes to the synchrophasor system. This can result in signal identification errors. Clearly, this will affect any display or analytical application since the signals are selected by name. In most cases, a mislabeled signal will be obvious, such as a frequency in place of a voltage, as it has very different values. However, it could be a current swapped for another current and lead to erroneous conclusions about the state of the system. It is not possible to predict every way this could occur, and therefore what the consequences would be. The best way to check is as mentioned before: Provide redundancy in measurements that are critical to the operation of the system. If one does not look right, check another before making a key decision.

2.4.7 Excessive or Inconsistent Latency

The first activity that latency affects is data collection. A PDC collecting data for time alignment will have to wait for the last frame corresponding to each timetag before completing the aggregated frame and forwarding it. Any data with excessive latency will be lost if the delay exceeds the PDC wait time. Similarly, even without a PDC, data sent to a display or analytical application will be unusable if the application does not wait long enough for it. Control center applications have to present information to operators within a time frame that is useful to them, that is usually in a time of seconds to a few minutes in some cases. Synchrophasor latency is generally a fraction of a second, but in a large network system could exceed many seconds. The user needs to assess how much delay a particular application can tolerate and allow a time delay up to that limit for data to be received. Latency can be assessed by monitoring when the data arrives using an accurate local clock. If some signals are commonly delayed close to the allowable limit, measurement and communication configurations have to be modified. Constant latency monitoring can alert users to developing problems, such as overloaded communication links or failing equipment.

Inconsistent latency can cause data to bunch up and be sent in blocks rather than one frame at a time at a regular interval. Display applications will seem jumpy in this condition, but as long as data does not overflow buffers, it does no real harm. Analytical applications may be prone to overruns and data loss. Jumps could also disrupt threshold triggers if a data burst had points that exceed the trigger followed by points below the trigger. However, these problems can be avoided with good application design, so inconsistent latency is not expected to cause problems in grid operations.

2.5 Summary

This chapter has discussed the impact of synchrophasor data quality on grid operations. Data quality is defined as aspects that affect its intended use. It has been broken down into six categories of errors that can occur in data. Each of these was defined and described. Error detection and control of errors were described, including procedures to assure better data quality. Finally, the impact of each of these types of errors on control center applications was discussed.

Data errors can occur at any time. They can be due to limitations of the measurement system, problems in the system, equipment failure and inadvertent problems created by system, and equipment changes. Continuous checking of data applications and system performance monitoring is essential to maintaining a high-quality system. Monitoring should include a timestamped record of detected problems as well as statistics on performance. Regular reporting shows changes in the system and alerts users to problems that may not be obvious from the use of the system. Detailed reports also help identify the root cause and location of a problem. Established maintenance and repair procedures complete the cycle to keep the system at a full performance level.

Synchrophasor measurements are increasingly being used in control center operations. They support overall grid awareness and alert operators to situations that they could not observe before, such as oscillations in the system and voltage or frequency instability. Maintaining data quality is essential to making these and other synchrophasor-based applications reliable and usable. Most data quality problems can be detected through monitoring and corrected before they affect the applications they support. With carefully planned and well-executed procedures, data quality can be kept at the level required for control center operation.

References

1. NASPI PMU data quality report, NASPI PMU applications requirements task force, NASPI-2017-TR-002 and PNNL-26313
2. IEEE Std. C37.118.1 & C37.118.1a, IEEE standard for synchrophasor measurement for power systems
3. Best practice recommendations for synchrophasor systems, synchrophasor data validation and conditioning project, phase 1 task 2 report, DOE project DE-AC03-76SF0098
4. IEEE guide C37.242, guide for synchronization, calibration, testing, and installation of phasor measurement units for power system protection and control
5. Quint R, Martin K, Nuthalapati S (Oct 2016) PMUs & critical infrastructure protection. In: Session 7 NASPI technical meeting. www.NASPI.org
6. IEC/ IEEE 60255-118-1, Standard for synchrophasor for power systems—measurements, 1st edn

Chapter 3
Testing and Validation of Synchrophasor Devices and Applications

P. Banerjee, S. Pandey, A. K. Srivastava and D. Lee

3.1 Introduction

Synchrophasor-based monitoring and control applications are being integrated into power grid control centers or being explored in pilot phase projects to realize the vision of real-time monitoring and control of bulk power system. The phasor measurement units (PMUs) are critical devices for implementing such advanced applications. Before deploying the PMUs and associated synchrophasor systems as well as applications in field, it is important to test and validate performance of the synchrophasor applications and associated devices to ensure high reliability and accuracy under different operating scenarios of the power system for specific applications.

The quality of synchrophasor data streams from PMUs of different vendors deployed in the power grid may not be the same. The inaccuracies and noises in the PMU data streams may impact the performance of synchrophasor applications. The different aspects of PMU data quality along with various tests that can eliminate bad data from PMU measurements are enlisted in this chapter. The impacts of PMU data quality on applications following the PMU Application Requirements Test Framework (PARTF) are also reviewed in this chapter. The software tool for data-driven event detection in real time has been used as one of the applications, and the results of the accuracy of event detection in the presence of bad data are presented in this chapter.

The synchrophasor application or wide area application mainly undergoes three stages before field deployment as shown in Fig. 3.1. The first stage is the algorithm development based on differential and algebraic equation (DAE) model of the power system. In this stage, the validation of the application is performed using off-line

P. Banerjee · S. Pandey · A. K. Srivastava (✉)
Washington State University, Pullman, WA, USA
e-mail: asrivast@eecs.wsu.edu

D. Lee
Southern California Edison, Rosemead, CA, USA

S. Nuthalapati (ed.), *Power System Grid Operation Using Synchrophasor Technology*, Power Electronics and Power Systems,
https://doi.org/10.1007/978-3-319-89378-5_3

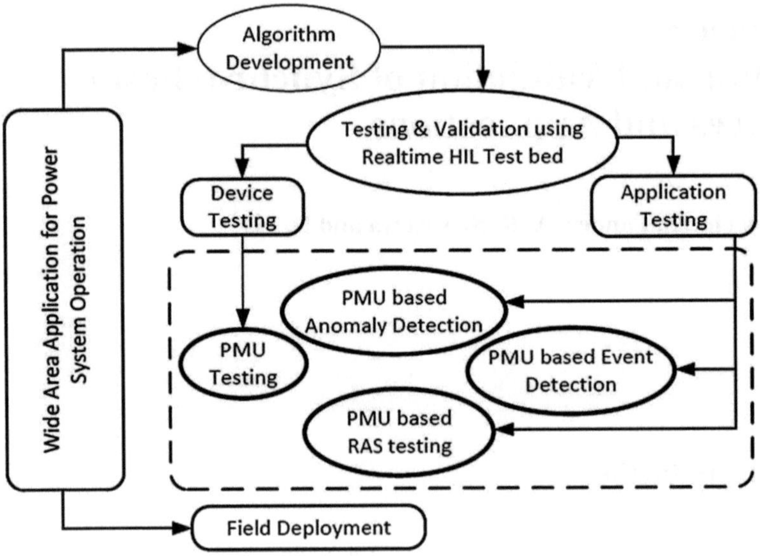

Fig. 3.1 Testing and validation of PMU and specific applications addressed in this chapter

simulators. The off-line simulators provide same simulation time increment for both the system and the controller model in the same simulation framework. The second stage is real-time testing and validation using hardware-in-the-loop (HIL) test bed. The application or the controller implemented in hardware setup cannot be validated using off-line simulators. The simulation time of the system model in the off-line simulators is not synchronized with the actual time used by the hardware controller, resulting in time skew of the controller output to the system or system response to the controller. The real-time simulators using HIL test bed simulate the test system in real time, and synchronization with the actual time of the hardware controller is guaranteed. The real-time HIL test bed emulated input–output equivalent of the true power system.

The synchrophasor-based control application consists of a large number of interacting components in the power system network. The overall application can degrade the system performance, even if one or more devices are individually tested and validated. Even each device may be validated satisfactorily, there are chances that the system consisting of one or more devices may fail to achieve the desired objective. Validation and testing of the application with one or more devices as a single entity is also desirable before field deployment. The requirement of end-to-end testing of such control application is discussed in this chapter. The Remedial Action Schemes (RASs) is considered as an example synchrophasor-based control application for performing end-to-end testing. The requirements of a RAS test bed are enlisted, and architecture of middleware called Erkios for RAS testing developed at WSU is presented in this chapter. The developed test bed is evaluated for single substation and multiple substation RAS.

Validation and conformance testing of PMU performed at South California Edison (SCE) using the PMU Performance Analyzer (PPA), software tool developed at Washington State University is presented in this chapter as a part of device testing. The specification of testing equipment and architecture of test bed for the PMU testing and associated applications are also presented in this chapter. PMU-based anomaly detection and PMU-based event detection are the two monitoring application presented in this chapter.

The results demonstrate that testing and validation of synchrophasor devices and applications is an iterative process to keep improving the performance of devices and applications involving all the stakeholders before the field deployment and even after field deployments for maintenance.

3.2 Review of Synchrophasor Technology and Phasor Measurement Unit (PMU)

The synchrophasor technology provides a real time-stamped magnitude and angle of voltage and current signals from different locations of the power system network. The two main components of the synchrophasor technology are the phasor measurement units (PMUs) used for estimating phasors and Global Positioning System (GPS) clock receiver used for time-stamping the phasors. The PMU estimates the voltage and current phasors, frequency and rate of change of frequency of the analog signal and transmits them to the local and the remote phasor data concentrator (PDC) for analysis and archiving as shown in Fig. 3.2. The PMUs are placed at all the five buses, which estimate the corresponding voltage magnitude and phase angle at each bus synchronized to the GPS clock signal. The five voltage phasors are transmitted to the PDC, located at the control center using fiber optic cable. The phasor

Fig. 3.2 Integrating synchrophasor technology in the power system

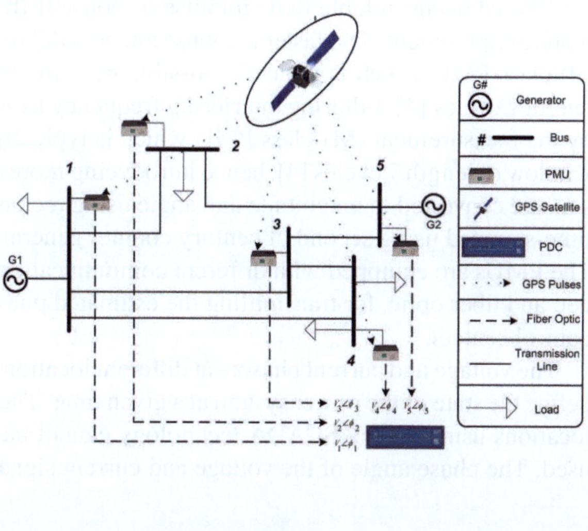

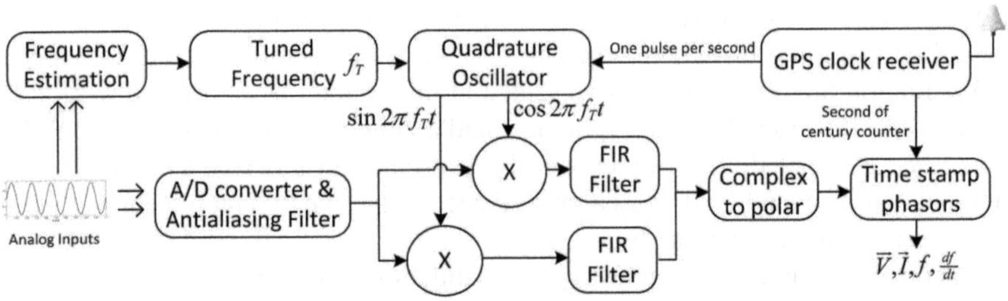

Fig. 3.3 Functional blocks of phasor measurement unit (PMU)

data received at the control center can be used for different applications, like state estimation, stability assessment, and emergency controls.

The internal functional blocks of the PMU are shown in Fig. 3.3. The scaled-down sinusoidal waveform of voltage and current signals is passed through an analog low-pass filter for removing the aliases by setting the filter cutoff frequency less than half the sampling frequency, thereby satisfying the Nyquist criteria. The analog-to-digital converter (ADC) generates microprocessor computable numerical values corresponding to the filtered analog signal. The number of bits of the ADC determines the quantization noise injected in the digital values of the converted signal. The ADCs of 12 bits and 16 bits are typically used for the PMU application. The tuned frequency f_T can be same as the nominal signal frequency of either 50/60 Hz or instantaneous signal frequency estimated using frequency estimator block. Interpolated discrete Fourier transform (IpDFT) [1], phase-locked loop (PLL) [2], and Prony [3] are some of the popular methods used for frequency estimation from the analog signal. The filtered signal is multiplied by sine and cosine signals which are synchronously generated by the quadrature oscillator based on one pulse per second (1PPS) signal from the GPS clock receiver. Multiplying of a signal with sine and cosine signal of frequency f_T is called hetrodyning. A pair of heterodyned signals are filtered using suitable finite impulse response (FIR) filter based upon the performance requirement. The faster response and less delay are required by the protection (P)-class PMU which is typically possible by using triangular window function of length 2 cycles [4]. Filtering interfering frequency from the analog signal is required by the measurement (M)-class PMU which is typically carried out using Hamming window of length 7 cycles [4], hence introducing more delay. The pair of filtered phasors are converted to magnitude and angle using rectangular to polar conversion and time-stamped using second of century counter generated by the GPS clock receiver. The PMUs are equipped with different communication interfaces, like serial, Ethernet, and fiber optic, for transmitting the estimated phasors to the PDC located at the control centers.

The voltage and current phasors at different locations in the power system network define the state of the power system at a given time. The measurements from different locations using legacy SCADA technology cannot be time-aligned as no GPS was used. The phase angle of the voltage and current signals is not possible to report by the SCADA. These shortcomings are addressed by the synchrophasor technology, which makes it a popular choice for input data for a wide range of power system monitoring and control applications.

3.2.1 Synchrophasor Applications in Power System Operation

The ability to measure phase angle and frequency of the positive sequence voltage signal from different locations of the power system network was the primary motivation for developing synchrophasor technology [5]. The NERC report [6] categorizes synchrophasor-based power system applications as real-time applications (frequency stability, power oscillations, voltage stability, event detection, state estimation, congestion management, outage restoration) and off-line applications (event analysis, model validation, load modeling, alarm settings, operator training). Synchrophasor applications are classified as [7] monitoring applications (frequency stability, state estimation), protection applications (backup protection, out-of-step protection, adaptive security–dependability protection), and control applications (damping controller, FACTS controller). A classification of synchrophasor applications, for example in which Angle frequency monitoring, Voltage stability, State estimation and Fault location are classified as Existing industry application while Restoration, Transient stability, Adaptive protection and Dynamic RAS are classified as Evolving application is provided in [8]. Hence, in a broad sense, the key applications of synchrophasor technology are [9]:

- wide area visualization,
- oscillation estimation,
- validation of generator parameter and model,
- islanding detection,
- voltage stability monitoring,
- potential malfunction identification of devices, and
- event analysis.

3.2.2 Need for Testing and Validation of Synchrophasors Devices and Applications

The synchrophasor monitoring and control application extensively use the measurements from the PMU, which may assist the system operator in the critical decision-making process and postmortem analysis after severe contingencies. The synchrophasor devices like PMU, PDC, and GPS clock are installed at the substation at the commissioning stage. These devices are expected to perform as per the standard, throughout their life cycle. However, the measurements from synchrophasor devices may be inaccurate due to the following reasons:

1. Different vendors use different phasor estimation algorithms, with variation in filters and window lengths.
2. The aging of insulation of CT/PT present inside PMU may result in attenuation of the signal and hence inaccurate phasor estimation.
3. A routine maintenance in substation may snap some of the PMU input connections that can remain undetected for a long time.
4. An internal bug in the phasor estimation process may appear after a long time due to incorrect handling of memory overflow or timer overrun.
5. Routine firmware upgrade of PMU devices is followed by complete PMU compliance testing.

3.3 Testing of Synchrophasor Devices

3.3.1 Review of Synchrophasors Standards and Compliance Testing of PMU

Different PMUs vendors in the market have adopted different algorithms for estimating the phasor values, and protocols for exchanging the phasor data to the control centers. The IEEE Standard for Synchrophasor Measurements for Power Systems [4] provides a guideline for the accuracy of the phasor estimation, and the protocol to be followed for the data exchange from the PMUs. The first edition of the synchrophasor standard was released as IEEE Standard 1344–1995 in 1995 [10]. This standard mainly laid the basis for the GPS clock synchronization accuracy requirement for the PMUs, located in the field. The accuracy of the GPS clock providing 1PPS signal is upto 1 μs. This standard, though mentioned some transient scenario, did not specify any test case for the performance evaluation of the PMUs.

Subsequently, IEEE Standard C37.118, in 2005 [11] superseded [10] and specified various test cases for the performance evaluation of the PMUs. The total vector error (TVE) was introduced for comparing the merits of the PMUs from different vendors. The TVE is the vector difference between the estimated phasor and the actual phasor, expressed as the percentage of the actual phasor. The TVE in percentage at any time instant is defined as

$$TVE = \sqrt{\frac{(X_r(n) - X_r)^2 + (X_i(n) - X_i)^2}{X_r^2 + X_i^2}} \times 100 \tag{3.1}$$

where $X_r(n)$ and $X_i(n)$ are the real and imaginary components of the estimated phasor, and X_r and X_i are the corresponding actual value of the phasors at any time instant, respectively. Apart from TVE, several other error metric are also used for compliance testing of PMU, which are frequency error (FE) and rate of change of frequency error (RFE) as defined below

$$FE = \left| f_{true} - f_{reported} \right| \tag{3.2}$$

$$RFE = \left| ROCOF_{true} - ROCOF_{reported} \right| \tag{3.3}$$

where $f_{reported}$ and f_{true} are the estimated frequency and the true frequency of the voltage signal, respectively, and $ROCOF_{reported}$ and $ROCOF_{true}$ are the estimated and the true rate of change of frequency (ROCOF) of the voltage signal, respectively.

The performance evaluation of the PMUs is based on the TVE reported for the different test cases specified in [11], which include off-nominal frequency, steady-state variation in amplitude and phase, and harmonic distortion along with out-of-band interference under steady-state operating conditions. The maximum value of the TVE stipulated for all the test cases, prescribed in [11], is 1%, which include summation of time synchronization error, instrumentation error, and processing error. This standard [11], however, did not cover the performance requirement of the phasor calculation under dynamic conditions.

The test cases for the dynamic signal variation are added in the revised version of the IEEE Standard C37.118.1 for synchrophasors, in 2011 [4]. The synchrophasor standard C37.118.1 [4] mentions different sets of tests for P-class and M-class PMUs. The P-class PMUs utilize shorter window length (in the range of one or two cycles) and provide less filtering of the phasors. P-class PMUs are typically used for faster operation with limited accuracy of the phasor measurements during dynamic and transients. The M- class PMUs use larger data window and larger filters to eliminate aliased signal from the fundamental component, and hence, [4] out-of-band interference test is required to be performed for M-class PMUs. To filter out interference signals from the fundamental, the required length of the data window is typically greater than 5 cycles of the fundamental component, which results in larger reporting delays but more accurate phasor measurements. The frequency error (FE) and rate of change of frequency error (RFE) are the additional performance criteria added in this standard for the performance evaluation of the PMUs. The dynamic signal variation added for testing the PMUs are amplitude and phase modulation, frequency ramp and step change of amplitude and phase. The response time, delay time, and overshoot for a typical step response test case are shown in Fig. 3.4. The maximum TVE specified is 1 or 3% for different test cases. The response time of FE and RFE for the PMUs of different classes is also separately specified for each of the test cases in the standard [4].

The IEEE synchrophasor standard [4] is amended in C37.118.1a.2014 [12], in which the RFE values for most of the test cases are revised. The oscillation test signal, which was previously carried out by using a signal having joint amplitude and phase oscillation, has been changed into two separate signals for the amplitude and the phase oscillations in the revised version of the standard [12]. The reporting latency of the M-class PMUs is also increased in the revised version of the standard [12].

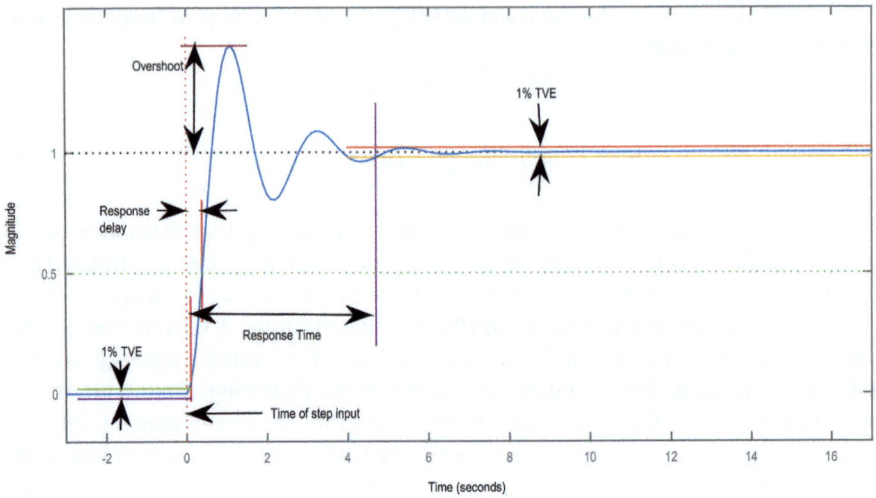

Fig. 3.4 Definition of various parameters for step response

3.3.2 Procedure and Hardware Requirement for PMU Testing

The IEEE TSS specifies the detail of the testing procedure for different test cases needed for testing a phasor measurement unit (PMU).

3.3.2.1 A Brief Summary of the PMU Testing Process is Presented Below

The procedure for PMU testing for magnitude range test:

1. Start the voltage and current at the lower magnitude and at the nominal frequency for 5 s.
2. Increase the input magnitudes by 10% of the nominal value and wait for 5 s.
3. Repeat Step 2 till the upper magnitude limit is reached.

The procedure for PMU testing for frequency range test:

1. Start the voltage and current at rated magnitude at the lower frequency limit for 5 s.
2. Increase the input frequency by 0.1 Hz of the nominal value and wait for 5 s.
3. Repeat Step 2 till the upper frequency limit is reached.

The procedure for PMU testing for harmonic distortion test:

1. Start the voltage and current at rated magnitude with 1% (for P class) or 10% (for M class) second harmonic at rated frequency for 5 s.

2. Increase the harmonic number by 1 and wait for 5 s.
3. Repeat Step 2 till the upper harmonic limit is reached.

The procedure for PMU testing for out-of-band interfering test:

1. Start the voltage and current at rated magnitude at rated frequency with 10% of rated magnitude of the first lower frequency limit of interfering signal for 5 s.
2. Increase the interfering signal frequency in exponential steps and wait for 5 s.
3. Repeat Step 2 till the interfering signal frequency reaches the first upper out-of-band limit.
4. Change the interfering signal frequency to the second lower out-of-band limit and wait for 5 s.
5. Increase the interfering signal frequency in exponential steps and wait for 5 s.
6. Repeat Step 5 till the interfering signal frequency reaches second upper out-of-band limit.
7. Repeat steps 1–6 for input signal frequency at $\pm 10\%$ of the nominal frequency of PMU reporting rate divided by two.

The procedure for PMU testing for step change of magnitude and phase test:

1. Start the voltage and current at rated magnitude and phase at the nominal frequency for 5 s.
2. The voltage and current magnitudes are changed to +10% and −10% of the initial value and remain for 1 s.
3. The voltage and current magnitudes are changed to nominal value after 1 s.
4. The steps 2 and 3 are repeated 10 times with the starting of the step shifted by PMU reporting time interval divided by ten w.r.t to the first step instant.
5. The steps 1–4 are repeated for +10 and −10° phase change keeping the magnitude at rated value.

The procedure for PMU testing for positive and negative frequency ramp test:

1. Start the voltage and current at rated magnitude at lower frequency limit for 5 s.
2. The frequency of the voltage and current is changed from lower limit to upper limit at 1 Hz/s.
3. Wait for 5 s at the upper frequency limit
4. Decrease the frequency from the upper limit to lower limit at −1 Hz/s.

The procedure for PMU testing for AM, PM, and joint AM–PM test:

1. Start the voltage and current with the modulation magnitude of 10% of nominal and modulation frequency of 0.1 Hz for 5 s.
2. Increase the modulation frequency by 0.2 Hz and wait for 5 s.
3. Repeat Step 2 till the upper modulation frequency limit is reached.
4. Repeat steps 1–3 for phase modulation keeping the magnitude at rated value.

3.3.2.2 Specification of the Equipment for PMU Testing Laboratory

Signal source: The PMU should be given input from a programmable voltage and current signal source.

1. The signal source should have at least 1 three-phase output for voltage and 1 three-phase output for current.
2. The voltage magnitude range of the signal source should be in the range of 12–600 V.
3. The current magnitude range of the signal source should be in the range of 1–15 A.
4. The signal source should generate the fundamental signal frequency from 45 to 65 Hz.
5. The magnitude of the voltage and current signals should be controllable using mathematical functions like level shifter, trigonometric functions, and algebraic equations.
6. The phase of the voltage and current signals should be controllable using mathematical functions like level shifter, trigonometric functions, and algebraic equations.
7. The signal source should also be able to inject atleast one harmonic with 10% of the rated voltage and current magnitudes. The single harmonic frequency should be variable from 100 to 3000 Hz.
8. The signal source should be controllable at required instances of the GPS pulses.

   GPS time source: the time input to the PMU and the signal source.

1. The GPS time source should be obtained from a GPS antenna.
2. The option to generate 1PPS, IRIG-B through copper or fiber should be present in the GPS clock receiver.

PMU measurement receiver: The phasor data concentrator (PDC) is typically used for archiving the C37.118 phasor data from a PMU.

1. The PDC should be configurable to receive C37.118 phasor data from Ethernet port or serial port.
2. The PDC should have sufficient memory to archive three-phase voltage, current, frequency, and rate of change of frequency data.
3. The PDC should be able to export the archived phasor data into computer processing data file like .xlsx or .csv files.

3.3.3 Testbed Architecture and PMU Performance Analyzer (PPA)

The PMU estimates the magnitude and angle of the voltage and current signal samples buffered for a window length. The PMU uses the GPS time stamp at the center of the data window to report the magnitude and angle. The true value of phasor at the center of the data window is required to report the accuracy of the estimated phasor by the PMU. The magnitude and angle of the voltage and current signals mandated by [4] for PMU testing may be constant with respect to time for static test cases or may be a simple function of time for dynamic test cases. In view of this, the PMU testing setup should have mechanism to obtain the GPS time and establish the true value of magnitude and phase which may be dependent on GPS time stamp, particularly for the dynamic test cases mentioned in [4]. The architecture of the test bed for PMU testing is shown in Fig. 3.5. The main components of the test bed are:

1. Programmable signal generator: The Real-Time Digital Simulator (RTDS) [13] or National Instruments CompactRIO (NI-cRIO) is used for generating programmable low voltage signal. Both the signal generators use GPS to generate time-varying dynamic voltage signals. The signals generated by RTDS are modeled in RSCAD, and the signals generated by NI-cRIO are modeled in LabVIEW software running in a computer.
2. Amplifier: The voltage and current amplifiers generate amplified voltage and current signals corresponding to low voltage signal from the signal generators.
3. PMUs: The amplified voltage and current signals are provided to the PMU under test. The GPS signal is also provided to the PMU to time-stamp the reported phasor.
4. PDC: The time-stamped phasor data from the PMU under test is aligned and archived by the PDC. The PDC may be a hardware type from any vendor or a software type like OpenPDC [14].
5. PMU performance analyzer: PMU performance analyzer (PPA) is a software tool, which imports phasor data from PDC, generates true phasor using the GPS time stamps from RSCAD or LabVIEW, and generates compliance report of the PMU under test.

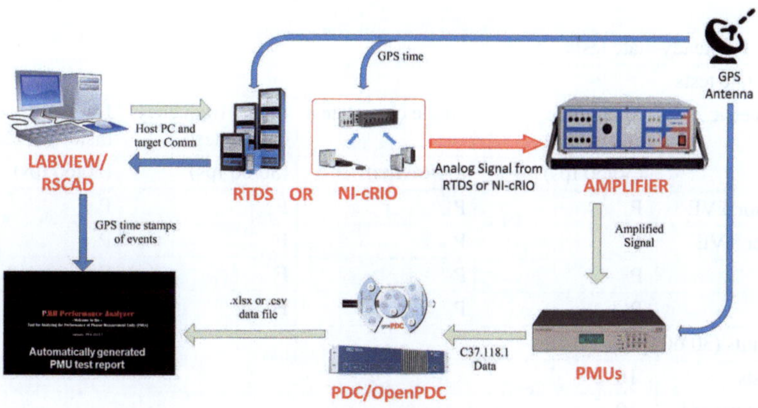

Fig. 3.5 Testbed architecture for PMU testing using PPA

3.3.3.1 Required Features of PPA

1. True value: It is assumed that the test PMU is provided with time-synchronized analog signal having parameter variation at different time instants. The signal generator changes one analog signal parameter at a known time instant and continues that parameter for a fixed time interval. The PPA should be able to generate true value of the phasor using the GPS time stamps of the phasor data.
2. The PPA should be able to process the exported .xlsx or .csv data files from the PDC.
3. The PPA should be able to calculate the total vector error (TVE), phasor magnitude error (ME) and phase angle error (PE), frequency error (FE) and rate of change of frequency error (RFE), and additional calculations for the dynamic step test results.
4. The PPA shall have a means of determining the time of arrival of PMU data messages and comparing that time against the message time stamp.
5. The PPA should be able to generate test result documentation and pie charts for the success rates of the PMU testing.

3.3.4 Example Results for PMU Testing at South California Edison

South California Edison (SCE) performed compliance testing of a DFR PMU according to the error metric mentioned in [4] for P-class PMU. The RTDS is used as programmable signal generator, and hardware PDC is used for archiving phasor data from the DFR PMU under test. The testing was done for different PMU phasor reporting rates of 30 and 60 fps. The steady-state test of magnitude change and angle change is complied by the PMU for all the error metric for both 30 and 60 fps, as shown in Table 3.1. The frequency change test is not complied by the PMU for all

Table 3.1 Steady-state tests

Steady-state tests				
Error metric	Magnitude change test	Angle change test	Signal frequency change test	Harmonic distortion test
	(30/60 fps)	(30/60 fps)	(30/60 fps)	(30/60 fps)
V Phasor TVE	P	P	F	P
I Phasor TVE	P	P	F	P
FE	P	P	F	F
RFE	P	P	F	F
Test results (30/60 fps)				
# of tests	16			
# of test passed	10			
Success rate	62.50%			

the error metric for both 30 and 60 fps. The harmonic distortion test is complied by voltage and current TVE for both 30 and 60 fps. The FE and RFE for 30 and 60 fps are also not complied for the harmonic distortion test. Out of the dynamic test of both positive and negative change in magnitude and angle, the PMU complied for only positive magnitude change of the response time for 60 fps. The overshoot metric for all the step change is complied for both 30 and 60 fps. The dynamic test of positive and negative frequency ramp is not complied for any of the error metric for both 30 and 60 fps, as shown in Table 3.2. The dynamic tests of amplitude modulation, phase modulation, and joint amplitude and phase modulation are complied for all the error metric for both 30 and 60 fps, as shown in Table 3.3. Hence, only a slight improvement (3.57%) on pass rate for dynamic tests with 60 fps was observed for this DFR-type PMU.

Table 3.2 Dynamic tests

Dynamic tests					
Error metric	Magnitude step change +10%		Magnitude step change −10%	Angle step change +10°	Angle step change −10°
	(30 fps)	(60 fps)	(30/60 fps)	(30/60 fps)	(30/60 fps)
Response time	F	P	F	F	F
Overshoot	P	P	P	P	P

Table 3.3 Dynamic tests

Dynamic tests					
Error metric	Positive frequency ramp	Negative frequency ramp	Amplitude modulation	Phase modulation	Joint amplitude and phase modulation
	(30/60 fps)	(30/60 fps)	(30/60 fps)	(30/60 fps)	(30/60 fps)
V Phasor TVE	F	F	P	P	P
I Phasor TVE	F	F	P	P	P
FE	F	F	P	P	P
RFE	F	F	P	P	P
Test results					
# of tests	28 (30/60 fps)				
# of test passed	16 (30 fps) 17 (60 fps)				
Success rate	57.14% (30 fps) 60.71% (60 fps)				

3.4 Testing and Validation of Synchrophasors-Based Monitoring Application

Validation of synchrophasor-based monitoring application is important as it affects the key decision criteria for the system operator. The PMU data quality severely affects the synchrophasor-based monitoring application. PMU data may incur errors due to several reasons. It is imperative to test the application with a clean set of data and with data set containing anomalies. Different applications can handle different thresholds of the percentage of bad data. An application that uses 300 data points might not be affected by 3 bad data points, whereas an application that uses 30 data points per window might give erroneous results if 3 data points are bad. Hence, it is important to test every application for the limit of data anomalies it can handle. It is also important to preprocess the PMU data before it is used for applications.

3.4.1 PMU Data Quality and Impact on Applications

Most of the sophisticated wide area monitoring and control applications for power system specify the availability of time-stamped phasor data at regular intervals whose data rate is mandated by the application requirement. The anomalies in PMU data can result in incorrect wide area visualization, maloperation of PMU-based control logic, and generation of incorrect alarms by the monitoring application. The large deployment of PMUs in the field is carried out in anticipation of high availability of synchrophasor measurements in the control center. However, the availability of good quality phasor data in the control centers is not guaranteed. In fact [15] presented that there is an increase in the percentage of unusable phasors in the eastern interconnection from 2009 to 2011, mainly due to the fact that the PMUs deployment in the field increases and the communication bandwidth is limited. Similar statistics are also observed in western interconnection, where CAISO reported that 18% PMUs are down or out of sync. The latency reported BPA goes as high as 200 ms for legacy communication network, which has been reduced to around 40 ms using recently upgraded communication network. The flow of the phasor data from the PMUs located in the field to the end application computer in the control center is shown in Fig. 3.6 [15].

The four of the total eight stages are identified to detect any possible errors in the PMU data. The first stage of error detection is the PMUs which can flag (a) data invalid due to large variation in input signal resulting in incorrect phasor estimation, (b) PMU error due to malfunction of any hardware component of the PMU, (c) synchronization error due to GPS satellite unavailability, and (d) time alignment error caused by inaccurate time tags of the phasors mainly due to unreliable GPS satellite.

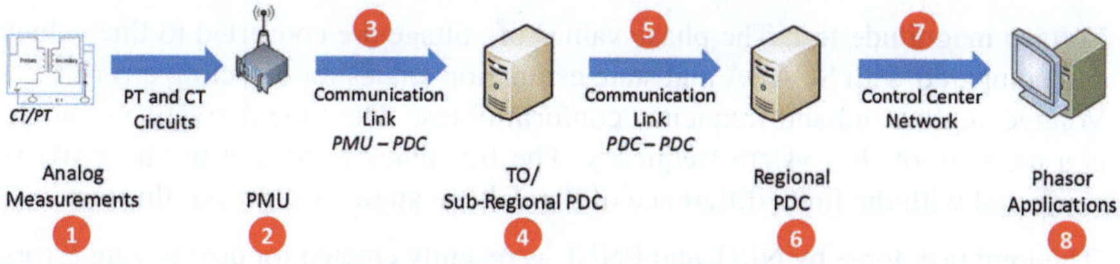

Fig. 3.6 Data flow in synchrophasor applications

The error detection of phasor data by subregional PDC and regional PDC can flag (a) data dropout error due to high latency in the communication network, (b) invalid time stamp, (c) unacceptable latency of the data packet, (d) invalid format of the data packet, and (e) parity and CRC error. The phasor application software in the control center can also detect several errors in the PMU data based on (a) sanity checks using SCADA data and (b) topology checks using the state estimator. [15] listed some of the scenarios of PMU data anomalies which arise from

- loss of data from one or several PMUs,
- loss of signals in a PMU,
- stale (non-refreshing) data,
- inconsistent data, data rates, and latencies,
- offsets in signal magnitude and phase,
- corrupted and drifting signals in a PMU,
- corrupted and drifting time reference in one or several PMUs,
- combination of several issues described above, and
- the failure of the topology processor and/or bad/incomplete topology information.

The reliability of the PMU is studied in [16] using hierarchical the Markov model. The proposed Markov model considered GPS signal and the PMU central processing unit module as the primary source of PMU data failure. The reliability model of PMUs using fuzzy sets is developed in [17] considering series model of the individual PMU components. The availability of the failure data of each component is difficult to obtain and fit in the series model. Reinhard [18] identified the primary sources of failure of PMU data are GPS signal, communication system, and combination of these for which the Markov model is developed. A comprehensive PMU data validation technique followed in ISONE is presented in [19]. The list of tests for PMU data validation in [19] is presented below:

- Reasonability test: The errors in time stamp, synchronization error, value range, and stale data are carried out in this test.
- Voltage angle test: The relative angle difference between the PMU measurements and the state estimation is monitored to be within the threshold value. The threshold value is determined from the typical difference in different loading conditions.
- Power flow test: Individual power flow from PMU and SCADA measurements is carried out, and errors between them are monitored for PMU data anomalies.

- Voltage magnitude test: The phase values of voltage are converted to line values and compared with SCADA and state estimation values for detecting errors.
- Voltage angle trend and frequency conformity test: The time derivative of angle is a measure of the system frequency. The frequency estimated by the PMU is compared with the finite difference of the voltage angle to carry out this test.

The joint task force by NIST and PNNL is recently created for quantifying errors caused by the network and phasor estimation algorithms and its impact on different WAMS applications. The wide area application performances are evaluated as a function of the different error axis. This is possible by carrying out multi-dimensional PMU error analysis. The possible list of PMU errors mentioned in [20] are (a) PMU measurement error characteristics (b) Clock error and weakness (c) Network and communications error and weakness The Frequency Response Analysis Tool (FRAT) is being specially developed for analyzing the frequency response of power system interconnection which is extended to quantify FRAT Application performance envelope. The communication network modeling using PDC data packets buffer and packet delay is also mandated by the task force.

3.4.2 PARTF Framework for Analyzing Impact of PMU Data Quality on Applications

The North American Synchrophasor Initiative (NASPI) released a document PMU Applications Requirements Task Force (PARTF) Framework. This document discussed about the data quality issue and tries to standardize the data attribute terminology and definitions. The PMU data can be used for applications which may use a single data points or a large data set and some applications might use the PMU data stream. There are several things that could lead to a bad PMU data quality. The document tries to categorise it into four types as given below:

1. At or within the PMU: It is possible that the CTs and PTS connected to the PMU have biases, the class of CT and PT Utilized (measurement or protection), differences or errors within the PMU and its filters and algorithms, or erroneous or missing timestamps associated with the measured grid conditions. These factors affect how the individual data point is created and whether it is accurate.
2. At any intermediate data aggregators: From point of the generation to point of use the data point can be stored at a number of places which may induce some error in the form of misalignment, erroneous compression, creation of duplicate data, or loss of data due to late deliveries. There can also be mislabeling or data losses at aggregators end.
3. Within the communications network: There can be a loss of communications network nodes, excess latency delaying delivery beyond the acceptable time window of the PDC, or data corruption introducing erroneous values.
4. At the point of use by the application: It can be a data storage issues or an insufficient application training period for an application that requires enough historical

data to show the application and its users the appropriate range of application solutions and exceptions.

These issues might corrupt the PMU data and make it unusable for PMU data-driven applications. Therefore, a PMU data preprocessing is required before using it for applications.

The PARTF framework also states the process of looking at the data from the applications requirements viewpoint and the impact of data issues on specific applications.

3.4.3 Preprocessing PMU Data for Event Detection

The PARTF framework outlines the data quality issues and application might not give the desired result if the data set contains bad data. Therefore, it is necessary to preprocess the PMU data, identify the bad data points and then it can be handled application specific.

An MLE-based unsupervised learning algorithm in combination with statistical, clustering and Prony analysis algorithm constitute a tool called SyncAD (synchrophasor anomaly detection). This tool is used to detect the bad data in PMU measurements. The architecture of the SyncAD tool is shown in Fig. 3.7.

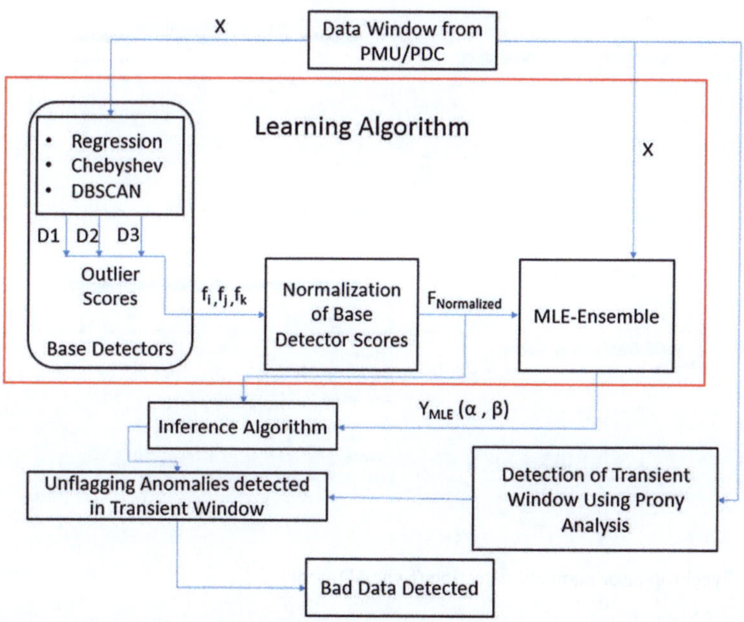

Fig. 3.7 Architecture of anomaly detection algorithm

Initially, linear regression, Chebyshev, and DBSCAN algorithm with loosely set threshold are applied on the PMU data window and compute the outlier score. It is important to analyze the outlier scores from the three base detector using the same frame of reference. Hence, the outlier scores are normalized using the Expectation Maximization (EM) algorithm. The normalized outlier scores are used by the MLE-Ensemble to compute the weight α and β. Subsequently, the Inference Algorithm uses the weights obtained from the MLE-Ensemble flags the data anomalies in the new data set. Finally, the Prony analysis is run to determine the transients in the data window. If the data anomalies are marked during transients, it is then unflagged and treated as good data. The reason behind this step is that the Anomaly Detection algorithm detects the extreme data points during a transient as anomalies. During an event, the fluctuations are high; therefore, linear regression, Chebyshev, and DBSCAN can flag them as anomalies.

The screenshot of the Synchrophasor Anomaly Detection (SyncAD) is shown in Fig. 3.8. PMU files are imported using the input browse button. The tool reads all the available PMU names. The User can select the PMUs to run the Anomaly detection algorithm and save the output file. The user can either replace the anomaly data point by the average or flag those instances. The SyncAD tool also shows the anomaly plot, original data plot, and the clean Data plot.

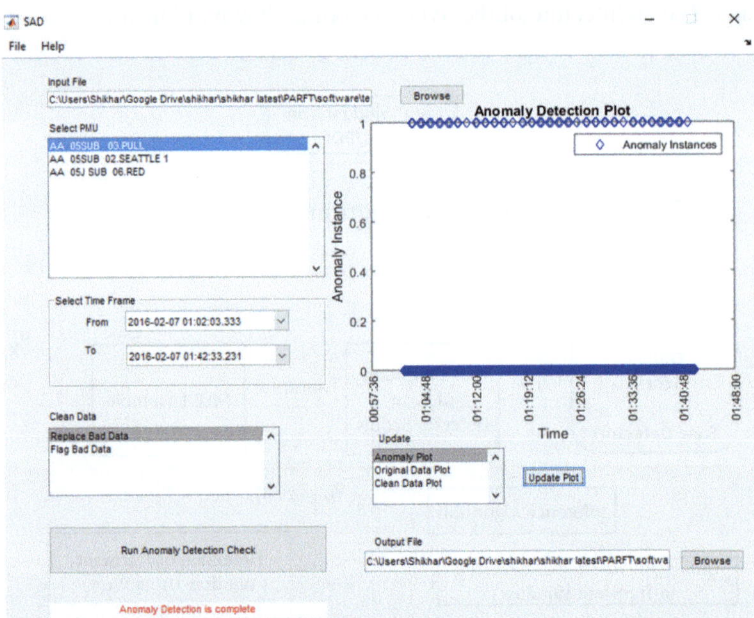

Fig. 3.8 Synchrophasor anomaly detection (SyncAD) tool

3.4.4 Synchrophasors-Based Event Detection

The typical refresh rate of SCADA measurements is every 4 s, which is used by the State Estimator (SE). The SE detects any topological and switching changes in the system. The high sampling rate of PMUs makes the real-time wide area view of the system possible. Researchers have used several techniques to detect events happening in the power system and have tried to classify them into Active Power and Reactive Power Events. A method to detect and classify the events into Active Power Event, Reactive Power Event and Fault Event has been presented in this chapter. This algorithm uses Density-Based Spacial Clustering With Noise (DBSCAN) to detect the operating point change Voltage, Current, Active power, and Reactive power changes during an event. The process is explained in the flowchart shown in Fig. 3.9.

These cluster change instances are used by the decision tree to classifies the events as shown in Fig. 3.10.

3.4.5 Testbed Architecture for Validation of Event Detection Application

Real-Time Digital Simulators (RTDS) were used to simulate the IEEE-14 bus in Smart Grid Demonstration and Research Investigation Lab (SGDRIL) at Washington State University (WSU) Pullman. The test bed architecture is shown in Fig. 3.11.

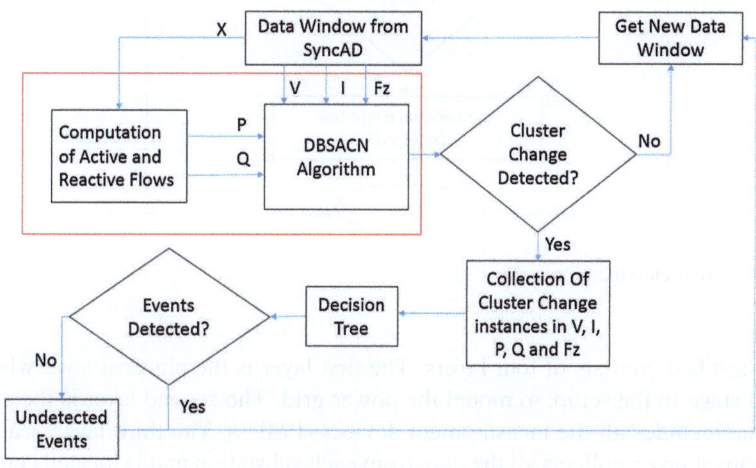

Fig. 3.9 Event detection algorithm architecture

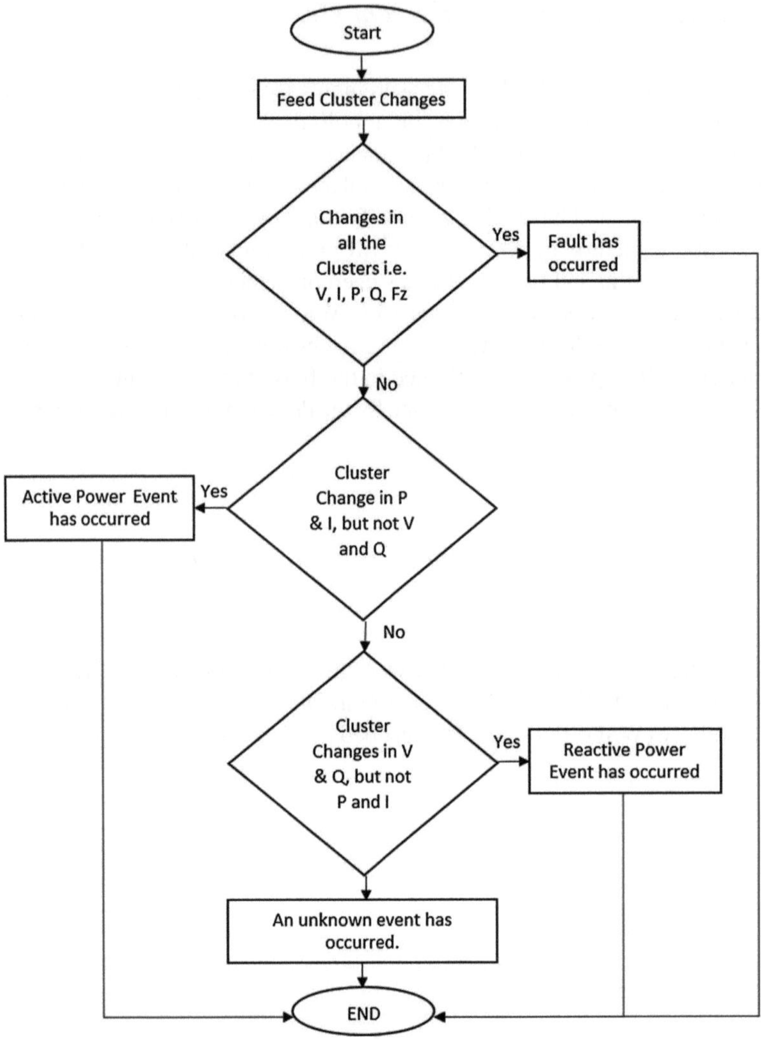

Fig. 3.10 Event classification

The test bed consists of four layers. The first layer is the physical layer which is the first stage in the setup, to model the power grid. The second layer is the sensor layer that includes all the measurement devices (PMUs). The third layer, real-time data archival layer, collects all the data from each substation and is located centrally in the control center. The anomaly detection algorithm in combination with event detection algorithm is deployed in the fourth layer.

Fig. 3.11 Test bed for performance evaluation of event detection

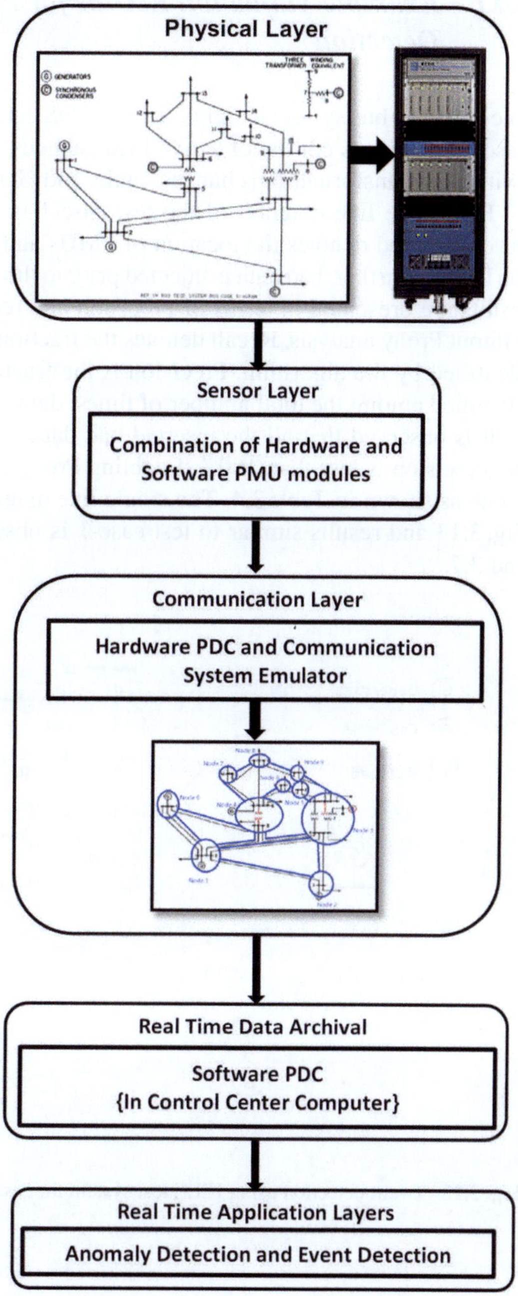

3.4.6 *Example Validation Results for PMU-Based Event Detection*

The IEEE-14 bus system is simulated in RTDS, and five PMUs are placed at different buses. The events consist of active load changes, reactive load changes, Cap Bank switching, transformer tap changes, faults and Generator outage.

The single line diagram of the test case-1 is shown in Fig. 3.12, in which the buses with red denotes the location of PMUs and the lines in red are monitored by the PMUs. Further, bad data is injected prior to the anomaly detection algorithm. The results are presented in terms of precision and recall with Prony analysis and also without Prony analysis. Recall denotes the fraction of the inserted bad data correctly identified by the algorithm. Precision is the fraction of inserted bad data correctly identified among the total number of times, data are identified as bad.

It is observed that all the inserted bad data are successfully detected; however, the precision is less than 100%. Enabling Prony analysis improved the precision to 100% as shown in Table 3.4. The single line diagram of the test case-2 is shown in Fig. 3.13 and results similar to test case-1 is observed as shown in Tables 3.5, 3.6 and 3.7.

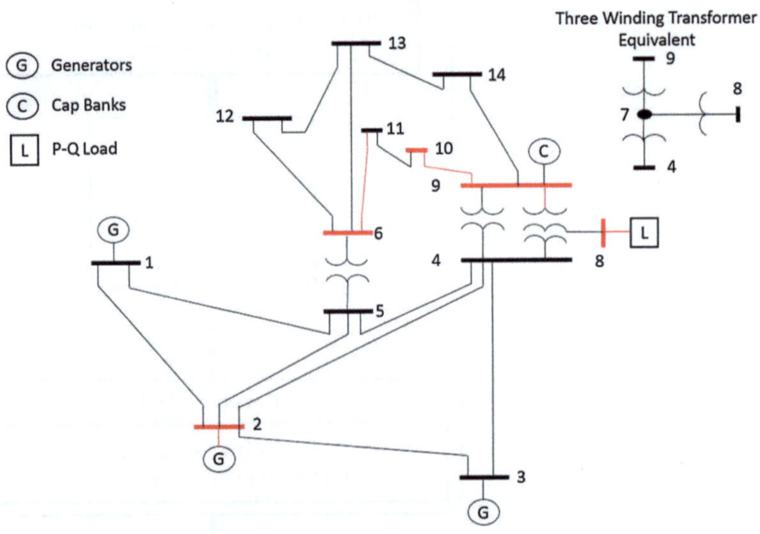

Fig. 3.12 Event detection using IEEE test system for case I

Table 3.4 Performance of anomaly detector case 1

S.No.	Without prony		With prony	
	Precision	Recall	Precision	Recall
1	0.9186	1	1	1
2	0.9080	1	1	1
3	0.9080	1	1	1
4	0.9186	1	1	1
5	0.9518	1	1	1

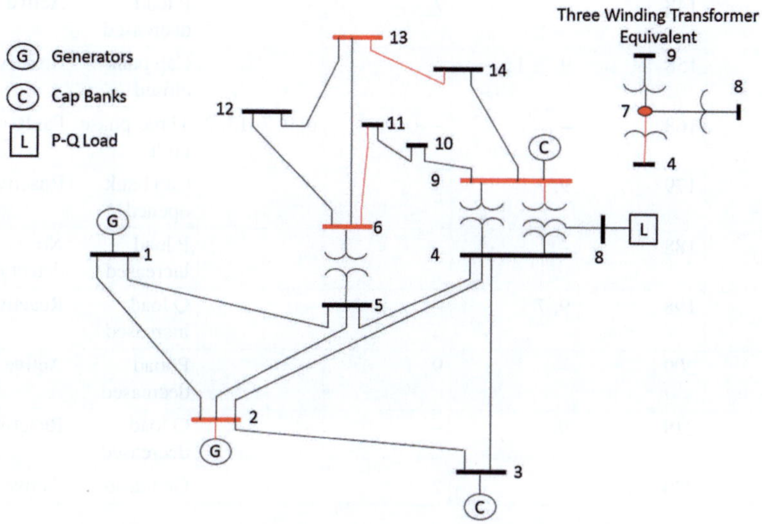

Fig. 3.13 Event detection using IEEE test system for case II

Table 3.5 Performance of anomaly detector case 2

S.No.	Without prony		With prony	
	Precision	Recall	Precision	Recall
1	0.8876	1	1	1
2	0.9404	1	1	1
3	0.9518	1	1	1
4	0.8876	1	1	1
5	0.9634	1	1	1

Table 3.6 Simulation result case 1

S.No.	Time (s)	Reactive event bus no.	Active event bus no.	Fault event bus no.	Actual events	Detected events
1	109	9, 7	–	–	Cap bank closed	Reactive
2	119	–	–	6, 9, 7, 13, 2	Three-phase fault	Fault
3	132	9, 7	–	–	Cap bank opened	Reactive
4	148	–	7	–	P load decreased	Active
5	158	9, 7, 13	–	–	Cap bank closed	Reactive
6	168	–	–	6, 9, 7, 13, 2	Three-phase fault	Fault
7	179	9, 7	–	–	Cap bank opened	Reactive
8	188	–	–	–	P load increased	No detection
9	198	9, 7	–	–	Q load increased	Reactive
10	209	–	9	–	P load decreased	Active
11	219	9, 7	–	–	Q load decreased	Reactive
12	229	–	2	–	Gen drop	Active

3.5 Testing and Validation of Synchrophasors-Based Control Application

The maturity of synchrophasor technology led to gradual adaptation of phasor measurements from PMU for control application. Real-time secondary voltage control compensating voltage fluctuation resulted from unexpected load disturbances is proposed in [21]. The direct observation of the phase angle difference in a power flow corridor using PMUs is one of the most important metric for real power flow control. Automatic feedback-based voltage and power flow control using phasor measurements are proposed by Marija in [22]. The proposed method demonstrates economic dispatch of resources to maintain (N1) reliability criteria. The phasor data from neighboring buses of a FACTS device like STATCOM have been used for generating control signal and control STATCOM in [23]. The reliability of the power system can be improved by preventing maloperation of distance relay in Zone 3

Table 3.7 Simulation result case 2

S.No.	Time (s)	Reactive event bus no.	Active event bus no.	Fault event bus no.	Actual events	Detected events
1	109	9, 7	–	–	Cap bank closed	Reactive
2	119	–	–	6, 9, 7, 13, 2	Three-phase fault	Fault
3	132	9, 7	–	–	Cap bank opened	Reactive
4	148	–	7	–	P load decreased	Active
5	158	9, 7, 13	–	–	Cap bank closed	Reactive
6	168	–	–	6, 9, 7, 13, 2	Three-phase fault	Fault
7	179	9,7	–	–	Cap bank opened	Reactive
8	188	–	–	–	P load increased	No detection
9	198	9, 7	–	–	Q load increased	Reactive
10	209	–	9	–	P load decreased	Active
11	219	9, 7	–	–	Q load decreased	Reactive
12	229	6	–	–	Tap down	Reactive
13	240	–	–	6, 9, 7, 13, 2	3 phase fault	Reactive
14	240	–	–	–	Cap bank closed	No detection
15	253	6, 13	–	–	Tap up	Reactive
16	263	–	2	–	Gen drop	Active

during power swing. A scheme of PMU-based blocking of Zone 3 operation of distance relay during power swing and not blocking during fault is presented in [24].

The operation of synchrophasor-based control applications depends on multiple components of the power system network. A typical set of components for such implementation is shown in Fig. 3.14. The power system layer consists of physical components of the system. The high voltage and current measurements are scaled to equivalent low voltage signal using the sensor and actuator layer. The sensor and actuator layer also consists of PMUs, which generate time-stamped digital phasor data from the low voltage analog signal. The data communication network consists of switches and fiber optic cables, which transmits the phasor data from the PMUs

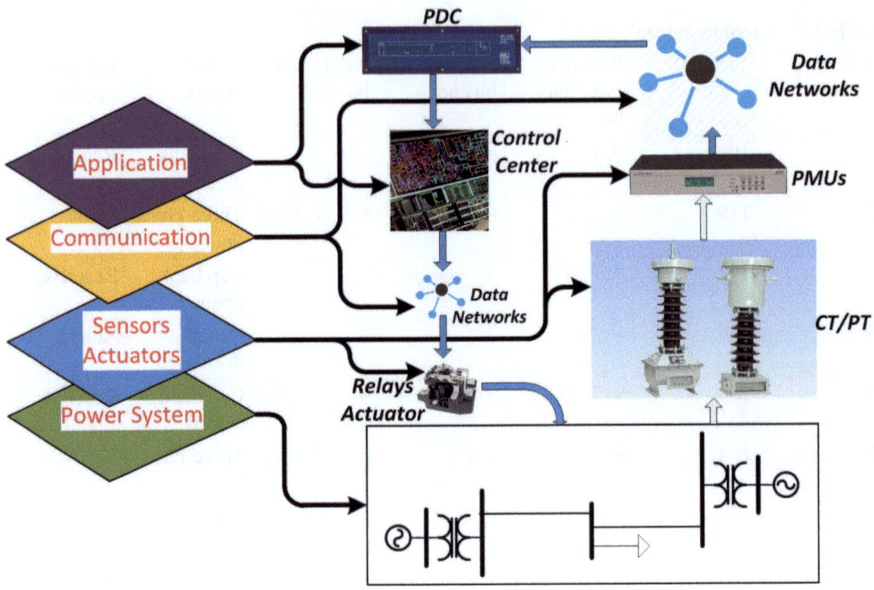

Fig. 3.14 Cyber-physical layers for synchrophasor control application

to the PDC located in the control center. The PDC and different control applications running in control center are part of application layer. The phasor data from the PDC are fetched by the control application and generate set points or actuation data. The control data from the application layer are again transmitted to the relays and actuators in the field through the communication layer. The relays and actuators in the field are part of sensors and actuator layer, which perform the control action in the power system layer.

Maloperation and failure are common for cause of performance degradation in a complex system like synchrophasor-based control applications. Most of the components in this architecture are tested and calibrated during installation of such devices. However, regular monitoring of individual devices and end-to-end testing is seldom carried out for the entire scheme. Hence, efficient and easy scheme for testing end-to-end synchrophasor-based control application is required.

3.5.1 Synchrophasors-Based Remedial Action Schemes

The PMU-based protection of power system equipment, particularly transmission lines are recently gaining interest in the industry. A study of the performance of special protection schemes (SPS) on Taiwan power system using PMU data is carried out in [25]. PMU data-based scheme of controlled separation of British Columbia and Alberta power systems has been demonstrated in [26]. Predicting instabilities using PMU data assisted with artificial neural networks (ANN) is proposed in [27]. The instability is mitigated by RAS which uses under frequency-based load shedding and islanding the system. A new special protection scheme (SPS) based on synchrophasor measurements from PMUs is proposed in [28]. The new RAS monitors the angles of the buses at large power plants in Alcan system and B.C. Hydro to prevent transient instability.

A summary of the survey of the System Integrity Protection Schemes (SIPS) across different continents is presented in [29]. The RAS deployed in North America constitute 65% of the global installations. The number of RAS installed in WECC system till 2012 was 190 [30], which increased to 282 till 2016 [31]. The reliability standard for RAS is drafted by NERC [32] in PRC0122, to ensure that RAS installation does not introduce unintentional or unacceptable reliability risks to the grid. The proposed standard [32] includes nine requirements generally organized in six areas:

1. R1 R3: Review process for new and modified RAS.
2. R4: Periodic Planning review of existing RAS.
3. R5: Review of actual RAS operations.
4. R6 R7: Corrective Action Plans to address deficiencies found in existing RAS via R4, R5, or R8.
5. R8: RAS periodic functional testing.
6. R9: RAS database.

The requirement R8 mention functional testing of typical RAS, every 6 years and limited impact RAS every 12 years. However, online testing of this important scheme is not possible without disturbing the normal operation of the system. The RAS logic is tested rigorously before commissioning in the field [33], but no further testing is performed during the lifetime of the RAS installation. Hence, correct working of RAS in case of a triggering event is uncertain after commissioning of RAS for 5–10 years. Therefore, in-field end-to-end remote testing of RAS is required to guarantee that it will actuate according to the schemes at all times and will not be blocked with any sort of software/hardware anomaly that may develop with time or any hidden failure that was not captured during design stage.

3.5.2 Testbed Architecture for RAS Testing

The following metrics of the RAS standard are monitored into the real-time test bed for reliable real-world validation.

1. Response time: The response time of RAS depends on the type of scenario for which it is deployed. Angular stability, short term voltage stability, and other similar problems require response time from a few cycles to a second. The slow stability problems require response time upto several seconds. Thermal overloading of transmission lines requires response time in the range from several seconds

to minutes based on the line rating and operating state. The response time is an important parameter to be validated for RAS testing. The delay caused by the test process and test network is an unavoidable operation of the test bed and, hence, should not be included in the evaluation. Further, the time taken for the actuator to react should not be included in the response time evaluation. The RAS system may include more than one utility which results in response time for the action can to include the time taken to communicate between different utilities. This issue is addressed by the NERC and WECC standards by stating the starting point to validate the RAS.

2. Communication Channels: The standards mandates that RAS should meet the Communication Systems Performance Guide for Protective Relaying Applications and Critical Communications Circuits-Guidelines for Design [34, 35]. This require that the RAS logic do not result in a false operation due to loss of channel, noise in the input signal or other failures. Similar communication design is also incorporated in the test bed for RAS testing.
3. Cyber Security: The NERC has adopted CIP-002-1 through CIP-009-1 standards for physical and electronic security of critical assets. The test bed is designed to follow these standards for performing end-to-end in-field testing.
4. Redundancy: The WECC standard states that, except for few devices like substation batteries, PT/CTs, breaker, and communication towers, all other devices installed in RAS logic and the RAS logic itself, should have redundancy. The standard also states that any misoperation of RAS should not have any adverse effect on the power system operation.

Erkios [36] is a middleware framework designed for in-field and end-to-end testing of RAS. The Erkios architecture is shown in Fig. 3.15, has four functional modules: Central Test System (CTS), Local Test System-Initiator (LTS-Initiator), Local Test System-Collector (LTS-Collector) and Test Wide Area Network (or Test WAN). The Central Test System acts as a coordinator during the RAS tests by generating the test signals required for testing and transmits the test signals to other modules. LTS-Initiator disables the PT/CT signals to PMU inputs and disables the RAS actuator output before starting the test. The LTS-Initiator enables the PMU inputs from PT/CT sensor and actuator for normal operation of RAS after the test is completed. LTS-Collector intercepts the remedial actions that the RAS has generated the emergency event of testing, and forwards it to the Central Test System. The Test WAN transmits data between Central Test System, LTS-Initiator, and LTS-Collector. The signal generator and amplifiers are used to generate the input signal sent by LTS-Initiator, suitable for PMU input. Fault tolerance techniques like heartbeats, timeouts, and self-testing features are programmed in Erkios to identify the failure of any RAS components or its own components. Failures in RAS logic can be categorized as omissive or assertive. Omissive failures occur when a component not functioning as expected, like damaged permanently, leading to a crash failure or it performs the required action late or too soon leading to a timing failure. Assertive failures occur when interaction between components results in incorrect outcome (also called syntactic failure) or a wrong value (also called semantic or value failure).

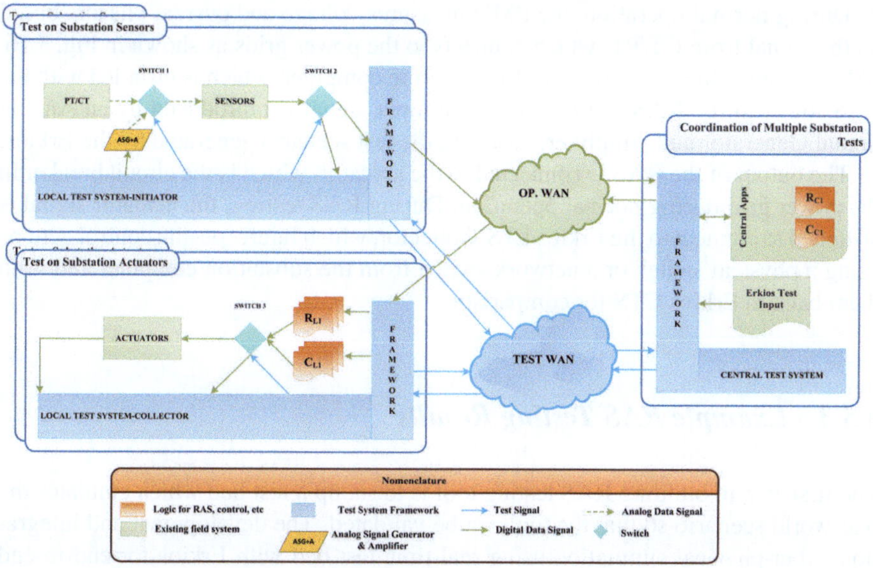

Fig. 3.15 Architecture of Erkios

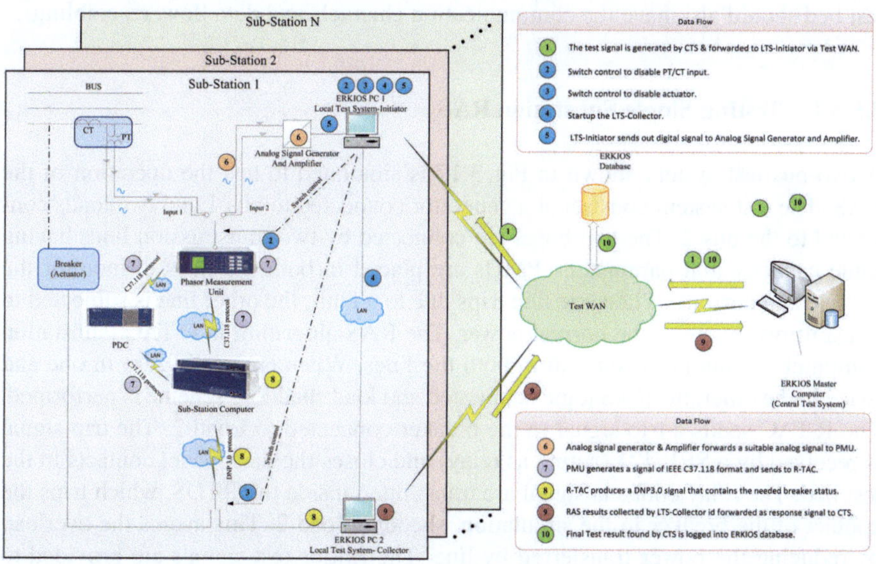

Fig. 3.16 Interfacing architecture with RAS at substation

During normal operation, the PMU measures voltage and current phasors based on the signal from CT/PT, which connects to the power grids as shown in Fig. 3.16. PMU transmits the measurements to substation computer, which is installed with the RAS logic. In the RAS testing mode, the input signal is reproduced by an Analog Signal Generator and Amplifier, based on the test scenario generated by the Erkios.

The output of the RAS is connected to the actuator, which is the circuit breaker in the power grid during normal operation. During RAS testing, the actuator signal is switched to connect to the Erkios LTS-Collector, which intercepts the control actions using a physical switch or a network switch from the substation computer and send them back to Erkios CTS for comparison.

3.5.3 Example RAS Testing Results

The first step in building RAS testing tool is to set up a test bed which emulates the real-world scenario so that the tool can be validated. The development and integration cyber-physical simulation using real-time test bed with Erkios for end-to-end in-field RAS testing is presented in this chapter. The test bed has a real-time power system simulator with the capability to interact with power system equipment like sensors (PMU—phasor measurement unit), relays, breakers in a closed loop. The test bed should also have the communication channels and data flow, resembling.

3.5.3.1 Testing Single Substation RAS

A two-bus test system shown in Fig. 3.17 is simulated to test the operation of the RAS. The test system consists of a generator connected to bus 1 and two loads connected to the bus 2. The two buses are connected by two transmission lines having same capacity and parameters. PMUs are placed in both the lines to monitor the voltage and current. When one line trips due to a fault, the other line is subjected to transmit two times of the normal power. The RAS algorithm in R-TAC (substation computer) monitors the current in both the lines. When current is zero in one and twice in the other, the RAS logic is initiated and load shedding scheme is performed. The R-TAC sends a trip signal to the breaker connected to Load 2. The trip signal is received by a SEL 421 (acting as relay) and closes the back panel contacts in the test bed. The relay contacts signal are transmitted inside the RTDS, which trips the contact of the breaker in the simulation, shedding load 2. This avoids the overload by reducing the power transferred by line. The Erkios test signals are provided to the PMUs and RAS logic senses the violation and generates the trip signal for load shedding. The trip signal is received by Erkios to analyze the results, but for this test bed the signal is sent back to the RTDS to visualize the response. The plots of current in transmission line 1 and line 2 is shows that the current magnitude in line 1 becomes zero when it is tripped after fault resulting in increase in current magnitude

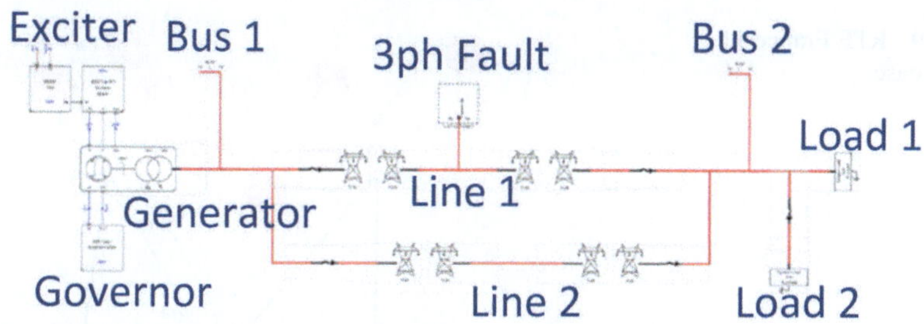

Fig. 3.17 Two-bus power system simulated for RAS testing

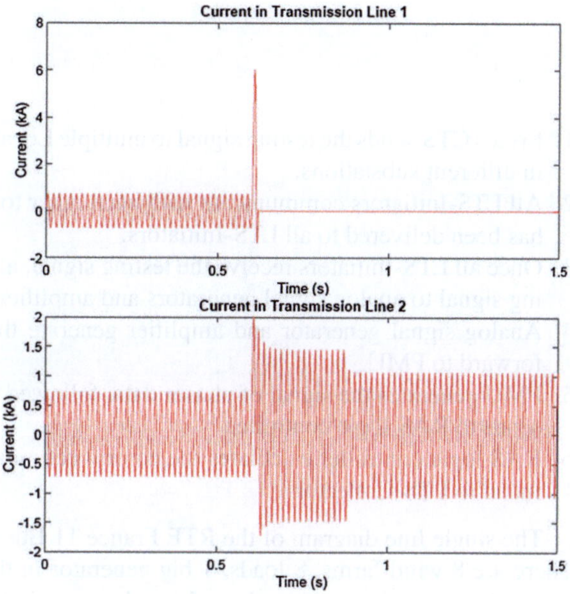

Fig. 3.18 Current in transmission lines 1 & 2

through Line 2. The RAS logic detects increase of current in line 2 and trips the load resulting in reduced current after 0.9 s as shown in Fig. 3.18.

3.5.3.2 Testing Multiple Substation RAS

In order to remotely test the performance of the multiple substation RAS, wind generation curtailment RAS for RTE France 11 Bus system is remotely tested by utilizing Erkios.

Comparing with the single substation RAS remote testing, the multiple substation RAS remote testing has one more important aspect needed to be considered, which is time synchronization. For multiple substation RAS remote testing, the testing process is shown as follows:

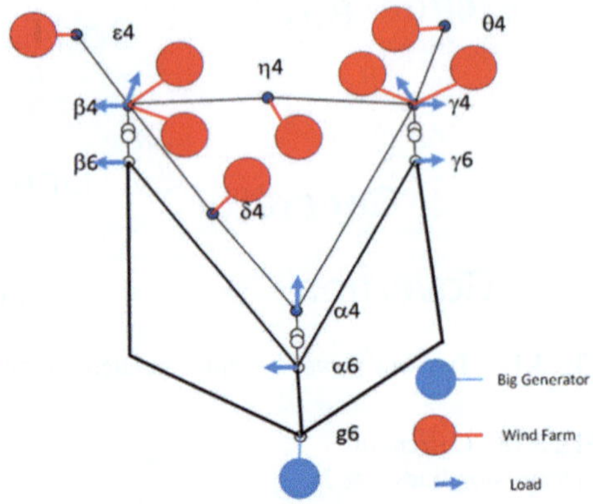

Fig. 3.19 RTE France 11 bus test case

1. Erkios CTS sends the testing signal to multiple Local Test System (LTS)-Initiators in different substations.
2. All LTS-Initiators communicate with each other to check whether testing signal has been delivered to all LTS-Initiators.
3. Once all LTS-Initiators receive the testing signal, all LTS-Initiators send the testing signal to analog signal generators and amplifiers at the same time.
4. Analog signal generator and amplifier generate the suitable analog signal and forward to PMU.
5. PMU generates the signal of phasor data, followed by IEEE C37.118 format and sends to RAS logic controller.
6. RAS logic controller calculates the control action and sends the controller command to LTS-Collector.

The single line diagram of the RTE France 11 Bus system is shown in Fig. 3.19. There are 8 wind farms, 8 loads, 1 big generator in the test case. With the change of weather condition, overload condition happens in the transmission line between substation $\beta 4$ and substation $\delta 4$. The wind generation curtailment RAS detects the overload condition and sends the control action with optimal curtailment value to the specific wind farm. In the remote testing, the Erkios CTS sends the testing signal to all the wind farm substation LTS-initiators. Within the threshold time, the LTS-collector successfully receives the control signal, which has the same optimal value with the MATLAB simulation, from RAS logic controller. Results indicate the successful operation of RAS.

3.6 Summary

This chapter presents the need and procedure for testing synchrophasor devices and applications before the field deployment. The specification of the testing equipment required for PMU testing laboratory is discussed in detail. The architecture of the test bed and the features of the software called PPA used for automatic PMU testing are presented. The successful implementation of PMU testing performed in the Southern California Edition (SCE) using PPA is presented. The result of the PMU testing shows that the compliance of the standard for 30 and 60 fps can be different for some of the test scenarios.

The streaming of PMU data through multiple layers of communication networks may result in data drop, latency, and corruption of data packets. Detecting events in the presence of bad data is performed using ensemble-based technique. Regression, Chebyshev, and DBSCAN are used as base detectors in the proposed ensemble-based method. The application performance of the anomaly detection and event detection is validated and presented in terms of the precision and recall metric in this chapter.

The RAS is considered as an example for testing synchrophasor-based control application. The test bed for end-to-end RAS testing is developed, and the architecture is presented in this chapter. The middleware for end-to-end testing of RAS is also discussed in this chapter. The results of single substation and multiple substation RAS show that the proposed test bed is capable of end-to-end testing of RAS.

Acknowledgements The authors would like to thank the Power System Engineering Research Center (PSERC), Rseau de Transport d'lectricit (RTE), France, and the US Department of Energy (DOE) for financially supporting this work. The authors would also like to thank Southern California Edition (SCE), National Institute of Standard and Technology (NIST), Pacific Northwest National Lab (PNNL), Schweitzer Engineering Lab (SEL), and Real-Time Digital Simulator (RTDS) for helping with various aspects of building this test bed and testing and validation of synchrophasor devices and associated applications. Additionally, the authors also appreciate the support and contributions from Patrick Panciatici, Yinghui Wu, David Bakken, and Arvind Mallikeshwaran.

References

1. Romano P, Paolone M (2014) Enhanced interpolated-DFT for synchrophasor estimation in FPGAS: theory, implementation, and validation of a PMU prototype. IEEE Trans Instrum Meas 63(12):2824–2836
2. Karimi-Ghartemani M, Ooi BT, Bakhshai A (2011) Application of enhanced phase-locked loop system to the computation of synchrophasors. IEEE Trans Power Deliv 26(1):22–32
3. de la O Serna JA, (2013) Synchrophasor estimation using prony's method. IEEE Trans Instrum Meas 62(8):2119–2128
4. IEEE standard for synchrophasor measurements for power systems (2011). IEEE Std. C37.118.1-2011 (Revision of IEEE Std. C37.118-2005)
5. Phadke AG, Thorp JS, Adamiak MG (1983) A new measurement technique for tracking voltage phasors, local system frequency, and rate of change of frequency. IEEE Trans Power Appar Syst PAS 102(5):1025–1038
6. Patel AE, Aivolaitas S (2010) Real-time application of synchrophasors for improving reliability. NERC Tech Rep
7. De La Ree J, Centeno V, Thorp J, Phadke A (2010) Synchronized phasor measurement applications in power systems. IEEE Trans Smart Grid 1(1):20–27
8. Lee H, Tushar, Cui B, Mallikeswaran A, Banerjee P, Srivastava AK (2017) A review of synchrophasor applications in smart electric grid. Wiley Interdiscip Rev Energy Environ 6(3):e223–n/a (Online). https://doi.org/10.1002/wene.223

9. Nuthalapati S, Phadke AG (2015) Managing the grid: using synchrophasor technology [guest editorial]. IEEE Power Energy Mag 13(5):10–12
10. IEEE standard for synchrophasers for power systems (1995). IEEE Std 1344-1995 (R2001)
11. IEEE standard for synchrophasors for power systems (2006). IEEE Std C37.118-2005 (Revision of IEEE Std 1344-1995) 1–57
12. IEEE standard for synchrophasor measurements for power systems-Amendment 1: modification of selected performance requirements (2014). IEEE Std C37.118.1a-2014 (Amendment to IEEE Std. C37.118.1-2011)
13. RTDS, "RTDS" (Online). http://www.rtds.com
14. OPENPDC, "OPENPDC" (Online). http://www.gridprotectionalliance.org
15. Synchro-phaor data quality (2011). Tech Rep. http://www.naspi.org/meetings/workgroup/2011February/presentations/techpanel/techpaneldataquality20110224.pdf
16. Murthy C, Mishra A, Ghosh D, Roy DS, Mohanta DK (2014) Reliability analysis of phasor measurement unit using hidden markov model. IEEE Syst J 8(4):1293–1301
17. Aminifar F, Bagheri-Shouraki S, Fotuhi-Firuzabad M, Shahidehpour M (2010) Reliability modeling of PMUs using fuzzy sets. IEEE Trans Power Deliv 25(4):2384–2391
18. Reinhard K (2012) On data quality and availability modeling of power grid phasor measurements. In: North American power symposium (NAPS). pp 1–5
19. Zhang Q, Luo X, Bertagnolli D, Maslennikov S, Nubile B (2013) PMU data validation at iso new england. In: Power and energy society general meeting (PES). IEEE, pp 1–5
20. Methodology for examining the impact of error and weaknesses in PMU data on operational applications (2011). Tech Rep. https://www.naspi.org/File.aspx?fileID=1572
21. Su HY, Liu CW (2013) An adaptive PMU-based secondary voltage control scheme. IEEE Trans Smart Grid 4(3):1514–1522
22. Liu Z, Il MD (2010) Toward PMU-based robust automatic voltage control (avc) and automatic flow control (afc). In: IEEE PES general meeting. IEEE, pp 1–8
23. Khan MT, Siddiqui AS (2016) Facts device control strategy using PMU. Perspect Sci Recent Trends Eng Mater Sci 8:730–732 (Online). http://www.sciencedirect.com/science/article/pii/S2213020916302117
24. Ouadi A, Bentarzi H, Chafai M (2016) A new PMU based power swing detector to prevent mal-operation of distance relay. Russ Electr Eng 87(10):572–578 (Online). https://doi.org/10.3103/S1068371216100059
25. Wang YJ, Liu CW, Liu YH (2005) A PMU based special protection scheme: a case study of taiwan power system. Int J Electr Power Energy Syst 27(3):215–223 (Online). http://www.sciencedirect.com/science/article/pii/S0142061504001425
26. Rahmatian M, Dunford W, Moshref A (2014) PMU based system protection scheme. In: Electrical power and energy conference (EPEC). IEEE, pp. 35–40
27. Hashiesh F, Mostafa HE, Khatib AR, Helal I, Mansour MM (2012) An intelligent wide area synchrophasor based system for predicting and mitigating transient instabilities. IEEE Trans Smart Grid 3(2):645–652
28. Palizban A (2015) Wide-area monitoring and control utilizing PMU measurements for a system protection scheme. Ph.D. dissertation, University of British Columbia
29. Madani V, Novosel D, Horowitz S, Adamiak M, Amantegui J, Karlsson D, Imai S, Apostolov A (2010) IEEE PSRC report on global industry experiences with system integrity protection schemes (sips). IEEE Trans Power Deliv 25(4):2143–2155
30. Vaiman M, Hines P, Jiang J, Norris S, Papic M, Pitto A, Wang Y, Zweigle G (2013) Mitigation and prevention of cascading outages: methodologies and practical applications. In: 2013 IEEE power energy society general meeting. pp. 1–5
31. Nerc standard PRC-012-2 remedial action schemes. Western electric coordinating council (2016) (Online). https://www.wecc.biz/Reliability/WECC-0126PRC-012-2WPRCPaperHennebergDavis.pdf
32. Remedial action schemes. NERC Standard PRC-012-2 (Online). http://www.nerc.com/pa/Stand/Prjct201005_3RmdialActnSchmsPhase3ofPrtctnSystmsDL/Clean_since_last_posting_PRC-012-2_02022016_final.pdf

33. Sykes J, Hu Y, Adamiak M, Apostolov A, Dac-Phuoc B, Deronja A, Ebrecht J, Henneberg G, Imai S, Madani V, Miller D, Quintana ADL, Vandiver B, Whittaker R, Zubair M, Ward S (2014) IEEE/PES PSRC report on design and testing of selected system integrity protection schemes. In: 2014 67th Annual conference for protective relay engineers. pp 738–742
34. (2013) Communications systems performance guide for electric protection systems. West Electr Coord Counc. (Online). https://www.wecc.biz/Reliability/CommunicationSystemPerformanceGuideforElectricProtectionSystems.pdf
35. (2010) Guidelines for the design of critical communications circuits. West Electr Coord Counc. (Online). https://www.wecc.biz/Reliability/GuidelinesfortheDesignofCriticalCommunicationsCircuits.pdf
36. Chvez LEO, Bakken DE, Bose A, Panciatici P (2015) Erkios: end-to-end field-based ras testing. In: 2015 IEEE power energy society innovative smart grid technologies conference (ISGT). pp 1–5

Chapter 4
Synchrophasor Technology at BPA

From Wide-Area Monitoring to Wide-Area Control

Dmitry Kosterev

Acronyms

BPA	Bonneville Power Administration
CERTS	Consortium for Electric Reliability Technology Solutions
CAISO	California Independent System Operator
COI	California–Oregon Intertie
DOE	US Department of Energy
EIM	Energy Imbalance Market
EPRI	Electric Power Research Institute
ERCOT	Electric Reliability Council of Texas
FCRPS	Federal Columbia River Power System
LBNL	Lawrence Berkeley National Laboratory
NASPI	North American Synchrophasor Initiative
NERC	North American Reliability Corporation
NREL	National Renewable Energy Laboratory
PNNL	Pacific Northwest National Laboratory
RAS	Remedial Action Scheme
TIP	(BPA) Technology Innovation Project
UVIG	Utility Variable generation Integration Group
WECC	Western Electricity Coordinating Council

Contributors:
BPA Transmission Planning
BPA Transmission Operations
BPA Transmission Engineering
BPA Transmission Innovation

D. Kosterev (✉)
Portland, OR, USA
e-mail: dnkosterev@bpa.gov

S. Nuthalapati (ed.), *Power System Grid Operation Using Synchrophasor Technology*, Power Electronics and Power Systems,
https://doi.org/10.1007/978-3-319-89378-5_4

4.1 History of the Synchrophasor Technology at BPA

Synchrophasors are precise time-synchronized measurements of power system quantities—voltages, currents, angles, frequency, active and reactive power, provided by phasor measurement units (PMUs). The primary benefits of the PMUs are (a) precise time synchronization (data points are time-stamped at the source at the moment of measurement) and (b) high data resolution (30–120 measurements per second compared to one measurement every 2 or 4 s by conventional SCADA). PMUs provide synchronized wide-area view of the power system dynamic state [1].

Bonneville Power Administration (BPA) was among the first adopters of the synchrophasor technology in the early 1990s. Initial PMUs were installed as stand-alone disturbance recorders at four substations collecting data locally at a rate of 30 times each second. The value of the synchrophasor technology was evident when the synchronized dynamic data enabled detailed analysis of July 2 and August 10, 1996 outages ([2, 3], also see Appendices A and B). Following the outages, BPA greatly expanded its PMU coverage to monitor large power plants, interties, and load centers. BPA also developed a network to stream real-time PMU measurements to its laboratory. In 2001, BPA started exchanging real-time PMU data with Southern California Edison (SCE), and later with California ISO (CAISO).

BPA also researched, developed, and prototyped several applications that use wide-area synchronized measurements for power system analysis. Power plant model validation using disturbance data was one of the early applications, providing a cost-effective alternative to staged generator tests. BPA's success with the approach influenced WECC Generating Unit Model Validation Policy developed in 2006, and disturbance-based model validation techniques have been adopted into NERC MOD Reliability Standards. Wide-area oscillation detection and analysis applications are enabled by the synchrophasor data. BPA collaborated with other grid operators and the research community to develop a suite of applications for analysis of disturbance ringdown, forced and ambient oscillations.

In mid-2000s, Vickie VanZandt, then Senior Vice President of BPA Transmission, sets the direction for the synchrophasor technology at BPA, saying, "It is time to move forward from wide-area monitoring to wide-area controls." BPA researched several analytical methods for using synchrophasor data for response-based voltage stability controls [4]. BPA also initiated a research project on developing control algorithms to dampen inter-area power oscillations. Brian Silverstein, then Senior VP of BPA Transmission, sets a clear guiding principle for synchrophasor-based controls—"First, do no harm".

4.2 BPA Synchrophasor Investment Project

BPA's original synchrophasor network was research-grade and was not suitable for real-time control center applications. In 2010, BPA started a capital project to build a secure, reliable, and production-grade synchrophasor infrastructure, as a participant in the Western Interconnection Synchrophasor Project (WISP) co-funded with federal Smart Grid Investment Grant funds. BPA's project included installations of dedicated "control" PMUs, high-capacity network routers, GPS units, substation telecommunication battery upgrades, and deployment of control center infrastructure and software.

Typical PMU racks in a substation

A typical PMU inputs two sets of three-phase voltages and six sets of three-phase currents, and calculates voltage and current phasors in polar form (magnitude and angle), bus frequency, line active, and reactive power. Digital statuses of circuit breakers and disconnects can be also recorded in a PMU message. BPA installed fully redundant "control" PMUs at 46 substations, measuring more than 4,000 power system quantities at a rate of 60 samples per second. "Control" PMUs are treated as Bulk Electric System (BES) Cyber Assets (CA). BPA "control" PMUs are fully redundant—they are connected to separate sets of potential and current transformers, and they stream data over redundant communications networks to separate BPA control centers.

BPA also deployed many non-redundant "data" PMUs, mainly at wind power plants in the Pacific Northwest. The "data" PMUs are used for engineering analysis, but not for real-time controls or operational decision making.

The interconnected nature of power system operations requires real-time data exchange among operating entities. BPA is a partner in the WISP, led by Peak Reliability Coordinator (RC), which deployed an unprecedented interconnection-wide network for real-time data exchange among operating entities in the West. BPA streams most of its synchrophasor data to Peak RC and exchanges a significant amount of real-time PMU data with 14 operating entities in the West.

BPA Synchrophasor Investment
<u>Project by Numbers (2016):</u>
46 "control" PMU sites
124 "control" PMUs
19 "data" PMUs
206,000 measurements per second
2GB of data archived every hour
18,000 meas. per sec. are used by analytical applications
14 operating entities that exchange real-time synchrophasor data with BPA

BPA implemented state-of-the-art control center architecture to handle huge volumes of streaming synchrophasor data and to perform real-time analytics. OSI Soft PI data historians are used to archive more than 200,000 measurements per second. The situational awareness application engine processes over 18,000 measurements per second. BPA deployed several real-time analytics described below. BPA technical staff developed visualization displays using PI Process Book to effectively present analytical information to BPA dispatchers. BPA established PMU data-based alarming for the oscillation detection application, with supporting operating procedures and training sessions. BPA dispatchers use a dedicated situational awareness PMU data historian to support real-time operations, and a similar historian with synchrophasor data and analytical results is available for BPA technical and engineering staff on a separate network. PMU data are also down-sampled to a 2-s SCADA rate and used in EMS applications such as state estimation.

Wide-area control was one of the key objectives of the BPA synchrophasor investment. BPA developed infrastructure to integrate streaming PMU data into a new Remedial Action Scheme (RAS) controller and developed an off-line environment for monitoring, testing, and validating the new PMU-sourced RAS controller's performance. BPA implemented a wide-area reactive power switching scheme using synchrophasors, which is currently in a test mode.

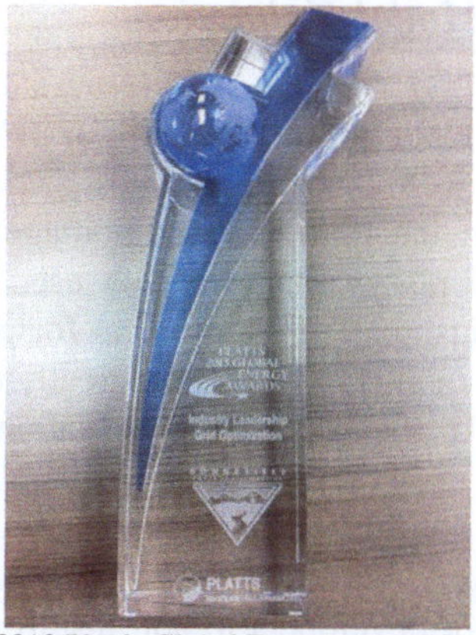

2013 Platt's Global Energy Award for Grid Optimization goes to the BPA synchrophasor project

In addition to the production system, BPA developed a synchrophasor application laboratory for researching, prototyping, and testing new analytical algorithms and applications. The laboratory has capabilities to extensively back-test algorithms using years of historic data as well as real-time streaming data, thereby accelerating application development. BPA has tested and validated its Mode Meter, oscillation detection, synchrophasor RAS, and voltage stability applications in the synchrophasor laboratory.

BPA synchrophasor investment project received Platt's Global Energy Award for Grid Optimization in 2013 [5].

BPA continues its commitment to the synchrophasor technology by having a multi-year plan for PMU deployment, implementing operating procedures for control room applications, and supporting research and demonstration of advanced applications using synchronized wide-area data.

4.3 Engineering Applications at BPA

The value of the synchrophasor investment project is unlocked through deployment of applications. BPA has successfully developed and deployed a variety of engineering applications over the past 15 years.

4.3.1 Power Plant Model Validation

Accurate power system models are required for reliable and economic power system operations, and power plants are the most critical part of the power system dynamic model. Following the 1996 cascading outages, which were attributable in part to poor system and power plant models used for system operations, the Western Interconnection instituted a requirement for periodic testing of generators for model verification. In the meantime, BPA greatly expanded its PMU coverage to power plants, with PMUs located at BPA substations looking at the power plant Point-Of-Interconnection (POI). BPA researched and successfully prototyped a disturbance-based model verification process in 2000, seeing the technique as an independent, cost-effective, and lower-risk option for model re-validation. BPA worked with grid simulator developers to add disturbance play-in functionality in their simulation packages. BPA's success influenced the WECC Generating Unit Model Validation Policy of 2006. The 2016 NERC MOD-026 and 027 Reliability Standards, which superseded WECC Policy, also recognize disturbance-based model verification as an acceptable method of compliance. To complement these Standards, the NERC Synchronized Measurement Subcommittee developed a Reliability Guideline on using PMUs for power plant dynamic model verification [6].

Today BPA has PMUs at the POIs of 13 conventional power plants, 33 interconnection points, and 133 generators, with over 21.5 GW of generating capacity. In addition, PMUs are now installed within BPA's service area at the POIs of 13 wind power plants with total capacity over 1.2 GW. As the number of power plants monitored grew, it became necessary to develop better ways to manage PMU data and power plant models. Through its Technology Innovation Program (TIP 52 and 274), BPA developed the power plant model validation (PPMV) application for model and data management. BPA also partnered with PNNL to implement advanced capabilities for managing data flow, comparing model performance, and reporting.

Today, the PPMV application is used in two modes of operation:

- Event analysis: BPA technical staff can compare the actual generator response with expected generator response during a system event for its entire generating fleet. This process helps to find generator performance abnormalities (see next section).
- Power plant model verification: BPA technical staff compares the actual responses (from recorded PMU data) with model responses for a specific generator over a span of several system events. If there is a good comparison between model and actual data, the model is verified, and the verification report can be used for compliance with the NERC MOD-026 and -027 Reliability Standards. Several BPA customers have used these BPA-provided reports to earn WECC plant model approval certificates.

If examination shows a consistent discrepancy between plant models and actual performance data, BPA will work with power plant operators to re-calibrate the model. BPA's research indicates that disturbance data can be used, with proper engineering judgment, to complement but not replace baseline model development and calibration.

Disturbance-based model verification and stress-testing are parts of an annual model checkup performed by BPA staff. When we started the process in 2011, about 60% of dynamic models were found to be deficient in their representation of actual plant responses to system events. Since then, the continued use of PPMV has materially improved the quality of power plant models in BPA service territory. Success stories of disturbance-based model verification are reported at NASPI and NERC technical conferences and IEEE publications [7, 8].

In 2012, BPA planning found significant differences between actual and simulated responses of the Columbia Generating Station to grid disturbances (left side). BPA worked with University Wisconsin through DOE CERTS program to successfully calibrate the model using PMU data (right side). The calibrated model matched very well a number of disturbance events from 2011 up- to-date, resulting in significant cost savings from not needing to re-test the nuclear power generator.

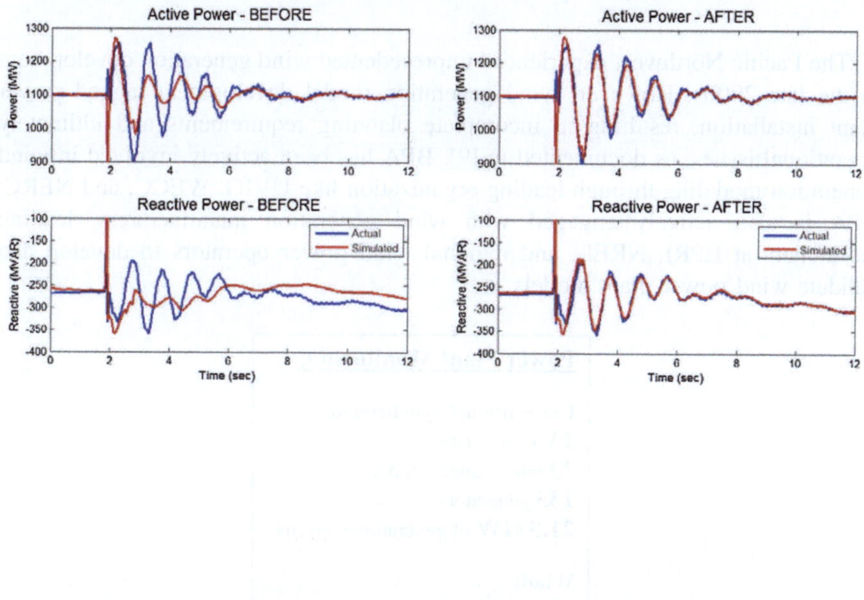

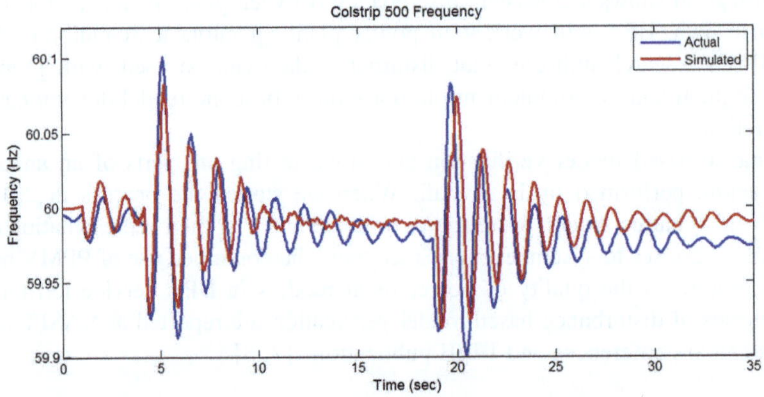

BPA performed comprehensive model review and validation using PMU data at the end of 2014. Numerous model revisions combined with some generator tuning resulted in significant improvements in the system dynamic performance. These helped BPA to make decision to increase its export capability to British Columbia from 2,000 to 2,500 MW during high wind- high water generation scenario in Pacific Northwest.

The Pacific Northwest experienced unprecedented wind generation development in the late 2000s. However, wind generation model development lagged power plant installation, resulting in incomplete planning requirements and ultimately operational issues, as documented in [9]. BPA has been actively involved in wind generation modeling through leading organization like UVIG, WECC, and NERC. BPA is also actively engaged with wind generation manufacturers, leading researchers at EPRI, NREL, and regional wind power operators to develop and validate wind power plant models.

Power Plant Monitoring:

Conventional-synchronous:
13 power plants
33 interconnection points
133 generators
21.5 GW of generating capacity

Wind:
13 wind power plants
1.2 GW of generating capacity

BPA was one of the first utilities in North America to require new power plants to install "data" PMUs as a part of their interconnection process. Several leading transmission operators in North America have since adopted similar requirements, including ERCOT and PJM Interconnection.

BPA has been also conducting large-scale industry outreach on model validation tools at NASPI, NERC, WECC, and EPRI forums, and also working with individual utilities on technology sharing. BPA is an active member of the EPRI modeling group and a user of EPRI's Power Plant Parameter Derivation tool.

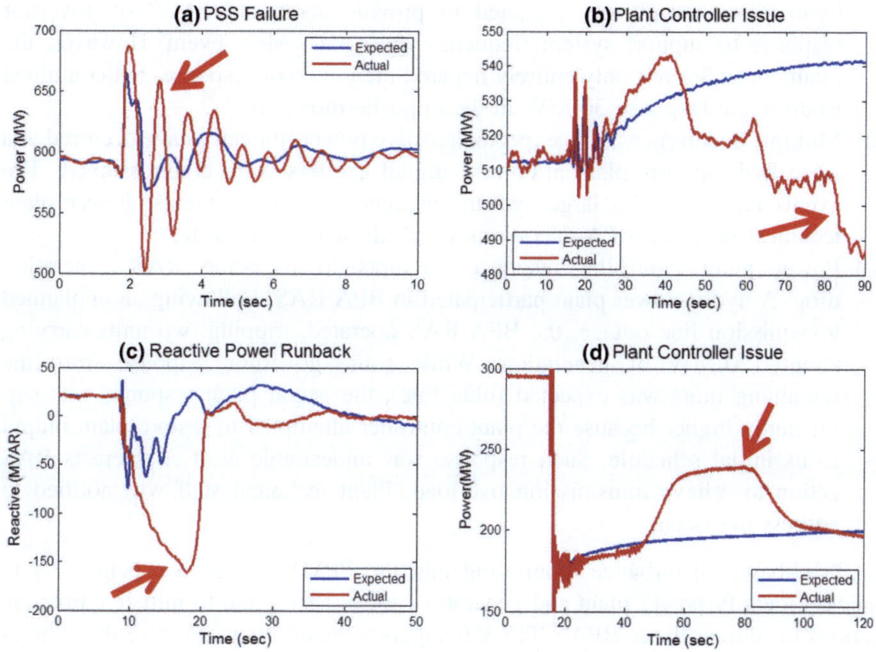

4.3.2 Power Plant Performance Monitoring and Analysis

Once a good model is developed, representing the desired power plant performance, the PPMV application can then be used as a "clinical" checkup of power plant dynamic performance. Over the years, BPA has used the PPMV application with PMU data to detect several abnormal and unexpected responses of power plants to grid disturbances (i.e., by spotting situations when a power plant model that has been shown to be accurate in several prior grid disturbances predicts plant behavior that does not match the plant's actual response in a new event).

Examples of PMU-detected control abnormalities are shown below:

(a) Power System Stabilizer (PSS) at a large power plant experienced an internal failure. The PSS status light was indicating normal operation. However, the plant's actual response to Chief Joseph brake application was much more oscillatory than expected. Further inspection done by plant engineering staff identified internal PSS failure.
(b) The power plant controller response overrides frequency pickup. A large hydro-power plant was expected to provide about a 30 MW of governor response to support system frequency during a system event. However, the plant controller not only entirely negated the governor response, but continued to drive the response 30 MW in the opposite direction.
(c) Multiple occurrences of unexpected reactive power runbacks have occurred at a large hydro-power plant after new digital exciters were commissioned. The events coincide with large system frequency deviation events. Power plant technical staff and BPA operations were advised on the issue.
(d) Power plant controller response attempts to over-ride RAS generation drop. A hydro-power plant participated in BPA RAS. Following an unplanned transmission line outage, the BPA RAS operated, tripping two units carrying about 120 MW of generation. While some governor response from the remaining units was expected (blue line), the actual plant response was significantly higher because the plant controller attempted to restore plant output to its initial schedule. Such response was undesirable as it counteracts RAS action to relieve transmission overload. Plant technical staff was notified to address the issue.

PMU-based disturbance monitoring and the PPMV application help BPA to quickly identify power plant and generator control issues and to initiate corrective actions to address them. BPA's TIP 350 supports the efforts to advance the suite of power plant monitoring applications.

4.3.3 System Model Validation and Event Analysis

The NERC MOD-033 Reliability Standard requires transmission planners to perform verification of system models used in planning studies. The verification is performed by comparing simulated disturbance events against the actual time-synchronized disturbance data. PMU technology is the best source of time-synchronized disturbance data required for system model validation.

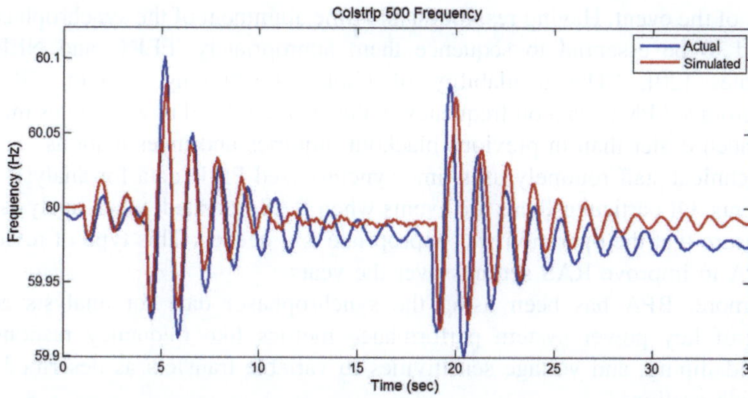

Model Validation studies for a transmission line switching in Montana in May 2015 show reasonably good agreement between simulated and actual responses

BPA and WECC have extensive experience with performing large-scale system model validation studies, including those for the August 10, 1996 outage, the August 4, 2000 oscillation, several large frequency disturbances during thermal governor model development in early 2000s, the June 14, 2004 Westwing event, and many RAS and large generation loss events [10, 21].

In recent years, WECC Staff, Peak RC, General Electric, BPA and other utilities in the West have made significant progress in bridging operational and planning models, working through the WECC Modeling and Validation Work Group. This team is developing a more complete mapping between operating cases and the dynamic model database, thereby enabling dynamic simulations performed using state estimator models. The group is also developing tools to map a state estimator snapshot onto a planning powerflow case, enabling verification of planning models as required by NERC MOD-033 Reliability Standard.

These improvements allow planning and operating engineers to perform model validations more frequently for a more diverse set of system events (large generation trips, RAS events, Chief Joseph brake tests, local faults), which helps engineers identify more accurate system operating limits [11].

4.3.4 Event Analysis

Wide-area visibility and time-synchronization of PMU data are also essential for reconstructing a sequence of events during complex system disturbances. Event analysis has been one of the earliest applications of the synchrophasor technology, greatly reducing staff time required to sequence the events and find causalities between them. For example, September 8, 2011, Pacific Southwest outage had more than 100 notable events occurred in less than 11 min, mainly clustered around four stages of the event. Having resolution and time alignment of the synchrophasor data turned to be essential to sequence them appropriately. FERC and NERC reports states [20]: "The availability of Global Positioning System (GPS)-time-synchronized PMU data on frequency, voltage, and related power angles made this task much easier than in previous blackout inquiries and investigations."

BPA technical staff routinely uses time-synchronized PMU data for analysis of system events. Of particular focus are events when RAS operated, when analysis is done to ensure that the operation was appropriate and needed. This type of review helped BPA to improve RAS settings over the years.

Furthermore, BPA has been using the synchrophasor data for analysis and baselining of key power system performance metrics like frequency response, oscillation damping, and voltage sensitivities to variable transfers as described in the following sections.

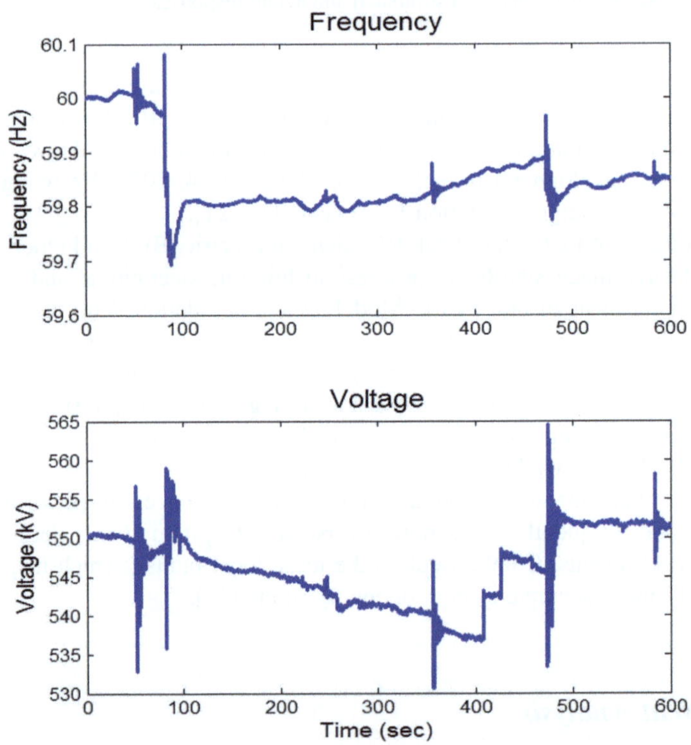

May 30, 2013 -Forest fires caused intermittent faults on a transmission line, eventually triggering operation of Remedial Action schemes, dropping several generators and inserting multiple reactive devices. Good PMU coverage enables efficient and accurate event analysis of this complex event.

4.3.5 Frequency Response Analysis

NERC BAL-003-1 Frequency Response Reliability Standard requires each interconnection to provide a certain amount of frequency response following a system frequency deviation event. The interconnection-wide frequency response requirement is then pro-rated among the Balancing Authorities (BAs), who are responsible for compliance with the Standard.

BPA has been using PMU data for frequency response analysis since late 1990s, and supported early attempts to develop Frequency Responsive Reserve requirements in the Western Interconnection through the 2000s. Supporting NERC BAL-003-1 Reliability Standard, BPA worked with the PNNL to develop a comprehensive set of tools and processes for frequency response monitoring and analysis. BPA's Technology Innovation Program, TIP 313, funded the development of the frequency response analysis tool (FRAT) [12].

The developed process encompasses (a) the assessment of frequency response at multiple levels—Western Interconnection, BPA Balancing Authority, and individual power plants, and (b) the impact of frequency response on transmission loading on major paths. BPA developed a comprehensive baseline of the frequency response performance of the Western Interconnection as a whole, and the BPA Balancing Authority in particular. The developed historic baseline is used for correlation analysis of the frequency responses relative to various system metrics, including total generation, hydro-generation, wind generation, time of day, and season. The FRAT application has helped BPA operations staff better understand BPA's historic performance as a Balancing Authority, the associated compliance risks, and make more informed decisions on the amount of frequency response available for marketing. NERC technical staff has adopted the FRAT application for their frequency response monitoring in support of BAL-003 Reliability Standard. BPA also has an application for verifying performance of power plants contracted to provide frequency response. BPA is now working on predictive tools to track its BA frequency response inventory based on generators online and their historic performance.

The geographic distribution of governor response following a generation loss or a RAS generation drop affects post-contingency power flows on transmission paths, and in turns affects post-contingency voltage stability. The California–Oregon Intertie (COI) and Pacific DC Intertie (PDCI) are stability-limited transmission paths affected by the post-transient governor response. BPA has been using PMU data to track power pickup on COI since the late 1990s. Governor power flow set-up used in planning and operating is tuned to reproduce the observed historic power pickup.

Appendix D provides step-by-step illustration of a frequency event detection and analysis performed by BPA engineering staff.

4.3.6 Oscillation Event Analysis

Oscillations are always present in power systems. Power oscillations are low amplitude and well-damped most of the time. But when the oscillations grow in magnitude, they may contribute to power system instability (such as occurred on August 10, 1996 in the Western Interconnection) or equipment damage (e.g., sub-synchronous control interaction that occurred in ERCOT in March 2009 [22]).

There are many types and causes of power oscillations. Oscillations can be grouped into two broad categories based on our approach in addressing them: (a) oscillations that we can predict, model, and mitigate by either control design/ tuning or by operating procedures, and (b) oscillations that result from equipment failure or forced, that are not generally predictable and have to be dealt as they develop.

Electromechanical oscillation is the most common type of oscillations that can be predicted, modeled, and controlled. Inter-area electromechanical oscillations are of high concern for BPA, and specifically North–South and Montana–Northwest (NW) modes [13]. North–South oscillations involve generators in British Columbia and Upper Columbia oscillating against generators in Desert Southwest and Southern California. The North–South modes are manifested in voltage and power oscillations on the California–Oregon Intertie. The North–South oscillations are typically well-damped; but the oscillation became unstable on August 10, 1996 leading to the large-scale power outage and was very lightly damped during the August 4, 2000 event. The Montana–Northwest oscillation has Colstrip power plant oscillating against generators in Pacific Northwest. Montana-NW oscillation can resonate with British Columbia-NW oscillation. This mode is manifested in power oscillations on the Montana Intertie and voltage oscillations in the Spokane area. The oscillation is typically well-damped, but could be become a problem under outage conditions.

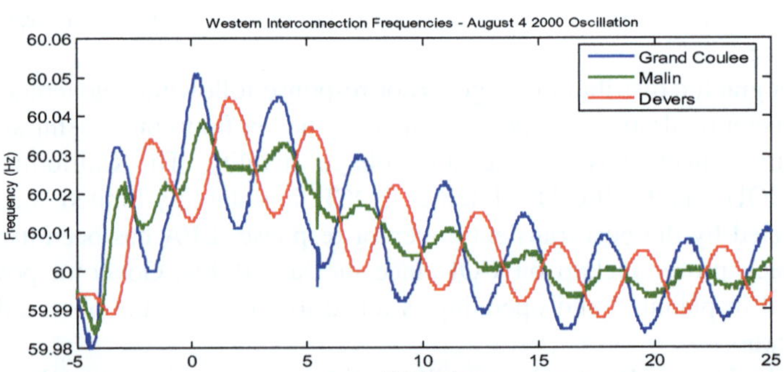

Wide-area synchronized measurements are necessary to understand which generators are oscillating against each other. North-South oscillation on August 4 2000: Grand Coulee – north, Devers – south, Malin – middle.

BPA has a long-established history of performing analysis of the inter-area power oscillations, going back to pioneering work by Dr. John Hauer. Analysis of oscillation ringdown events is used for baselining oscillation damping versus system conditions and contingencies, benchmarking simulated damping against actual power system damping, developing operating procedures to address low damping oscillation risks. BPA advanced the state of "ringdown" analysis tools under TIP 050. BPA's current set of "ringdown" tools include interactive applications developed by PNNL and the University of Wisconsin, and a batch-processing application developed by Montana Tech University. BPA's planning and operations performed extensive large-scale simulations of a wide range of system conditions and contingencies to identify scenarios when low damping can occur. These studies provided a technical foundation for the development of the operating plans to address low damping conditions.

BPA performs annual system tests to stimulate power oscillations under controlled conditions. The tests usually include insertion of a 1,400 MW Chief Joseph braking resistor to initiate an oscillation ringdown and signal injections in the Pacific HVDC Intertie. Under TIP 349, BPA is developing a baseline of the system performance during these tests to continue improving understanding of the factors impacting inter-area power oscillations.

Chief Joseph braking resistor

BPA also participated in DOE-led research on applications to estimate damping of the inter-area power oscillations using ambient data, an application known as the Mode Meter. The goal of the application is to recognize when low damping conditions indicate high system stress. Such application could be used as an early warning of potential oscillation problems. The Mode Meter application has been implemented in the BPA synchrophasor application server and is now in a test mode. Our experience indicates that it is technically challenging to distinguish low damping conditions due to system stress versus low damping condition caused by forced oscillation. BPA is working with researchers at Montana Tech University,

University of Wyoming, University of Wisconsin, and PNNL to advance this technical issue with funding from US Department of Energy.

Forced low energy oscillations could provide early indication of equipment health issues, mistuned controllers, or bad operating points. A PNNL team developed signal processing techniques to pick up these low energy oscillations, which are otherwise not visible just by looking at the signals. BPA worked with the PNNL team to prototype these methods in the BPA's synchrophasor laboratory under TIP 305 and is now developing a structured environment around the analytics and alarming under TIP 350.

Engineering analysis of power oscillation paved the foundation for the control room applications like oscillation detection and Mode Meter described in the later session, including development of operating procedures for dispatchers on how to respond to oscillation events and low damping conditions.

4.3.7 Voltage Fluctuations due to Variable Transfers

Large-scale integration of renewable generation in the Western Interconnection can greatly benefit from expanding the footprint of generation resources used for energy balancing. Over the next decade, the need for flexible generating capacity is expected to increase significantly to support the integration of large amounts of solar generation [14].

The developing western Energy Imbalance Market (EIM) aims to expand the balancing resources by pooling generation resources across multiple Balancing Authorities, thereby reducing the need to build new peaking capacity, lowering operating costs, and improving efficiencies. One of the technical challenges is to make sure that the balancing services can be delivered reliably over the transmission network. Transmission network controls and business processes were designed for relatively slow and predictable power ramps. Faster ramps caused by variable energy resources like wind and solar require greater automation of business processes, more nimble controls, heavier duty switching, and better situational awareness and system assessment tools. BPA transmission is taking steps toward automated arming of its Remedial Action Schemes (RAS) and has researched concepts of advanced voltage controls and situational awareness under its Technology Innovation Program (TIPs 51, 348, 355, 370).

BPA Technology Innovation supported research to identify the system impacts of variable transfers, including development of methodologies and simulation tools under TIPs 237 and 281. This research found that the acceptable magnitude of voltage fluctuations has been the most limiting factor for variable transfers. BPA Transmission Planning and Operations have used PMU data from disturbances and power ramps to independently validate the results of the powerflow studies performed to establish limits on variable transfers.

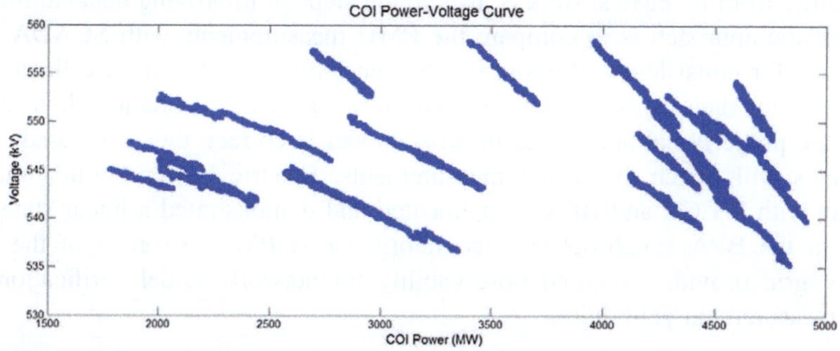

A family of Power-Voltage curves for disturbance events and ramps showing voltage changes at Malin due to power transfer variations on the California – Oregon Intertie. The slope of voltage fluctuations becomes much steeper under higher power transfers.

Current research under TIP 348 is evaluating measurement-based predictive algorithms using wide-area synchronized measurements, and TIP 370 plans take the concepts from monitoring to controls.

4.3.8 State Estimation

BPA's synchrophasor system sends 2-second down-sampled data to its Energy Management System (EMS) applications including state estimation. PMU-measured bus voltages, line active and reactive power measurements typically have better resolution than the existing SCADA measurements. Bus voltage phasor angle is a direct measurement of the power system state. BPA uses a state estimator from GE Grid Solutions (formerly Alstom, Areva, and ESCA) which is capable of using phase angles in its solution.

BPA's PMU data have very high availability, and data quality is monitored and properly flagged. In contrast, many neighboring transmission owners are sending PMU data with frequent data dropouts, or unflagged bad data. While accurate phasor angles from external system can improve BPA state estimator solution, inaccurate phasor angles can also cause solution issues. Better data quality monitoring is required to ensure that bad measurements are flagged and eliminated from state estimation.

4.3.9 Data Quality Monitoring

BPA deployed applications for monitoring data quality, including data dropouts, data latency, and loss of synchronization. In general, BPA data have very high data availability from its PMUs. BPA is also taking steps in improving data calibration. A simplistic approach is to compare the PMU measurements with SCADA measurements for consistency. A more comprehensive approach is to use linear state estimation for data calibration and system network model validation. If measurements are properly calibrated and network model is correct, the "estimated" measurements will match the actual measurements. Electric Power Group, under a contract with WECC and BPA, implemented and demonstrated a linear state estimator in the BPA synchrophasor laboratory. Good PMU coverage of the BPA 500-kV grid provides required observability for network model verification and PMU measurement calibration.

4.4 Control Room Applications at BPA

Bringing synchrophasor technology into the control room was one of the main objectives of BPA's original 2010 investment project. BPA developed several operations applications and visualization displays and deployed these on a control room video-wall as early as October 2013. BPA has since held regular dispatcher training sessions, improved the displays, and prepared technical reference documents for these applications. Finally, BPA developed formal operating procedures for the oscillation detection application in June 2016.

4.4.1 Oscillation Detection

BPA implemented and deployed the oscillation detection application in its control room in October 2013. The application's primary purpose is to detect unanticipated oscillations that result from control system failures, local power plant instability, forced oscillations, excitation of inter-area modes, or a generating unit in an unstable operating region. The oscillation detection application scans multiple signals (power, frequency, voltages) across the grid for indications of growing or sustained high energy oscillations. It monitors large power plants, including BPA wind hubs, load centers, AC interties with Canada, California and Montana, and controlled elements like the Pacific HVDC Intertie and Static VAR Compensators.

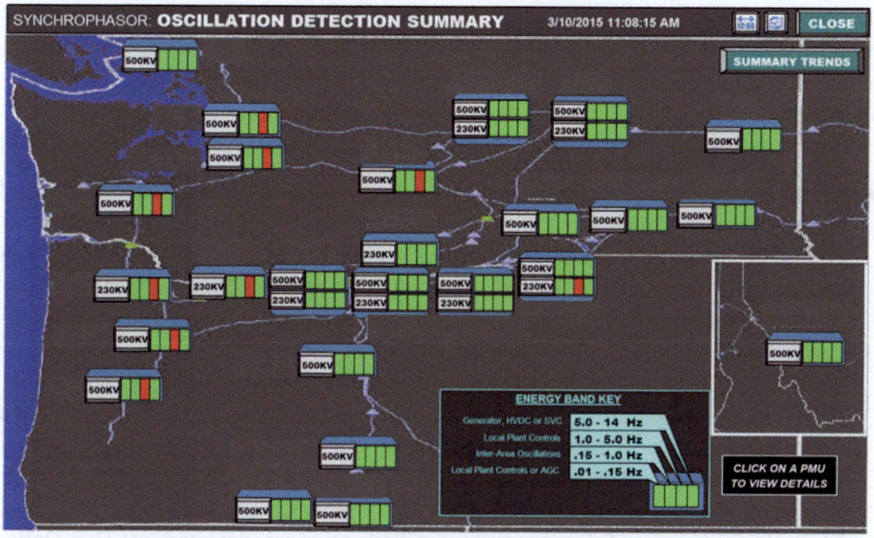

A picture of the Oscillation Detection display in BPA control center

The application uses algorithms developed by Dr. Dan Trudnowski at Montana Tech University and Dr. John Pierre at the University of Wyoming. BPA staff developed visualization displays to present the information to dispatchers, planning and operations staff. The oscillation detection display is very effective in identifying a type and magnitude of an oscillation event. Each PMU has four oscillation frequency bands. These bands correspond to various types of oscillations: (i) AGC or local plant controls, (ii) inter-area oscillations, (iii) local plant oscillations or control problems, (iv) voltage control problems at generators and power electronic devices. Forced oscillations can be present in any of the frequency bands. In the display, a status indicator changes color from green to red when an oscillation alarm is triggered for that frequency band, which has proven to be an effective visual indicator of the type of oscillation present. The display also informs a dispatcher whether an oscillation is local or wide-area. If an alarm is indicated at only one site, then the oscillation is local. But if many sites are affected, then the oscillation is wide-area, or it is a high energy oscillation that propagates through the grid (see figure). A dispatcher can click on a site to get further details on the oscillation—such as real-time trends of monitored signals, calculated oscillation energy, and alarm status. This information helps to better diagnose the problem, locate the oscillation source, and ensure appropriate corrective actions are taken.

BPA developed operating procedures for dispatchers to address the oscillation events, developed technical reference documents, and performed several training sessions for BPA dispatchers. The operating procedures and alarming went into effect in June 2016. BPA published a technical paper through a US CIGRE conference on its experience with the development, implementation, and initial operational experience with the oscillation detection application [14].

Here are examples of oscillation events detected, their causes, and actions taken:

- Multiple instances of high-frequency reactive power oscillations were detected at a large wind generation hub in 2013 and early 2014. These oscillations correlated with high-wind generation output. The oscillation frequency was about 13–14 Hz, close to the resonant frequency of an adjacent series-compensated line. BPA was concerned about the risk of a sub-synchronous control interaction, should a wind power plant become radially isolated on the series-compensated line (similar to an event that occurred in ERCOT in 2009). BPA engineering staff approached the wind power plant owner/operator, and they notified the wind-turbine generator manufacturer. The manufacturer upgraded the control firmware in April 2014, and the oscillations have diminished since.
- Multiple instances of local power oscillations were detected at The Dalles hydro-power plant in late summer—fall of 2015. Because of drought, hydro-generation was low, transmission loading was low, and system voltages were high. As a result, hydro-generators were absorbing a large amount of reactive power. The oscillation events correlated with times when generators were operating close to their Under-Excited Limiters. Further analysis done by the US Army Corps of Engineers identified control interactions between a generator's Power System Stabilizers and Under-Excitation Limiters as the cause of the oscillation. The US Army Corps of Engineers retuned the Under-Excitation Limiter controls in January 2016, and performed system tests to confirm that the new controls settings provided a stable response.

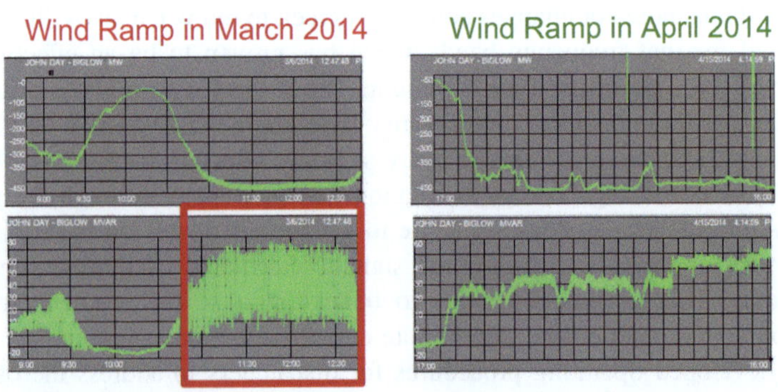

Wind ramps prior to a controller upgrade in April 2014 showed a 70 MVAR oscillation in reactive power (left). Wind power ramps after the upgrade show oscillation activity greatly diminished (right).

- An event of sustained operation in a rough zone was recorded at John Day hydro-power plant in October 2014. While hydro-power generators regularly cross a rough zone during loading and unloading, sustained operation with oscillations is highly undesirable. The problem was caused by the plant controller, which erroneously dispatched the generator to operate in the rough zone. The BPA's dispatcher called the power plant and notified a plant operator about the oscillation. The plant operator re-dispatched the unit to a stable operating point; the plant controller error was also fixed.
- In April 2016, equipment failure near the Sylmar terminal of the Pacific HVDC Intertie caused large active and reactive power variations at the BPA's Celilo terminal. The oscillation propagated through the Pacific Northwest network. Events like this demonstrate the interconnected nature of the power system.
- In May 2016, a "rogue" power controller activity was detected at John Day powerhouse line#1. It caused random changes in active power for about an hour. The behavior was attributed to plant controller issues in the John Day plant.
- The oscillation detection application has detected several local generator instability events caused by transmission losses. Because such oscillations are usually high energy, they propagated several substations away from the source (cause) of the oscillation.

4.4.2 Frequency Event Detection

BPA developed a frequency event detection application and deployed it in the control room in 2014 for monitoring purposes. This application is very practical and useful for dispatchers, technical operations staff, and engineers. It detects an onset of a system frequency deviation from nominal. The application finds the bus frequency measurement with the largest rate of change, typically corresponding to the PMU closest to the location of generation loss. The information is then displayed on a video-wall—the largest dot corresponds to the PMU assumed to be closest to the location of the generation loss. The smaller dot sizes show the propagation of the event through the system.

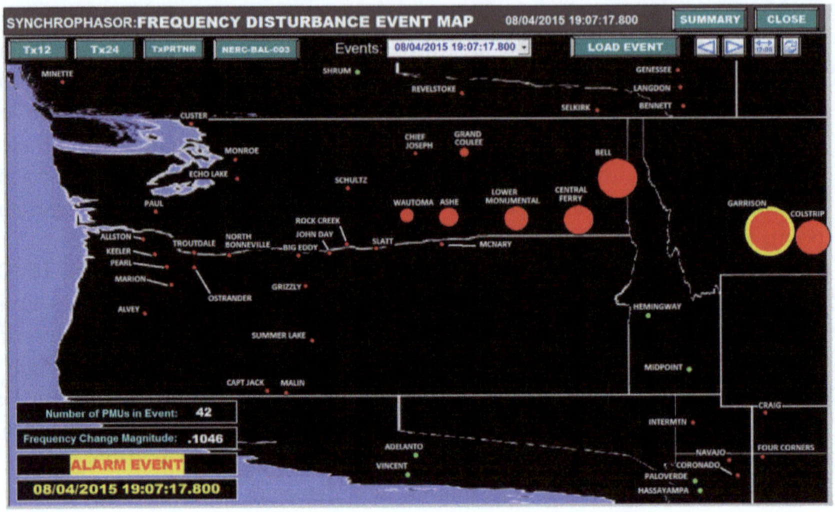

Frequency Event Detection display for a generation loss in Montana. Size of the red dots represents frequency propagation through the system

The trending display includes system frequency as well as power pickup on the interties with Canada, Montana and California. As discussed in the engineering applications section, post-transient frequency response can lead to transmission overloads; therefore, a synchronized view of system frequency and path flows improves situational awareness for the operations staff.

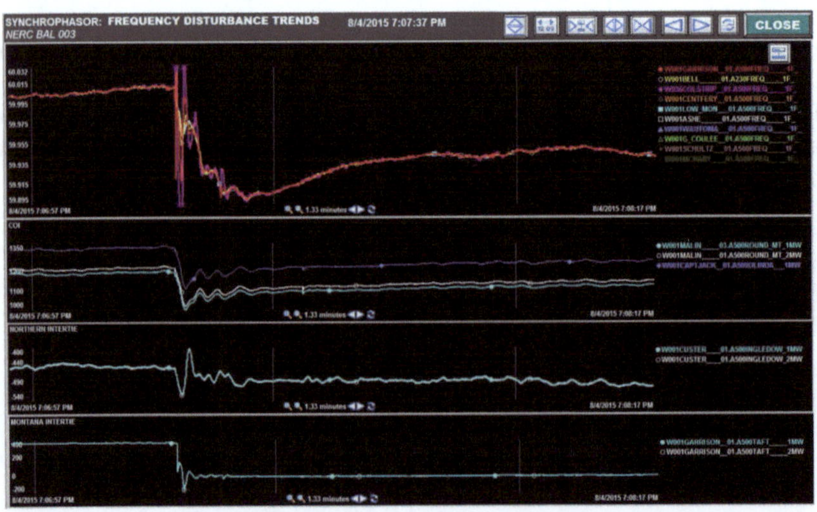

Frequency disturbance trend display provides a synchronized view of the system frequency and active power flows on BPA-operated interties during a frequency deviation event

Appendix D describes integrated frequency response analysis deployed at the BPA.

The detection of a frequency event also triggers an automatic update of a MW flow display, showing the dispatchers how MW flow on the grid (based on phase angle differences on paths between PMUs) get redistributed as result of the event that caused the frequency to change (e.g., a Gen drop).

Following the event, a dual bar chart is shown next to each path that had a voltage angle change of at least 0.5°. The top bar chart shows the relative phase angle difference/MW flow on the path before the event (relative to an historical max), and bottom bar chart shows the relative angle difference/MW flow after the event. It is easy to see how flows on individual paths/lines increased or decreased.

4.4.3 Islanding Detection

Generators in the Western Interconnection operate synchronized, and large-scale power system breakups and islanding events are very infrequent. However, when breakups occur, fast restoration of power service is vital. The objective of the BPA Islanding Detection application is to provide information on power system islands, breakup cut-planes, their frequencies, and angular separation between islands, which dispatchers and technical operations staff can use during the system restoration process. Because of BPA PMU coverage, the phase angle display can be also used during regular line reclosing to ensure the line closing angle is safe.

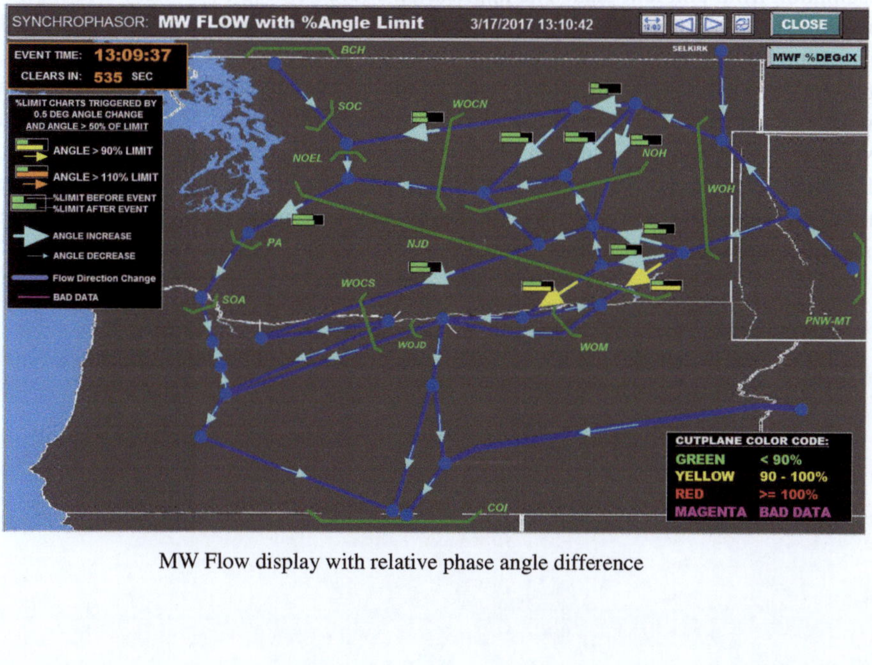

MW Flow display with relative phase angle difference

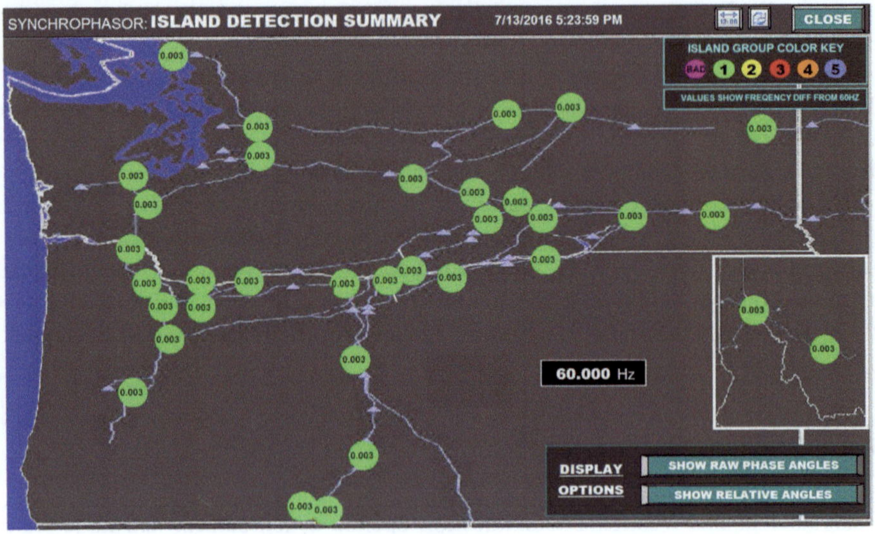

Island detection summary display, the Northwest grid is synchronized.

4.4.4 Mode Meter or Low Oscillation Damping Detection

The Mode Meter application estimates damping of the grid's natural modes of oscillation from ambient data. The application could be used as an early warning of potential oscillation problems due to high system stress. BPA deployed analytical algorithms co-developed by Montana Tech University, University of Wyoming, and PNNL. The application is implemented for estimating damping of the inter-area modes of oscillations—North–South Mode A, North–South Mode B, Montana Northwest, and British Columbia Northwest. The application calculates frequency, damping, and energy for each oscillation mode, and the results of analytics are archived in the BPA historian. BPA is developing a composite alarm and operating procedures for the application.

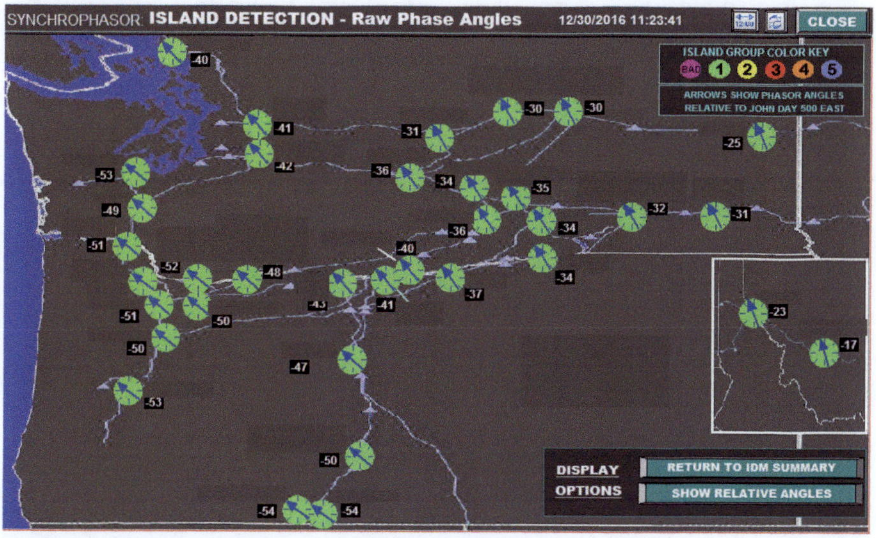

Phase angle display

Lightly damped oscillations can develop either due to high system stress (e.g., August 10, 1996 and August 4, 2000 on COI) or due to forced oscillations (e.g., November 30, 2005 Nova Joffre event). This distinction matters because dispatchers take different actions depending on the cause of an oscillation. The current algorithm does not differentiate when low damping conditions are due to system stress versus low damping condition caused by forced oscillations. BPA is working with researchers at Montana Tech University, University of Wyoming, University of Wisconsin, and PNNL to advance this technical issue with support from DOE.

4.5 Synchrophasor-Based Controls

BPA's synchrophasor system was designed and built to enable wide-area stability controls. BPA implemented a synchrophasor-based wide-area voltage stability controller as a part of the investment project, also known as "synchrophasor RAS". The control scheme and its implementation have been approved as a safety net by the WECC RAS Reliability Subcommittee. The controller was tested in monitoring mode for two years before becoming operational in May 2017.

BPA researched several algorithms for voltage stability controls in the early 2000s, and prototyped them in its laboratory. This experience was useful and informative for the work done under BPA TIP 051 voltage stability controls, where BPA developed and performed extensive modeling, simulations, and back-testing of new algorithms. BPA also engaged an industry-wide expert panel to review its work.

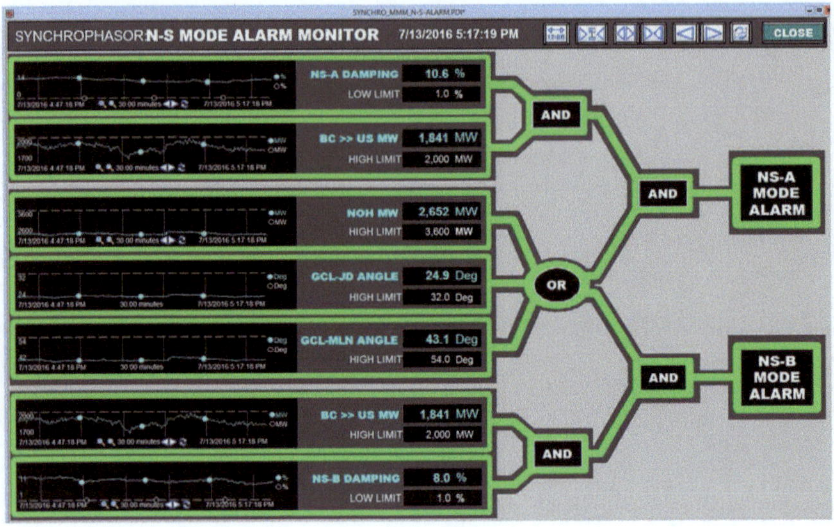

Composite Low Damping Alarm concept:
damping estimates are paired with flows and phasor angles as indicators of system stress

The synchrophasor RAS has a hierarchical architecture. The master controller receives wide-area synchrophasor measurements to perform a real-time assessment of transient voltage stability risks. The implemented controller includes three phase-plane algorithms to identify potential stability risks; if any of the three algorithms is triggered, the RAS sends a control signal to several BPA substations to trigger local voltage controllers. Should the local voltage drop below a pre-defined threshold, the local controller switches shunt capacitors and reactors with a minimum time delay.

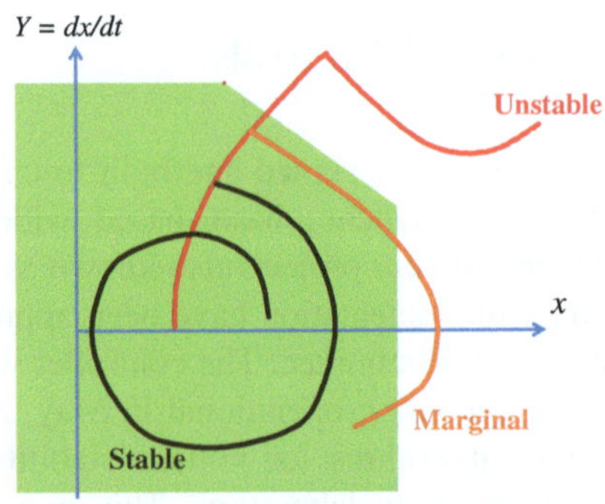

Illustration of phase-plane algorithms: an algorithm is triggered once a trajectory gets outside the "green" zone for a pre-defined time period.

BPA designed the synchrophasor-based control system up to WECC's RAS standards. BPA's synchrophasor RAS is fully redundant, including redundant measurements, PMUs, and network equipment (routers, switches, etc.) streaming data to both BPA control centers. BPA also deployed a diversity principle to account for potential data dropouts or bad measurements. Each signal has its primary, secondary, and tertiary options of "like"-measurements, so that if the primary signal is not available or flagged as "bad", the algorithm will switch to the next available signal. Timing is a critical part of the synchrophasor RAS, as BPA planning required the total response of less than half a second. Synchrophasor RAS receives measurements every cycle to perform voltage stability assessment. BPA has measured statistics on data latency, including signal processing by a PMU, network latencies, RAS controller throughput, and telecom delay from RAS controller output to BPA substations and relaying, to verify that the incoming data speeds do not impede the half second RAS response required.

BPA implemented an innovative approach for testing and verification of the synchrophasor RAS controller. RAS controllers are tested annually or any time when a controller change is made, so it is essential to have automated tools and procedures for efficient RAS testing. Most conventional RAS algorithms are event-driven by line-loss logic, and their testing is event-based. The challenge with testing the synchrophasor RAS is that its algorithm is driven by a time-sequence of analog measurements. BPA had to develop a special RAS test automation unit that streams synchrophasor data and compares actual to expected controller outputs.

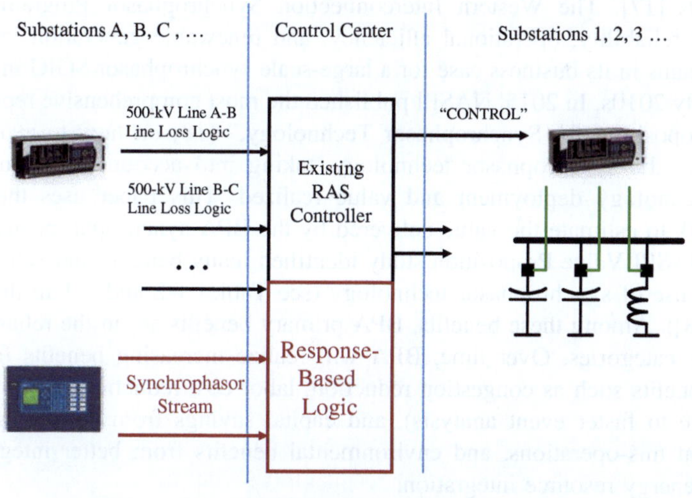

Hierarchical architecture of the synchrophasor RAS controller

BPA prototyped the RAS controller in its synchrophasor laboratory. This environment allowed BPA to back-test algorithms using a wide range of historic data events, such as large generation trips, line faults and outages, power oscillations, and also bad data cases. This testing helped further refine the algorithm performance.

BPA has a replica of the RAS controller algorithm in its situational awareness application engine. It runs the PMU data through the RAS algorithms every 5 s to determine whether the algorithms would have triggered. The results are archived in the BPA historian, and available to BPA operations and planning staff for further analysis and performance assessment.

4.6 Value Realized from the BPA Synchrophasor Project

BPA's current synchrophasor project investments total about $50 M, comparable to the synchrophasor technology investments made by Pacific Gas and Electric, Southern California Edison, and Peak Reliability Coordinator. These costs include:

- Cost of field installations (PMUs, telecom, substation upgrades)—about $30 M
- Cost of control center network, servers, and applications—about $8 M
- Technology innovation projects—about $10 M

This section describes and quantifies many of the benefits realized from the synchrophasor project investments.

California Energy Commission was one of the first one to perform a study to develop a business case for the synchrophasor technology deployment in mid-2000s [17]. The Western Interconnection Synchrophasor Program (WISP) identified reliability, operational efficiency, and renewable generation integration value streams in its business case for a large-scale synchrophasor SGIG investment in the early 2010s. In 2015, NASPI published the most comprehensive report, "The Value Proposition for Synchrophasor Technology," [18] on how to evaluate the benefits of the synchrophasor technology, taking into account experiences with actual technology deployment and value realized. This paper uses the NASPI framework to estimate the value delivered by the BPA synchrophasor investment.

The NASPI Value Proposition study identified many benefits and value metrics from the use of synchrophasor technology (see Tables 1-2 and 3-1 in the NASPI report [18]). Among these benefits, BPA primary benefits are in the reliability and resiliency categories. Over time, BPA will realize increasing benefits from economic benefits such as congestion reduction, labor cost reductions (as from hours saved due to faster event analysis), and capital savings from early diagnosis of equipment mis-operations, and environmental benefits from better integration of variable energy resource integration.

The value of the synchrophasor investment projects extends beyond BPA. BPA streams its real-time data to Peak Reliability Coordinator, where it is used for the interconnection-wide situational awareness, system performance assessment, and system model validation.

Table 4.1 Benefits realized through deployment of the synchrophasor applications

	Status	Improved grid reliability	Improved grid throughput and usage	Better economics and cost savings	Better environmental impacts/ enabling renewables
Control room					
Oscillation detection	Since 2013	X			X
Frequency event detection and MW flow	Since 2014	X			
Islanding detection	Since 2013	X			
Phase angle display	Since 2013	X			
Synchrophasor RAS	Implemented in 2016, in test mode	X	X		X
Engineering analysis					
Power plant model validation	Fully deployed in 2014	X	X	X	X
Power plant performance monitoring	Under development	X			
System model validation applications	Under development	X		X	
Event analysis	Original application	X		X	X
Frequency response analysis	Deployed since 2014	X		X	
Oscillation detection and analysis	Deployed since 2014, development continues	X			X
Tools for voltage sensitivity due to variable transfers	Studies performed in 2015	X	X		X

We will use the framework developed by NASPI to highlight examples of value delivered by BPA's synchrophasor investment. Table 4.1 provides an overall summary of the how applications relate to specific value streams. Specific applications and their benefits are reviewed below:

- *Oscillation Detection*—BPA's oscillation detection application was first implemented in October 2013, and operating procedures were developed and went into effect in June 2016. BPA has since detected numerous oscillation events, and resolved them either operationally or through follow-up control fixes. Oscillations could cause either direct grid instability (as in WECC on August 10, 1996), or transmission and generation equipment damage (as with ERCOT's 2009 wind oscillation event). Detecting oscillations and acting on them in timely manner improves overall system reliability and equipment safety. In the case of oscillations caused by wind generation controls, early identification of the oscillation and analysis of its causes can reduce the extent of wind plant curtailments.

- *Frequency Event Detection and MW Flow* —Frequency response is a key reliability metric for the bulk power system. Frequency response impacts other reliability metrics, as the governor response to frequency deviations affects transmission line loading and voltage stability. An example of this occurred on June 14, 2004, when a delayed cleared fault in Arizona resulted in more than 4,000 MW of generation tripped, pushing the Western Interconnection into under-generation and close to the stability edge [12]. After the system survived the initial power swing, the new challenge emerged: power flow on the COI was far in excess of its SOL, while the system frequency was abnormally low. Navigating the system through such complex events requires comprehensive wide-area situational awareness tools. BPA's frequency event detection (FED) application provides better situational awareness to BPA dispatchers for similar situations in the future. The FED display provides clear indications where an under-frequency event was initiated, and MW flow display shows how the governor response for all of the generators online affects power transfers on interties and flow gates. This integrated view allows operators to quickly diagnose and mitigate under-frequency problems before they cause a stability collapse.
- *Islanding Detection*—Islanding events, which occur when a grid collapse ends up with an portion of the grid in balance between load and generation, are very rare. The existence and boundaries of an electrical island can be determined in real time using PMUs. Streaming PMU data is used for islanding detection. PMU data and phase angle monitoring can be used to re-synchronize and speed system restoration should an islanding event occur in the Pacific Northwest, reducing customer outage time. BPA islanding detection was implemented in the control room in October 2013.
- *Phase Angle Display*—BPA phase angle displays were implemented in the control room in October 2013. The application displays either absolute or relative phase angles, selectable by a user, for BPA substations. This information can be used for transmission line reclosing or re-synchronizing parts of the system during an islanding event. After a major grid collapse, operators have to piece the system back together, figuring out a carefully choreographed sequence that balances available generation and load across a wide region. Wide deployment of PMUs across the BPA main grid makes this much easier because operators can use phase angle matching for faster and safer synchronization and

line reclosing, reducing total customer outage time with less risk to costly electrical assets.

- *Synchrophasor RAS*—BPA's synchrophasor RAS was approved by WECC in 2015 as a safety net to protect the grid against unplanned contingencies and extreme events, such as June 14, 2004 event described above. Voltage stability is one of the limiting factors for dynamic transfers when COI flows are near its operating limit, or as much as 20% of time as estimated from historic data. BPA is considering whether the synchrophasor RAS can be used to mitigate voltage stability limitations and to enable a wider operating range for dynamic transfers. Synchrophasor RAS helps to advance the environmental policies in the Western Interconnection toward reliable integration of renewable generation by protecting against extreme events and expanding the operating range of dynamic transfers. BPA estimates that using this RAS scheme increases COI voltage stability limit by 100–150 MW without the need to invest additional capital funds.
- *Power Plant Model Validation* —BPA's PPMV application was initially researched and developed in 2000. Numerous enhancements in model and data management have been implemented over the years, and the current application has been in active use since 2014. Regular use of power plant model validation application results in multiple generator model revisions, which ultimately improves grid reliability since the models are used to study operating limits and make grid investments. Thanks to PPMV, better generator models across the Northwest have improved simulated system performance, enabling BPA to increase its export limit on the Northern Intertie from 2,000 to 2,500 MW in 2015 without any incremental capital investment. Higher export capability could become useful during high-water, high-wind, light load conditions, similar to those during 2011 spring runoff, so that Northwest could export power instead of curtailing wind generation.

  With synchrophasor-based model validation, BPA is realizing time and money savings by reducing the need for physical generator testing and modeling, with no lost revenue from plant downtime, or risk to equipment damage during staged tests. BPA helped several transmission customers to comply with NERC MOD-026 and -027 by performing PMU-based model verification.

  Accurate wind generation models are required to make appropriate system investments and to avoid curtailments due to instability events. Because initial wind generation models were incomplete, BPA and wind developers were unable to foresee voltage stability issues under weak network conditions in some parts of their system. When local instability events occurred, BPA had to curtail wind generation power output until appropriate reinforcements or operational improvements were made. With the use of the PPMV application, BPA and industry experts have improved dynamic models of wind power plants in Pacific Northwest, thereby making sure appropriate reinforcements and controls are in place to reduce the risk of wind curtailments.

- *Power Plant Performance Models Monitoring*—With automated PPMV and ongoing plant monitoring, BPA can identify potential equipment failures before they occur, supporting condition-based maintenance that should reduce catastrophic equipment failures and direct maintenance costs. Faster identification of generator problems creates facility efficiencies by reducing equipment downtime and protecting generator revenues. Reducing equipment failures can reduce crew labor costs, including travel time and overtime, and enable more cost-effective equipment acquisition and inventory management.
- *Event Analysis*—PMU-based event analysis provides greater insight in the root cause than SCADA data. PMU-based event analysis significantly reduces labor costs and time required to complete an after-event sequence of events and root cause analysis.
- *System Model Validation*—Accurate power system models are required for reliable system operations. Periodic model validation ensures that the models are accurate and up to date. NERC MOD-033 Reliability Standard requires Transmission Planners to conduct model validation studies for their footprint. Models are compared against wide-area disturbance data provided by PMUs. BPA collaborated with industry to develop tools to streamline system model validation process. Leveraging its experience with system model validation and event analysis, WECC and BPA developed interactive tools to greatly reduce the staff hour required to prepare model validation cases.
- *Frequency Response Analysis*—BPA's frequency response analysis tool (FRAT) was developed in 2014. BPA uses FRAT for frequency response analysis at the interconnection, BPA BA and power plant levels. The tool allows BPA to assess its historic performance and to identify which generator contributing to its frequency response. BPA used the frequency response baselining in its filings with FERC during the development of NERC BAL-003-1 Frequency Response Standard. BPA also uses FRAT for compliance with NERC BAL-003-1 Reliability Standard, and helping BPA to make decisions on how much of frequency response surplus make available in the power market.
- *Oscillation Analysis Tools*—Damping of inter-area power oscillations is a necessity for grid security. BPA has used oscillation analysis tools and wide-area synchronized PMU data to identify and better understand the modes of inter-area oscillations in the Western Interconnection and recognize the conditions when these oscillations can put the system at risk. BPA developed its strategy to address the oscillation damping risks. Forced oscillations could be an indicator of impending control failure or improper tuning. BPA worked with equipment owners to correct forced oscillation issues caused by malfunctioning equipment at wind power plants and hydro-power generators.
- *Voltage Analysis*—BPA has developed several tools to analyze voltage fluctuations due to variable energy transfers, as on the California–Oregon Intertie. Variable transfers are limited primarily by voltage fluctuations and secondarily by voltage stability. PMU data were used in baselining historic fast power ramps

to validate study results used to set the dynamic transfer limits on California–Oregon Intertie. Increasing dynamic transfer capabilities on COI and PDCI enables the Balancing Authorities in the West to expand balancing footprint to accommodate a larger amount of renewable generation.

4.7 Technology Innovation Pipeline

BPA has a robust pipeline of technology innovation projects to further unlock the value of the BPA investment in the synchrophasor technology.

4.7.1 Synchrophasor Infrastructure

As more solar and wind generation in the West becomes electronically coupled to the grid, those plants' dynamic response is affected by phase imbalances and point-on-wave phenomena during system faults. The resolution and attenuation of phasor quantities may no longer be adequate for analysis of disturbance performance of electronically coupled generators. A 500-kV fault in Southern California on August 18, 2016 resulted in tripping more than 1,000 MW of electronically coupled generation, even though phasor quantities were within disturbance ride-through envelopes. BPA recognized the need for digital oscillography capabilities at its wind power plants back in 2009, and installed extended digital fault recorders with PMUs at wind power plants. Moving forward, BPA is evaluating multi-functional devices capable of streaming out synchrophasor data at 60 samples per second for real-time applications, and local recording of digital oscillography for engineering event analysis.

BPA uses OSI Soft PI historian to support operational and event analysis functions. However, traditional databases are generally not well suited for big data analytics. BPA is researching advanced computational platforms based on distributed cluster computing for efficient analysis of the high volume of synchrophasor data. BPA has been collaborating with PNNL researchers to deploy a big data analytical system using the Apache Spark cluster computing framework at the BPA synchrophasor lab.

BPA exchanges real-time synchrophasor data with several operating entities in the Western Interconnection. The current method of data exchange is point-to-point, but greater efficiencies could be gained by using a publish–subscribe method instead. BPA is monitoring an ongoing NASPI-PNNL study developing a proposal for next-generation information technology and communications architecture for synchrophasor data exchange.

4.7.2 Engineering Analysis

BPA will continue incremental improvements to the existing tools deployed for ringdown analysis, model validation, and frequency response analysis. As we move forward, the focus is shifting toward system dynamic performance baselining and monitoring. BPA is developing baselining applications for oscillation analysis (TIP 349) to correlate observed damping of power oscillations with system conditions. TIP 313 has developed frequency response baselining capabilities at the interconnection, BPA Balancing Authority, and power plant levels.

Equipment monitoring has emerged as one of the unexpected success stories of the synchrophasor technology, providing capabilities to detect equipment failures and identify control mis-operations and cyber intrusions. BPA will continue research and demonstration of analytics for detecting equipment abnormalities, building on initial work under TIP 305 and TIP 350. This work also ties to developing the capabilities for big data analytics at BPA.

BPA will continue deployment of tools for compliance with NERC MOD Standards. Applications for power plant model verification (MOD-026, -027) are fairly mature and have been in use at BPA for several years. The focus is now shifting on applications for MOD-033 system model validation Reliability Standard. One of such applications is called linear state estimator (see NASPI workshop on linear state estimation [20]). Through collaboration with WECC and Peak RC, BPA has installed and successfully tested a linear state estimator in its synchrophasor laboratory. The application can be used to assess model-data integrity, identify persistent measurement errors, validate network model data, and condition data for use by downstream analytical applications. BPA is also monitoring similar research efforts at other utilities, primarily Dominion Virginia Power.

4.7.3 Control Room Applications

BPA made synchrophasor-based oscillation detection operational in June 2016. The application is simple and effective in detecting and alarming on events of high energy power oscillations. While the oscillation cause and source can be identified in most events, there have been cases when the oscillation origin was not obvious. The problem becomes particularly challenging in instances when a forced local oscillation excites an inter-area mode. BPA is partnering with the research community and other operating entities to improve analytical methods for location of the oscillation source. BPA is also monitoring ISO New England work on locating oscillation source using wide-area measurements.

The Mode Meter application was developed under the DOE-WECC Western Interconnection Synchrophasor Project. The application has run in monitoring mode since October 2013. BPA's objective is to use the application to identify low damping conditions due to the system stress. As noted earlier, the current version of the application does not differentiate between low damping due to system stress or forced oscillations. BPA is collaborating with the research community on the algorithm improvements.

BPA continues to advance research and demonstration of voltage stability applications, using both model and measurement-based approaches. BPA TIP-325 demonstrated analytical methods for voltage stability assessment using real-time state estimator models and augmenting them with real-time synchronized measurements. Similar applications are deployed by Peak RC and ISO New England. BPA TIP-348 is evaluating measurement-based approaches for predicting voltage changes on the power grid due to reactive switching or variable transfers.

Phasor Angle Monitoring is another application BPA had for several years. BPA installed its first Phasor Angle Monitor in 1997 following the recommendations from August 10, 1996 power outage report. Poor data availability with the original research PMUs made the monitor unusable for real-time operations. BPA deployed a revised version of phasor angle displays in its control room as a part of the synchrophasor investment project in October 2013. Currently, there are no alarming or operating procedures for dispatchers associated with the phasor angle display. BPA is planning to initiate a study to determine on how to make phasor angle information actionable.

4.7.4 Wide-Area Controls

BPA's wide-area synchrophasor RAS was implemented and approved by WECC as a safety net. BPA continues monitoring the RAS control performance with a goal of making the scheme operational in 2017. The new RAS could be used to remove some of the existing voltage stability constraints on dynamic transfers. BPA will continue researching other algorithms to improve voltage security in Pacific Northwest load centers and interties.

A team of technical staff from Sandia National Laboratories, Montana Tech, and BPA has developed and demonstrated a control system for damping of north–south oscillations in the Western Interconnection. The control system acquires measurements from PMUs located across the Pacific Northwest, and uses them as feedback information in constructing a power command signal. This control signal is used to modulate active power flow on the PDCI by as much as +/– 125 MW (of less than 4% its 3,220 MW-rated capacity). Extensive system studies were done to demonstrate the controller effectiveness in damping power oscillations (see Fig. 4.1), as well as robustness with respect to a wide range of generation patterns and contingencies. A major milestone was achieved in September 2016 when the control system was first tested in closed-loop operation. Tests were done under controlled conditions when the oscillation damping was already good. Still, the controller has shown more than 4% improvement in damping of the primary north–south oscillation mode when the control system was active (see Fig. 4.2). Test results were

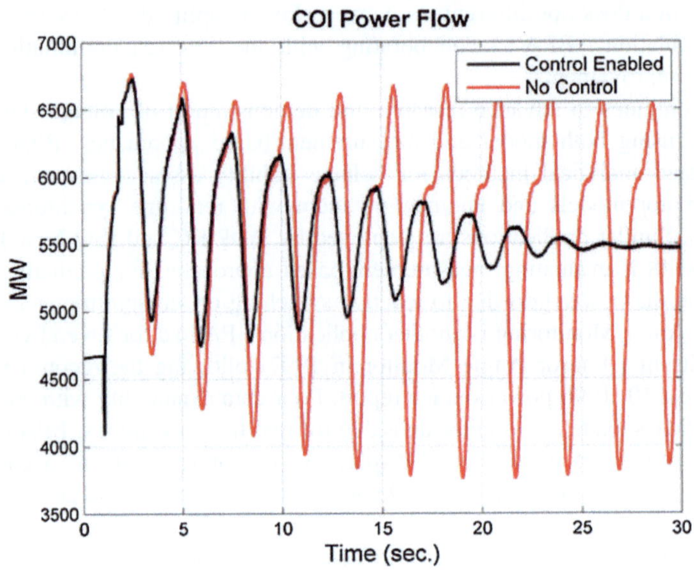

Fig. 4.1 Simulations show significant improvement in damping of the north–south oscillation due to the controller action in a high-stress system

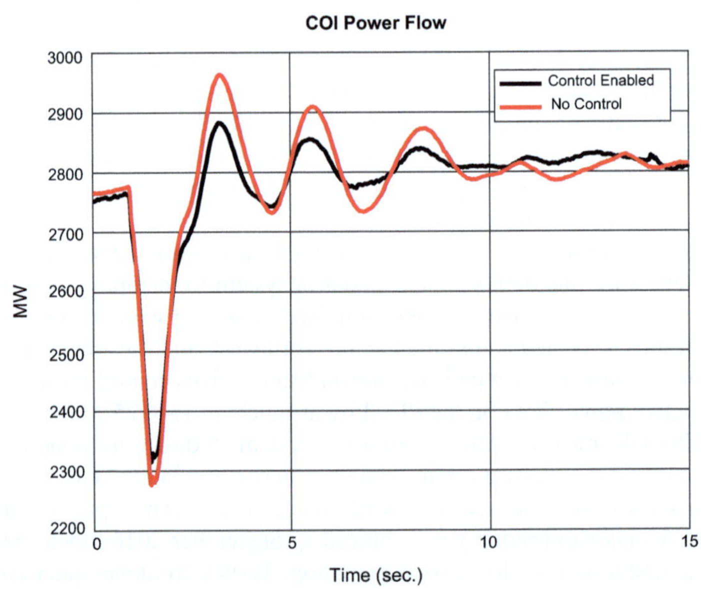

Fig. 4.2 System tests show the actual improvement in the north–south oscillation damping due to the controller action in an unstressed system

consistent with the study work performed, validating the effectives and robustness of the developed control system.

This is the first time that wide-area synchronized PMU measurements have been successfully demonstrated in real-time feedback control on a large-scale power grid in North America. Significant engineering effort went into making sure that the controller causes "no harm" to the grid under all possible operating conditions. A "watch dog" supervisory system was developed and implemented to monitor grid conditions, signal quality and latencies, PDCI status, and the controller hardware status. This project was supported by BPA Technology Innovation and the DOE Office of Electricity Transmission Reliability and Energy Storage programs.

Automated Generation Control (AGC) performs generation-load balancing functions. AGC systems in general do not recognize transmission system constraints. BPA suspends AGC during transmission contingencies that initiate RAS operation. Similarly to its synchrophasor RAS, BPA needs to consider adding condition-based rules to suspend/modify its AGC action. On August 10, 1996, AGC action to replace tripped generation at McNary with generation at Upper Columbia contributed to increased transmission system stress [2]. Phasor angles between Upper Columbia and interties were increasing well beyond safe operating levels. Suspending AGC when phasor angles exceed safe operating limits is being considered.

4.7.5 *Collaboration and Technology Outreach*

US Department of Energy and North American Synchrophasor Initiative have played a vital role in the advancement and deployment of the synchrophasor technology and its applications in North America. The DOE program funds development of new synchrophasor hardware and applications concepts, while NASPI coordinates industry learning, problem-solving, and strategy development to foster synchrophasor technology adoption and value realization. WECC's Joint Synchronized Information Subcommittee and recently, NERC's Synchronized Measurement Subcommittee have complemented DOE and NASPI with more hands-on, peer-to-peer synchrophasor deployment advice. BPA is committed to supporting DOE, NASPI, and NERC mission in advancing the synchrophasor technology outreach for grid reliability, resiliency, and adaptability to meet the challenges of large-scale renewable generation integration in the West. BPA will also continue playing a regional role in providing technical support to Northwest utilities with technology implementation and value proposition.

Over the years, BPA has established collaborative relationships with operating entities outside the Western Interconnection, specifically ISO New England, ERCOT, Dominion Virginia Power, and Oklahoma Gas and Electric. This type of collaboration helps BPA to appreciate new diverse perspectives and to learn from other's operating experiences.

4.8 Synchrophasor Project Team

Executive Sponsors:

- Jeff Cook, VP of Transmission Planning
- Mike Miller, VP of Transmission Engineering
- Michelle Cathcart, VP of Transmission Operations
- Terry Oliver, Chief Technology Officer

PMU Deployment Project Management:

- Lawrence Carter

Control Room Applications:

- Pete Raschio (Manager), Nick Leitschuh (PM), Vincent Minden, Bill Wagner, Jeff Anderson, Kliff Hopson

Synchrophasor RAS team:

- MayMay Adolf, Carldez Thomas, John Kerr, Gary Swoboda, Ziyuan Zhang, Kerry Bilyeu, Abebe Masho, Karl Knoll

Synchrophasor Laboratory:

- Greg Stults, Tony Faris, Steve Yang

Technology Innovation:

- Gordon Matthews

Transmission Engineering:

- Steve Laslo, Caitlin Martin, Grisel Mendez, Joe Andres

Transmission Operations:

- Margaret Albright (Manager), Jim Burns, Dan Goodrich, Ashley Donahoo, John Anasis, Eric Heredia, Bart McManus, Rian Sackett

Transmission Planning:

- Melvin Rodrigues (Manager), Dmitry Kosterev (Editor), Jeff Cutter, Gordon Kawaley, Hamody Hindi, Dave Cathcart, Sergey Pustovit, Christina Lee, Sam Hirsi, Kalin Lee, Jeff Johnson.

4.9 Relevant Technology Innovation Projects

TIP-050—Oscillation Damping Controls
TIP-051—Voltage Stability Controls
TIP-052—Power Plant Dynamic Monitoring
TIP-274—Development and Demonstration of Applications for FCRPS Compliance with Model Validation Standards
TIP 281—Impacts Due to Dynamic Transfers (Maxisys)
TIP 289—Wide-Area Damping Control Proof-of-Concept Demonstration (Sandia, Montana Tech)
TIP 299—Synchrophasor Linear State Estimator and PMU Data Validation & Calibration
TIP 305—Data Integrity and Situational Awareness Tools (PNNL)
TIP-313—Power Frequency Controls
TIP 325—Real-Time System Operating Limits (SOL) Computation and Visualization for BPA (V&R Energy)
TIP-348—Measurement-Based Voltage Stability
TIP-349—Oscillation Baselining Application
TIP-350—Power Plant Monitoring Center
TIP-352—Development and Demonstration of a Phasor-Driven Tool for Adaptive Stability Model Calibration using GE PSLF (PNNL, General Electric)

Acknowledgements We gratefully acknowledge BPA engineers who paved the road to the deployment of the synchrophasor technology at BPA. Bill Mittelstadt, John Hauer, Carson Taylor, Jules Esztergalyos, and Ken Martin were among the first visionaries to recognize the value of the technology, develop the first synchrophasor network, research and prototype leading-edge applications. Support from John Haner, Susan Wiess and Don Watkins helped BPA to move through early stages of technology deployment.

BPA executives and management provided strong support and inspiration for the research, development, and deployment of the synchrophasor technology at BPA, including Vickie VanZandt, Brian Silverstein, and Larry Bekkedahl.

BPA synchrophasor investment project was unprecedented for the Agency with a very aggressive timeline. Jim Dow developed the initial project plan, schedule, and cost estimates. Scott Lissit served as a Project Manager during the critical first phase of the project. The successful execution of the synchrophasor investment project is attributed to BPA Project Management Organization.

A project of this complexity and magnitude requires collaboration with industry organizations, other utilities, national laboratories, and researchers. We want to acknowledge Alison Silverstein, NASPI Project Manager, for her leadership in the technology outreach and vision for the technology potential. Phil Overholt at DOE, Jeff Dagle at PNNL, and Joe Eto at LBNL have been supporting research and development of the synchrophasor technology over the past two decades, taking concepts and ideas and converting them into production-quality applications. Ryan Quint and Bob Cummings at NERC have been the visionaries on how the technology can be applied for improving system reliability. Damir Novosel has been the technology champion for the past two decades, leading work on standards, technology solutions, and value proposition. Armando Salazar, Bharat Bhargava, Jim McIntosh, and Jim Hiebert were among the first pioneers to bring the technology in a control room environment. We greatly value collaboration with many operating entities in the West and their technical staff.

We are certainly grateful for the opportunity to work with the leading researchers in the industry and collaborate with them on application development and deployment—Dan Trudnowski, Matt Donnelly, John Pierre, Dave Schoenwald, Bernie Lesieutre, Chris DeMarco, Joe Chow, Pouyan Pourbeik, Henry Huang, Jim Follum, Pavel Etingov, Frank Tuffner, and Ning Zhou, and many others.

We are especially thankful to Alison Silversten, NASPI Manager, for her extensive contributions to this publication.

Appendix A: July 2, 1996 Western Interconnection Outage

On July 2, 1996, a large-scale power outage occurred in the Western Interconnection. It started with a loss of two 345-kV transmission lines from Wyoming to Idaho. Thermal overloads cascaded into lower-voltage lines, which resulted in voltage collapse in the Boise, Idaho area. But the disturbance did not stop there. Voltage collapse propagated into Southern Oregon and caused fast voltage decline on the California–Oregon AC Intertie (COI) as shown in Fig. 4.3. Line protection opened three COI lines, resulting in uncontrolled system breakup and loss of load and generation. Total of 2 million customers were affected, and 11,850 MW of power interrupted.

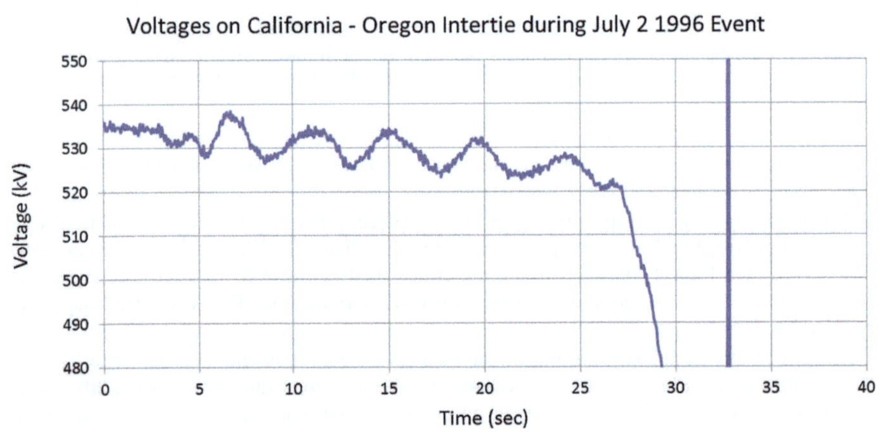

Fig. 4.3 Voltages on California–Oregon Intertie during July 2, 1996 Western Interconnection power outage

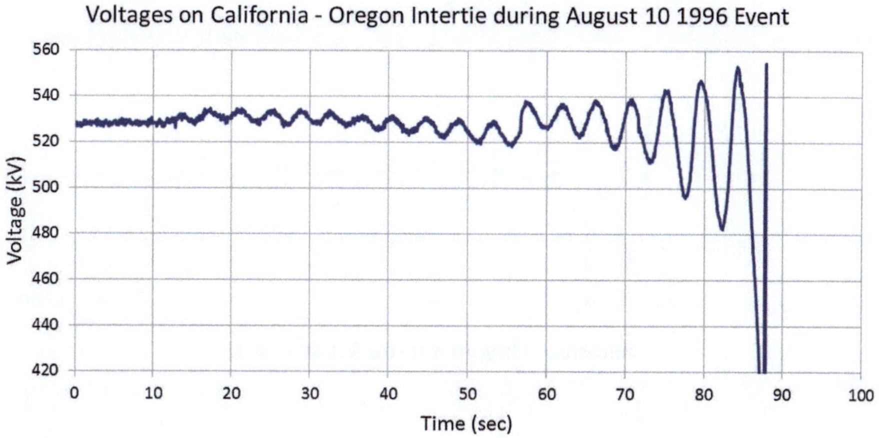

Fig. 4.4 Voltages on California–Oregon Intertie during August 10, 1996 Western Interconnection power outage

Appendix B: August 10, 1996 Western Interconnection Outage

On August 10, 1996, another large-scale power outage occurred in the Western Interconnection. The interconnection broke into four islands, interrupting service to 7.5 million customers mainly in California and the Southwest. The sequence of events was initiated by the loss of a Keeler–Allston 500-kV line near Portland, Oregon. The line loss resulted in overloads on lower-voltage transmission lines in the area and depressed grid voltages. Several minutes later, two lower-voltage lines in Portland area tipped due to thermal overloads, and further depressing voltages. McNary generators were already boosting full reactive power to support grid voltages, and started sequentially tripping by over-excitation protection. McNary tripping started growing voltage and power oscillations on California–Oregon Intertie as shown in Fig. 4.4. COI lines were opened by line protection about 90 s from the time oscillations began.

Appendix C: June 14, 2004 Generation Outage in the West

A delayed cleared fault in Arizona caused generation loss in excess of 4,000 MW, which in turns drove system frequency down to 59.5 Hz, and created large power swing and voltage drop at California-Oregon Intertie [21]. Once the system settled, BPA operators were presented with a dilemma: system frequency was abnormally low at 59.75 Hz, while COI power flows were significantly above its system operating limit (Fig. 4.5).

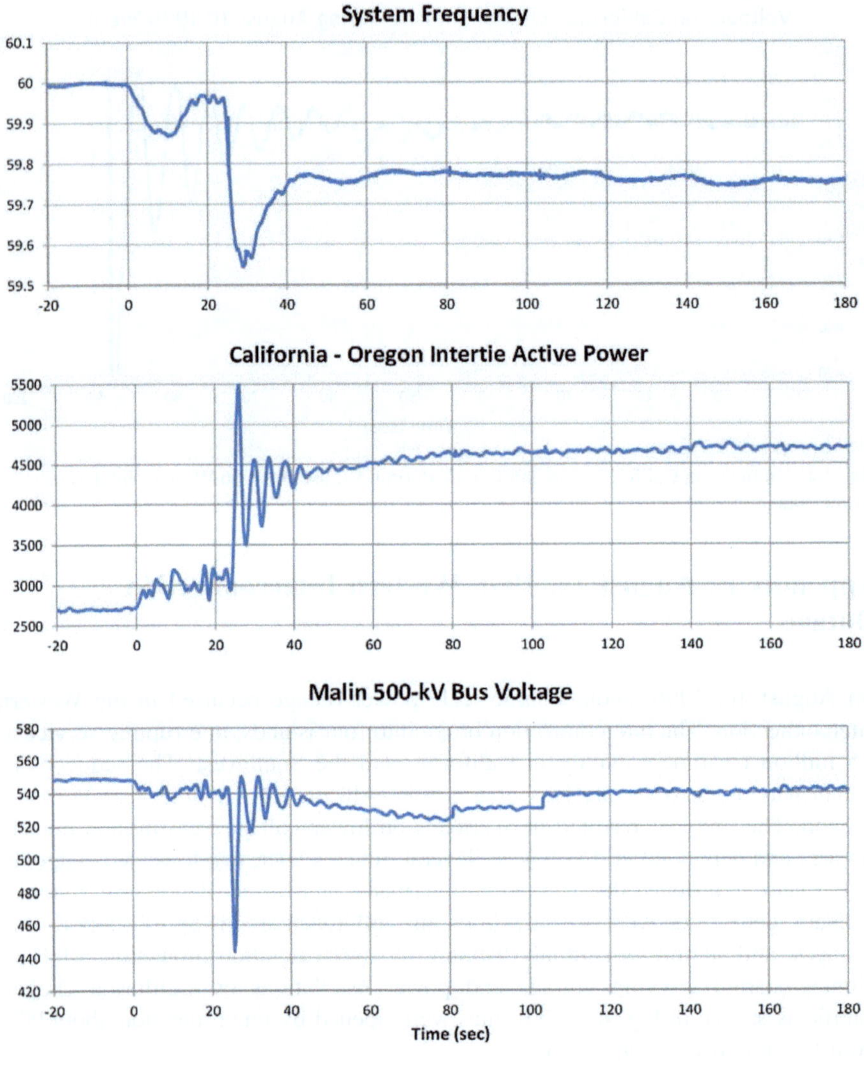

Fig. 4.5 Frequency, power, and voltages during June 14, 2004 event

Appendix D: Frequency Response Analysis at BPA

Event Detection

BPA runs the frequency event detection application in real time. Both BPA and partner data are used by the algorithm. The application takes a derivative of voltage phasor angles to calculate bus frequency. Although PMUs report bus frequency, there is a difference between frequency filtering algorithms used by various PMUs. An onset of frequency deviation is detected by magnitude of frequency deviation and its rate of change. The application uses rate of frequency change to locate the source of frequency disturbance, as the bus frequency will decline the fastest closer to the event location.

Notification

The application sends out a notification of the event occurrence as well as information about the magnitude of frequency deviation and PMU which had the fastest rate of change of frequency.

> The Synchrophasor Frequency Deviation Module (FDM) detected an alarm event at 9/1/2015 10:30:42 AM Pacific Daylight Time (GMT-07:00:00)
>
> Rank1 PMU = W066NAVAJO____01
>
> Number of PMUs Affected = 44
>
> Frequency Change Magnitude = 0.07672414

Visualization

BPA technical staff has access to visualization displays similar to ones used by BPA dispatchers in the control room.

An overview display shows Western Interconnection map (with zoom in on Pacific Northwest). The size of the dot corresponds to rate of frequency change at that PMU location. In this case, it is evident that a frequency disturbance originated in Arizona.

BPA technical staff can also pull out 90-s plots of system frequency during the event, as well as power flows on California–Oregon Intertie, Northern Intertie, and Montana Intertie.

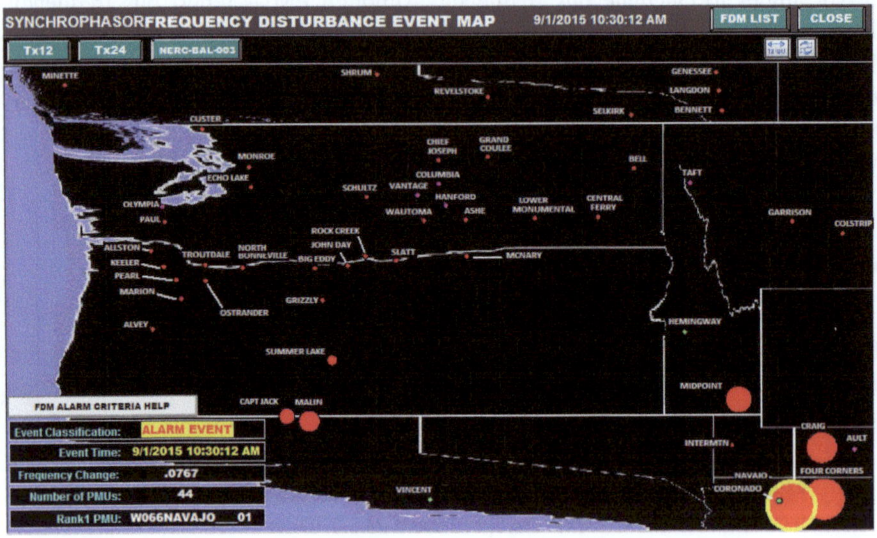
SYNCHROPHASOR FREQUENCY DISTURBANCE EVENT MAP
9/1/2015 10:30:12 AM
FDM LIST
CLOSE
Tx12
Tx24
NERC-BAL-003
MINETTE
SHRUM
REVELSTOKE
GENESSEE
LANGDON
SELKIRK
BENNETT
CUSTER
MONROE
ECHO LAKE
CHIEF JOSEPH
GRAND COULEE
BELL
TAFT
COLUMBIA
SCHULTZ
VANTAGE
HANFORD
OLYMPIA
WAUTOMA
ASHE
LOWER MONUMENTAL
CENTRAL FERRY
GARRISON
COLSTRIP
PAUL
ROCK CREEK
JOHN DAY
ALLSTON
TROUTDALE
NORTH BONNEVILLE
BIG EDDY
SLATT
MCNARY
KEELER
PEARL
MARION
GRIZZLY
OSTRANDER
ALVEY
HEMINGWAY
SUMMER LAKE
MIDPOINT
FDM ALARM CRITERIA HELP
CAPT JACK
MALIN
CRAIG
AULT
INTERMTN
NAVAJO
FOUR CORNERS
CORONADO
VINCENT
Event Classification: ALARM EVENT
Event Time: 9/1/2015 10:30:12 AM
Frequency Change: .0767
Number of PMUs: 44
Rank1 PMU: W066NAVAJO___01

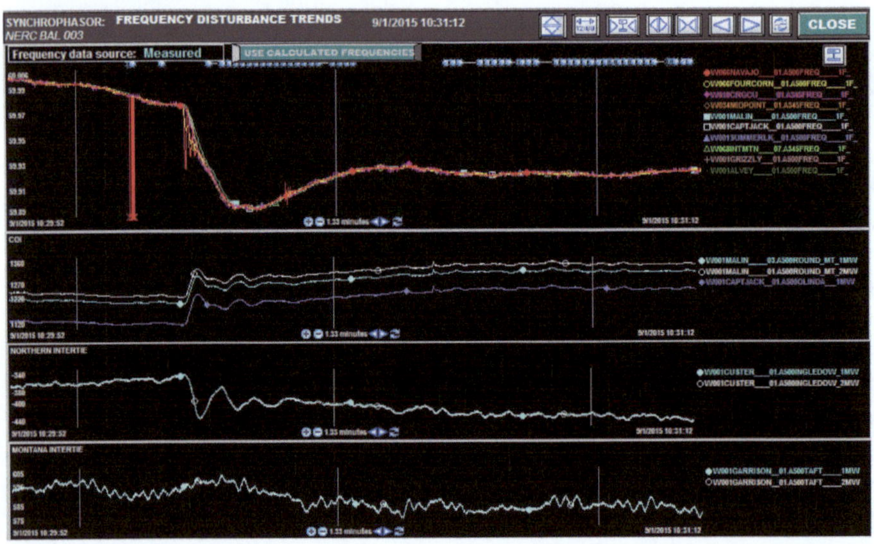
SYNCHROPHASOR: FREQUENCY DISTURBANCE TRENDS
NERC BAL 003
9/1/2015 10:31:12
CLOSE
Frequency data source: Measured
USE CALCULATED FREQUENCIES
COI
NORTHERN INTERTIE
MONTANA INTERTIE

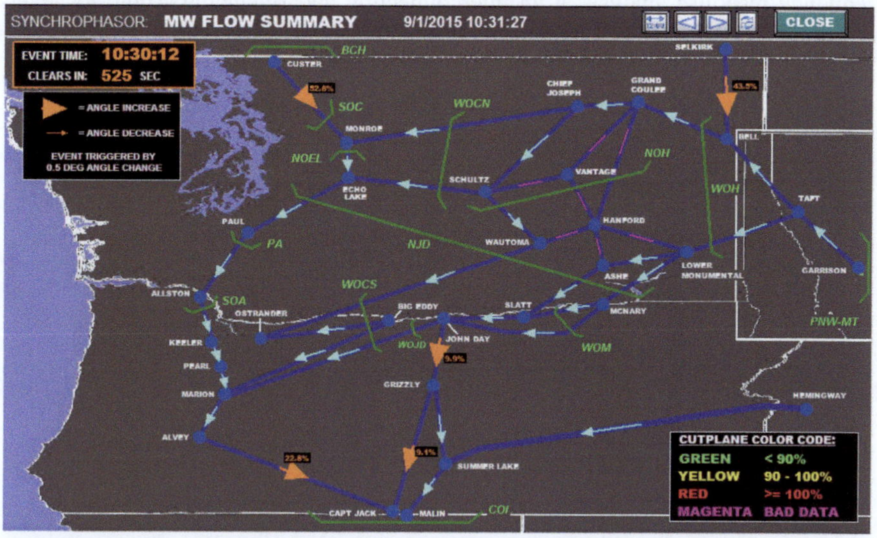

"MW Flow Summary" display shows power pick-up on transmission lines due to governor response.

Data Extract

BPA has several applications to extract data from its OSI Soft PI servers. One of the advantages of OSI Soft PI architecture at BPA is that an application can simultaneously access both SCADA and PMU PI historians. BPA uses high-resolution measurement of bus frequencies, voltages, phasor angles, intertie flows, and generation from PMU historian, and 2-s AGC quantities from its SCADA historian. SCADA data are time-synchronized at control center, so that there is a time lag between PMU and SCADA measurements, in our experience at least SCADA scan rate of 2 s. The data extract applications can time-shift SCADA quantities for better alignment with synchrophasor measurements.

Analysis

BPA uses the frequency response analysis tool (FRAT), co-developed with PNNL, for its frequency response analysis. FRAT is an interactive application to perform frequency response analysis at an interconnection and Balancing Authority levels according to NERC BAL-003-1 methodology.

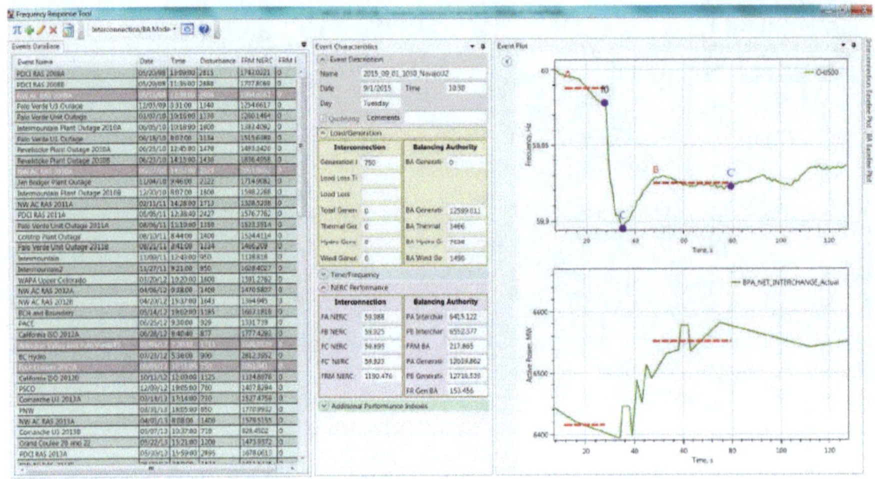

Baselining

The FRAT database has an extensive system event log, going back to 2008. This log creates the opportunity to evaluate frequency response trends over many years. BPA has been using the FRAT baseline to support its arguments in filings with Federal Energy Regulatory Commission (FERC) during the development of NERC BAL-003-1 Frequency Response Reliability Standard.

First, we can trend the Western Interconnection frequency response and compare against the requirement set forth by NERC BAL-003-1 Reliability Standard. Trending WECC frequency response is important for tracking the impact of the changing generation resource mix has on the frequency response.

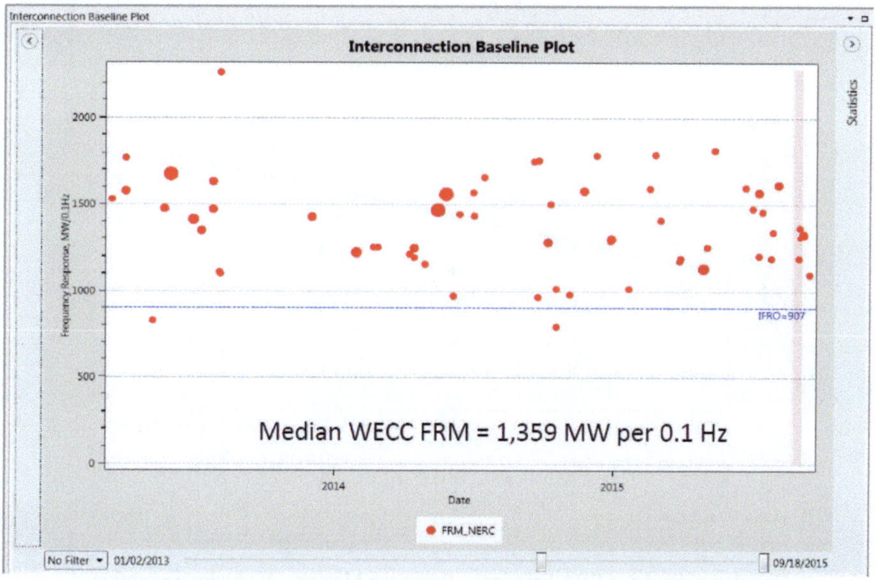

Similar plots can be made for BPA's Balancing Authority. Ultimately, a Balancing Authority is an entity responsible for compliance with NERC BAL-003-1 Standard. Such analysis is very helpful for BPA to understand its inventory of frequency responsive reserves, and factors affecting its performance. BPA can perform correlation analysis between observed frequency response and generation resource mix—hydro, wind, etc. For BPA, there is a strong correlation between available hydro-generation capacity and its frequency response performance.

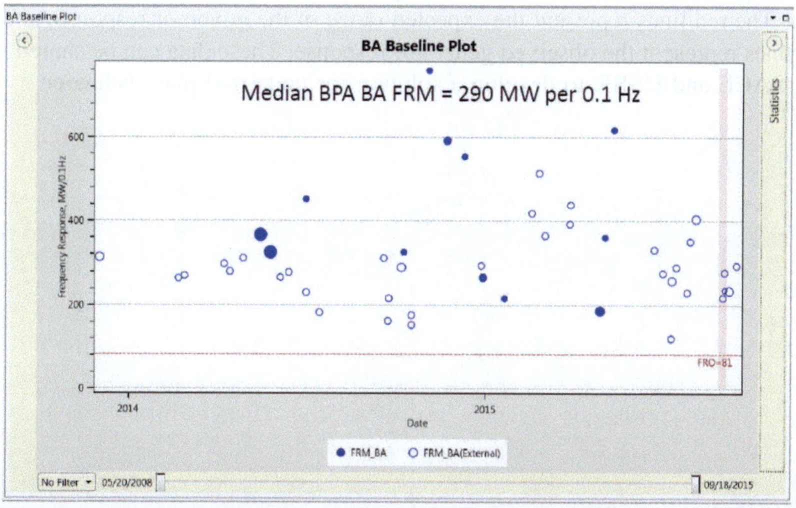

Historic trend of BPA Frequency Response Measure

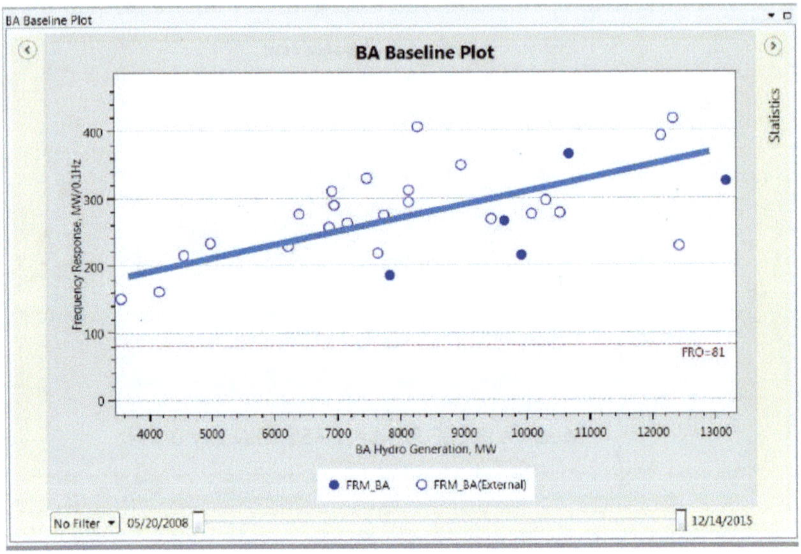

Correlation between BPA Frequency Response Measure and hydro generation

Generating Fleet Performance Analysis

BPA developed a MATLAB application for analysis of generator performance during a frequency disturbance event. An application plots observed versus expected range of generator responses. The application is very useful in identifying generator response abnormalities.

Example below shows examples of good and questionable responses to the event. The red lines represent the expected range of the governor response, and the blue lines represent the observed generator response. These data can be shared with the USACE and USBR to develop a solution for undesired plant behavior.

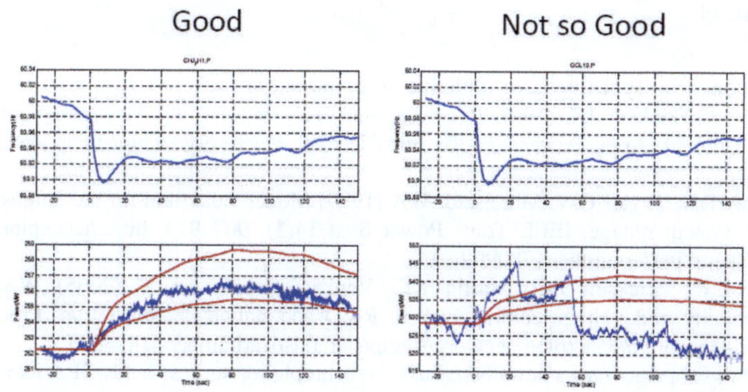

Power Pickup Analysis

Frequency response is not only about arresting system frequency decline and recovery following a resource loss. Governor response to frequency deviation can increase power flows on transmission system and thereby create voltage stability risks. This is particularly relevant to California–Oregon Intertie which can be post-transient voltage stability limited by generation outages in California and Desert Southwest. BPA has been baselining governor pickup on California–Oregon Intertie since late 1990s to make sure that actual pickup is accurately represented in power system studies.

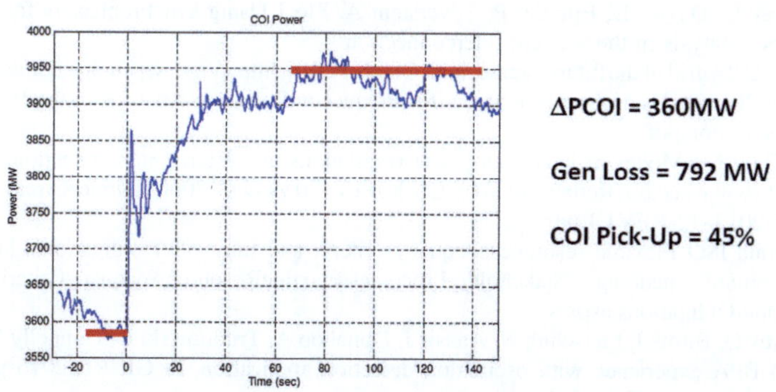

References

1. North American synchrophasor initiative. www.naspi.org
2. Taylor CW, Erickson DC (1997) Recording and analyzing the July 2 cascading outage. IEEE Trans Comput Appl Power 26–30. http://ieeexplore.ieee.org/stamp/stamp.jsp?arnumber=560830
3. Kosterev DN, Taylor CW, Mittelstadt WA (1999) Model validation for the August 10, 1996 WSCC system outage. IEEE Trans Power Syst 14(3): 967–979. http://ieeexplore.ieee.org/stamp/stamp.jsp?arnumber=780909
4. Taylor CW, Erickson DC, Martin KE, Venkatasubramanian V (2005) WACS—wide area stability and voltage control system: R&D and online demonstration. Proc IEEE 93 (5):892–906. http://ieeexplore.ieee.org/stamp/stamp.jsp?arnumber=1428005
5. https://www.bpa.gov/news/newsroom/Pages/Synchrophasor-success-lands-BPA-its-first-Platts-Award.aspx
6. NERC (Sept 2016) Reliability guideline, power plant dynamic model verification using PMUs. http://www.nerc.com/pa/RAPA/rg/ReliabilityGuidelines/Reliability%20Guideline%20-%20Power%20Plant%20Model%20Verification%20using%20PMUs%20-%20Resp.pdf
7. Overholt P, Kosterev D, Eto J, Yang S, Lesieutre B (2014) Improving reliability through better models. IEEE Power Energy Mag 44–51. http://magazine.ieee-pes.org/files/2014/04/12mpe03-overholt-2301533.pdf
8. North American Synchrophasor Initiative (2015) Model validation using phasor measurement unit data: NASPI technical report, March 2015. [Online]. Available: https://www.naspi.org/documents
9. Federal Energy Regulatory Commission (2012) Comments of the Bonneville power administration under AD12–10, May 2012. http://elibrary.ferc.gov/idmws/file_list.asp?document_id=14023794
10. Kosterev D, Davies D (2010) System model validation studies in WECC. IEEE Power Eng Soc Meet. http://ieeexplore.ieee.org/xpls/abs_all.jsp?arnumber=5589797
11. Kincic S, Davies, D, Kosterev D, Member H, Zhang B, Thomas M, Vaiman J, Ramanathan RW (2016) Bridging the gap between operation and planning models in WECC. IEEE Power Eng Soc Meet. http://ieeexplore.ieee.org/stamp/stamp.jsp?arnumber=7741091
12. Kosterev D, Davies D, Etingov P, Silverstein A, Eto J Using synchrophasors for frequency response analysis in the western interconnection
13. GIGRE 2014 grid of the future symposium, October 2014, http://cigre.wpengine.com/wp-content/uploads/2015/06/Using-Synchrophasors-for-Frequency-Response-Analysis-in-the-Western-Interconnection.pdf
14. WECC (2013) Modes of inter-area power oscillations in western interconnection Nov 2013. https://www.wecc.biz/Reliability/WECC%20JSIS%20Modes%20of%20Inter-Area%20Oscillations-2013–12-REV1.1.pdf
15. California ISO Flexible resource adequacy criteria and must-offer obligations. http://www.caiso.com/informed/Pages/StakeholderProcesses/FlexibleResourceAdequacyCriteria-MustOfferObligations.aspx
16. Kosterev D, Burns J, Lietschuh N, Anasis J, Donahoo A, Trudnowski D, Donnelly M, Pierre J (2016) BPA experience with oscillation detection application, in GIGRE 2016 grid of the future symposium, Oct 2016
17. IEEE Power System Dynamic Performance Committee (2012) Identification of electromechanical models in power systems. Technical Report PES-TR15, June 2012. http://sites.ieee.org/pes-resource-center/files/2013/11/TR15_Modal_Ident_TF_Report.pdf
18. Novosel D, Madani V, Bhargava B, Khoi V, Cole J (2008) Dawn of the grid synchronization. IEEE Power Energy Mag 49–60, Jan/Feb 2008. http://ieeexplore.ieee.org/stamp/stamp.jsp?arnumber=4412940

19. NASPI Technical Report (2015) The value proposition for synchrophasor technology—itemizing and calculating the benefits from synchrophasor technology use, Version 1.0, Oct 9, 2015. www.naspi.org
20. NASPI technical workshop on synchrophasor data and state estimation, Mar 25, 2015. www.naspi.org
21. FERC and NERC Report (2011) Arizona-Southern California outages on Sept 8, 2011. https://www.ferc.gov/legal/staff-reports/04-27-2012-ferc-nerc-report.pdf
22. Agrawal B, Kosterev D (2007) Model validation for a disturbance event that occurred on June 14 2004 in the western interconnection, presented at 2007 power engineering general meeting. Tampa, FL
23. Adams J, Carter C, Huang S-H (2012) ERCOT Experience with sub-synchronous control interaction and proposed remediation. IEEE PES transmission and distribution conference and exposition, May 2012. http://ieeexplore.ieee.org/stamp/stamp.jsp?arnumber=6281678

Chapter 5
Use of Synchrophasor Measurement Technology in China

Shi Bonian

5.1 Introduction

In developing smart power grid, the synchrophasor measurement technology plays a crucial role for power system analysis and control. Phasor measurement unit (PMU) and wide-area measurement system (WAMS) are becoming the critical measurement infrastructures of transmission and generation systems [1]. As the mid of 2017, approximately 4100 PMU sets have been deployed in power grids in China, covering all 500 kV substations and some important power plants and 220/110 kV substations. In addition, more than 30 WAMS center stations are in service, providing important dynamic information about the power system operation. In this chapter, the recent and emerging development of PMU/WAMS, communication, and synchronization network in China will be briefly summarized first, and then some major advanced applications utilizing synchrophasor measurement technology will be presented.

5.2 Development and Applications of PMUs in China

5.2.1 Status of PMU Deployment

Power transmission grid in China is run by two major transmission system operators (TSOs), i.e., State Grid Corporation of China (SGCC) and China Southern Power Grid (CSG). Both have deployed large numbers of PMUs as part of

S. Bonian (✉)
Beijing Sifang Automation Company, Ltd, Beijing, China
e-mail: shibonian@sf-auto.com

S. Nuthalapati (ed.), *Power System Grid Operation Using Synchrophasor Technology*, Power Electronics and Power Systems,
https://doi.org/10.1007/978-3-319-89378-5_5

Table 5.1 Amount and proportion of PMUs in power plants and substations at over 110 kV level

Voltage level (kV)	1000	±800	750	±660	500	330	220	110
Amount	7	6	39	26	723	165	861	200
Proportion (%)	0.35	0.30	1.92	1.28	35.67	8.14	42.48	9.87

nationwide mandate that requires all substations on the 500 kV networks and above and generators larger than 100 MW should be monitored by PMU.

To take SGCC as the example, by the end of 2013, the number of PMU sets installed at the substations and power plants is 2027 [2]. Sorted by the voltage level, the statistical installation of PMUs is shown in Table 5.1.

As the mid of 2017, approximately 4100 PMU sets have been deployed in power grids in China, covering all 500 kV substations and some important power plants and 220/110 kV substations within SGCC and CSG.

Chinese standards on PMUs are being updated to conform to the recently released IEEE standards C37.118.1 and C37.118.1a. Much effort has been devoted to improve the measurement accuracy based on latest released PMU standards.

5.2.2 PMU Supporting IEC 61850 Protocol

With the enhancement of synchronized phasor measurement technology, current major practice for PMU development in China includes two aspects: support smart substation technique and improve the measurement accuracy based on latest released standards on PMU.

Approximately 180 PMUs supporting IEC 61850 have been commissioned in recent years. With the fast development of smart substation technologies, over 3500 smart substations have been put into operation, covering voltage range from 110 to 750 kV. Most of these smart substations are located in low voltage levels; however, approximately 180 high-voltage substations are required to possess the synchronized phasor measurement function. Phasor measurement and communication functions realized in smart substation structure are displayed in Fig. 5.1.

In Fig. 5.1, PMU takes inputs from merging unit and switchgear control unit via process buses, in the form of IEC 61850-9-2 SV and GOOSE protocols. After phasor calculation, the phasors can be transmitted, at a slow rate, to a monitoring server via station buses, in the form of Manufacturing Message Specification (MMS). At the same time, phasor can be transmitted, at a fast rate of up to 100 frames per second, to a substation phasor data concentrator, in accordance with IEEE C37.118 standard, and then to remote dispatching center.

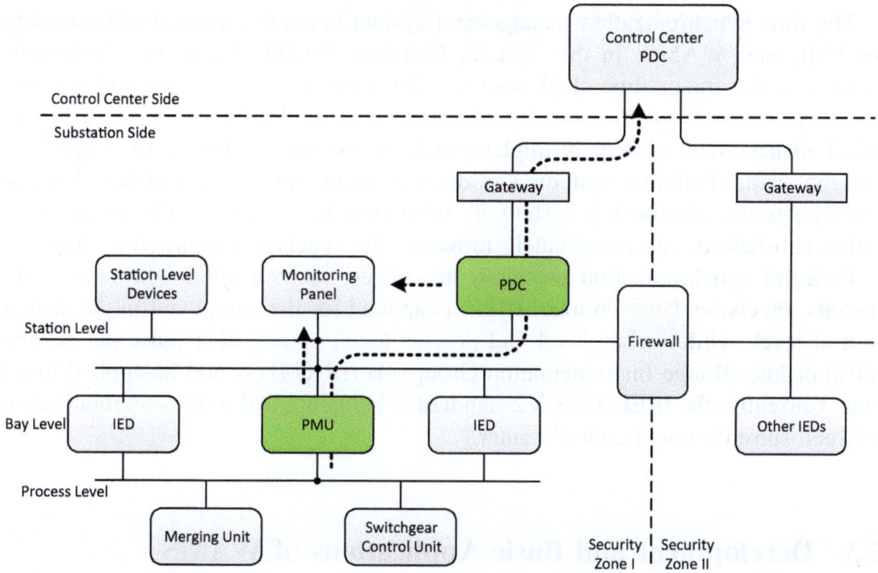

Fig. 5.1 PMU functions realized under smart substation environment

5.2.3 Data Transmission Network and Time Synchronization Networks in Chinese Power Grids

In consideration of power transfer security and dispatching automation, the applications based on state grid dispatching digital network (SPDnet) have a high demand for reliability, real time, and safety. Therefore, key performance requirements such as network time delay, convergence time, and safety must be guaranteed. Currently, according to the principle of dual independent backbone networks and multiple layers integration, an exclusive dispatching data network based on exclusive channels (Synchronous Digital Hierarchy + IP + Optical Fiber) has been applied, which has the URL stratification structure and alternate route function. Highly reliable core equipments together with double master controls are adopted in this network. Three-level border gateway protocol route is used to realize a multi-protocol label switching virtual private network, through which different types of network tasks can be isolated and the safety of dispatching tasks can be ensured. Quality of service strategy is employed to ensure the transmission bandwidth and real-time demand of data flow for important tasks.

Several standards have been established for real-time data application of power systems, including DL/T 476-2012 for real-time data communication among dispatching centers, DL/T 634/5 (104) for real-time data communication between central station and substation, DL/T 860 7-2 (IEC 61850 compatible) for data communication within substation. In addition, IEEE C37.118 is used for WAMS/PMU phasor data communication.

The time synchronization management system is another critical infrastructure for PMU and WAMS. In this system, Compass Satellite Navigation System is adopted as the major time clock source. Global Positioning System (GPS) is the auxiliary one, and synchronous digital hierarchy is the backup solution. A hierarchical management system is implemented for the monitoring of time synchronization status. Different systems and devices, including power grid dispatch and control system, monitoring system in substation level, and synchronized IEDs within substations, run coordinately to realize the synchronization monitoring.

Different synchronization protocols have been adopted within the smart substations. Precision Time Protocol (PTP) is applied for the equipment in the station control level, while in bay level and process level, optical fiber pulse per second (PPS) or Inter-Range Instrumentation Group—B (IRIG-B) would be applied to set time. Currently, the IEEE 1588 V2 standard is being applied to the communication between substation and control center.

5.3 Development and Basic Applications of WAMS in China

5.3.1 Architecture of WAMS

All of the national, inter-regional, and provincial dispatching centers have been deployed with WAMS central stations. The total number is 39 until the end of 2013. In addition, some of the control centers in metropolitan power grids, such as Guangzhou and Shenzhen, also designed and are constructing their own WAMS.

Two comprehensive dispatching supporting platforms have been developed, i.e., D5000 in State Grid of China and OS2 in China Southern Power Grid, to integrate different systems deployed in control room, such as energy management system (EMS), dynamic security assessment (DSA), dispatcher training system (DTS). WAMS is one of the core subsystems in these platforms as well. The framework of a typical WAMS is demonstrated in Fig. 5.2. In addition to the measurement terminals (PMUs) and the communication network, the WAMS central station is composed of several servers, including front-end communication, WEB, historical database, real-time database, and advanced application servers. Among these servers, the real-time database and advanced application servers are the key components.

The real-time database techniques oriented to power system equipment have been developed based on Common Information Model (CIM) criterion, wide-area distributed storage, and direct location methods. A substantial increase in access efficiency is observed relative to the previous system. The scope of data access is extended from local to remote area, which solves the problem of real-time data share in the entire network with high efficiency. Several techniques including fixed time duration, equal data time interval, and hash table index have been used to

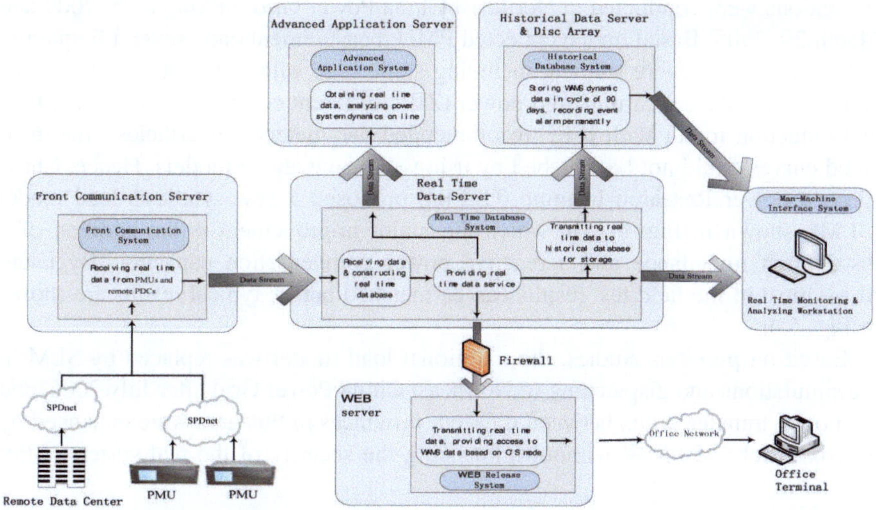

Fig. 5.2 Typical framework of WAMS

satisfy the requirements of storage and retrieving of massive data. The processing capability of the real-time database reaches the level of 5 million events per second.

Many successful applications based on PMU data or WAMS have been implemented in the D5000 or OS2 platform. Some of these applications will be introduced in the following section.

5.3.2 Power System Model Parameter Identification and Validation

Much work has been done in the field of parameter identification for line and transformer model, which can be considered as static-type model expressed with linear algebra equations. PMU-recorded steady-state data is used for identification. For dynamic type model with differential and algebra equations, such as generator and motor load, PMU-recorded data during dynamic process is used for identification, which will be introduced in this section.

5.3.2.1 Load Model and Parameter Identification

Many disturbance data recorded by PMU show significant differences between the simulation results and real system responses. Errors in the system models and parameters are the main causes, especially the load. To validate the simulation accuracy, four field tests of artificial three-phase-to-ground short circuit in 500 kV substations were conducted in Northeast China Power Grid on March 25, 2004 and March 29, 2005. Based on the collected PMU measurement data, several frequently used load models were studied, including static load with different percentages of constant impedance, current, and power (ZIP), different combinations of static load and induction motor with IEEE recommended parameters. Nevertheless, the measured curves could not be matched by using all of the above models. Hence, China Electric Power Research Institute (CEPRI) proposed a new synthesis load model (SLM), shown in Fig. 5.3, in which the major improvement is the addition of a distribution impedance and a reactive power compensation capacitor. By using SLM, most of the field test results can be matched better; typical results are shown in Fig. 5.4.

Based on previous studies, the traditional load model was replaced by SLM in the simulations and dispatching of Northeast China Power Grid after July 2006, and the power transfer limits between different provinces in this area were enhanced by approximately 450 MW without influencing the security of the real system. After

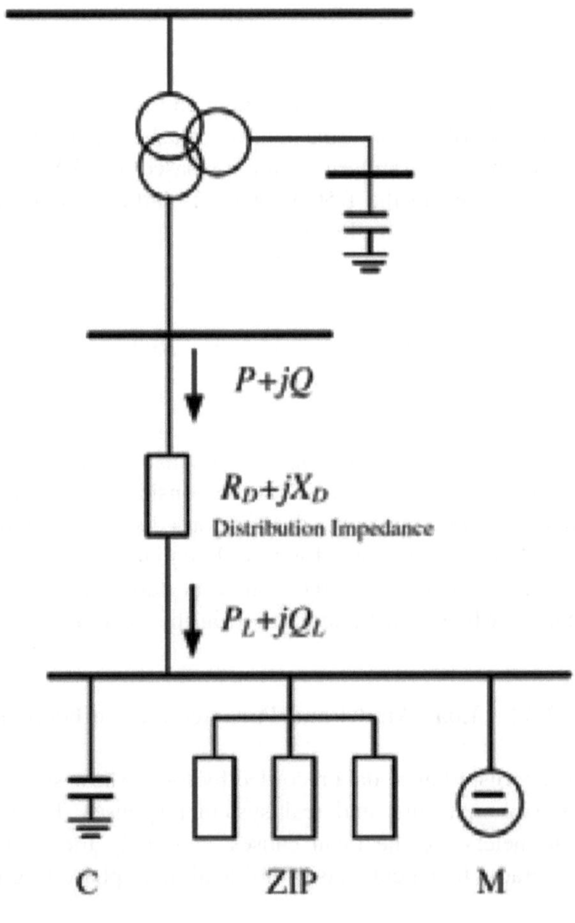

Fig. 5.3 Configuration of synthesis load model

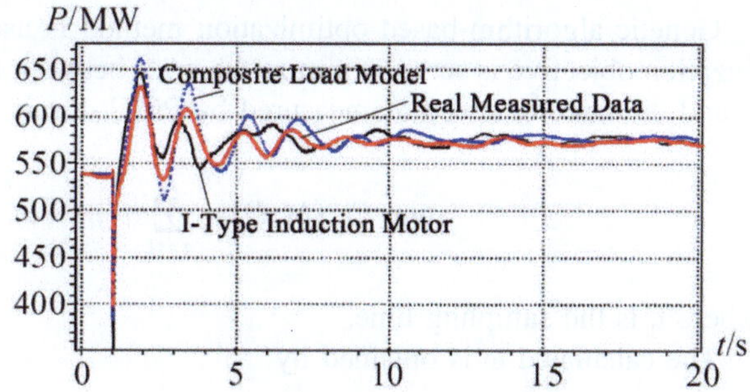

Fig. 5.4 Comparisons between the recorded PMU data and simulated data

2007, SLM model and parameters were also validated by recorded PMU data under many disturbances in North China and Central China regional power grids and some provincial grids.

5.3.2.2 Identification of Generator's Moment of Inertia

Generator parameters may vary greatly in different operating conditions. Some identification or parameter optimization techniques have been employed to realize generator parameter identification. Generally, the input PMU data are the generator terminal voltage phasors, excitation voltages, and excitation currents, aiming to minimize the error between the simulated generator current and the current measured by PMU [3]. The identified generator parameters include synchronous reactance, transient reactance, subtransient reactance for direct and quadrature axes, and their corresponding time constants.

The moment of inertia is one of the most important parameters that determine the dynamic performance of a generator unit [4]. It is combination of all the rotatory components such as generator, turbine, and rotatory exciter. Traditionally load rejection tests are used to obtain the combined MOI at the field of plants, but may have damage to generator unit. Thus, to identify the MOI parameter from PMU-recorded event data is another feasible choice.

Based on the movement equation of generator rotor

$$M\times\frac{d\omega}{dt}+D\times\omega=\frac{N_p\times P_m}{\omega}-\frac{N_p\times P_e}{\omega} \tag{5.1}$$

where M is the moment of inertia, ω is the angular speed, D is the damp ratio, N_p is the number of the poles, P_m is the mechanical power, P_e is the electromagnetic power.

Here, the variables to be solved is the moment of inertia M and the damp ratio D. Mechanical power P_m can be treated as constant during first 1–2 seconds, while the initial value can be assessed from the steady-state electromagnetic power P_e measured by PMU.

Genetic algorithm-based optimization method is used to get MOI, as the optimization objective is set to minimize the error between the calculated angular speed ω and the angular speed ω_r measured by PMU during a frequency disturbance,

$$\min\ Err(M,D) = \sum_{t=1}^{K} \omega(t_i) - \omega_r(t_i))^2 \tag{5.2}$$

where t_i is the sampling time.

The calculated ω is obtained by

$$\omega(t_n) = \frac{P_m - P_e(t_{n-1}) - D \times \omega^2(t_{n-1})}{M \times \omega(t_{n-1})} \Delta t + \omega(t_{n-1}) \tag{5.3}$$

where Δt is the time interval.

The recorded generator output active power and frequency of the event used for optimization are shown in Fig. 5.5a with one minute length, and zoomed transient data in Fig. 5.5b with 4 s length.

The recorded power and frequency data are used as input to Eq. (5.3), and after 126 iterations of adopted genetic algorithm, the error converges and the final result for JR generator #2 with M = 102,648 kg m^2 and D = 557 kg m^2/s. The convergence process is displayed in Fig. 5.6.

5.3.3 Disturbance Recognition and Location

When disturbance occurs in power systems, the operators of control centers can receive many warnings and sequence of events (SOEs) information. However, these information usually only contains breaker trips and protection relay actions. The operators cannot directly know what kind of faults lead to these actions of breaker or protection relays. The action reasons usually can be obtained by analyzing the fault recording wave files after the fault events. It will be time consuming and not be benefit for fault process. Based on real-time dynamic data of PMUs, online identification methods for different kinds of faults or disturbances have been designed and implemented. These fault or disturbance types that can be identified include short-circuit faults, generator breaker trips, phase shift failures of DC converters, islanding. An example of detection of short-circuit fault is displayed in Fig. 5.7.

As the PMU distribution points have not completely covered the entire power grid, in case that faults happen at locations that without install PMU, or within power grids with relatively lower voltage levels, it will only be able to detect the occurrence of disturbance instead of directly locating the fault points. At the very

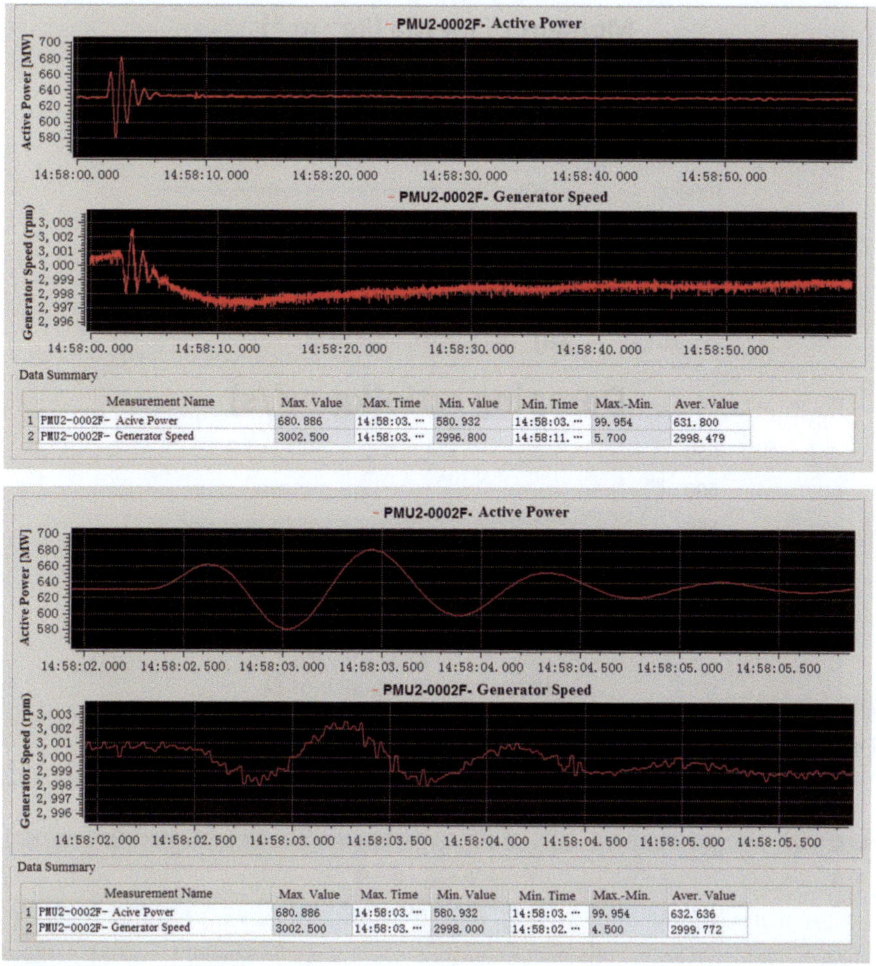

Fig. 5.5 Recorded power and frequency for power plant JR #2 on January 7, 2014 event

moment of fault occurrence, such electrical quantities as voltage and frequency will change abruptly, with the incremental quantity being related with the distance from the fault point. As a result, in accordance with the electrical incremental quantity of each measurement node at the time of disturbance, it is able to draw the voltage contour map upon the geographical wiring diagram, thus approximately locating the region with faults. An example is displayed in Fig. 5.8.

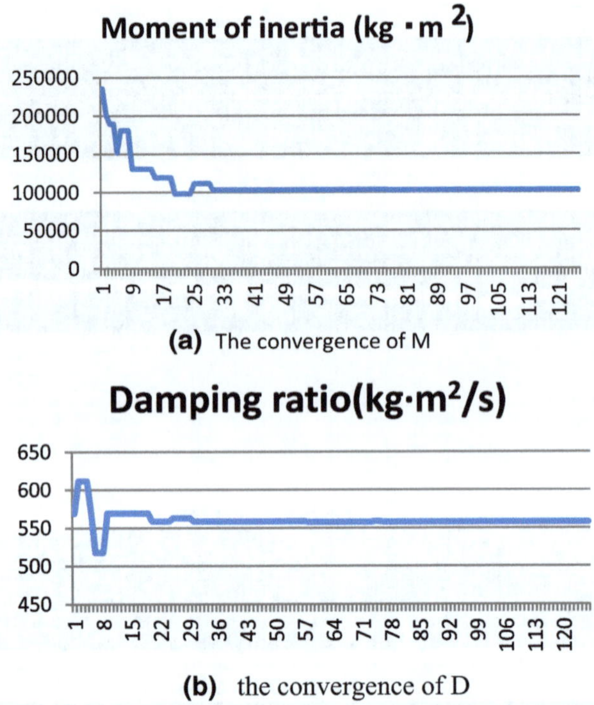

Fig. 5.6 Identified parameter of JR generator #2

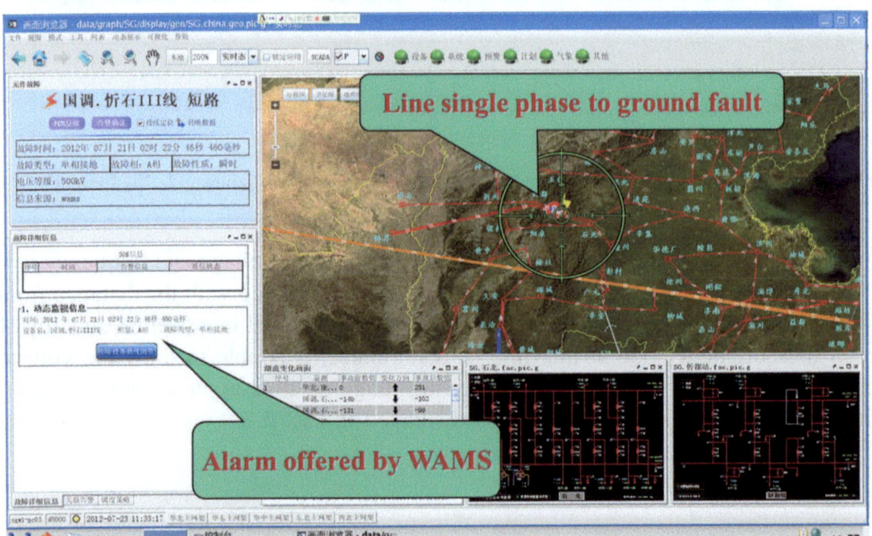

Fig. 5.7 Short circuit fault detected by WAMS

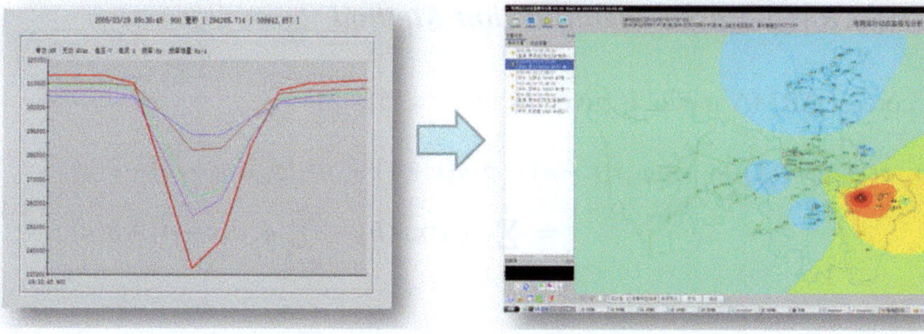

Fig. 5.8 Voltage contour visualization for fault area tracking

5.4 Estimation of Electromechanical Modes from Ambient PMU Data

5.4.1 Classification of PMU Data

In general, field PMU data fits two classifications: typical and non-typical data. Typical data carries system mode information and is describable by the model structure used by an identification algorithm. In contrast, non-typical data cannot be described by a general linear model and is commonly encountered as outliers, which are values that significantly deviate from normal values. After PMU data is collected at the dispatching center, the first step is to classify the data into typical and non-typical data according to some prediction error model.

Typical data mainly includes ambient data and ringdown oscillations. Ambient data is produced when the power system is working under an equilibrium condition, and the major disturbance is from small-amplitude random load changes. Ringdown oscillation data occurs after large disturbances, such as a line tripping out of service, where the resulted oscillations are observable. The ringdown oscillation data carries the highest level of information density and the mode estimation converges fast to the true values.

However, it is valuable to identify oscillation information from ambient data, since there are huge numbers of ambient data versus only few ringdown records. To check the existence of ringdown oscillation in the typical data, some fast oscillation detection method is adopted, such as the waveform detection method. If ringdown is checked, then some real-time alarming message can be delivered.

If the features of the data satisfy the assumptions of ambient definition, then the ambient data analysis method will be applied. If weak or negative damping ratio is derived from ambient data analysis, then some online pre-alarming message can be delivered to warn the dispatcher for some necessary regulation control.

5.4.2 ARMA-Based Identification Method

The core of ARMA model is given as follows [5],

$$\begin{aligned} x(\kappa) &= \varphi_1 x(\kappa-1) + \cdots + \varphi_n x(\kappa-n) - \phi_1 a(\kappa-1) - \cdots - \phi_m a(\kappa-m) + a(\kappa) \\ &= \sum_{h=1}^{n} \varphi_h x(\kappa-h) - \sum_{p=1}^{m} \phi_p a(\kappa-p) + a(\kappa) \end{aligned} \tag{5.4}$$

(k = 1,2,…,N), where $x(\kappa)$ is the system response, $a(\kappa)$ is the stochastic noise input, $\varphi_1,\ldots,\varphi_n$ are the coefficients of AR part, $\phi_1,\ldots,\phi_m$ are the coefficients of MA part, N is the length of data.

The eigenvalues λj and λj* can be calculated by solving the AR polynomial. Therefore, the parameters of low-frequency oscillation modes can be obtained,

$$\begin{cases} f_j = \dfrac{\sqrt{\ln \lambda_j \cdot \ln \lambda_j^*}}{2\pi\Delta} \cdot \sqrt{1-\varsigma_j^2} \\ \varsigma_j = -\dfrac{\ln|\lambda_j|}{\sqrt{\ln \lambda_j \cdot \ln \lambda_j^*}} \end{cases} \tag{5.5}$$

where f_j is the mode frequency, ς_j is the mode damping ratio, and Δ is the sample time interval.

Before using ARMA method to analyze the ambient data at different PMU sites, some preprocessing work is necessary to make use of the measured PMU data. First, the outliers are removed, and then, the data window is detrended using a linear model [6]. After that, the detrended data is passed through a low-pass anti-aliasing filtering and downsampled finally. In this research, the original PMU data collected in the dispatching center is recorded at the rate of 25 Hz. After a 2.5 Hz cutoff low-pass FIR filter (order = 50), the input data is downsampled to the rate of 5 Hz.

In this research, via some parameter selection procedures, the order of AR is set as 14, and the order of MA is set as 13. The data window is 5 min, and the window is updated at the step of 10 s.

Since at different PMU sites, the measured ambient data may suffer some different measurement noise levels, which will introduce some variation on ARMA output, although, theoretically an interarea mode frequency should be almost the same for different measurement locations within the same power grid. So, in order to make the estimation process at the dispatching center be an unsupervised mode meter, it is necessary to know a priori an estimate of the mode frequency to be tracked. For each interarea mode, a frequency range should be given in advance.

Then, the ARMA outputs from multiple PMU sites should first be categorized into different oscillation groups, according to the given frequency range. This procedure is a kind of clustering in nature, where some a priori knowledge is necessary. For each oscillation mode group, the average center of different frequency-damping ratio sets, i.e., f_j, ς_j, is calculated. The calculated average center denotes the system electromechanical oscillation mode characteristics.

5.4.3 Application Case in CSG

China Southern Power Grid locates in a vast area of the southernmost part of China, comprising five geographical regions: Guang Dong (GD), Guang Xi (GX), Gui Zhou (GZ), Yun Nan (YN), and Hainan (HN), as shown in Fig. 5.9. It is also closely connected to Hong Kong (HK). In the year of 2011, a ± 800 kV HVDC transmission line from the west (YN) to the east (GD) was put into operation as the backbone between YN to GD. The hydraulic power is distributed in the west, and the load centers are concentrated in the east. AC/DC parallel transmission corridors deliver bulk power load usually over 10 GW level. With the fast growth of CSG interconnection, low-frequency oscillation becomes one of the bottlenecks of power transmission capacity.

The CSG interarea oscillation modes and their frequency ranges are presented in Table 5.2. The first two interarea oscillation modes are well damped, but in some special operating conditions their damping can be reduced. The damping ratio of the third interarea oscillation mode in some conditions can be somewhat weak.

Above proposed estimation algorithm based on ambient PMU data for CSG is realized in the dispatching center under the CSG data mining project [7].

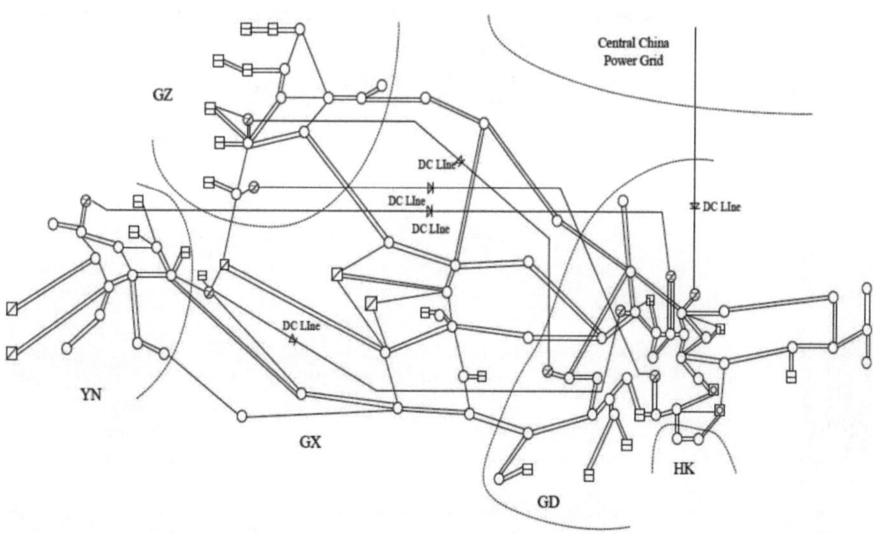

Fig. 5.9 Single-line diagram of studied system

Table 5.2 CSG interarea oscillation modes

Mode	Osc freq range (Hz)
YN/GZ-GD	0.30–0.50
YN-GZ	0.45–0.70
HN-GD	0.70–0.85

Figure 5.10 is the real-time display interface of the data mining and visualization system, where the specific or the total three interarea oscillation mode information is displayed in real time, including the variation trend of oscillation frequency and damping ratio in curve form, and the phase plane plot of damping ratio versus frequency.

In Fig. 5.10, at the time of screen snapshot, the characteristics of first two interarea oscillation modes, that is, YN/GZ-GD mode (0.4 Hz, 16.72%) and YN-GZ mode (0.58 Hz, 8.4%), are quite consistent with the some offline simulation analysis results.

But the third interarea mode, that is, HN-GD mode (0.76 Hz, 0.33%), appears abnormally, where the estimated damping is near zero under normal operation. The reason for this phenomenon is that there exists some forced oscillation component in the tie-line active power measurement, which is taken as the ARMA model input. Under such condition, the assumption for the ARMA method to identify oscillation mode under ambient noise, that is, random input to the linear system model, cannot be satisfied anymore. More detail on explanation is given as follows:

To take the classical second-order differential equations of single-machine infinite bus system, shown in (5.6) as the example,

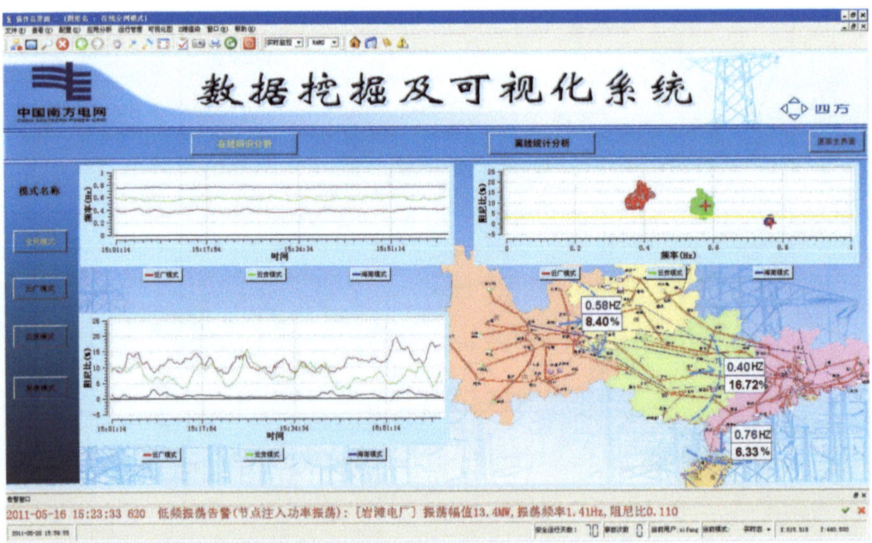

Fig. 5.10 Real-time display interface of data mining system

$$\Delta \ddot{\delta} + 2\varsigma\omega_n\Delta \dot{\delta} + \omega_n^2\Delta\delta = \frac{1}{2*H}(\Delta P_m - \Delta P_e) \tag{5.6}$$

where ς is the damping ratio and ω_n is the natural frequency of the oscillation.

Under the ambient noise assumption, (5.6) is transformed to:

$$\Delta \ddot{\delta} + 2\varsigma\omega_n\Delta \dot{\delta} + \omega_n^2\Delta\delta = n(t) \tag{5.7}$$

where n(t) is white noise.

The output signal of (5.7) is used to identify the oscillation mode (ω_n,ς).

However, if forced oscillation exists, the right side of (5.6) will be turned into:

$$\Delta \ddot{\delta} + 2\xi\omega_n\Delta \dot{\delta} + \omega_n^2\Delta\delta = R_0\cos(\omega t) \tag{5.8}$$

where R_0 is the equivalent external input amplitude.

The solution to (5.8) will be

$$\Delta\delta(t) = \Delta\delta_1(t) + \Delta\delta_2(t) \tag{5.9}$$

where

$$\Delta\delta_1(t) = A_0 e^{-\xi w_n t}\cos(\omega_n\sqrt{1-\varsigma^2}t + \phi_1) \tag{5.10}$$

$$\Delta\delta_2(t) = \frac{R_0}{\sqrt{(\omega_n^2 - \omega^2)^2 + 4\xi^2\omega_n^2\omega^2}}\cos(\omega t - \varphi_2) \tag{5.11}$$

In the expression (5.9), the second item is the dominant component, which represents non-decaying cosinusoidal curve, with oscillation frequency ω and zero damping ratio.

When the $\Delta\delta(t)$ of (5.8) is input to the ARMA model, the calculated damping ratio only reflects a pseudo-external input oscillation mode, instead of power grid intrinsic oscillation mode. The identified result, especially the damping ratio, deviates from the actual power grid intrinsic oscillation mode. This fact reveals the restriction of the ambient data analysis technique, when the power grid suffers some forced oscillation. However, this phenomenon can be used for judging the oscillation type, while the oscillation event occurs, which will be discussed in the Sect. 5.4.4.

Figure 5.11 is the offline statistics display interface of the data mining and visualization system, where the history of specific or the total oscillation mode information is displayed, including the variation trend over long time of oscillation frequency and damping ratio in curve form, and some statistical characteristics are also presented. Then, the range and average of certain mode can be easily observed over certain period, and some operation rule can be summarized.

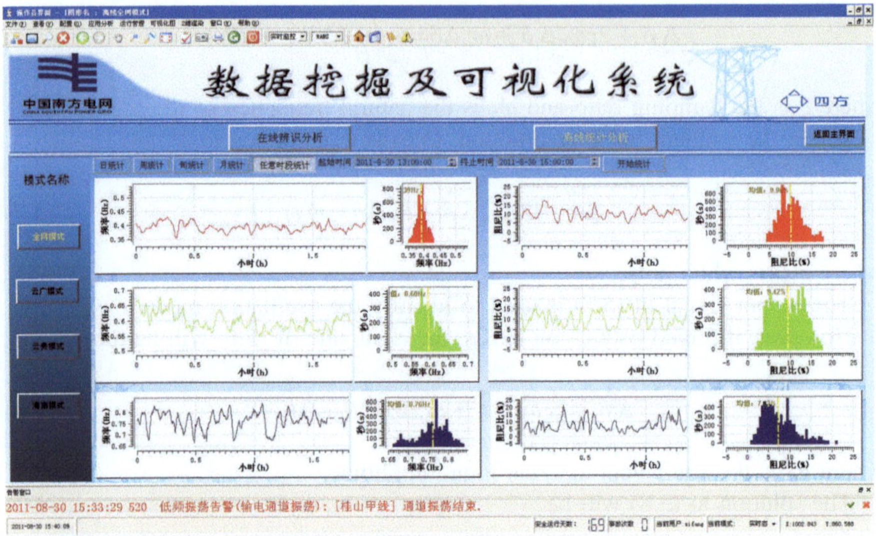

Fig. 5.11 Offline statistics display interface of data mining

5.4.4 Low-Frequency Oscillation Mechanism Analysis

On the April 21, 2008, CSG experienced an oscillation event, which lasted for over 6 min and peak-to-peak amplitude of oscillation on certain 500 kV tie-line was 91 MW as well as 66 MW at certain generator output. The oscillation phenomenon was detected on tie-lines between YN and CSG's backbone grid. Investigated result revealed that the oscillation was caused by generator valve operation and was damped out after the generator tripping.

Figure 5.12 shows the recorded data on one typical tie-line, line Luo-Bai, during oscillation event on April 21, 2008, comprising the ambient data, ringdown data, and some outliers also, as indicated in the figure.

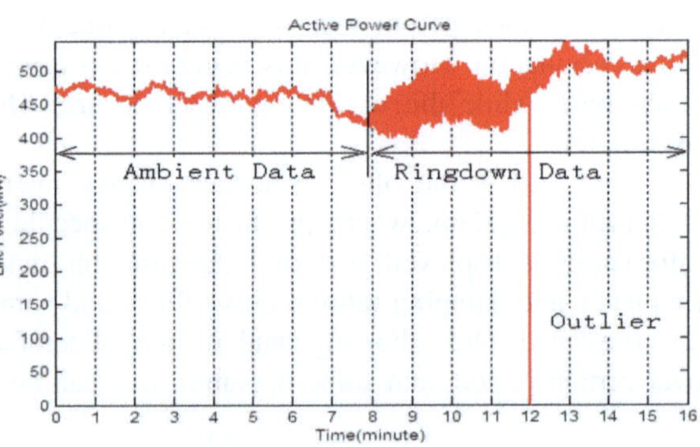

Fig. 5.12 Active power on Luo-Bai tie-line during oscillation event on April 21, 2008

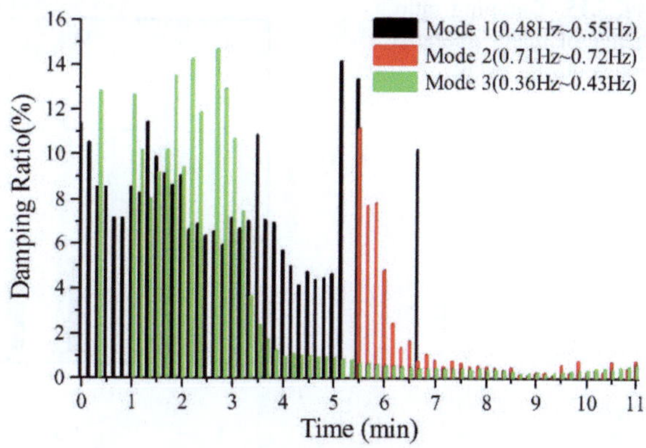

Fig. 5.13 ARMA output of active power on Luo-Bai tie-line

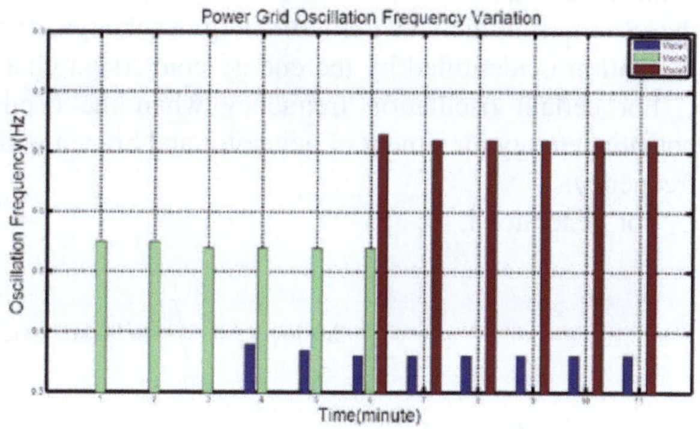

Fig. 5.14 Oscillation frequency change of major interarea modes during event

By utilizing procedures introduced in Sect. 5.4.2, such as, outlier removal, detrending, filtering and down sampling, and ARMA calculation, the ARMA output result is shown in Fig. 5.13.

By adopting clustering method introduced in Sect. 5.4.3, the estimated oscillation frequency and damping ratio results for different oscillation modes are displayed in Figs. 5.14 and 5.15.

In Fig. 5.14, before the ringdown, the only main mode is YN-GZ mode (0.54 Hz), and the YN/GZ-GD mode was not observable due to limited PMU records at that time. But during the ringdown, only the YN/GZ-GD (0.36 Hz) and its second harmonics (0.72 Hz) are observable. In Fig. 5.15, before the ringdown, the YN-GZ mode's damping ratio is about 9%, but during the ringdown, the YN/GZ-GD's damping ratio suddenly drops to the value of near zero.

Comparing the estimated oscillation modes both from ambient data, i.e., from pre-disturbance condition, and from the ringdown data, i.e., during disturbance, the two results are found to be quite different, especially when then the damping ratio of ringdown data suddenly drops to the value close to zero, then it indicates the forced oscillation type.

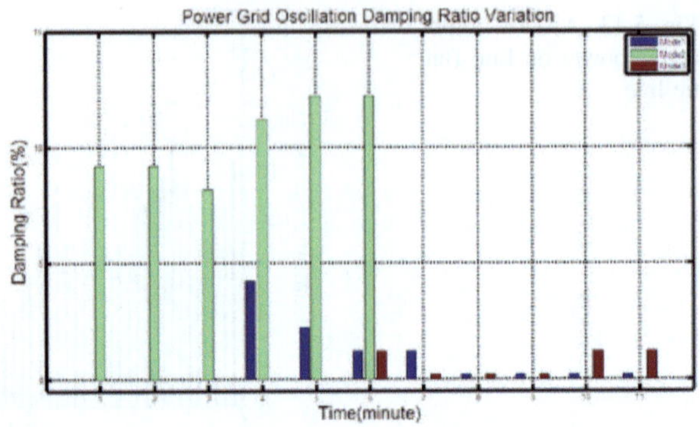

Fig. 5.15 Damping ratio change of major interarea modes during event

Further, network distribution of external input energy method is adopted to confirm the existence of forced oscillation [8]. Based on the energy distribution of the system from the viewpoint of energy exchange, disturbance source of the forced oscillation is identified by the energy conversion characteristics.

For certain oscillation frequency when the oscillation occurs, calculate the potential energy increment of generator and branch with measured active power and frequency.

For generator i,

$$\Delta V_{PEi}(t) = \int_0^t \Delta P_{ei} \Delta \omega_t \omega_0 dt \tag{5.12}$$

where ΔV_{PEi} is the potential energy increment of generator i, ΔP_{ei} is the active power increment of generator i compared with steady-state value, $\Delta \omega_i$ is the frequency deviation of generator i, and ω_0 is the nominal frequency.

For branch i-j connecting node i,

$$\Delta V_{PEi}(t) = \int_0^t \Delta P_{ij} \Delta \omega_i \omega_0 dt \tag{5.13}$$

where ΔV_{PEi} is the potential energy increment of branch i-j connecting node i, ΔP_{ij} is the active power increment of branch i-j compared with steady-state value, $\Delta \omega_i$ is the frequency deviation of node i, and ω_0 is the nominal frequency.

Using the same PMU-recorded event data, here the active power of RH #2 generator, active power sum of tie-lines L-B I and L-B II, active power of MW #1 generator are selected as input for calculation of potential energy increment. The active power curves are displayed in Fig. 5.16.

By using Eqs. (5.12) and (5.13), the potential energy increment of the generators and branches is calculated, as shown in Fig. 5.17.

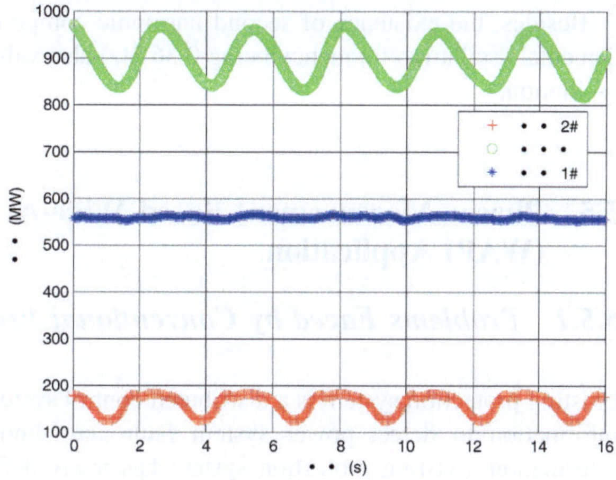

Fig. 5.16 Active power oscillation during April 21, 2008 event

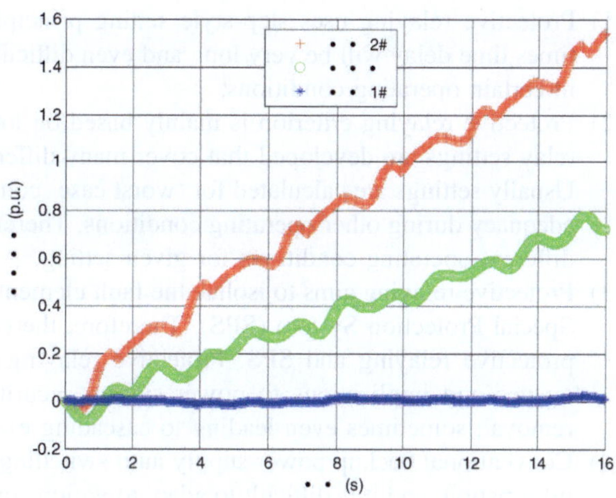

Fig. 5.17 Potential energy increments during April 21, 2008 event

Although the active oscillation amplitude on the tie-lines L-B I and L-B II is larger than that of the HH 2# generator, but the non-periodic component of the potential energy increment for HH 2# generator is significantly larger than the non-periodic component of potential energy increment for tie-lines L-B I and L-B II. At the same time, according to the sign of non-periodic component of the calculated potential energy increment, it indicates that the potential energy increment is injected from the HH 2# generator to the major power grid. Also, from the sign of non-periodic component of the calculated potential energy increment from tie-lines L-B I and L-B II, it shows that the potential energy increment is injected from S/S LP to S/S BS and further propagated to GD power grid. The results demonstrate that the oscillation type is forced oscillation and the oscillation source is located in the HH power plant, which is consistent with the conclusion of the actual accident investigation.

Besides, the existence of second harmonic component (0.72 Hz, with the fundamental oscillation frequency being 0.36 Hz) also validates the existence of forced oscillation.

5.5 Phasor Measurement-Based Wide-Area Protection (WAP) Application

5.5.1 Problems Faced by Conventional Protection

Existing protection system is a distributed control system, which mainly using local information to detect power system fault and abnormal state. Due to limited information, existing protection system has many deficiencies, mainly in the following aspects [9, 10]:

(1) Protective relaying uses step-style setting principle to coordinate, and sometimes time delay will be very long and even difficult on setting and cooperation in certain operating conditions.
(2) Protective relaying criterion is mainly based on local measurement. Protective relay settings are developed that cover many different operating arrangements. Usually settings are calculated for 'worst case' conditions and then checked for adequacy during other operating conditions. Therefore, it is difficult to adapt to different operating conditions for given setting.
(3) Protective relaying aims to isolate the fault elements, and it is independent with Special Protection System (SPS). Therefore, there is no coordination between protective relaying and SPS. Protective relaying cannot consider the consequence and implications to power system security and stability due to fault removal, sometimes even leading to cascading events.
(4) Conventional backup power supply auto switching is based on local-substation information, and it is difficult to adapt to various operating conditions of power systems. It cannot realize remote restoration operation, and some overload effect caused by auto switching operation cannot be considered.

Based on above issues, it is necessary to develop a new protection and control systems that based on wide-area information, which is called wide-area protection (WAP) system.

5.5.2 Architecture of WAP System

Wide-area protection system can be described as: Getting real-time information of regional power grid through communications network, and with wide-area information to determine fault location, and remove the fault components from power system quickly to selectively, and has certain security and stability control ability.

Typical wide-area protection system consists of one (or two for redundancy) master station and several subsidiary control stations, as shown in Fig. 5.18.

(1) Master station mainly issues protection and control functions that require multi-substation information which is collected through communication network by control stations. Master station can analyze power grid operation mode and status, determine fault location, and send corresponding commands to control stations.
(2) Subsidiary control station is responsible for acquisition of electrical and state quantities and issues the protection and control functions that only require local-substation information. Control station sends information that master station required to master station through communication network. At the same time, control station can receive, verify, and execute commands issued by the master station.

As far as protection function is concerned, master station and subcontrol station both act as backup protection.

WAP should take full advantage of current differential protection to improve selectivity and sensitivity of backup protection when synchronization signal is available and healthy and to improve system reliability with wide-area redundancy information.

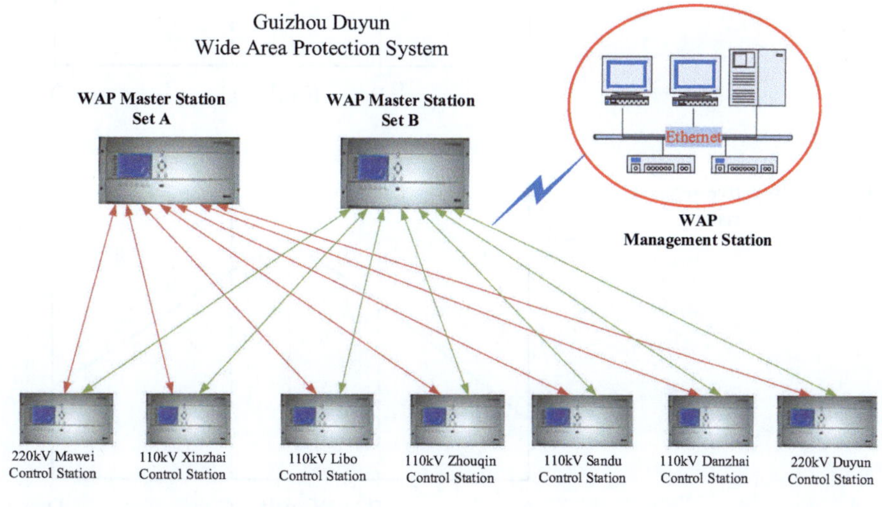

Fig. 5.18 Architecture of WAP system

5.5.3 *Proposed WAP Functions*

5.5.3.1 Enhanced Current Differential Protection

Differential protection of buses, transformers, and generators is a well-established protection principle that has no direct counterpart in protection of long transmission lines.

With synchronization signal, which may get from GPS or Compass satellite, current differential protection can be used. In order to prevent mal-operation of current differential protection due to synchronization signal abnormal, sequence voltage characteristic permit criterion is used. For symmetry fault, positive-sequence voltage is increased from fault point to power sources, and Fig. 5.19 shows positive-sequence voltage characteristic.

For asymmetry fault, negative-sequence voltage is decreased from fault point to power sources, and Fig. 5.20 shows negative-sequence voltage characteristic.

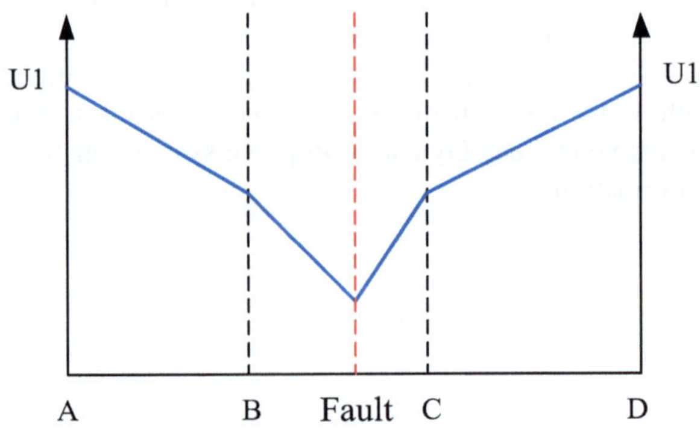

Fig. 5.19 Positive-sequence voltage characteristic

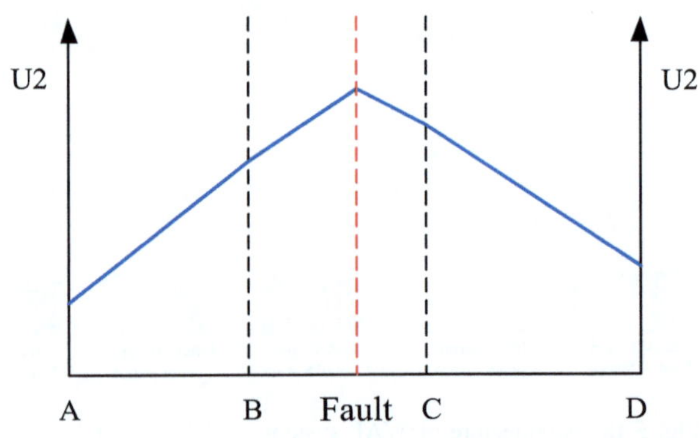

Fig. 5.20 Negative-sequence voltage characteristic

5.5.3.2 Pilot Direction Protection

In the case of lost synchronization signal, pilot direction protection is used as a backup for current differential protection. Direction element can use zero-sequence, negative-sequence, or impedance directions. In order to improve direction element reliability and validity, a comprehensive direction relay is brought out, which is described as:

$$\begin{aligned} Dc &= 2D_1 + D_2 + D_0 \\ Dc &\geq 2 \end{aligned} \tag{5.14}$$

where D_1 represents impedance direction, D_2 represents negative-sequence direction, and D_0 represents zero-sequence direction. When direction is positive, it takes 1 and 0 otherwise. When Dc is greater than or equal to 2, then a fault occurred. D_2 and D_0 reflected the system impedance of behind protective relaying installation, which are related to power system operating status. However, D_1 is independent of power system operating status. Therefore, improve D_1 weight to 2.

Considering the data reliability during current differential protection switch to pilot direction protection, and power backward situations, pilot direction protection requires a certain delay time.

5.5.3.3 Protection for CB Failure

Current differential protection and pilot direction protection cannot provide backup protection for adjacent equipment. In case of CB failure, the fault cannot be cleared; CB failure protection must take effect and trip adjacent CBs. Two criteria can be used:

(1) Overcurrent criterion. For a given fault line, WAP sends trip command to CBs of fault line. If there are still large current in CB after a certain delay time, then WAP regards this CB as failure and sends trip commands to adjacent CBs to the failure CB. In order to prevent unexpected mal-operation, remote trip commands are must be verified by local criterion.
(2) Regional current differential criterion. Figure 5.21 is an example diagram. Line L1 faulted, and QF1 occurred CB failure. WAP uses CT5, CT2, CT3, CT4 current to make a region current differential criterion. When differential current is large, then WAP regards QF1 as failure and sends remote trip to QF5, QF2, QF3, QF4, and QF1 (send trip command to QF1 again). For protection of CB failure, the fault clearance time can be decreased from 1.0–2.0 to 0.2 s.

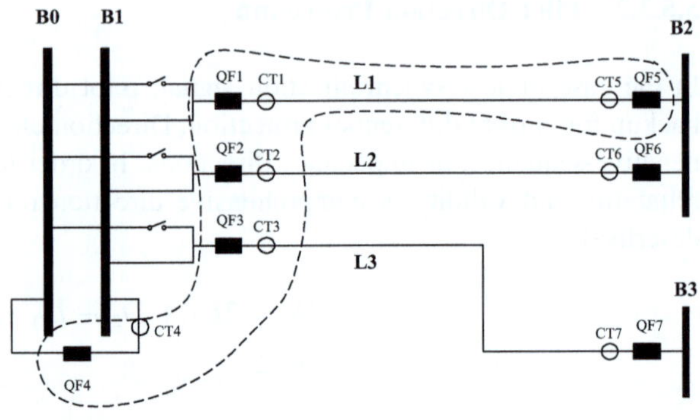

Fig. 5.21 Protection criterion for CB failure with regional current differential

5.5.4 Application Case Study

5.5.4.1 Brief Description of Duyun WAP System

Duyun WAP project was implemented on seven 110 kV substations in Duyun district of Guizhou Province and was put into operation in December 2011, as shown in Fig. 5.22. The communication structure takes the ring form.

Some major features of this system are shown as below:

(1) Using current differential protection to detect fault when synchronization signal is OK and using pilot direction as backup protection of current differential protection in the case of synchronization signal lost;

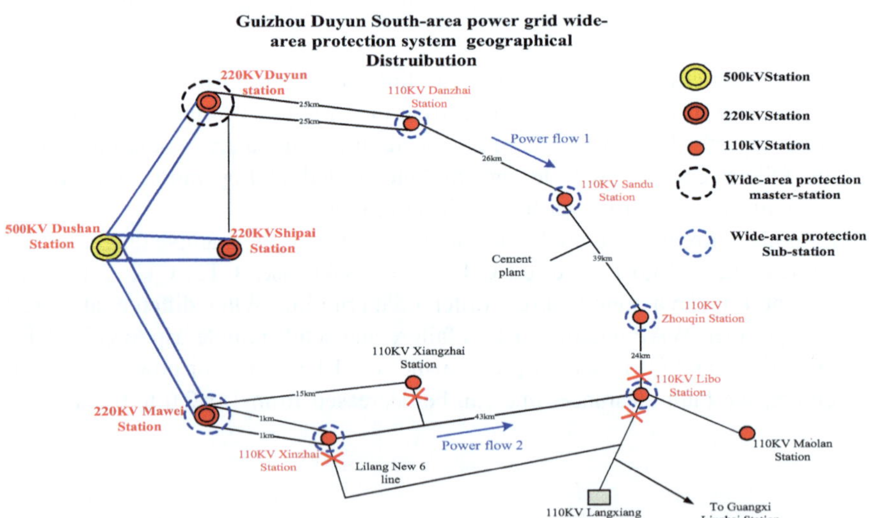

Fig. 5.22 Overview of Duyun power grid

(2) Accurate fault location distance based on synchronized voltage and current phasor at terminal of transmission line;
(3) Remote trip with local criterion verification for CB failures;
(4) Auto restoration for substation without power supply under wide-area protection platform.

5.5.4.2 Recorded Event with Wide-Area Backup Protection in Operation

During WAP system operation, one three-phase-to-ground short-circuit fault occurred on X-L line. Quickly after the occurrence of this fault, the Duyun WAP system identified the fault correctly and sent the trip command in time, as shown in Fig. 5.23.

Recorded fault voltages and fault currents are shown in Fig. 5.24.

Based on post-event analysis, the calculated fault is located at the end of line L-Z and falls within the range of protection zone 2 area. The length of line L-Z is 31.1 km, and the fault distance calculated from recorded data is 27.255 km, i.e., 87.6% of the line length. Under these circumstances, the local distance protection did not trip for this case. Finally, the fault was cleared by wide-area differential protection. After checking the relevant messages and fault recording data, the period of time from fault onset to complete fault isolation was 88 ms, which verified the fast and correct operation of wide-area differential protection.

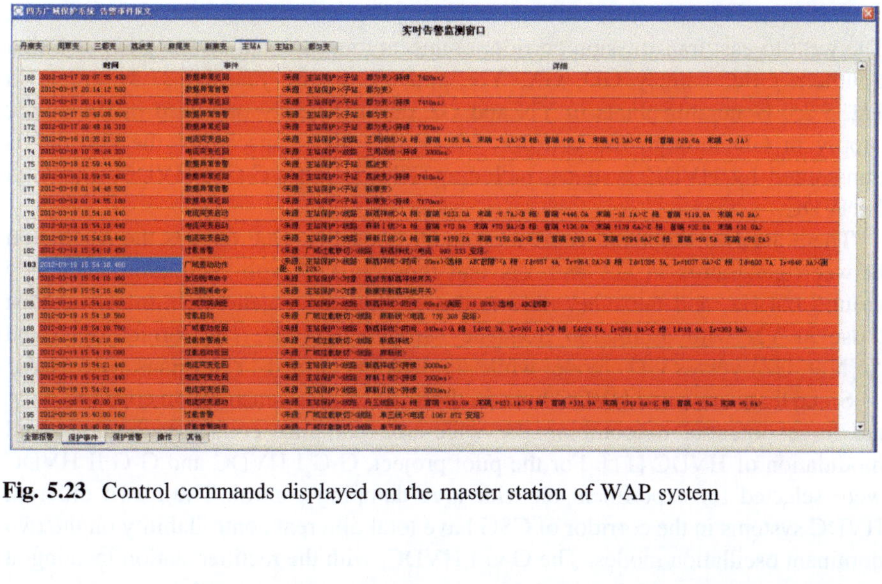

Fig. 5.23 Control commands displayed on the master station of WAP system

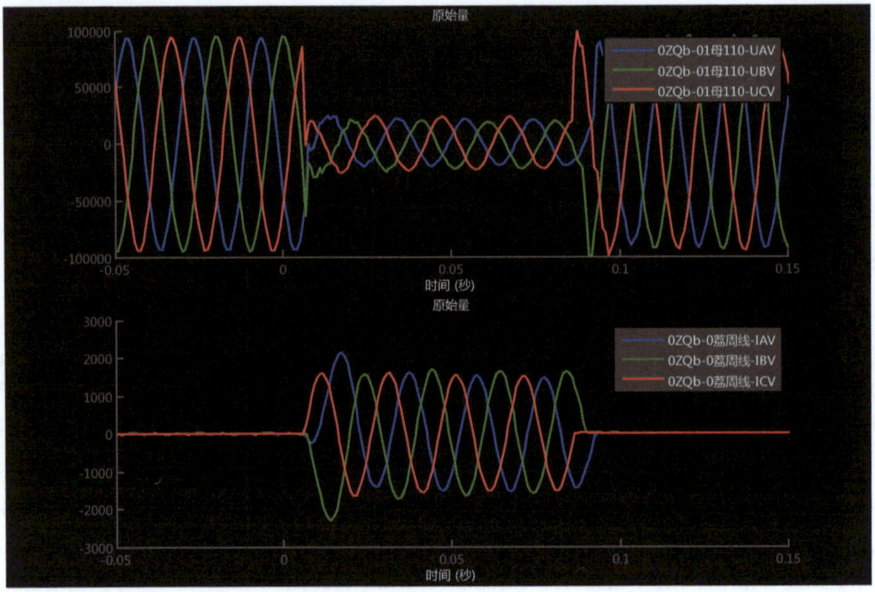

Fig. 5.24 Recorded fault voltages and fault currents

5.6 Wide-Area Damping Control Utilizing HVDC Modulation

5.6.1 Research Background

The west-to-east transmission corridor connects four regional power grids of China Southern Grid, namely GD, GX, YN, and GZ in the year 2008, as shown in Fig. 5.25. Generator plants in YN and GZ provide nearly one-third of the power supply in GD through the corridor. More than 60% capacity of the corridor is transmitted by HVDC systems, including TSQ HVDC, G-G I HVDC, and G-G II HVDC.

There are two dominant oscillation modes in CSG. One reflects the oscillation between generators in GZ and YN and those in GD with oscillation frequency around 0.4 Hz, and the other reflects the oscillation of generators in GZ against those in YN with oscillation frequency around 0.57 Hz. The former is termed '(YN + GZ) versus GD' mode, while the latter is termed 'YN versus GZ' mode.

Since there are some HVDC links in the west-to-east transmission corridor, there are huge potential to combine the wide-area damping control technology and modulation of HVDC [11]. For the pilot project, G-G I HVDC and G-G II HVDC were selected for experiment on wide-area damping control. These two different HVDC systems in the corridor of CSG have total different controllability on the two dominant oscillation modes. The G-G I HVDC, with the rectifier station locating at

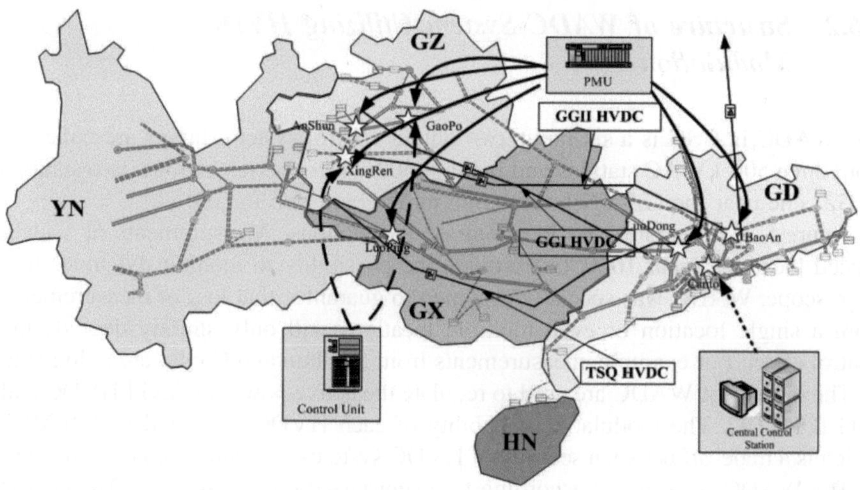

Fig. 5.25 System structure of studied power grid

GZ, shows good performance in damping oscillation mode of YN versus GZ. On the other hand, the G-G II HVDC, with the rectifier station near the middle of intertie of YN and GZ, has more impacts on (YN + GZ) versus GD mode than others.

Hereby, utilizing modulations of G-G I and G-G II HVDC links for different damping purposes, respectively, is reasonable in the design of WADC, i.e., G-G I for YN versus GZ mode and G-G II for (YN + GZ) versus GD mode. The principle for input signal selection is then clear: The input signal for G-G I should offer good observability of the YN versus GZ mode, while input signal for G-G II offers good observability of (YN + GZ) versus GD mode.

Advanced WAMS/PMU technology in CSG allows selecting inputs from numerous candidates. Signals such as intertie power flow, voltage phase, and frequency were compared and discussed thoroughly by simulation studies. The frequency of AC system can be directly related to rotor speed of generators nearby; thus, frequency difference between two regional grids reflects interarea oscillations effectively. Furthermore, the output of damping controller with input of frequency deviation has strong physical implication of the well-known damping torque concept. In the final scheme of WADC, frequency measurements in GZ, YN, and GD are used.

5.6.2 Structure of WADC System Utilizing HVDC Modulation

The WADC in CSG is a six-input–two-output control system. Inputs are collected from three 500 kV AC stations and three ± 500 kV converter stations: two stations in GZ, one near the GZ-YN border, one in YN, and two in GD.

Figure 5.25 illustrates the locations of the stations. Measurements at widely spaced locations (near 1000 km) provide the possibility to monitor dynamics in a large scope. WADC was specially designed to guarantee that loss of measurements from a single location or even multiple locations will only slightly degrade the control effect. For example, measurements from AnShun and GaoPo are redundant.

The outputs of WADC are used to regulate the active power of G-G I HVDC and G-G II HVDC. The modulation capability of each HVDC is limited to 300 MW, which is a trade-off between security of HVDC systems and damping performances.

The WADC is designed according to a centralized control scheme. The central control station is located at control center of CSG in Guangzhou, as shown in Fig. 5.25. Central station plays the core role of the whole system. Its functions include receiving, validating, and aligning data stream from PMUs, detecting unusual events inside power grids, monitoring the status of communication network, calculating real-time commands based on inputs, sending commands to control units, and checking status of control units. The entire architecture of the WADC system is shown in Fig. 5.26.

The control block design is shown in Fig. 5.27, consisting of a washout, two lead–lag blocks, a proportional block, and a limiter [12]. The controller is designed using pole placement technique offline initially. Parameters of the lead–lag block and the proportional block are tuned according to an algorithm based on fast online Prony analysis.

An online Prony algorithm was developed to monitor changes of oscillation mode in real time. If the current oscillation frequency is too much different from the center frequency of filter, the parameters of filter and lead–lag blocks will be adjusted according to the adaptive algorithm as soon as possible. The time delay in

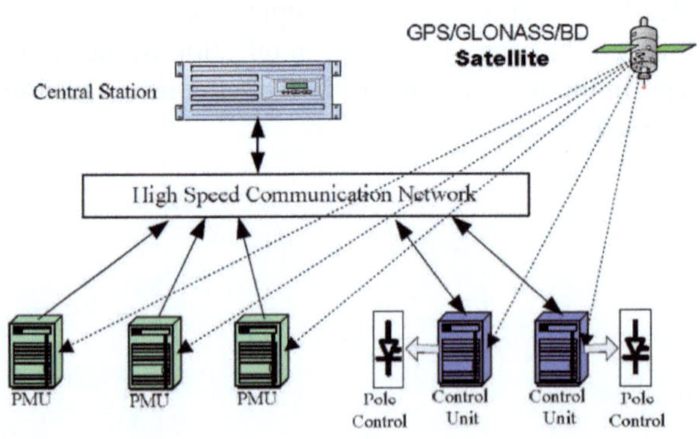

Fig. 5.26 Architecture of WADC system with HVDC modulation

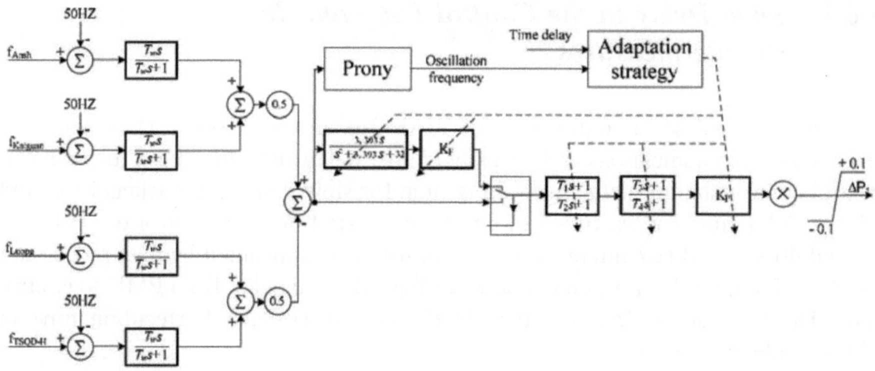

Fig. 5.27 Control block design of WADC

control loop is also monitored online, and phase compensation will follow the change of time delay.

Since time delay for data transmission is inevitable in wide-area control system, some lead–lag blocks and a band-pass filter in central control station are adopted to mitigate the negative effect of time delay, which will be introduced in Sect. 5.6.3.

Besides, it is necessary to add dead band to continuous controllers, especially for wide-area control schemes, to prevent inadvertent actions during steady state, which may be caused by measurement noise and other small disturbances. A dead band with the range of ±9 MW has been implemented in the WADC control loop.

The interface of WADC to HVDC was developed with collaboration from HVDC vendor Siemens. The interface receives binary and analog signals from control units and interprets them to Pole Control of HVDC. The interface can block the input from control unit as soon as external signals are implausible.

The HVDC links operate in constant power control mode under normal condition: the rectifier in constant current control and the inverter in constant voltage control. Signal ΔP is generated from the wide-area controller described above and added to the reference signal *Pord* to regulate the active power of HVDC link during oscillations. Figure 5.28 shows the configuration of the pole control system.

Fig. 5.28 Configuration of the pole control system

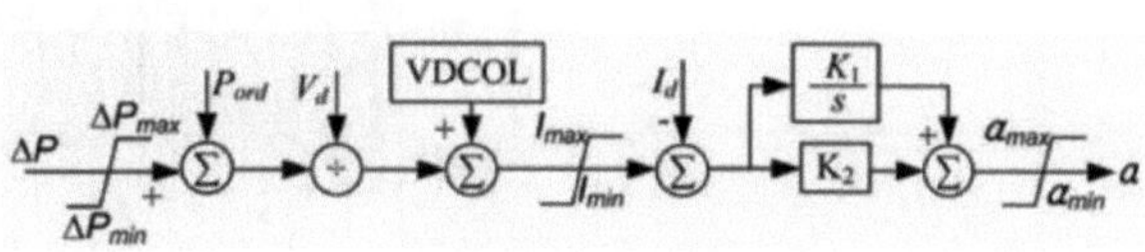

5.6.3 Time Delay in the Control Loop and Its Countermeasures

The total delay time in control loop of WADC is around 110 ms. Delay time for fiber-optic communications is less than 15 ms; delay time for data processing in central control station including waiting time for slowly arriving packets is around 15 ms; delay time for execution of control command in Pole Control of HVDC is around 40 ms, and remaining delay time of 40 ms is consumed by data processing in PMU. Figure 5.29 shows the field recordings of time delay from PMU to control unit. The time delay does not include the control command execution time in HVDC pole control.

For the oscillation control with the target modes in the range of 0.3–0.6 Hz, time delay of 110 ms will introduce a phase lag of 12–24°, which can be compensated by lead–lag blocks in central control station.

With the oscillation frequency got from Prony analysis and the time delay measured in section one, phase shift for time delay compensation can then be calculated from

$$\theta = 2\pi f \cdot \Delta t \tag{5.15}$$

where θ is the phase shift due to the time delay in radian, f is the dominant oscillation frequency in Hz, and Δt is the total time delay in seconds.

To a lead–lag block $(1 + sT_{lead})/(1 + sT_{lag})$, define $\alpha = T_{lag}/T_{lead}$, then α can be got from:

$$\alpha = \frac{1 - \sin\theta}{1 + \sin\theta} \tag{5.16}$$

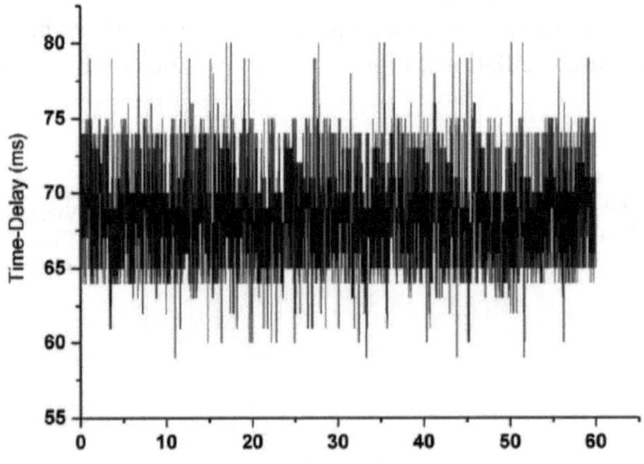

Fig. 5.29 Delay time in control loop of WADC

Then, the parameters of the lead or lag block can be obtained from:

$$T_{lead} = \frac{1}{2\pi f \cdot \sqrt{\alpha}} \tag{5.17}$$

$$T_{lag} = \alpha \cdot T_{lead} \tag{5.18}$$

As each lead–lag block compensation is limited to a maximum of 60° for practical reasons, two lead–lag blocks are used here. Each block compensates half of shift phase.

5.6.4 Operational Experience of WADC

The WADC system was committed into operation in 2008. On June 14, 2009, some field tests were performed to validate the effect of damping control under different operation modes.

Figure 5.30 presents some of the typical results during the experiment. For the measured angle difference curves, Prony analysis is further performed, and the result from Prony analysis is listed in Table 5.3.

In Fig. 5.30a, the relative power angle between YN and GZ reflects the YN versus GZ oscillation mode. From Table 5.3, it is evident that since G-G I HVDC WADC is designed to damp this swing mode, the damping ratio can be improved greatly from 7.6 to 20.5%.

Also from Table 5.3, it can be seen that as G-G II HVDC WADC is designed to damp the oscillation between (YN + GZ) and GD, it has no influence on the YN versus GZ mode. However, Fig. 5.30b shows that G-G II HVDC WADC can increase the damping ratio of (YN + GZ) versus GD mode from 10.9 to 30.9%. The solid lines in both Figs. 5.30a and 5.30b demonstrate that the coordinated WADCs can provide effective damping to both interarea oscillation modes.

On August 29, 2016, with the commissioning of the Luxi Back-to-Back MMC HVDC, YN power grid and China Southern main power grid became two asynchronous grids interconnected with HVDCs, and the WADC system with HVDC modulation ends its role for damping either YN versus GZ mode or (YN + GZ) versus GD mode.

However, a new form of very low-frequency oscillation appears in YN power grid, since the generation in YN grid is composed of many large hydro power plants. Due to improper setting of governor control parameters and the characteristics of hydro turbines, the very low-frequency oscillation appears for some times, with the oscillation frequency around 0.05 Hz. In 2017, some research has been done in utilizing HVDC modulation function to damp this kind of oscillation after accurately detecting the onset of very low-frequency oscillation. Currently, RTDS tests are completed and the effectiveness of the HVDC modulation control with oscillation detection integrated is verified.

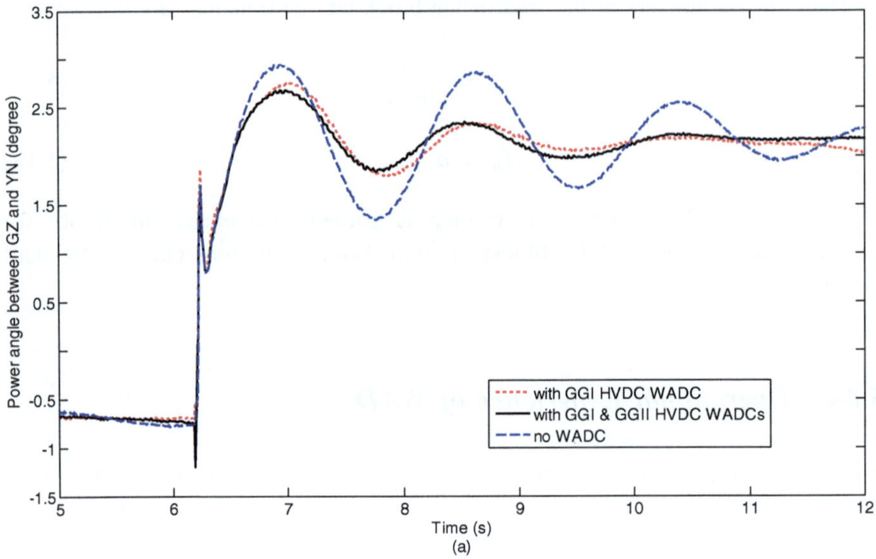

(a) Power angle between YN and GZ

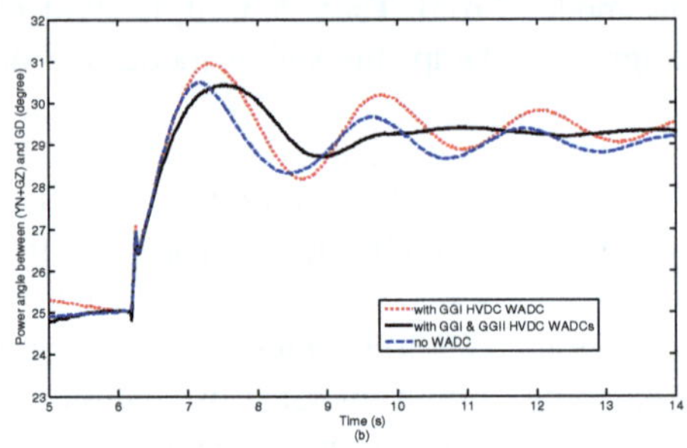

(b) Power angle between (YN+GZ) and GD

Fig. 5.30 Field test results of the WADC system with HVDC modulation

Table 5.3 Prony analysis results of the angle difference during the tests

	YN versus GZ mode		(YN + GZ) versus GD mode	
	Osci. freq. (Hz)	Damp. ratio (%)	Osci. freq. (Hz)	Damp. ratio (%)
No WADC	0.583	7.551	0.455	10.078
WADC with G-G I	0.588	20.459	0.423	10.893
WADC with G-G I and G-G II	0.584	19.137	0.309	30.854

5.7 Monitoring and Assessment on Integrated Wind Farm

5.7.1 *Intelligent Alarm for the Cascading Tripping of Wind Turbines*

Wind turbines in China suffer from cascading tripping induced by the internal or external faults of a wind farm. When groups of wind farms are electrically located close to one another, cascading tripping may occur. This event puts considerable pressure on utility companies. However, utility companies have no access to monitoring data on wind turbines because of commercial secrecy. Using PMUs installed at step-up substations, an intelligent alarm system can detect the occurrence of cascading tripping events, trace the entire process, and subsequently assess the impact. Basic approaches to conduct online alarm processing include:

- Detecting the inception of disturbance based on sudden large changes of voltages, and subsequent power losses and voltage swells are needed to justify the detection;
- Recording power losses of different wind farms in real time; number of tripped wind turbines in each wind farm is estimated according to the operating condition before disturbance;
- Sudden restoration of voltages can be regarded as a sign of adjustment of reactive power compensation devices and ending of the disturbances.

Moreover, changes of voltage differences are often used as an effective way to detect faults:

$$\Delta^2 U = \left| |u_i - u_{i-1}| - |u_{i-1} - u_{i-2}| \right| \tag{5.19}$$

where $\Delta^2 U$ is the second order of voltage change, and u_i is the voltage measured at time instant i.

One example for using the above method is given in Fig. 5.31.

5.7.2 *Event Analysis for Subsynchronous Interaction Between Wind Farms and AC Networks*

The one-line diagram of the target system is illustrated in Fig. 5.32, which covers most of the main grid in Xinjiang Uygur Autonomous Region. As shown in the diagram, there are many wind farms located within this system. The generated wind power is collected through 35/110 kV lines to substations A, B, and C. With little local load near the wind farms, their power is then transmitted to substation D via long-distance transmission lines. Through 220 kV double circuit lines, the wind power is delivered to substation F, where the voltage is further stepped up to 750 kV. Thus, the wind power is next injected into the 750 kV main grid, in which

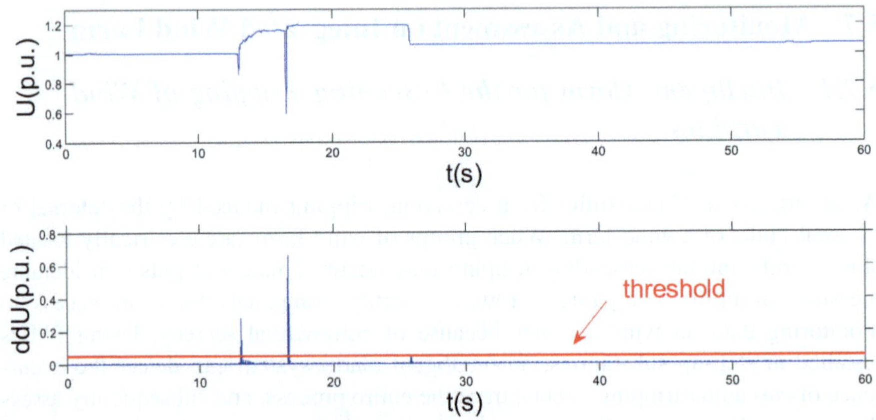

Fig. 5.31 Detection of voltage dips during faults

several hub substations, namely E, F, G, H, and L, are also identified in the diagram. Two thermal power plants, namely plants M and N, are connected to substation H via 500 kV transmissions. Under normal conditions, the power flow is usually delivered from substations E and G to substation F, and then to substation H. Both the power and the thermal power from plants M and N is supplied to substation L. Besides, part of the power is transmitted to Central China Power Grid through a ±800 kV ultra high-voltage DC (UHVDC) transmission line.

As shown in Fig. 5.32, most of wind farms are located near substations A and B. In these wind farms, many identical 1.5 MW direct-drive PMSGs have been deployed. These large-scale wind farms are far from the main power grid. According to the data provided by the grid company, the short-circuit capacities at substation B and A are about 2000 MVA and 1450 MVA. However, the total installed wind power capacity at substation A and B is about 1100 MVA at the end

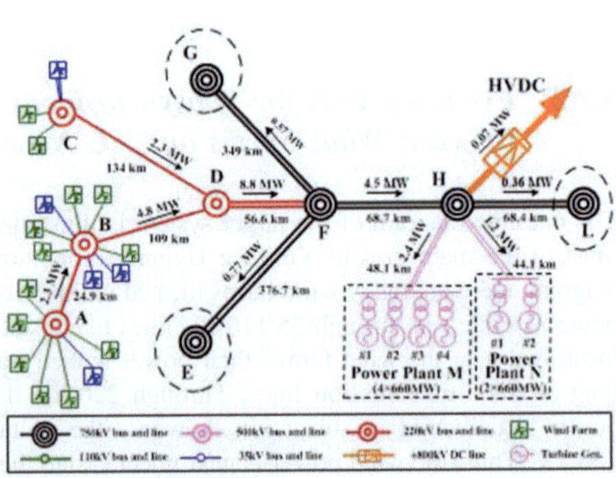

Fig. 5.32 One-line diagram of the practical system suffering unstable SSI

Table 5.4 Torsional frequencies of generators

Generator name	Torsional mode 1 (Hz)	Torsional mode 2 (Hz)	Torsional mode 3 (Hz)
#1/#2/#3/#4 (power plant M)	15.4	25.3	30.8
#1/#2 (power plant N)	18.3	31.5	39.5

of 2014 and is still growing. Thus, the SCR is about 1.3 or even less. In other words, these PMSG-based wind farms are radially connected to a relatively weak AC network.

Power plant M is equipped with four identical units, and plant N has two identical units. Due to different shaft parameters for the units in plants M and N, torsional frequencies are different and listed in Table 5.4.

With the installed capacity of these wind farms increasing, many sustained power oscillation events have occurred in these farms since June 2014. The oscillation frequency of active power varies in a very broad range from 20 to 40 Hz, which are within the subsynchronous frequency range. Initially, power oscillation only appeared in the power grid near wind farms. However, as the number of running wind turbines increases, the oscillation amplitude grows under some undesirable conditions. In the worst case, active power oscillation would even spread to the 750 kV main grid and excite severe torsional vibration in turbogenerators that are hundreds of kilometers away from the wind farms.

On July 1, 2015, a severe SSI event occurred in the aforementioned system. Figure 5.33 shows the recorded active power of a wind farm at two time intervals during the SSI event.

The sustained power oscillation induced by these PMSG-based wind farms excited intense torsional vibration on shafts of units #1, #2, and #3 in power plant M (unit #4 was under maintenance at that time). During this event, their protection devices that were installed originally to counteract the possible subsynchronous torsional interaction (SSTI) between units and HVDC tripped the three units successively, causing a total power loss of about 1280 MW. It led to a dramatic drop of HVDC power from 4500 to 3000 MW. Moreover, this event made the frequency of this regional grid decrease from 50.05 to 49.91 Hz.

By analyzing the recorded PMU data, the flow of the subsynchronous power in the system was reconstructed and their magnitudes were illustrated also in Fig. 5.32. Obviously, such subsynchronous power started from the PMSG-based wind farms and flowed along the path from substations A/B/C, sequentially to substation D, F, H, and finally to the power plants of M/N. Note that the exchange of subsynchronous power between HVDC and substation H is almost zero. This situation indicates that HVDC did not participate into such SSI event. Therefore, it is safe to conclude that the sustained subsynchronous power oscillation was caused by the SSI between multiple PMSGs and the weak AC network. The SSI further spread to the main grid and stimulated intense torsional vibration on shafts of

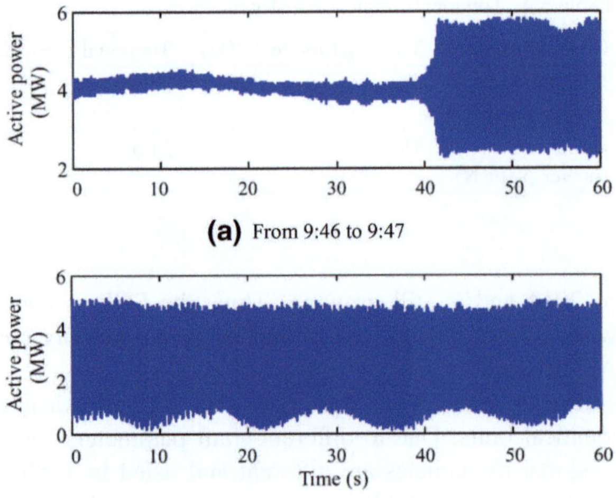

(a) From 9:46 to 9:47

(b) From 11:52 to 11:53

Fig. 5.33 Field measured active power output of a wind farm

turbogenerators that unfortunately have a torsional mode coinciding with the SSI frequency. Once the magnitude of the excited vibration exceeds the threshold of the protection relays, those affected generators would be tripped immediately.

By performing spectrum analysis on recorded active power, as Fig. 5.33, it can be observed that the oscillatory active power (with a frequency of 26.95 Hz) started to diverge at 09:46:40 and soon entered a state of sustained oscillation. From 11:52, the oscillation frequency changed to 30.66 Hz. This implies that the characteristics of SSI in a practical system vary as system conditions change. The SSI event lasted for about 3 h and 20 min (9:46–13:05), and the variation of SSI frequency is displayed in Fig. 5.34.

From Fig. 5.34, it can been seen there were four occasions that the frequency of SSI crossed the torsional modes. The number one to three crossings were experienced in such a short period that the torsional vibration on the corresponding generator shafts had not be sufficiently excited. In other words, the generators were safe during those three occasions. However, the fourth crossing lasted for about

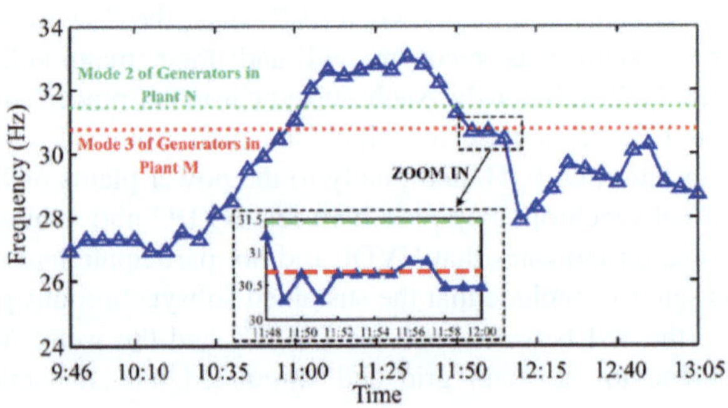

Fig. 5.34 Variation of the SSI frequency and associated torsional frequencies

6 min (as zoomed-in curves in Fig. 5.34). As a result, intense shaft torsional vibration at mode 3 was excited, which led to triggering of generators in plant M,

5.7.3 *Online Wide-Area SSR Monitoring for Wind Farms Integrated with Weak AC Networks*

5.7.3.1 Synchronous Wide Frequency Range Measurement Unit

Based on the mature synchrophasor measurement technology for fundamental frequency defined phasor, the SSR detection function is expanded on the present PMU device. The computation model covering wide frequency range measurement is illustrated in Fig. 5.35, where both fundamental component and subsynchronous components are calculated.

Besides normal synchrophasor calculation function, the wide frequency range measurement unit calculates the instantaneous power from sampled data after ADC. Further frequency spectrum analysis is performed on the instantaneous power, and the existence of the SSR component can be detected. When the SSR component is detected, the wide frequency range measurement unit will trigger the data recording to record the sampled data during the SSR occurrence. Meanwhile, the SSR occurrence tag and the magnitude of calculated SSR component are sent to WAMS in the control center via synchrophasor data frame in real time. The format of data frame is the same as defined by IEEE C37.118, which includes:

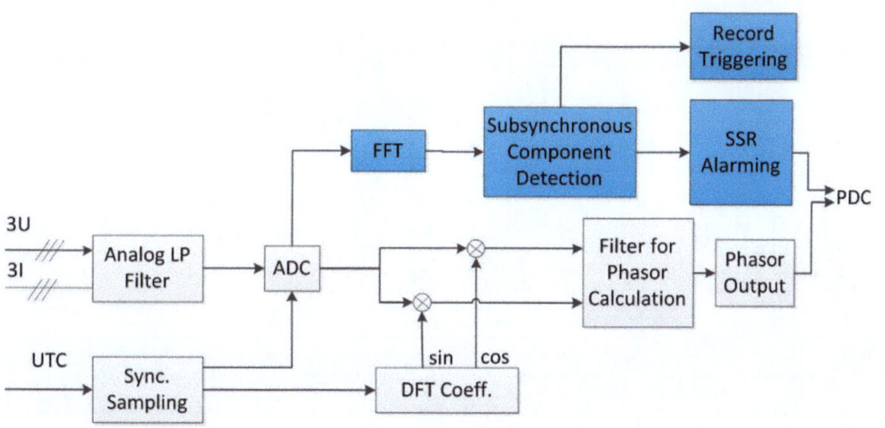

Fig. 5.35 Computation model for wide frequency range measurement

(1) The status word (STAT), indicating whether there is a SSR component;
(2) The SSR occurrence sign (SSO), binary value, indicating whether this measurement unit has detected the SSR;
(3) Magnitude of SSR (SSP), analog value, the magnitude of detected SSR component;
(4) Frequency of SSR (SSF), analog value, the frequency of detected SSR component.

5.7.3.2 Online SSR Monitoring and Alarming

To integrate online wide-area SSR monitoring function with the current WAMS, which is already deployed in the control center, the extended architecture of WAMS is shown in Fig. 5.36, with added blocks in blue color.

The online wide-area SSR monitoring and alarming system is integrated with current D5000 platform. The major added modules include the management module for SSR monitoring, SSR analysis module, and visual display interface. The management module for SSR monitoring is further divided into SSR detection submodule, field recording trigger submodule, and field recording sample data management submodule. SSR analysis module includes field recording sample data file accessing submodule and the SSR analysis tool. Visual display interface includes harmonic analysis geographical display submodule, SSR event reporting and exporting submodule.

The online wide-area SSR monitoring system has been constructed and applied in some provincial power grids in Northwest China since 2016.

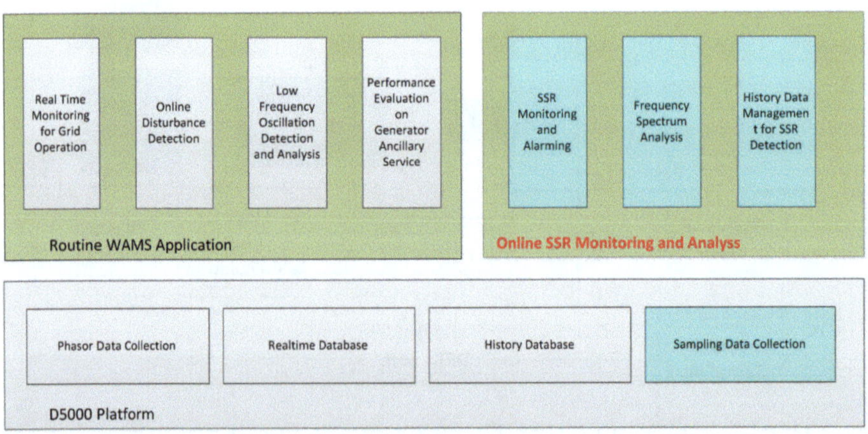

Fig. 5.36 Extending architecture for online SSR monitoring

References

1. Phadke AG, Thorp JS (2008) Synchronized phasor measurements and their applications. Springer, New York
2. Chao L, Bonian S, Wu X, Sun H (2015) Advancing China's smart grid: phasor measurement units in a wide-area management system. IEEE Power Energy Mag 13(5):60–71
3. Huang Z, Guttromson R, Hauer J (2004) Large-scale hybrid dynamic simulation employing field measurements. IEEE Power Eng Soc Gen Meet 2004:1–7
4. Wall P, Gonzalez-Longatt F, Terzija V (2012) Estimation of generator inertia available during a disturbance. IEEE Power Energy Soc Gen Meet 2012:1–8
5. Wies R, Pierre J, Trudnowski D (2003) Use of ARMA block processing for estimating stationary low-frequency electromechanical modes of power systems. IEEE Trans Power Syst 8(1):167–173
6. Zhou N, Trudnowski D, Pierre JW, Sarawgi S (2008) An algorithm for removing trends from power-system oscillation data. IEEE Power Energy Soc Gen Meet 2008:1–7
7. Shi B, Jingtao W, Yang D, Liu Y (2013) Estimation of electromechanical modes from ambient PMU data in China southern power grid. J Int Counc Electr Eng 3(3):228–233
8. Chen L, Min Y, Chen Y (2014) Evaluation of generator damping using oscillation energy dissipation and the connection with modal analysis. IEEE Trans Power Syst 29(3):1393–1402
9. Benmouyal G, Schweitzer EO, Guzmán A (2002) Synchronized phasor measurement in protective relays for protection, control, and analysis of electrical power systems. In: Western Protective Relay Conference, Oct 22–24. Spokane, Washington
10. IEEE power system relaying committee working group C6. Wide area protection and emergency control. IEEE Report, 2008
11. Smed T, Andersson G (1993) Utilising HVDC to damp power oscillations. IEEE Trans Power Deliv 8(2):620–627
12. Chao L, Xiaochen W, Jingtao W, Li P, Han Y, Li L (2012) Implementations and experiences of wide-area HVDC damping control in China southern power grid. IEEE PES Gen Meet 2012:1–8

Chapter 6
Identification of Signature Oscillatory Modes in ERCOT by Mining of Synchrophasor Data

Prashant C. Palayam, Sidharth Rajagopalan, Bill Blevins and Sarma Nuthalapati

6.1 Introduction

The presence and impact of modes of oscillation in the power system are well-established [1]. Much work has been done on the estimation of these modes using both theoretical—linearizing the system equations about an operating point [1]—and measurement-based [2, 3] approaches. With the advent of synchrophasor technology, which allows visualization of such oscillations in real-time environments [4, 5], the application of measurement-based techniques for estimation and detection of oscillatory modes as they happen is possible [6].

Increasing amounts of wind generation coming on-line across the nation brings different forms of control systems and strategies to the power grid. This poses many operating challenges for system operators [7]. One of the challenges is the presence of high-/low-frequency oscillations from the wind farm control systems. Fast responding wind controllers with bad settings and/or design introduce control system oscillations which are either intermittent with high energy or consistent with low energy. Such oscillations can plausibly drive nearby wind farms to oscillate at the same frequency and/or interact with other conventional generation units on the grid causing them to oscillate in turn. This can cause significant forced damage to mechanical parts such as turbine blades and generator shafts. These forced oscillations have also been known to cause voltage fluctuations at the distribution level

P. C. Palayam
Electric Power Group, Pasadena, USA

S. Rajagopalan · B. Blevins (✉)
Electric Reliability Council of Texas (ERCOT), Taylor, TX 76574, USA
e-mail: bill.blevins@ercot.com

S. Nuthalapati
Peak Reliability, Vancouver, USA

S. Nuthalapati (ed.), *Power System Grid Operation Using Synchrophasor Technology*, Power Electronics and Power Systems,
https://doi.org/10.1007/978-3-319-89378-5_6

leading to tripping of motors and pumps [5]. One example of such an event was recorded on the ERCOT system in January 2014 [4].

With growing evidence of such oscillations in power systems [4–9], it became essential to develop a process for identifying the signature oscillations that characterized the system. A data mining effort (using PMU data) was undertaken as a part of the Discovery Across Texas (DAT) Project funded by the US Department of Energy (DOE) and awarded to the Center for Commercialization of Electric Technologies (CCET) [10].

In this chapter, a generalized approach to identify signature oscillations is presented. Two metrics to classify the detected oscillations are proposed and illustrated using archived data from the ERCOT Synchrophasor Network [4, 5, 11]. The details of the project and a brief description of the data mining tool are presented in Sect. 6.2. The proposed metrics used to classify the oscillatory modes are explained in Sect. 6.3, and the generalized approach to identify signature oscillations using these metrics is presented in Sect. 6.4. Section 6.5 illustrates the proposed approach using archived data in the ERCOT system, and the concluding remarks are presented in Sect. 6.6.

6.2 Data Mining for Oscillations on the ERCOT System

There have been several documented cases of oscillation events on the ERCOT system being analyzed using PMU data. This section presents some discussion on two selected cases of such oscillations, before proceeding to a brief description of the data mining effort.

6.2.1 Sustained, Poorly Damped Oscillations

In April 2013, the ERCOT system saw an oscillation event caused by incorrect settings being applied to a wind farm control system during a system update.

The plot in Fig. 6.1 shows voltage magnitude recorded by a PMU located near the wind farm. These were sustained oscillations that lasted several hours and had a mode frequency of 2 Hz. The oscillations ceased when the wind farm source of the oscillations was tripped offline.

6.2.2 Rapid, Un- or Negatively Damped Oscillations

In October 2013, a wind farm on the ERCOT system, operating to a constrained output (40 MW max.) due to weak grid conditions, tripped. The wind farm had implemented a primary frequency response (PFR) module which responded to a

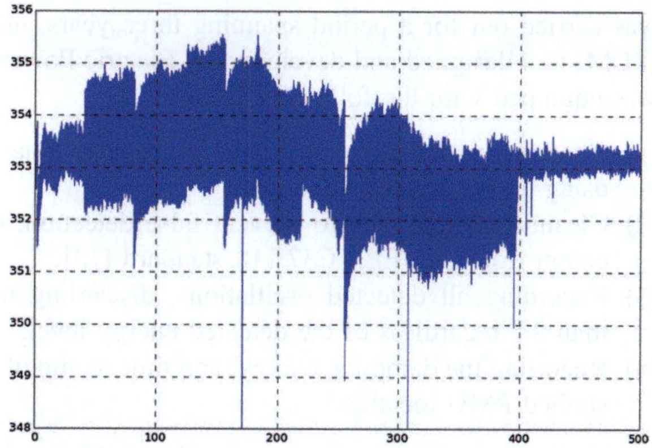

Fig. 6.1 Plots of sustained oscillations in voltage magnitude recorded on the ERCOT system

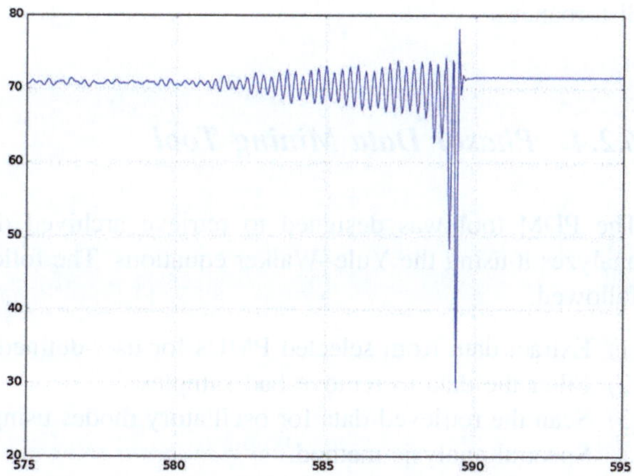

Fig. 6.2 Plots of negatively damped oscillations in voltage magnitude recorded on the ERCOT system

drop in frequency following a loss-of-generation event. The output of the unit was ramped up well beyond 40 MW leading to negatively damped oscillations of $\sim$ 3.2 Hz, which caused the wind farm to be tripped by its own protection systems.

Figure 6.2 shows a plot of the negatively damped oscillations in voltage magnitude recorded by a PMU located near the wind farm. Unlike the case is shown in Fig. 6.1, the duration for these oscillations was extremely short, lasting less than 10 s in total.

6.2.3 Data Mining for Oscillations

The Data Mining Project was undertaken to mine synchrophasor data for oscillation incidents (such as the ones described in Sects. 6.2.1 and 6.2.2 above) and classify detected modes into historical patterns of occurrence. The detection of oscillations was carried out for a period spanning three years, using the Phasor Data Mining (PDM) tool designed and developed by Electric Power Group (EPG) [12]. The tool was equipped with the following features:

(1) Continuous, automatic scanning of archived data from chosen PMU locations using a user-defined window.
(2) Cleaning of bad data to prevent false detection, using the Stat Word specification from the IEEE C37.118 standard [13].
(3) Recording all detected oscillations, discarding modes with damping greater than 8% regardless of the detected energy level.
(4) Reporting the damping, energy, and time-stamp of selected oscillations for each studied PMU location.

A value of 8% was chosen as a threshold for the damping as that was the minimum observed on the system for damped oscillations in response to disturbances.

6.2.4 Phasor Data Mining Tool

The PDM tool was designed to retrieve archived data for selected PMUs and analyzes it using the Yule–Walker equations. The following general algorithm was followed:

(1) Extract data from selected PMUs for user-defined time window.
(2) Filter the data to remove bad samples.
(3) Scan the retrieved data for oscillatory modes using the Modified Yule–Walker Spectral analysis method.
(4) Create final selection of oscillatory modes for each PMU.
    a. Discard modes with damping >8%
(5) Report time-stamp, energy, and damping for all selected modes for every PMU.
(6) Repeat, starting with step '1' until all the data has been analyzed.

The PDM tool performed several functions such as data loading and filtering, mode scanning and exporting results into a database. The approximate time taken to completely process one minute of data from 6 PMUs was 4 s. Table 6.1 shows the performance of the tool for 1 min batches of data extrapolated for 1 h, 1 day, 1 week, and 1 month.

The tool was configured to carry out multiple threads of its process in parallel to expedite the mining activity.

Table 6.1 Phasor data mining tool performance

Task	Time
Data loading (1 min of data, All PMU signals)	3 s
Data filtering	1 s
Mode solving	
Result exporting	
Total (All frequency bands)	4 s
Ideal example	
1 h	4 min
1 Day	1.6 h
1 week	11.2 h
1 month (31 days)	2.06 days

6.2.5 *Post-processing*

The PDM algorithm identified many oscillations. It also frequently reported more than one set of results for a PMU measurement signal within a 1 min window. Hence, it was important to differentiate between 'significant' (high energy) modes, weaker (low energy) modes, and modes produced due to numerical inaccuracies.

The following rules combined using 'and' logic were applied. A mode was discarded as weak if:

(1) its energy was less than 20% of that of the mode with maximum energy, and
(2) its occurrence (time in minutes) was less than 20% of that of the mode with maximum occurrence.

6.2.6 *Mode Identification*

Two new metrics were developed as discussed in Sect. 6.3. These metrics were used to identify signature modes in ERCOT system.

Six different PMUs at four locations on the ERCOT grid were used in the study. These locations were chosen to be close to wind generation, but these PMUs did not directly monitor the output of the local wind farms. The archived PMU data sampling rate was 30 samples per second, and the quantities recorded by the PMUs used in the study were frequency, voltage (magnitude and phase angle), and current magnitude. Bad data in the PMU measurements was removed using the PMU status word, before performing the scanning procedure outlined in Sect. 6.2.4. A PMU located close to a load center was used as reference for angle difference calculation.

The study results were used to determine patterns of occurrence and the factors to be configured for monitoring and detecting these oscillations. Patterns of occurrence fell into two broad categories:

(1) Intermittent and
(2) Consistent.

Those oscillatory modes driven by bad settings in their controllers were intermittent, meaning that they were not continuously present and were excited in response to system conditions/events. As will be discussed in Sect. 6.5, the energy of these modes showed a good correlation with levels of local wind production.

On the other hand, the consistently present modes of oscillations had low-energy level irrespective of the level of wind production but were present all the time.

The mining metrics used for the study helped identify patterns of occurrence for each oscillatory mode. Based on the results, four factors to be addressed for purposes of real-time monitoring were determined:

(1) Location and signal
(2) Mode frequency range
(3) Minimum energy
(4) Damping.

6.3 Metrics for Classification of Oscillations

While looking for signature of oscillations in a system, it is important to somehow classify the oscillations based on measurable metrics. The mode frequencies form the characteristics of the oscillations. Accordingly, oscillations may be ranked based on the following two factors for any frequency:

(1) Highest energy and
(2) Highest occurrence.

The metrics based on these factors were defined as the *Monthly Highest Energy* and the *Monthly Mode Occurrence*.

6.3.1 Monthly Highest Energy

Monthly Highest Energy (MHE) is the maximum energy calculated for an oscillatory mode during the month. This metric provides the maximum level of energy reached by the mode in that month. The greater this value is, the more significant the mode becomes.

6.3.2 Monthly Mode Occurrence

Monthly Mode Occurrence (MMO) is the ratio of the total time in minutes reported for a particular mode to the total time in minutes in the month being analyzed.

$$MMO\,(\%) = \frac{t_m}{t_{all}} \times 100 \tag{1}$$

where

t_m total time reported for mode *m* and
t_{all} total time in the month

This metric provides the fraction of time in the month for which the mode is present. More the value is, more is the presence, and hence the mode is of higher significance.

6.4 Generalized Approach for Identification of Signature Oscillations in a System

This section presents a generalized approach to identify signature oscillations in a system, developed as part of the data mining project as detailed in Sects. 6.1 and 6.2. It is the authors' belief that this approach can be applied to any power system with available PMU data, for which signature oscillations are to be determined. Once identified, the modes may be configured using wide area visualization tools, as alerts/alarms to aid the system operators.

Step 1: Identify key locations with installed PMUs whose data will be used for the analysis. The selection of these locations could be based on various factors as suited to the study at hand—local electromechanical modes, inter-area oscillations, proximity to wind generation, etc. In addition, the following are important:

(1) Frequency, voltage, and current magnitude quantities will all be used in the analysis.
(2) A reference PMU for angle difference calculations shall be assigned. This reference could be a PMU which is close to a load center in the system or chosen to be near the center of inertia of the system.
(3) Bad data will need to be removed (see Sect. 6.2.3).

Finally, select the period of data used for mining such that it covers different seasons and conditions of the system. This period would be influenced by the system and the PMU data archived.

Step 2: Scan the data at periodic intervals—every minute is recommended—for oscillatory modes with low damping regardless of detected energy level. Gather time-stamp, energy, and damping for all detected modes for data from every PMU selected in Step 1.

Step 3: Calculate MMO and MHE mining metrics for each oscillatory mode for all PMU data under study. Discard weaker modes with low occurrence and low-energy level.

Step 4: Compare the MMO for each of the remaining oscillatory modes with its MHE and other parameters according to the factors that informed the choice of PMUs (e.g., wind production levels). The following determining factors may be used:

(1) *Correlation* between MMO, MHE, and the chosen parameters for the study, e.g., wind production, will identify whether the parameter influences the particular mode.

(2) *Intermittency versus Consistency* for MMO and MHE will identify the behavior of the mode on the system, e.g., a low-energy background oscillation that is consistently present across all studied time periods.

Step 5: Based on the results from Step 4, determine if applicable, four factors for each identified mode for purposes of real-time monitoring.

(1) *Location and Signal* are the PMU and corresponding measurement signal with highest MMO for that mode.

(2) *Mode frequency range* is the band-pass filter range enveloping the oscillatory mode frequency. A range of up to ±1.5 Hz is recommended.

(3) *Minimum energy* is the threshold(s) at which warnings are to be generated. From empirical evidence and operational experience, two settings at 20% and 50% of maximum MHE are recommended for 'alert' and 'alarm,' respectively, signifying increasing criticality.

(4) *Damping* is the threshold beyond which warnings are to be generated. A conservative value of at least 5% is recommended to allow system operators time to take action on poorly damped oscillatory modes.

6.5 Illustration on the ERCOT System

A summary of the ten identified oscillatory modes in the ERCOT interconnection for the time period from 2012 to 2014 is shown in Table 6.2.

Table 6.2 Occurrence of oscillatory modes over three years

#	Mode (Hz)	2012	2013	2014	Oscillation type
1	0.6	Till March	Absent	Absent	Local
2	0.9	Present	Present	Present	Wind production related
3	1.5	Only in April	Absent	Absent	Control systems
4	1.7	4 months	Absent	Absent	Control systems
5	2.0	Absent	April	Absent	Control systems
6	2.7	Present	Present	Absent	Wind production related
7	3.2	4 months	Absent	Only in Jan	Control systems
8	5.0	Present	Present	Present	Control systems
9	5.4	Present	Present	Present	Control systems
10	6.0	Present	Present	Present	Control systems

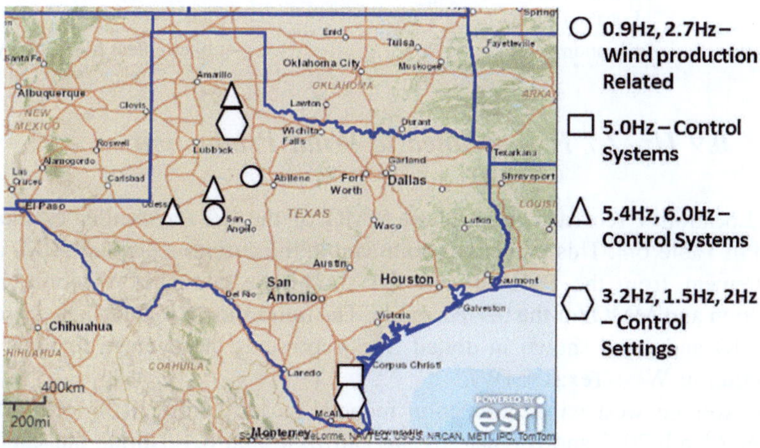

Fig. 6.3 Location of identified oscillatory modes in ERCOT

Figure 6.3 shows the location of oscillatory modes in the ERCOT region. The 0.9 and 2.7 Hz modes were dominant in West Texas, whereas the 5.0 Hz oscillatory mode, marked in green, was dominant in the Rio Grande Valley region of ERCOT. The 5.4 and 6.0 Hz oscillatory modes were dominant in the Panhandle and Far West regions of ERCOT. The 1.5, 2, and 3.2 Hz modes were found in the Valley and the Panhandle.

Table 6.3 shows the list of ten oscillatory modes with the PMU location that saw the highest MMO of that particular mode. The pattern of occurrence was also compared with the MHE trend for each mode. As shown in Table 6.3, the 0.9 and 2.7 Hz modes, identified earlier as having a close relationship to wind production, were also found to have high energy under conditions of high wind production. On the other hand, the 5.0 and 5.4 Hz modes were low energy and present irrespective of wind production levels.

Table 6.3 Pattern of energy level of oscillatory modes

#	Mode (Hz)	Nearest PMU	Related to wind production	Highest energy level
1	0.6	Location 2	No	Low energy and flat
2	0.9	Location 1	Yes	High energy and tracking occurrence
3	1.5	Location 3	No	Low energy
4	1.7	Location 4	No	High energy
5	2.0	Location 3	No	Low energy
6	2.7	Location 5	Yes	High energy and tracking occurrence
7	3.2	Location 2	No	High energy
8	5.0	Location 3	No	Low energy and remained flat
9	5.4	Location 2, Location 5	No	Low energy and remained flat
10	6.0	Location 2	No	Intermittent high energy

6.5.1 0.9 Hz, 2.7 Hz—Related to Wind Production

PMU Location 1 recorded the highest MMO of the 0.9 Hz oscillatory mode as shown in Table 6.3. This mode was dominant in the current magnitude (IM, amps) measurement from the PMU. Figure 6.4 shows the comparison between wind production and MMO for the 0.9 Hz mode. The trend shown in thick line represents the MMO and trend shown in dotted lines represents the average regional wind production in West Texas (MW).

The average west wind production reduced steadily from 3100 to 1600 MW between March 2012 and August 2012 as shown in Fig. 6.4. During the same time period, the MMO of 0.9 Hz oscillatory mode reduced from 48 to 34% indicating a direct relationship to wind production. Similarly, for the period from August 2012

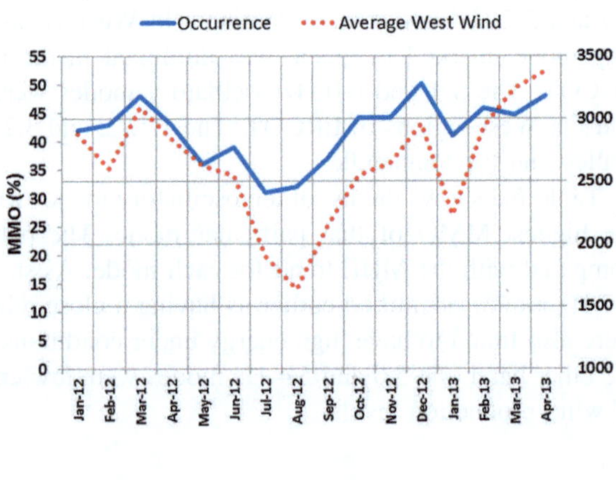

Fig. 6.4 Comparison between 0.9 Hz mode occurrence and average west wind production

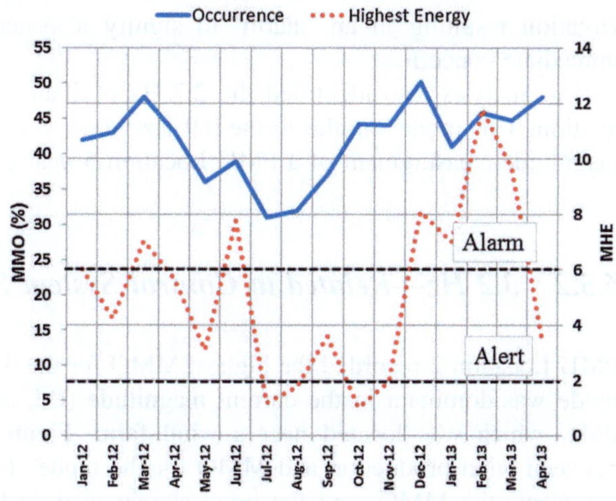

Fig. 6.5 Comparison between 0.9 Hz mode occurrence and highest energy

to December 2012, the average wind production steadily increased from 1600 to 3000 MW and correspondingly, the MMO of the 0.9 Hz mode increased from 34 to 50%, following the pattern of average wind production.

This comparison shows that wind generation in West Texas had a strong correlation to the 0.9 Hz oscillatory mode reported by PMUs in the region.

Figure 6.5 shows the comparison between the mining metrics MHE and MMO, for the 0.9 Hz mode in the current magnitude signal from PMU Location 1. The blue trend represents the MMO, and red trend represents the MHE for the mode.

As can be seen, the peaks and valleys of the derived metrics matched very well, providing a second indication that the 0.9 Hz mode is related to wind production. As the plot in Fig. 6.5 suggests, when the MMO increases, the MHE also increases showing that the 0.9 Hz oscillatory mode is strongly present when the average west wind production is high and vice versa. For instance, the energy of 0.9 Hz oscillatory mode was below 2 in August 2012 when the average west wind production was 1600 MW. On the other hand, the energy of 0.9 Hz oscillatory mode was 12 when the average west wind production was close to 3000 MW.

The study recommended monitoring of the 0.9 Hz mode in real time. Based on the four factors for configuration detailed in Sects. 6.2.6 and 6.4, the current magnitude measurement from PMU Location 1—which had the highest MMO—was labeled as the critical measurement. Frequency range recommended was between 0.8 and 1.2 Hz to capture the oscillatory mode. To prevent aggressive alarming at low-energy levels and high damping, the recommended minimum damping threshold for real time was 5%, with an alert to be generated if the threshold was violated. Similar thresholds were proposed based on MHE—the maximum recorded MHE for this mode was 12 in February 2013 as shown in Fig. 6.5, while the minimum MHE was close to zero. Hence, as a conservative estimate based on operational experience, the minimum threshold for an alert based on the energy was set at 2. A second threshold of 6 was also incorporated, with a violation resulting in an 'alarm' to signify a system condition of greater and immediate concern.

The analysis also identified the 2.7 Hz oscillatory mode related to wind production. This mode, similar to the 0.9 Hz mode, was also dominant in the current magnitude measurement of a PMU Location 5 near a wind farm.

6.5.2 3.2 Hz—Related to Control System Settings Changes

PMU Location 2 recorded the highest MMO for the 3.2 Hz oscillatory mode. This mode was dominant in the current magnitude (IM, amps) measurement from the PMU which was located near a wind farm. Figure 6.6 shows the comparison between wind production and MMO for the mode. The trend shown in thick line represents the MMO, and the trend shown in dotted line represents the average regional wind production in MW for the north region.

The 3.2 Hz mode MMO trend provides two indications. First, this oscillatory mode does not occur consistently and is intermittent. Second, the pattern of occurrence does not match the average wind production data indicating no causal relationship between the two for this mode. The intermittent occurrences suggest that this mode was more likely to be excited by a temporary change in either the system state or the control systems of the local wind farm.

Figure 6.7 shows the comparison between the mining metrics for the 3.2 Hz mode in the current magnitude signal from PMU Location 2. The trend in thick line represents the MMO, and the trend shown in dotted line represents the MHE for the same mode. The peaks and general trends of the derived metrics matched well. Taken together with the conclusions from Fig. 6.6, this suggests that there was a weak correlation to regional wind production.

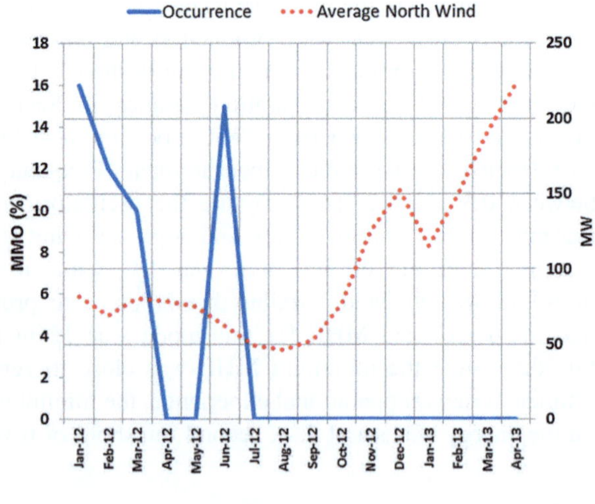

Fig. 6.6 Comparison between 3.2 Hz mode occurrence and average north wind production

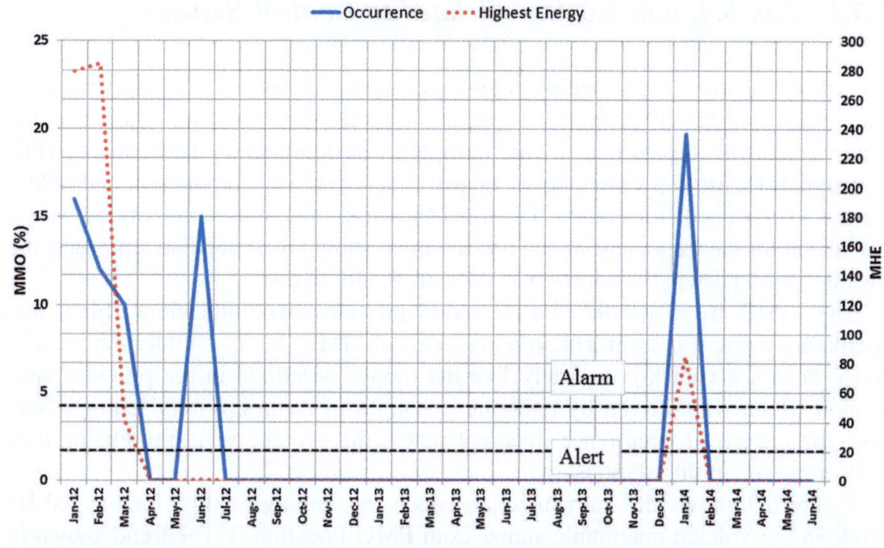

Fig. 6.7 Comparison between 3.2 Hz mode occurrence and highest energy

The third highest intermittent MMO was 20% in January 2014. This is in good agreement with the oscillation incident on January 10, 2014 [3]. The highest energy during the same month was approximately 80. Similarly, the 3.2 Hz mode occurrence touched 15% in June 2012, however with low energy. This difference in energy between the two cases was found to be related to the MW output of the local wind farm. Hence, it was concluded that this intermittently occurring oscillatory mode was driven by temporary changes in control system settings in the local wind farm and occurred either with high or low energy depending upon the local wind generation.

The study recommended monitoring of the 3.2 Hz oscillatory mode in real time with specific configuration. The current magnitude measurement from PMU Location 2 which had the highest MMO was labeled as the critical measurement. Frequency range recommended was between 2.5 and 4.5 Hz to capture the oscillatory mode. The recommended damping threshold was 5% below which an alert would be generated. As with the 0.9 Hz mode, two levels—alert and alarm—were recommended for the energy level with the thresholds set conservatively at 20 and 50, respectively.

The analysis had also identified other oscillatory modes 1.5, 1.7, and 2 Hz related to control system settings changes. These additional oscillatory modes also had similar characteristics of intermittent occurrence and varying energy depending upon the local wind generation.

6.5.3 5.0, 5.4, and 6.0 Hz—Related to Control Systems

The PMU Location 3 recorded the highest occurrence of 5.0 Hz oscillatory mode as shown in Table 6.3. This mode was present in all the measurements taken from the PMU which was located near a wind farm, but was dominant in the frequency (FR, Hz) signal. Figure 6.8 shows the comparison between wind production and MMO for the 5.0 Hz oscillatory mode. The trend shown in thick line represents the MMO calculated for the frequency signal, and the trend shown in dotted line represents the average regional wind production in MW in South Texas.

The MMO trend for the 5.0 Hz mode provides two indications. First, this oscillatory mode is present with varying levels of MMO, but is not intermittent. The mode does not vanish completely like the modes described in the previous subsection which were driven by temporary changes in control system settings. Second, the pattern of occurrence does not match the average wind production data indicating no causal relationship.

Figure 6.9 shows the comparison between the mining metrics for the 5.0 Hz mode in the voltage magnitude signal from PMU Location 3. The trend shown in thick line represents the MMO in the voltage signal, and the trend shown in dotted line represents the MHE for the same oscillatory mode.

The MHE remained very low for the 5.0 Hz oscillatory mode and, as shown in Fig. 6.9, the trend remained almost entirely flat suggesting no relationship with regional wind production. However, the varying levels of MMO for this mode suggest a relationship to the operation of local wind farms and the specific design of their control systems. The PMU locations that recorded the highest occurrence of the 5.0 Hz mode were near wind farms employing type-3 wind turbines. The mode was more dominant in the Rio Grande Valley region when compared to other regions in EROCT like the Panhandle and the Far West.

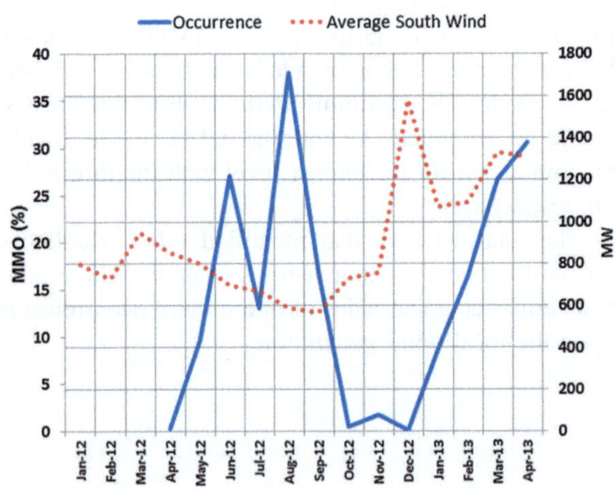

Fig. 6.8 Comparison between 5.0 Hz mode occurrence and average south wind production

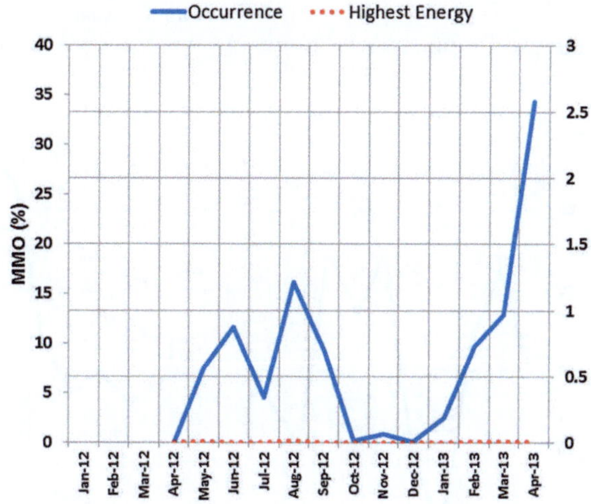

Fig. 6.9 Comparison between 5.0 Hz mode occurrence and highest energy

The PMU Location 2 in the Panhandle and PMU Location 5 in the Far West region of ERCOT recorded the highest occurrence of the 5.4 Hz oscillatory mode. This mode was present in all the measurements from the PMU but dominant in the voltage signal (VM, kV). The 5.4 Hz oscillatory mode had similar characteristics as 5.0 Hz oscillatory mode, but was more dominant in the Panhandle and Far West regions of ERCOT rather than the Rio Grande Valley. The energy level remained low and flat and did not suggest any relationship to regional wind production. The PMU locations were near wind farms that were employing type-3 wind turbines.

PMU Location 2, in the Panhandle, also recorded the highest occurrence of the 6.0 Hz oscillatory mode shown in Table 6.3. This mode was present in all the measurements from the PMU but dominant in the current magnitude signal (IM, amps). The 6.0 Hz oscillatory mode was in general found to be no different from the other oscillatory modes like the 5.0 and 5.4 Hz modes driven by the operation of the wind farm and the specific design of their control systems.

Measurements from PMU Location 2 in the Panhandle region of ERCOT were chosen as critical and to be monitored for both the 5.4 and 6.0 Hz oscillatory modes. Figure 6.10 shows a comparison of the MMO in the current magnitude signal at Location 2 for both 5.4 and 6.0 Hz oscillatory modes. The thick line and dotted line trends represent mode occurrence of 6.0 Hz and 5.4 Hz, respectively.

6.5.4 Summary

The 0.6 Hz oscillatory mode was present only in the first three months of 2012 and did not appear thereafter. Also, it was noticeable only in Location 2. This was a local mode found to be related, not to wind production, but topology changes

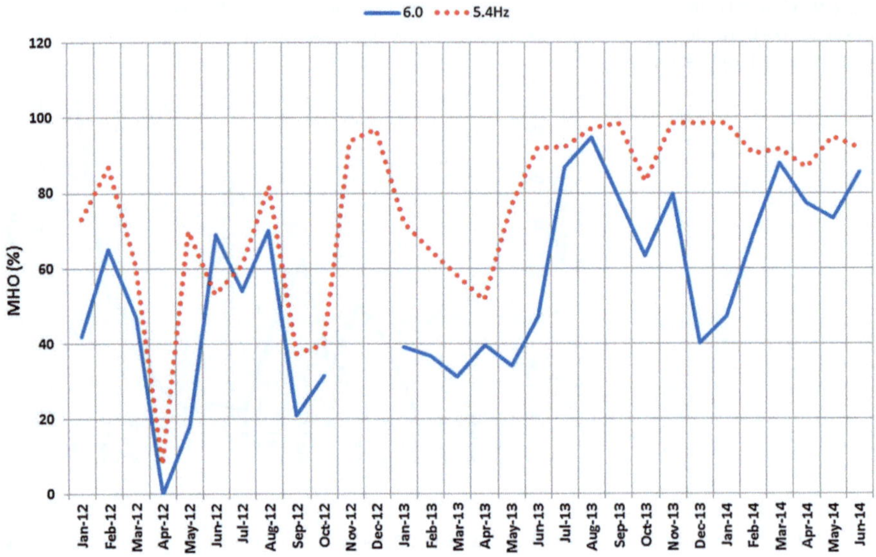

Fig. 6.10 Comparison between the 5.4 and 6.0 Hz mode occurrence at Location 2

around a nearby wind farm. The energy level of this oscillatory mode was low and flat.

The 0.9 and 2.7 Hz oscillatory modes appeared every month of every year in 2012 and 2013. They were both high energy modes which tracked the local wind production. However, in 2014 the 2.7 Hz mode was absent while the 0.9 Hz mode continued to be present. This coincides with the energization of the Competitive Renewable Energy Zones (CREZ) transmission upgrade project.

The 1.5, 1.7, 2.0, and 3.2 Hz oscillatory modes were intermittent modes, driven by temporary changes in the control systems settings in local wind farms. The occurrence of 3.2 Hz oscillatory mode in January 2014 as shown in Table 6.2 was validated against the previously mentioned oscillation incident (see Sect. 6.1) [3] on January 10, 2014 with good agreement.

The 5.0, 5.4, and 6.0 Hz oscillatory modes occurred every month in all three years. However, these modes were of low energy and continuously present, driven by design of control system and operation of local wind generation.

6.6 Conclusion

This chapter presented generalized approach for identifying signature oscillations in a system. It also presented two new metrics to classify oscillations in the systems. The approach was illustrated using ERCOT PMU data. The results are based on the data mining study conducted by ERCOT and EPG as part of the Discovery Across

Texas demonstration project. The study identified ten unknown and unmonitored oscillatory modes. The PMU location and measurement with the highest occurrence of an oscillatory mode were chosen as a critical location/measurement for additional monitoring in real time. The study recommended specific configurations of mode frequency, damping, and energy ranges and thresholds for each oscillatory mode to aid in monitoring, diagnosis, and operator action in real time for prevention of system and ensuring safe operation of the electric grid.

The wind production was found to closely track the occurrence of two particular modes. Several additional modes related to control system setting changes from other wind farms were also identified by the study. The discovery of specific oscillations belonging to specific wind farms provided an opportunity to review specific oscillation events with wind farm owners/operators for model validation. Further studies aimed at re-tuning wind farm control systems that are known to generate low-/high-frequency oscillations will benefit system operations and integration of wind generation into the electric grid.

References

1. Kundur P (1994) Power system stability and control. McGraw-Hill, New York, NY, USA
2. Pierre JW, Trudnowski DJ, Donnely MK (1997) Initial results in electromechanical mode identification from ambient data. IEEE Trans Power Syst 12(3):1245–1250
3. Trudnowski DJ, Pierre JW, Zhou N, Hauer JF, Parashar M (2008) Perfromance of three mode-meter block-processing algorithms for automated dynamic stability assessment. IEEE Trans Power Syst 23(2)
4. Koellner KM, Burks S, Blevins B, Nuthalapati S, Rajagopalan S, Holloway ML (2015) Synchrophasors across Texas. IEEE Power Energy Mag 13(5):36–40
5. Chen J, Shrestha P, Huang S-H, Sarma NDR et al (2012) Use of synchronized phasor measurements for dynamic stability monitoring and model validation in ERCOT, July 2012. IEEE PES General Meeting, San Diego, USA
6. PES magazine DOE Article
7. Huang S, Schmall J, Conto J, Adams J, Zhang Y, Carter C (2012) Voltage control challenges on weak grids with high penetration of wind generation—ERCOT experience. IEEE PES General Meeting, San Diego, USA, July 2012.
8. Ballance J, Bhargava B, Chen H, Palayam PC, Hiebert J (2014) Analysis of power systems oscillations in WECC system using synchrophasor technology. CIGRE 45th General Session, Aug 2014
9. Diagnosing Equipment Health and Mis-operations with PMU data. NASPI Technical report. https://www.naspi.org/File.aspx?fileID=1417. Accessed Mar 2015
10. Center of Commercialization of Electric Technology (2015) Final Technical report on technology solutions for wind integration in ERCOT, Feb 2015. https://www.smartgrid.gov/document/technology_solutions_wind_integration_ercot
11. Rajagopalan S, Nuthalapati S, Mortensen T, Blevins B (2014) Post-event analysis of a compound event in the ERCOT system using synchrophasor data. In: CIGRE grid of the future symposium, Houston, USA
12. Electric Power Group. http://www.electricpowergroup.com
13. IEEE Std C37.118.1-2011—IEEE Standard for Synchrophasor measurements for power systems. https://standards.ieee.org/findstds/standard/C37.118.1-2011.html

Chapter 7
Oscillation Detection in Real-Time Operations at ERCOT

Sidharth Rajagopalan, Patrick Gravois, Bill Blevins, Wei Liu and Sarma Nuthalapati

7.1 Introduction

The Electric Reliability Council of Texas Inc. (ERCOT) serves as the independent system operator (ISO) for most of the state of Texas. The ISO manages delivery of electric power to 25 million customers in Texas, representing 85% of the state's electric load and 75% of Texas land area. The ERCOT electric grid consists of around 41,500 miles of transmission and more than 570 generating units [1]. The total installed generation capacity for peak demand in ERCOT is 78,000 MW, and its recorded peak load is 71,110 MW. There is more than 19,000 MW of wind generation capacity installed and operating in the ERCOT market, and in March 2014, the total wind generation reached a record 16,141 MW, which was 50% of the load at the time [2].

Most of the generated power in ERCOT is consumed in industrial and urban load centers in East, Central, and South Texas, while most of the wind energy is installed in the northwest part of the state because of abundant wind resources there. This separation between resources and load centers has created some unique challenges in the integration of renewable resources, and in meeting the growing demand for power while also effectively ensuring the reliability of the grid [3–5].

Recognizing the potential benefits of synchronized phasor measurements, or synchrophasors, ERCOT started a collaborative effort in 2008 to implement this technology in its operations and planning processes. This effort, coordinated by the Center for the Commercialization of Electric Technology (CCET) under a grant

S. Rajagopalan · P. Gravois · B. Blevins (✉) · W. Liu
Electric Reliability Council of Texas, Taylor, TX 76574, USA
e-mail: Bill.Blevins@ercot.com

Present Address:
S. Nuthalapati
Peak Reliability, Vancouver, WA 98662, USA

S. Nuthalapati (ed.), *Power System Grid Operation Using Synchrophasor Technology*, Power Electronics and Power Systems,
https://doi.org/10.1007/978-3-319-89378-5_7

from the Department of Energy, has engaged various entities, including ERCOT transmission service providers (TSPs) and a software application provider [6].

As PMUs have been installed across the ERCOT grid, analyses using the incoming synchrophasor data have indicated that synchronized phasor measurements can greatly improve both ERCOT operations procedures and planning studies. This chapter covers in detail three oscillation events that were detected by Operations personnel as they were occurring on the system. The issues in all three situations were found to be incorrectly operating controllers and/or weak grid conditions [7]. The control systems in both cases were fixed and the behavior verified using PMU data in future operations.

7.2 The ERCOT Phasor Measurement Task Force [8]

During the latter half of 2013, the ERCOT Board of Directors adopted a 2014–2018 Strategic Plan. This plan acknowledged "the greater use of synchrophasor measurement of real-time grid status" as a key trend within ERCOT and the industry. In conjunction with this, the ERCOT Reliability and Operations Subcommittee (ROS) developed a goal for 2014 to "evaluate and determine Synchrophasor requirements and address any needed changes in Protocols and Guides." This goal was intended to be met, in part, by the formation of the Phasor Measurement Task Force (PMTF) which kicked off in November 2013.

At the PMTF kickoff meeting, representatives from ERCOT, TOs, Generator Owners (GOs), the North American Synchrophasor Initiative (NASPI), and CCET were in attendance. NASPI and CCET representatives presented on the status of synchrophasor technology within the broader electric industry, as well as the success of the "Discovery Across Texas" (DAT) smart grid demonstration project which employed synchrophasor technology and advanced analytic tools to monitor wind dynamics, enhance transmission operations management practices, and improve grid reliability. The group elected its leadership and confirmed the ROS-identified charter which consisted of eight key issues as the PMTF's main scope of work.

In order for the PMTF to complete its work, ERCOT stakeholders and industry experts were invited to attend and present material the group could learn from. Technical topics include: PMU hardware, PDC design, situational awareness software, and model validation techniques. Efforts within other NERC regions were also reviewed in order to discover best practices. Ultimately, the group narrowed its focus to the three most beneficial "use cases": oscillation detection and alarming, generator model validation, and post-disturbance analysis and reporting. These use cases were chosen, in part, based on ERCOT's prior experience with the synchrophasor infrastructure installed as part of the DAT project.

To support the identified use cases, the PMTF worked on three key deliverables over the course of 2014: the ERCOT Synchrophasor Technology Assessment, the ERCOT Synchrophasor Communications Handbook, and draft Nodal Operating

Guide language which is binding language within ERCOT. The assessment answers the key questions posed by ROS and summarizes the status of this technology within ERCOT. The impact of related NERC reliability standards and FERC rulings was also considered. The communications handbook provides ERCOT stakeholders with technical details about how to record and transmit synchrophasor data to ERCOT. The binding language outlines requirements for the installation of PMUs by TOs and GOs. Each of these documents was published and presented to ERCOT ROS toward the end of 2014.

7.3 Detection of Oscillations in Real Time

In ERCOT region, most renewable generation projects are located remote from load centers and conventional synchronous generators. As a result, weak grid conditions (lack of sufficient system strength, in terms of short circuit ratio, SCR) are one of the key challenges to accommodating high renewable penetration. Under such conditions, voltage becomes more sensitive and in turn, more precise and robust controller settings and design can be crucial to maintain stable operation. Most oscillatory responses observed in recent years were caused by either improper controller tuning or limited controller functionality in the renewable generation voltage and/or frequency control.

7.3.1 Controller Parameter Degradation

During a real-time event in late February 2014, ERCOT operations engineers observed oscillations from a PMU close to a hydro resource (see Fig. 7.1). It was also observed that the oscillations ceased as shown in Fig. 7.2, when the resource went offline and that a second resource at the same site did not produce these

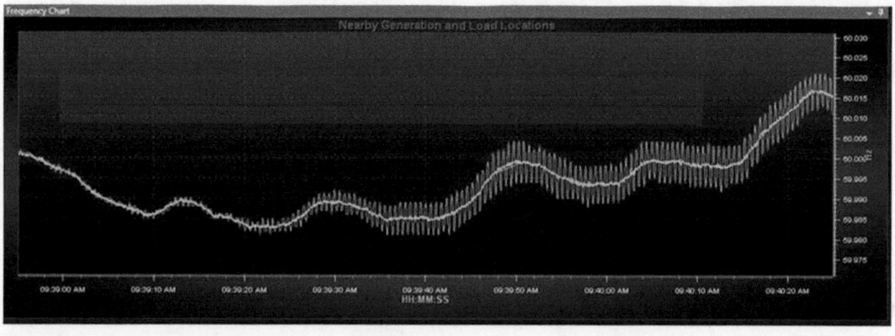

Fig. 7.1 Start of the oscillations (late February 2014)

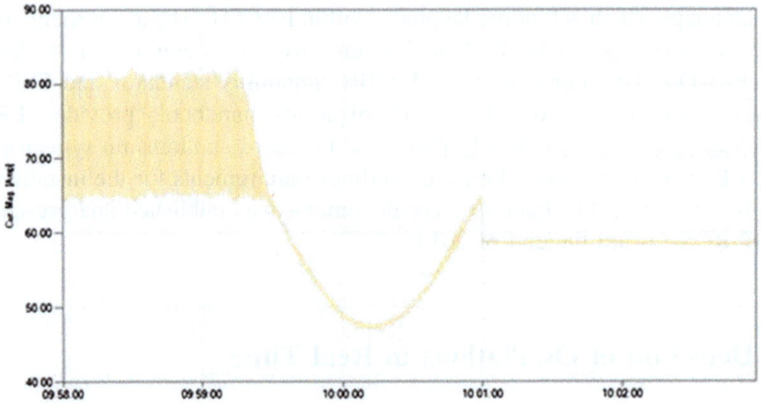

Fig. 7.2 Oscillations cease when unit goes offline

oscillations when it came online. The dominant mode of these oscillations was 1.8 Hz.

Discussions with the power plant operators confirmed that the oscillations were indeed originating from the first resource. The likely source was determined to be degraded parameters in one the control cards for the problematic unit and together with the plant operator, this was to be tested.

Usually, the resources operate one at a time. ERCOT was going through severe weather at the time, and the plant operators decided to operate only that unit identified as not causing the oscillations until the severe weather conditions passed. So, in March 2014, the resource determined as the source of the oscillations received new control cards and was brought online. ERCOT engineers monitored the live PMU data stream as the resource came online and were able to confirm that the oscillations were significantly reduced (Fig. 7.3), to the point of nonexistence.

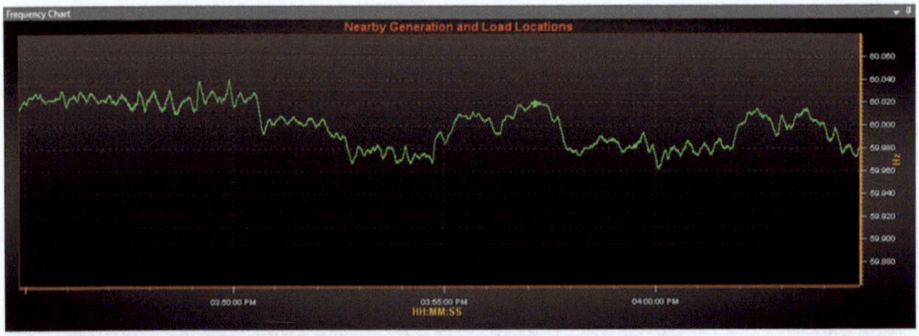

Fig. 7.3 No oscillations in March after controller replacement

7.3.2 *Weak Grid Oscillations*

During a real-time event in January 2014, ERCOT operations engineers observed oscillations from a single PMU located near a wind plant (Fig. 7.4). The wind plant, due to controller settings, had been susceptible to oscillations in the past under weak system conditions (outage of a nearby line). However, with the plant generating around 56 MW and no outages in the area, the oscillations were unexpected [8].

Removing the wind farm AVR from service did not help mitigate the oscillations, so in order to address the situation ERCOT advised the wind farm to curtail output to 45 MW. This reduced the oscillation magnitude but did not remove them completely. The unit was further constrained to 40 MW, and the oscillations ceased thereafter (Fig. 7.5).

The oscillation frequency was 3.3 Hz with damping below 1% (Fig. 7.6). This was in agreement with earlier studies, indicating a relation to controller settings at the plant. The plant operator confirmed a system update had been installed

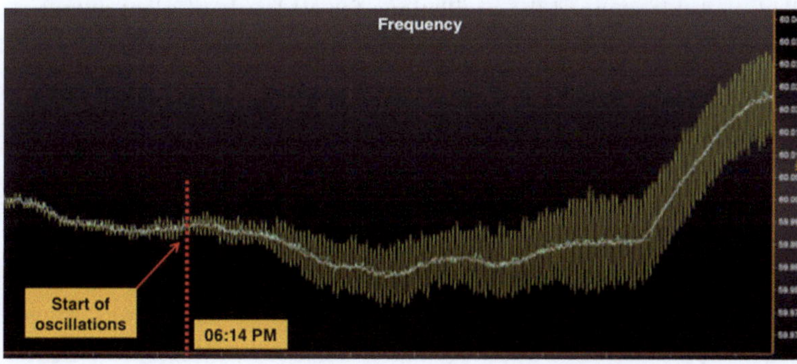

Fig. 7.4 Start of oscillations

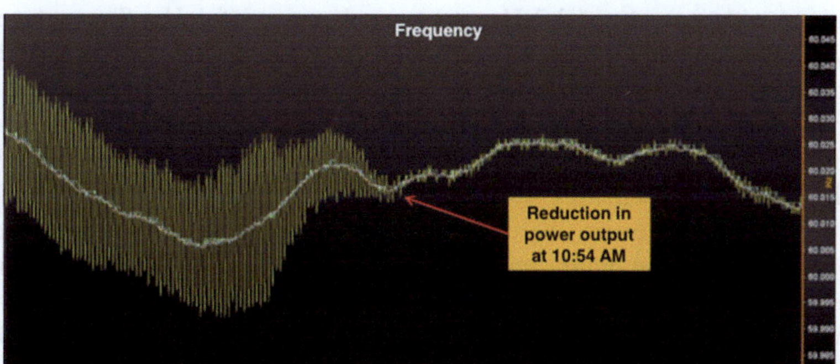

Fig. 7.5 End of oscillations after 40 MW constraint

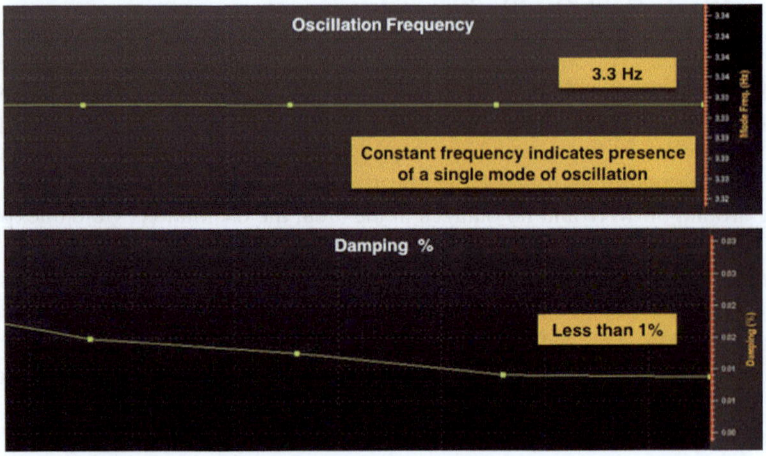

Fig. 7.6 Oscillation frequency and damping

(to improve performance) around the time the oscillations began. The updates were uninstalled, and this stopped the oscillations for all levels of plant output.

7.3.3 Controller Parameter Settings

During an event in October 2013, ERCOTs synchrophasor data analytics software recorded oscillations from a single PMU located near a wind plant (Fig. 7.7). Unusually, the oscillations recorded were negatively damped in both the voltage magnitude and phase angle. The rising magnitude of oscillations triggered an out-of-step relay at the interconnecting transmission substation, removing the wind plant from service. The mode of oscillation was 3.8 Hz, and the damping was −0.1%.

In the days preceding this event, the PMU had recorded periods of sustained oscillations with a mode of 3.7 Hz and poor damping of 1 Hz. The outage of a nearby transmission facility had weakened the system in the area, and the wind farm output had been curtailed to 40 MW. Information from the resource indicated that a primary frequency response (PFR) controller at the wind farm had been enabled. The PFR controller was driving the output of the wind farm over the 40 MW set point leading to the oscillatory behavior (see Sect. 7.3.2 above).

At the time of the event, the conditions described above were exacerbated by low load (early morning hours) and a loss of generation event. The resulting drop in frequency engaged the PFR control and drove the wind farm output from a curtailed 37 MW to a maximum of 48 MW in less than 4 min. The low load made the system weaker than usual, and the sharp ramp caused the wind farm to enter a

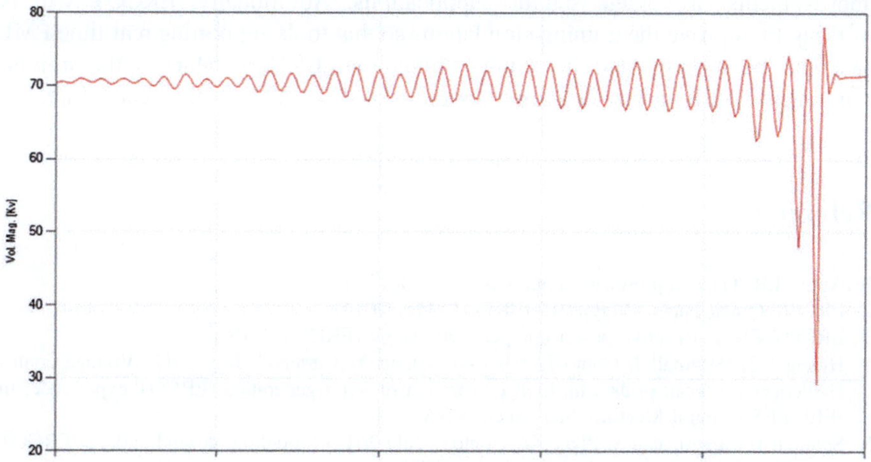

Fig. 7.7 Voltage magnitude in kV showing the negatively damped oscillations

period of negatively damped oscillations which lasted 10 s, before the loss of the wind farm via out-of-step relay activation.

ERCOT directed the wind farm to disable the PFR for the duration of the outage. This ensured that the wind farm was able to meet the curtailed set point, and no further oscillations were recorded by the PMU that originated from the PFR controls.

7.4 Conclusions

In this chapter, we covered three oscillation events that were detected by ERCOT Operations personnel as they were occurring on the system. In all the cases, controller parameters, weak grid conditions, or both were found to be the motivating factors, and in each case, the issue was addressed in real time with directives to the resource(s) identified.

ERCOT has found benefit in its existing PMU network and collaborated with its stakeholders to codify rules to support three primary use cases—oscillation monitoring, model validation, and post-event analysis. PMUs have provided benefit beyond these use cases but as no other tools existing today addressed these three use cases as well as PMUs, and it was decided to focus on those tasks. Other possible benefits which ERCOT will be considering would be linear state estimator/hybrid state estimator, Inertia monitoring and analysis and a backup ACE calculation using PMUs. An area of potential research going forward is to consider if PMUs can be used to calculate grid stability index for weak grid areas and possible improvements in voltage stability applications. Additionally, ERCOT will be working to improve the training simulations so that tools supporting real-time PMU data can be incorporated in routine simulation. ERCOT believes the training component will be a key in getting adoption of these tools in the near future.

References

1. About ERCOT. (http://www.ercot.com/about/index.html)
2. ERCOT Quick Facts—May 2014 (ERCOT, May 2014)
3. ERCOT CREZ Reactive power compensation study (ERCOT, 2010)
4. Huang S-H, Schmall J, Conto J, Adams J, Zhang Y, Carter C (July 2012) Voltage control challenges on weak grids with high penetration of wind generation: ERCOT experience. In: IEEE PES General Meeting. San Diego, USA
5. Schmall J, Rajagopalan S, Zhang Y, Conto J (July 2012) Modeling of wind parks at ERCOT. In: IEEE PES general meeting. San Diego, USA
6. CCET (Sept 2013) Technology solutions for wind integration in ERCOT. DOE Report
7. Chen J, Shrestha P, Huang S-H, Sarma NDR et al (July 2012) Use of synchronized phasor measurements for dynamic stability monitoring and model validation in ERCOT. In: IEEE PES general meeting. San Diego, USA
8. Koellner KM, Burks S, Blevins B, Nuthalapati SN, Rajagopalan S, Holloway ML (Sept–Oct 2015) Synchrophasors across Texas: the deployment of phasor measurement technology in the ERCOT region. IEEE Power Energy Mag 13(5), 36–40

Chapter 8
Oscillation Detection and Mitigation Using Synchrophasor Technology in the Indian Power Grid

P. K. Agarwal and Chandan Kumar

8.1 Introduction

Any electrical grid is susceptible to low-frequency oscillations as these are inherently present in the system. However, such low-frequency oscillations are non-observable due to high positive damping. These oscillations are observed only if some power system control interact in a way causing excitation of the oscillation modes by reducing their damping. Indian grid has also been experiencing low-frequency oscillations since early days. However, their observability is now better with the introduction of Synchrophasor technology [1, 2]. In earlier days, these oscillations were known to the system operator only after being reported by any generating station. For example, in Western Regional (WR) grid of India, oscillations between Vindhyachal and Korba regions were observed when the transmission linkage between the Eastern and Western part of the Western grid was weaker. Further, during large grid synchronizations, the Thyristor Controlled Series Compensators (TCSCs) were used in order to suppress the inter-area oscillations after synchronization. For example, synchronization of Eastern Regional (ER) grid and Western Regional (WR) grid in 2003 was done using tie links having TCSC for damping the low-frequency oscillations [3]. Similar was the case while synchronizing the Eastern Regional grid with the Northern Regional grid wherein TCSC installed tie lines were used to damp out the inter-area oscillations [4, 5]. Oscillations were also observed after synchronization of NEW (North-East-North East-West) grid with the Southern Regional grid since December 2013 [6–9]. Long inter-regional synchronizing tie lines require TCSCs to suppress the oscillations as

P. K. Agarwal (✉) · C. Kumar
Power System Operation Corporation Ltd., New Delhi, India
e-mail: pkagarwal@gmail.com

C. Kumar
e-mail: chandan@posoco.in

S. Nuthalapati (ed.), *Power System Grid Operation Using Synchrophasor Technology*, Power Electronics and Power Systems,
https://doi.org/10.1007/978-3-319-89378-5_8

used in ER-WR and ER-NR grids synchronization. Since NEW grid and SR grids were synchronized over a 208-km-long 765 kV 2*S/C lines, no FSC plus TCSC was envisaged in the planning horizon of synchronization of NEW and SR grids.

In recent years, Indian system operator has to operate power systems closer to its stability limits because of rapid growth in demand. Furthermore, large regions being connected by few lines has resulted in lightly damped modes becoming prominent and observable to the system operator. Technology and operating philosophy are being sought for increasing the power transfer over existing (possibly weak) interconnections, thereby reducing the damping of the already lightly damped modes which increases the operator's challenges. Consequently, ensuring that the damping of modes of rotor oscillation in power systems provides adequate margins of stability has been and still is of concern to system planners and operators. In order to damp oscillations, firstly there is a need to classify the modes under various categories based on the observation and literature.

A. Classification of Oscillation

Based on the literature survey [10–17], the low-frequency oscillations can be broadly classified into the following categories:

1. Inter-area mode: 0.05-0.3 Hz
2. Intra-area mode: 0.4–1.0 Hz
3. Inter-plant mode: 1.0–2.0 Hz
4. Intra-plant mode: 1.5–3.0 Hz
5. Control mode: not defined
6. Torsional mode: 10–46 Hz.

The associated oscillations in the above classifications are the characteristic or natural modes of the system. The frequency and damping of such modes generally change with the operating conditions, i.e. changes in the system configuration and load/generation in the grid. Such changes in operating condition may cause the electrical power system to drift or be forced towards a small-signal stability limit. These conditions result in negative damping of modes which is seen as growing oscillation in various electrical parameters. Such negative damped oscillations have huge repercussions on the power system and can even cause blackouts.

B. Impact of Low-Frequency Oscillation

It is well known to the power system engineers that the oscillations have played an important role in the world famous blackouts [18]. These include the following:

1. 1996 WSCC Blackout
2. 2003 Blackout in USA and Canada
3. 2003 Blackout in Denmark–Sweden
4. 2003 Italy Blackout
5. 2011 Chile Blackout

So, there is a need of systematic study and planning required for small-signal stability in power system to avoid such blackouts. Inter-area oscillation causing System Protection Scheme (SPS) triggering due to negative damping has also been observed in the Indian power system. The observability of oscillation has placed new dilemmas with regard to the type, nature, source and impact of the oscillations in front of the operator. In due course of time, the operator gradually is now able to comprehend more about the oscillations using Synchrophasor measurements along with conventional SCADA.

C. Observability of Low-Frequency Oscillations in the Indian Grid

Earlier, the SCADA data available with the system operator and/or the data from generators were used for the study of oscillations which had its own limitations due to low resolution. With the introduction of Synchrophasor measurements units in Indian grid, the oscillations in the grid can now be easily monitored with precision at national control room and each regional control room in real time. This has helped the system operator in giving feedback to the planners regarding weak interconnections where LFO is observable. Based on the various observations and analysis of the data from the Synchrophasor based Oscillation Monitoring System (OMS), baselining of the modes in present Indian power system has also been done and is given below [1]:

- Mode 0.2–0.25 Hz: Southern grid and NEW grid
- Mode 0.7–0.75 Hz: Eastern, North-Eastern and Western grid
- Mode 0.6 Hz: Eastern, North-Eastern and Western grid
- Mode 0.5 Hz: North-Eastern grid with rest of the grid.

8.2 Cases of Low-Frequency Oscillation in the Indian Grid and Measures Taken/Required for Their Improved Damping

Various cases of low-frequency oscillation and few cases of sub-synchronous oscillation have been observed in the Indian grid during last few years. Such high observability of oscillation was possible only with the availability of Synchrophasor data at the desk of system operator [1]. Among these incidents, the cases of sub-synchronous resonance observation at HVDC terminal, oscillation in the area of southern part of Eastern grid, oscillation in the North-Eastern grid, etc., are of interest. These oscillation cases analysis based on the Synchrophasor data, SCADA data and other data sources is presented in this chapter. However, a summary of these case studies is given below in order to familiarize the readers.

- The sub-synchronous resonance (SSR) and sub-synchronous torsional interaction (SSTI) phenomenon in the electrical grid can be due to the proximity of a fixed series capacitor or an HVDC link to a generating station. Both the situations are disastrous for the generator rotor which gets damaged after experiencing sub-synchronous oscillations. There are very few cases in Indian system on SSR and SSTI out of which one has been monitored through Synchrophasor and the same has been explained in this chapter.
- Oscillation cases in the Eastern and North-Eastern regions are prominent. The areas in Eastern Region near to Southern Regional grid boundary have shown highest observability in terms of oscillations. Significant inter- and intra-plant oscillations have been seen in North-Eastern region which have resulted in tripping of units and lines [19]. One case from North-Eastern system has been explained in this chapter in order to show how the low-frequency oscillation affects the system stability and protection operation.
- Based on experience, it is observed that oscillations when excited by a single unit can be easily monitored and concluded and after this the remedial action can also be planned to take suitable measures. During such cases, oscillations have either initiated when these units were brought in service or when they were taken out or have tripped abruptly during oscillations. Further, the oscillations have been observed to die out in a short span of time. These cases have provided various lessons during system operation to avoid such oscillations. Few such case studies have been included in this chapter for the benefit of the readers.
- The interesting observation during many of the cases of oscillations was their observability aspect. Even though in many cases oscillation frequency lies in the category of inter-area or intra-area oscillation, their observability was local in nature.

Few case studies have been presented below to understand how Synchrophasor data has helped operators in understanding the very nature of low-frequency oscillations (LFO) and planning of various remedial measures.

A. Low-Frequency Oscillations at Generating Plant due to Switching of a Nearby High Capacity Transmission Line

Low-frequency oscillation in a generating plant may get excited due to the malfunction of its controller when there is a change in the grid parameter by virtue of any nearby switching activity. In one of the cases, it was found that generating unit started hunting and caused oscillation in the grid during the switching of a 765 kV long line which was connected from the adjacent substation near to the generating station. The observed low-frequency oscillation has subsided after the tripping of the unit which was hunting. The detailed description of the event is given in the next paragraph.

One of the 500 MW units of the plant was just synchronized with the grid, and it was generating 98 MW. The unit was on speed control mode. The controller mode of the generating unit was being changed from speed control mode to load control mode of the turbine. At this moment, a nearby long 765 kV line was opened by the system operator as a part of planned outage. Subsequently, the unit started hunting

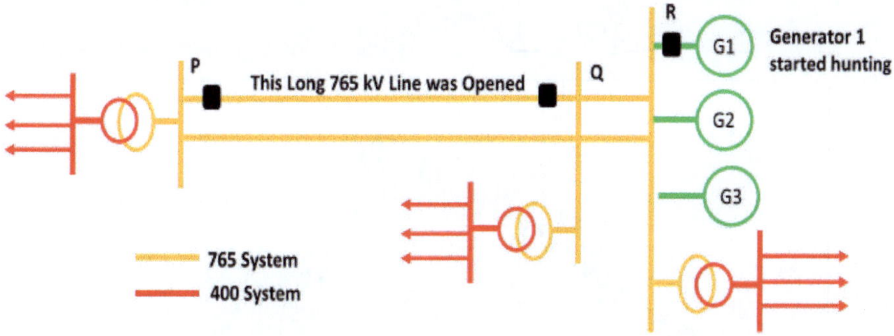

Fig. 8.1 Network schematic for the illustration of the case study 1

with wide fluctuation in its active as well as reactive power. The schematic diagram for better visualization of the event is given in Fig. 8.1.

Prior to analysing the actual cause of the oscillation, it is to be understood that the turbine prime mover control system usually has three modes of operation, i.e. speed control, load control and pressure control. Prior to synchronization, i.e. in start-up phase, any generating unit prime mover is kept on speed control mode. In this mode by adjusting the control valves, the total amount of steam flow entering steam turbine is changed, which is directly proportional to speed of rotation. Once the generator is synchronized to the grid, control system is in either load or pressure mode. In the load control mode, control error that represents input to the controller is formed as a difference between set point and actual load.

In the above case, it was found that the control was being changed from speed to load control mode to increase the power of the generating unit after synchronization. During this time, a nearby long 765 kV line was being opened by the system operator as a part of planned outage. As the line was manually tripped, it has led to 7–8 kV change in the bus voltage of the generating station. Due to this voltage dip during the line outage, there was a change in load feedback. This has resulted in toggling of turbine control between speed control to load control and eventually resulted in hunting of the real power of the unit between 17 and 250 MW causing severe oscillations in the grid. Along with real power, reactive power of the unit was also oscillating.

Figures 8.2 and 8.3 show the respective frequency and voltage observed at various nodes during the excitation of inter-area oscillation due to the local oscillation at the generator. This incident of oscillation has provided a lesson to the system operator to not perform switching near to substation while a generating unit is being synchronized. Further, it has also provided a generating station to check for planned switching operation near to its generating station which is being given by system operator on day-ahead basis to all constituents.

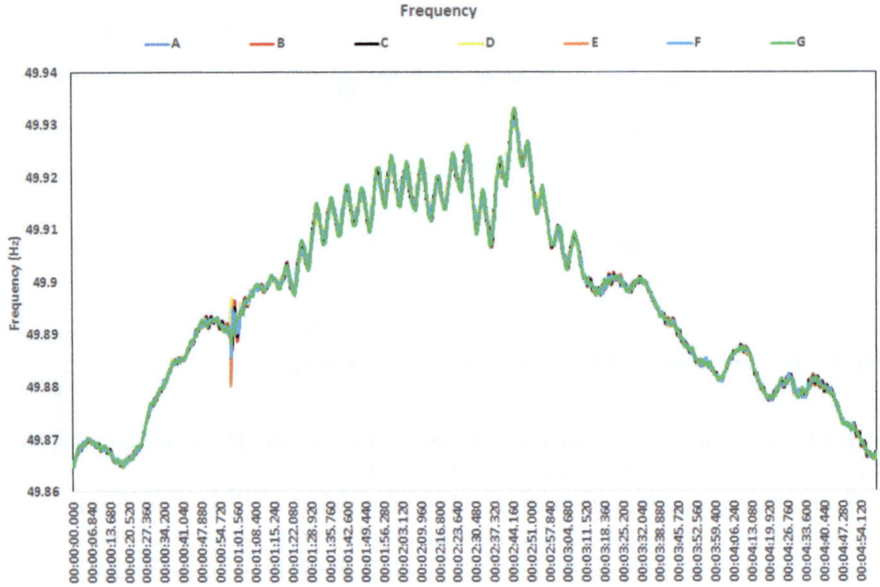

Fig. 8.2 Frequency of nearby nodes in the grid during the oscillation

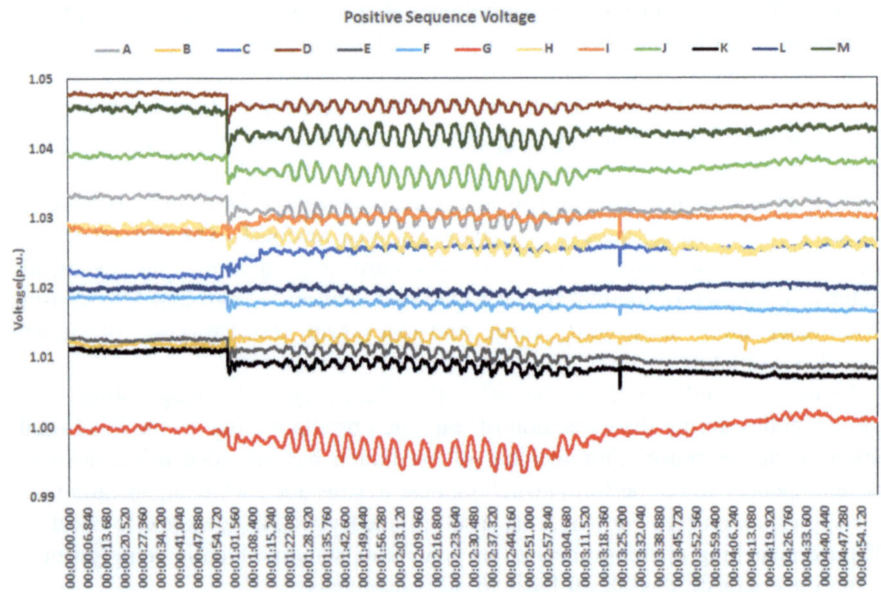

Fig. 8.3 Positive sequence voltage of nearby nodes in the grid during the oscillation

B. Sub-synchronous Torsional Interaction Near to HVDC Converter Terminal

The sub-synchronous torsional interaction (SSTI) is observed when an HVDC converter terminal is located near to a generating complex. SSTI is observed on two different operating conditions, i.e. low generation at the plant and multiple commutation failures in the HVDC due to fault on the line. A brief overview of the SSTI to understand the phenomenon is provided in the next paragraph.

The phenomenon of sub-synchronous oscillations is basically associated with synchronous machines and can be of three types:

1. The classical form of subsynchronous oscillation is subsynchronous resonance (SSR) and it occurs when a natural frequency of a series compensated transmission system interacts with a natural frequency of the turbine prime mover.
2. The second type of subsynchronous oscillation is called subsynchronous control interaction (SSCI). Such SSCI oscillations are a result of control interaction and not from the electrical resonance from SSR. The problem in such cases is primarily due to the use of mechanically derived speed signals in the feedback control system.
3. A third concern with regard to torsional modes is of sub-synchronous torsional interactions (SSTI) which originate because of other transmission equipment such as SVC, HVDC.

The power transmitted by HVDC system is in general constant current or constant power control scheme, which is independent of frequency. Therefore, the HVDC rectifier tries to maintain the constant value of the current through current controller amplifier (CCA) function. The current error of HVDC and accelerating power of generator are related to each other's closed loop control. The HVDC current order may change due to fault on converter end or fault on inverter end. To keep the error in current at HVDC terminal to be zero, the generator has to accelerate/decelerate accordingly. Any disturbance in the system resulting in current error will have a wide frequency range, and so it includes frequency in SSR range. If this frequency in SSR range is excited in the HVDC control system, then the accelerating power reflected to the generator rotor is also in the same frequency. If this frequency matches with the resonance frequency or the natural modes of the turbine shaft, then it will result in SSR. Therefore, HVDC current control can cause sub-synchronous torsional interaction with rectifier end connected generators. The gain of the constant current regulator will increase when the control angle (α) increases, so generators near to HVDC are prone to sub-synchronous oscillation at larger α, for example, when HVDC is running at lower voltage.

Two oscillation cases on sub-synchronous torsional interaction were observed in a generating plant connected to the converter terminal of HVDC system under two different scenarios. The first case was when multiple commutation failures occurred in HVDC during faults near to the inverter substation under inclement weather. These faults were reflected on the converter end of the HVDC as dip in the voltage and rise in the current. With these commutation failures, HVDC power flow was also varying and the current controller was trying to keep the power constant. As explained above, the HVDC current order would change due to fault on either converter end or on inverter end, and to keep the error in current at HVDC terminal to zero, the generators have to accelerate/decelerate accordingly. During these faults, voltage dip is observed, and in order to keep the current same, α has to be increased, and with this, SSTI was observed in the grid near to converter terminal from PMUs.

Figure 8.4 shows the converter end AC side voltage where the generator is connected which have experienced the sub-synchronous oscillation. Subsequently, the SSTI damping controller got activated; however, the extent of SSTI was high and even after damping controller activation it has resulted in tripping of both poles as per the design with subsequent SPS operation leading to tripping of some of the designated units. In principle, damping of the sub-synchronous oscillations due to SSTI can be achieved by modulation of the current order or the firing angle, which gets activated after detection of SSTI after recovery of the fault. Figure 8.5 shows the zoomed plot of the SSTI as measured from the digital fault recorder device from the field at the inverter end of the HVDC terminal. The frequency of the sub-synchronous oscillation in general ranges from 10 to 40 Hz. In this case, it was observed to be around 6.69 Hz from the RMS values of electrical parameter from the disturbance recorder. Further from the PMU installed at the converter terminal of the HVDC, SSR frequency comes out to be 6.69 Hz as shown in Fig. 8.6. This mode was observed in frequency, voltage and current measured from PMU.

A thorough analysis on the available data was done by the HVDC team and system operator for this event and some interesting facts came out. The SSR frequencies deduced from the AC side instantaneous voltage by the damping controller were 15.55 Hz and 11.15 Hz. This can be seen in Figs. 8.7 and 8.8. These two frequencies were superimposed on each other, and thus, the average frequency observed was 13.35 Hz. However, the same was observed as 6.69 Hz frequency when measured on the RMS values of parameters from fault recorder as shown in Figs. 8.9 and 8.10.

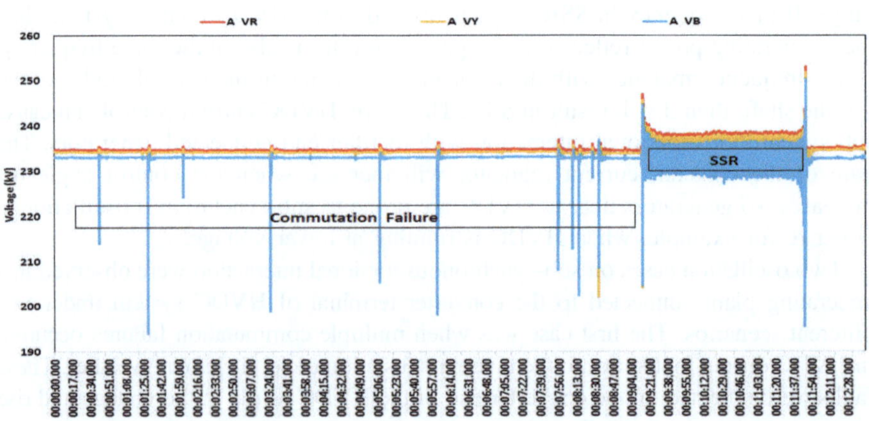

Fig. 8.4 Voltage of AC side on the HVDC converter end. Multiple commutation failures followed by SSR can be observed

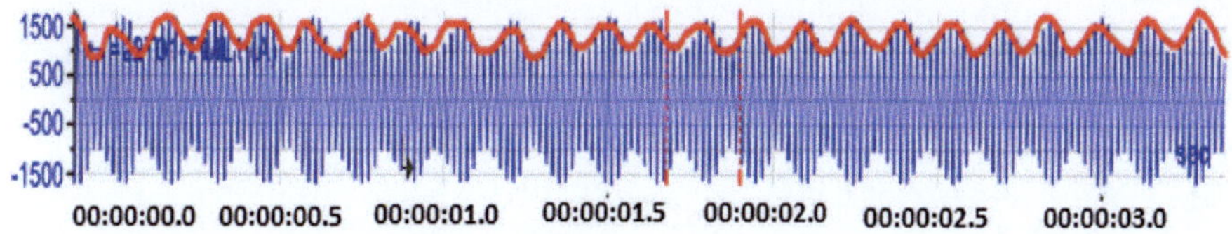

Fig. 8.5 DFR from the inverter terminal of HVDC which provides 6.69 Hz as SSR frequency

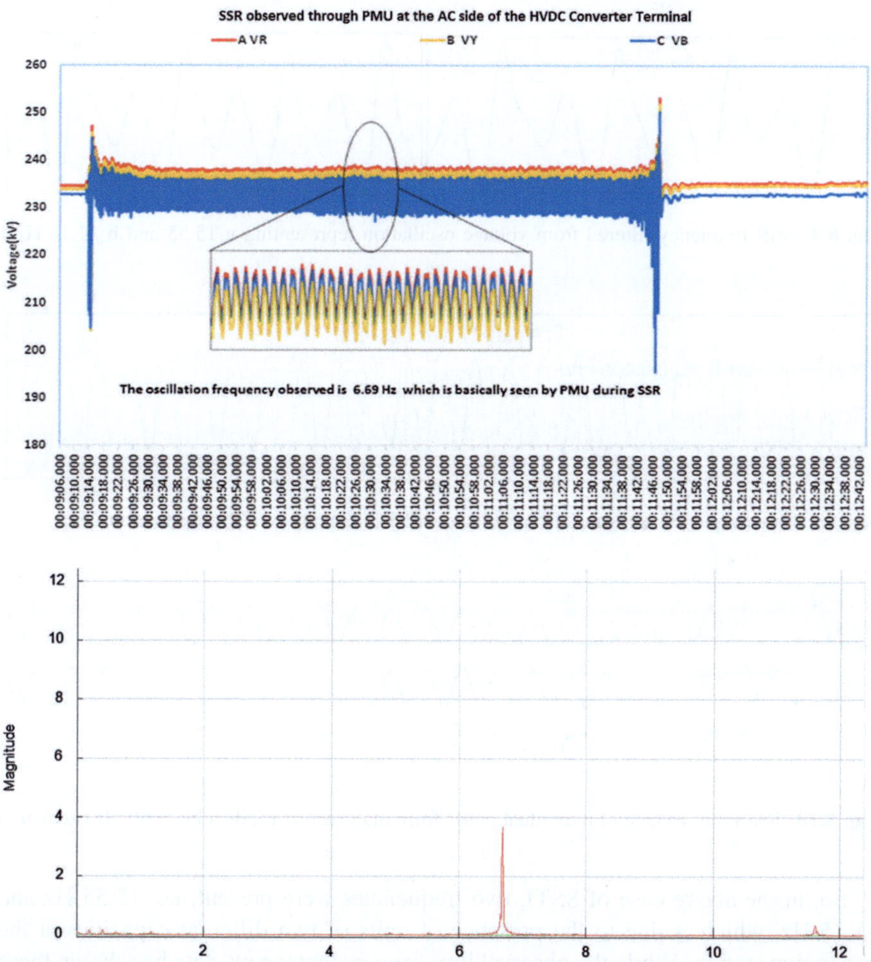

Fig. 8.6 Voltage of AC bus of converter terminal of HVDC and its Fourier transform indicating 6.69 Hz oscillation

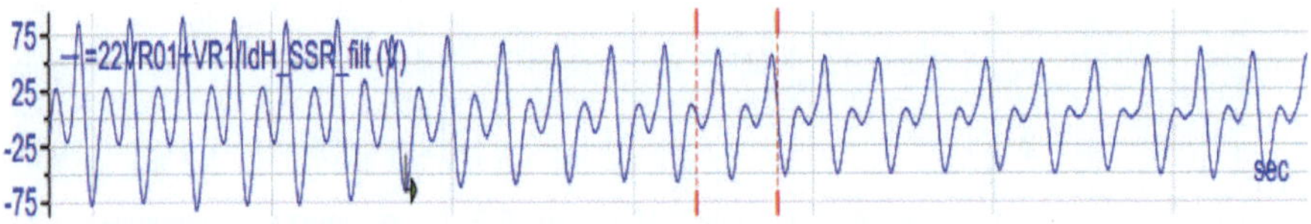

Fig. 8.7 SSR frequency filtered from voltage oscillation indicating presence of two frequencies superimposed on each other

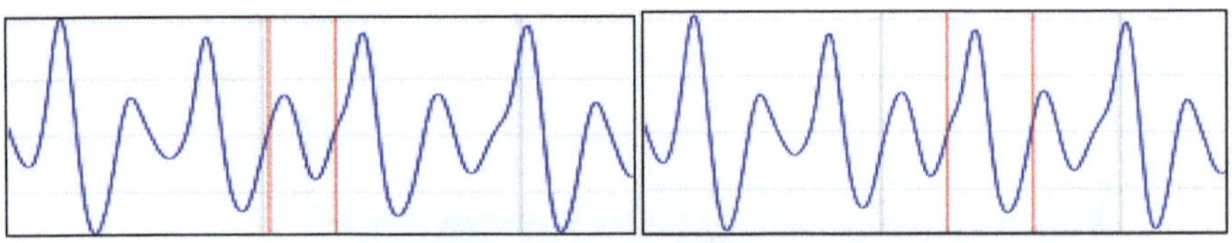

Fig. 8.8 SSR frequency filtered from voltage oscillation representing **a** 15.55 and **b** 11.15 Hz

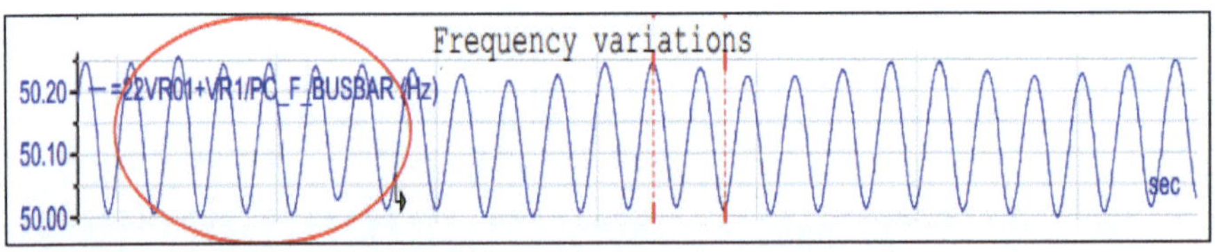

Fig. 8.9 Bus frequency measured from fault recorder indicating 6.69 Hz oscillation

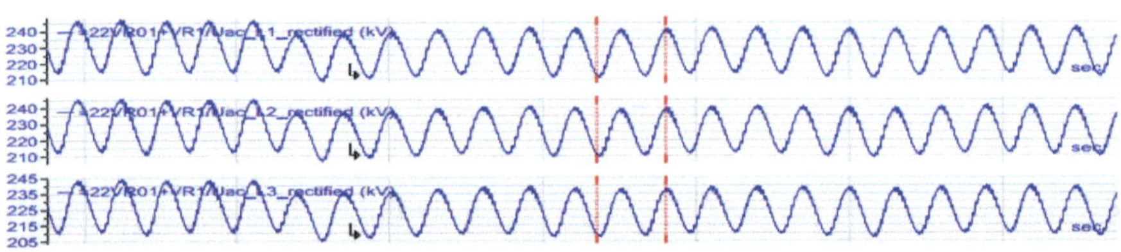

Fig. 8.10 RMS bus voltage of individual phase from fault recorder indicating 6.69 Hz oscillation

So, in the above case of SSTI, two frequencies were present, i.e. 15.55 Hz and 11.15 Hz, which is due to the presence of units of two different capacities at the generating station. While the observability from instantaneous data has shown these two frequencies, the RMS quantities were showing only one frequency which was half of the average frequency of these two modes. The same frequency was measured from fault recorder and PMU data also. Such type of measurement-related changes and their reason need to be further analysed from the signal processing techniques.

The second case of oscillation appeared when HVDC was running at reduced power order (1000–1200 MW), and there was a low generation of 300–400 MW at

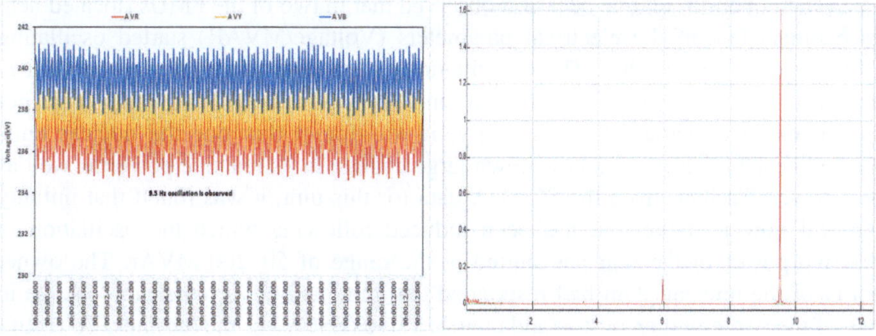

Fig. 8.11 Voltage of AC bus of converter terminal of HVDC and its Fourier transform indicating 9.5 Hz oscillation

the generating station with only one type of unit running out of the two different capacities of units present there. During this period, sub-synchronous torsional oscillations were observed on several occasions. The reason for sub-synchronous oscillation during low generation without any power system fault could not be ascertained. The oscillation frequency which was observed from Synchrophasor data was around 9.5 Hz (Fig. 8.11). The frequency of actual oscillation can be either 9.5 Hz or higher frequency like 15.5–34.5 Hz based on aliasing observed in Synchrophasor. It can be seen that 15.5 Hz oscillation was present in the earlier case also. This suggests the presence of 15.55 Hz oscillation in present scenario when only one type of unit is running.

Learnings from the above two cases of SSTI can be summarized as follows:

- From the above two cases, it was found that two different SSR frequencies are present when two different capacity units are running at the generating station. However, only one frequency is present when only one type of unit is running at the station. The oscillatory mode corresponding to one type of unit is 15.5 Hz and for other 11.15 Hz.
- Whether PMU can measure SSTI or SSR oscillation is a matter of discussion. PMU, which is reporting at 25 samples per second, can measure the oscillation up to 12.5 Hz as per the Nyquist criteria. Any higher frequency beyond 12.5 Hz will be reported as 25-F_o in the Synchrophasor due to aliasing. So a SSR having frequency of 15 Hz will be observed in PMU as 10 Hz.

C. Low-Frequency Oscillation at Nuclear Power Station

The oscillation due to the malfunction of emergency stop and control valve of nuclear power plant has been observed in the Indian grid in the past [1]. It was attributed to the failure of the relay governing the valve. Further, one more case of low-frequency oscillation was observed at a nuclear station where such behaviour was observed. This had resulted in wide variation in the power flow of the nuclear unit. This case gives a good overview of observability aspect from the Synchrophasor during the oscillation in the grid.

System operator on one occasion observed that in two of the PMUs (located near each other) few of the electrical parameters (Voltage/MVAR) started oscillating with respect to each other. These oscillations were observable only in the voltage and reactive power at these two locations. Real power and frequency have not shown much of variation. On enquiry, it was found that one the nuclear units located near to these nodes has shown large change in its reactive power as well as real power. Further, from the SCADA data for this unit, it was found that initially the real power of the unit has been reduced following which the oscillation in reactive power of the unit has started in the range of 50–200 MVAr. The owner entity of the nuclear plant had responded that they had observed steam leakage in one of the two emergency stop and control valve (ESCV) of the turbines. To rectify the steam leak process, isolation of this ESCV was required. Hence, turbine output was reduced to 350 MW in order to isolate the ESCV where leakage was observed. After isolation of ESCV, the entire steam load was shifted to the other ESCV, and with this, hunting has started in the unit. This hunting was damped after increasing the generation to 440 MW.

Figures 8.12 and 8.13 show the four parameters, i.e. voltage, frequency, active power and reactive power during the oscillation. It can be seen that the observability of this mode was more in voltage and reactive power. In addition, the observable oscillatory frequency was 0.2 Hz (an inter-area mode) but observable at only two nearby nodes to generators only. Therefore, the controller behaviour in the nuclear unit has excited the 0.2 Hz which is a global mode but the observability of this mode was only local.

D. Low-Frequency Oscillation in the North-Eastern Grid

Cases of low-frequency oscillations due to switching operation of transmission elements have been observed in the Indian grid, prominently in regions with weak interconnection and having small generating units (50 MW and below). The low-frequency oscillation modes in these regions get excited on account of nearby

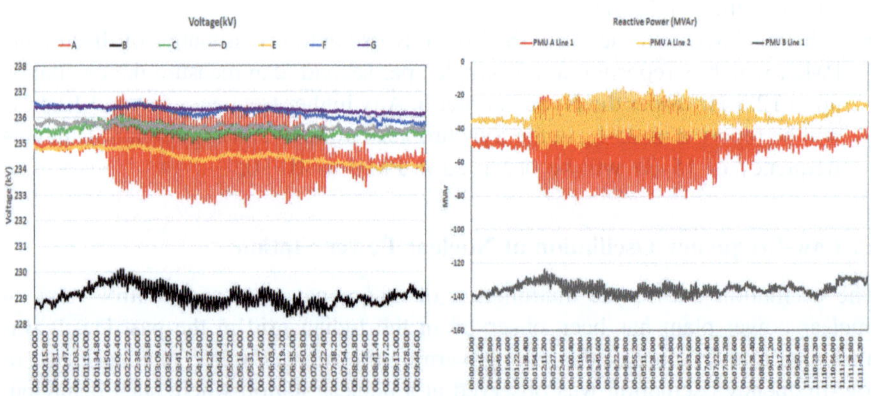

Fig. 8.12 Voltage of nodes in the grid and reactive power of lines from the nodes where maximum magnitude in voltage oscillation are observed

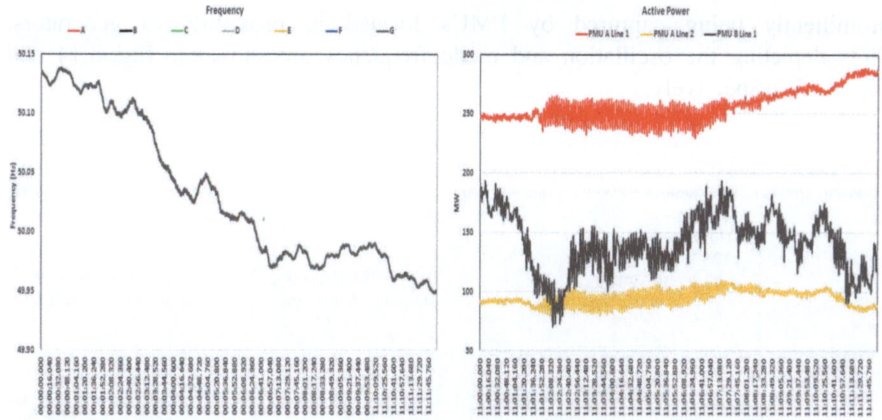

Fig. 8.13 Frequency of nodes in the grid and real power of lines from the nodes where maximum magnitude in voltage oscillation are observed

switching operation and resulting malfunction of generator controls. In absence of PSS tuning, the oscillations do not die out fast due to insufficient local damping. In the first case study, it was shown that switching operation can have effect on generation control and result in controller malfunctioning. Such oscillations on account of switching operations have been prominently observed in the North-Eastern Regional (NER) grid of the Indian Power System. A description of one such event observed in North-Eastern grid in India is explained in this case study.

Small hydro-generators having unit sizes of 50 MW and below are connected through 132 kV tie lines to load centres in the North-Eastern region having low fault levels. The area surrounding these units and the load centres is known to be prone to bad weather conditions, resulting in tripping of one or more transmission elements. Further, damage to protection equipment on account of the lightning surge and high earth resistance on account of prevailing soil conditions has also been reported from this area.

The hydro-generating plants A, B, C and D were generating 44 MW, 96 MW, 39 MW and 19 MW, respectively, where generating plants C and D are connected to the same bus. At this point in time, a nearby 132 kV transmission line supplying radial load developed a fault and the fault had a delayed clearing. This incident triggered low-frequency oscillations modes around 1.2 Hz (inter-plant) and its harmonic mode at 2.4 Hz (intra-plant). The units at generating stations A, C and D started hunting prominently, and the phenomenon was visible to power system operators at generating stations as well as regional control centre.

The oscillations were prominent in frequency and active power flows rather than voltage and reactive power flows. The oscillations continued for a total duration of about 8 min in the grid and were captured by all PMUs of NER grid, most prominently being captured by PMUs located in proximity to generators. Plots depicting the oscillation and mode frequency are shown in Fig. 8.14 and Fig. 8.15, respectively.

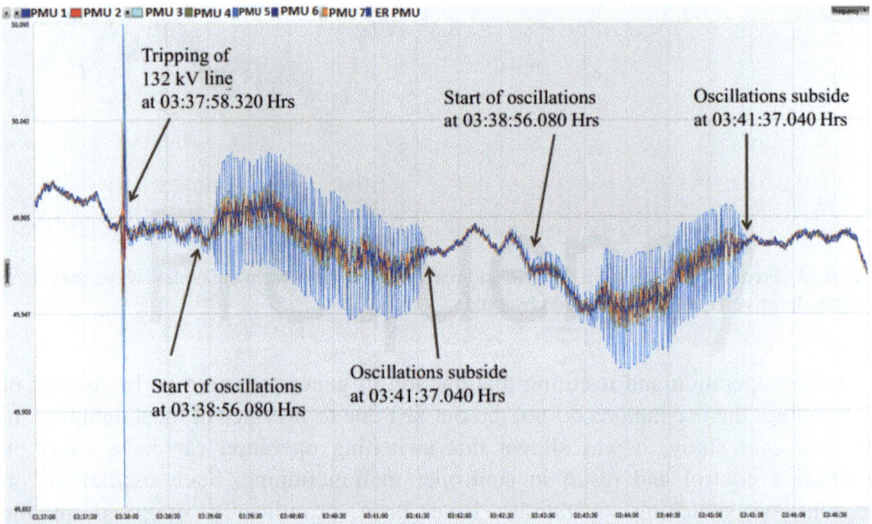

Fig. 8.14 Frequency plot from PMUs showing oscillation from PMUs of NER and ER

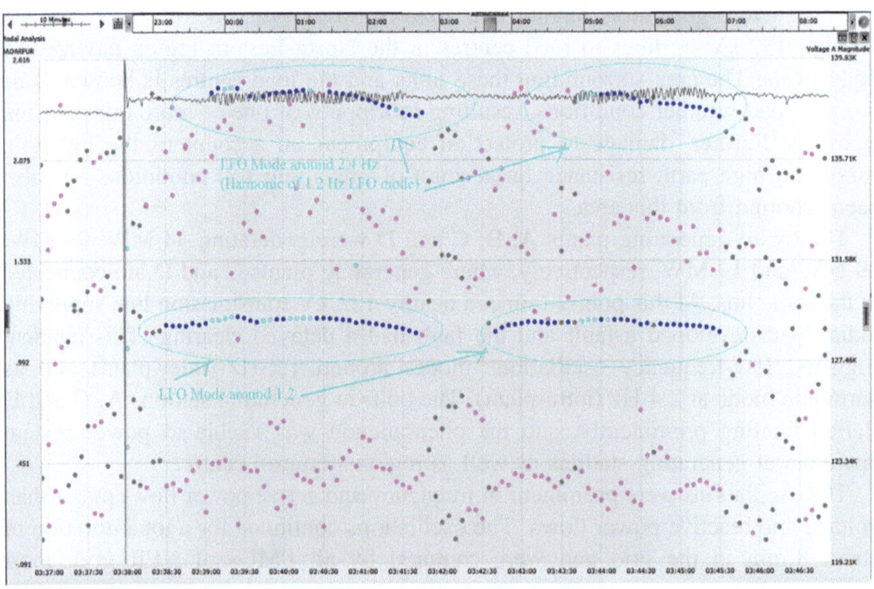

Fig. 8.15 Modal analysis plot from PMU at 132 kV node near to generating station in NER

These oscillations have resulted in tripping of units at generating stations A and C, after which the oscillations continued between units at generators B and D. This was also captured from SCADA plots shown in Fig. 8.17. In Fig. 8.15, good observability of oscillations in active power flows on the lines in NER can be observed. While in Fig. 8.17 it can be observed from SCADA plot that even the oscillating units at generating stations A and C got tripped, but oscillations kept continuing between units at generators B and D.

While the exact reason behind these oscillations is yet to be indicated by respective generators, it is clear that in the absence of adequate damping torque in the grid has led to prolonged oscillations which got damped out only after some of the participating units had tripped causing change in the system equilibrium point (Fig. 8.16).

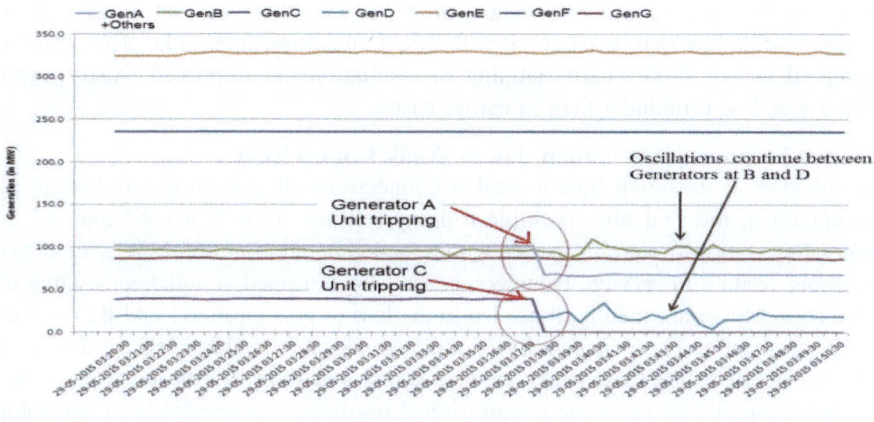

Fig. 8.16 Plot of generation from SCADA

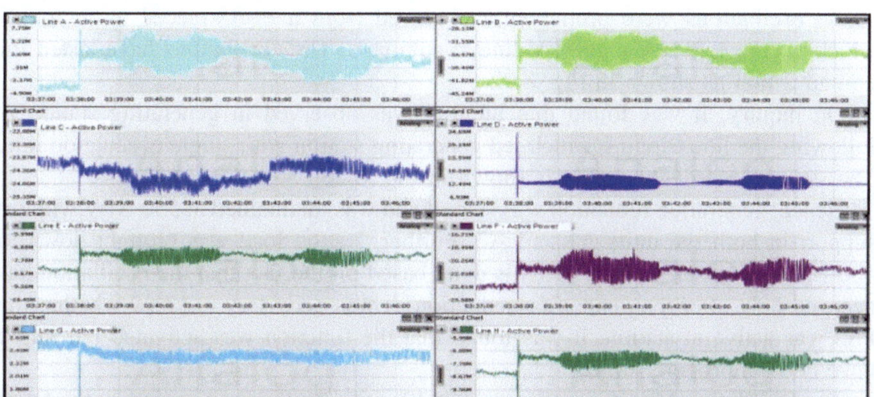

Fig. 8.17 Plot of active power flows from NER PMUs

E. Low-Frequency Oscillation and Impact of PSS Tuning

In this case, the power swing and small-signal stability issues were addressed for a generating station where units were connected at two different voltage levels. The interconnecting lines from these voltage levels were having large difference in electrical angles. The schematic diagram of the generating station is shown with the help of Fig. 8.18.

In case of contingency arising due to loss of evacuation lines from one voltage level, large swings were observed on evacuation lines at other voltage level based on Synchrophasor data from nearby PMUs supported by disturbance recorder from the station. Further, Synchrophasor measurement also provided that during such conditions, oscillations were being observed in system frequency which were inter-plant in nature as shown in Fig. 8.19.

Based on this input, SPS scheme was modified to take care of power swing criteria and PSS of generating units was tuned to take care of small-signal stability. After the PSS tuning and SPS enhancement, it was observed that under contingency, oscillation damped out very fast and had low amplitude. This can be observed in Fig. 8.20 where damping of oscillation has improved. Also, power swing was less prominent than in earlier cases.

F. Low-Frequency Oscillation due to Weak Connectivity

In this case, it is shown how a weaker connectivity of generator could result in oscillation in the grid after multiple tripping/contingencies. Units A1 and A2 of generating station A were in service with a total generation of 865 MW with all six outgoing feeders in service. However, several lines tripped simultaneously due to protection trip; consequently, it has led to the loss of generators B1 and B2 on over speed protection. The connectivity of the stations prior to this event is shown in Fig. 8.21. After these lines which are connecting generator A with the grid, i.e. A–C double circuit and A–D one circuit tripped resulting in connectivity of generator through A–D single circuit only which is also shown in Fig. 8.21.

Immediately following the line tripping which left generating station A with single evacuation line, oscillations were observed throughout the regional grid. These oscillations were prominently observed from a PMU placed at station G which is connected by a 400 kV line to station E. The oscillations have lasted for around 2 min as shown in Fig. 8.22.

On inquiry, it was found that hunting was observed in generating station A following the line tripping which led generating station with single evacuation line. The station recorder snapshot is shown in Fig. 8.23.

Hence, generation reduction was started and was maintained at 550 MW with oil support in both the units A1 and A2. Further, station load was brought down to 400 MW for better stability. This is done based on the fact that the oscillation has decayed only after the generation reduction. It was necessary to know whether the PSS was active at station. It was found that the function was activated but might have been improperly tuned.

The Oscillation Monitoring System (OMS) utilizing the multi-prony method was used to analyse the above oscillations and to identify the dominant modes using the

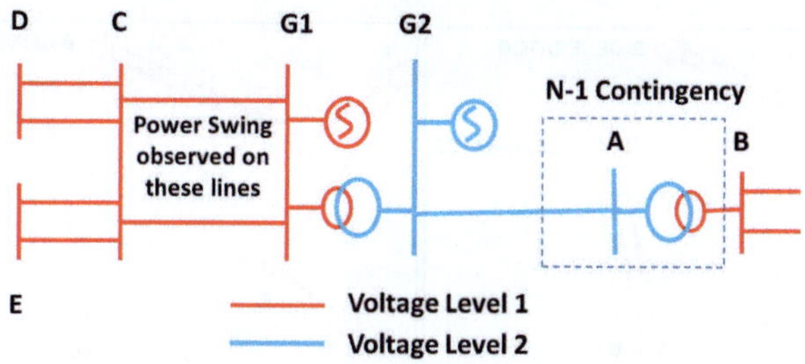

Fig. 8.18 Schematic diagram of the N-1 event causing power swing and oscillation

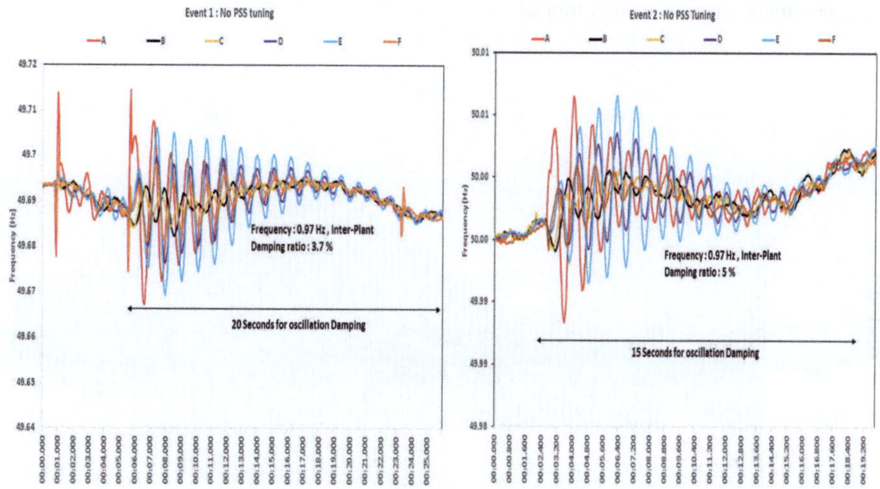

Fig. 8.19 Inter-plant oscillation at thermal station due to inadequate damping in the system

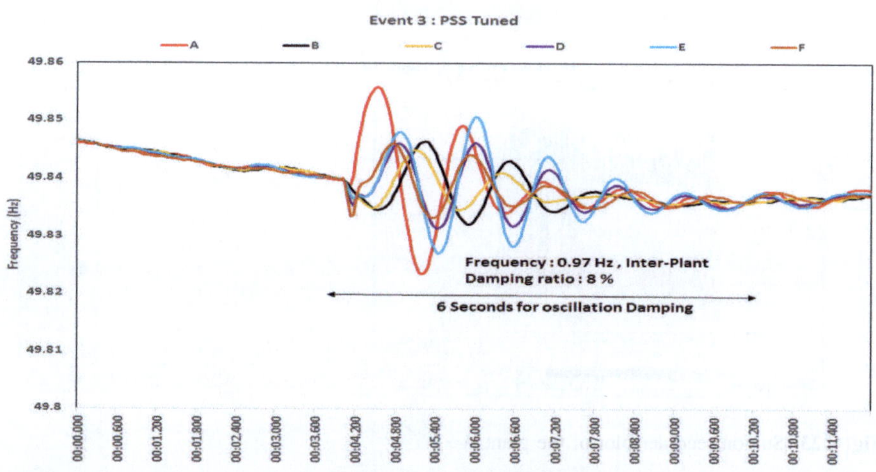

Fig. 8.20 Inter-plant oscillation damping improvement after PSS tuning

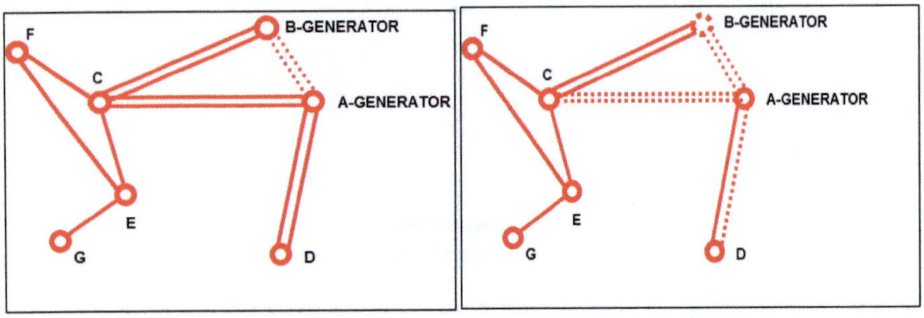

Fig. 8.21 On the left is the connectivity when units at generator B tripped, while on the right is the connectivity when units at A tripped

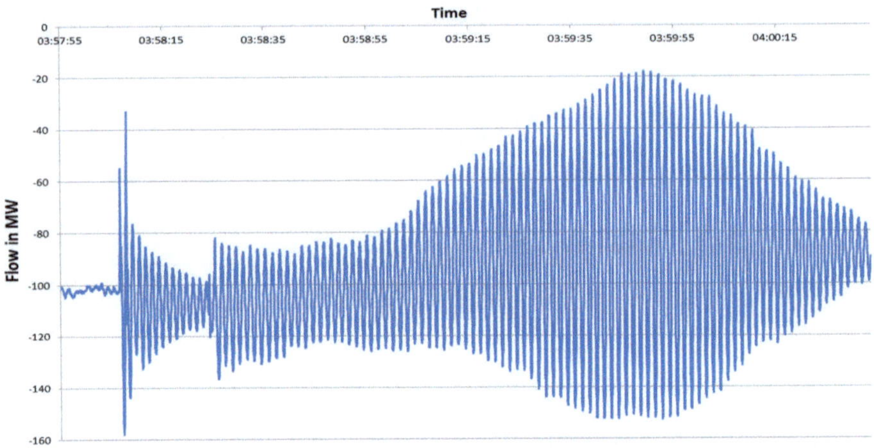

Fig. 8.22 PMU plot of active power flow in the 400 kV line G–E

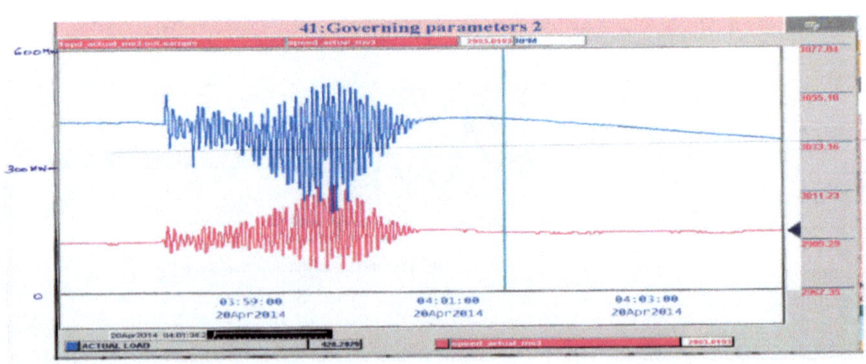

Fig. 8.23 Station recorder plot of the plant A

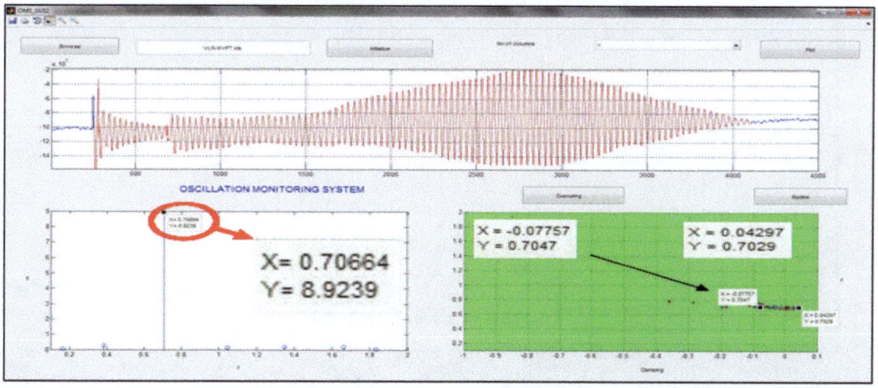

Fig. 8.24 Snapshot of OMS engine during the event indicating presence of 0.7 Hz oscillatory mode

Synchrophasor data [1]. The analysis was performed on the PMU data from station **G** and the frequency of the dominant mode of oscillation was 0.7 Hz which can be observed in Fig. 8.24. The damping ratio changed from −1 to +1.7% over the period.

The above case study has shown that a weak connectivity with grid and lack of PSS tuned response may result in oscillation in the generating the station. Further, it also provides a feedback to check for such connectivity issues at all generators in the Indian grid where a similar event could occur. The recommended action in order to mitigate such oscillation is to tune the PSS and implementing better SPS scheme for multiple contingencies.

G. Very Low-Frequency Oscillation due to Hydro-Power Plant

The case of oscillation arising during start–stop sequence in a hydro-unit has also been observed in Indian grid. This is one unique case where oscillation in hydro-units were observed by the system operator, analysed and then corrected with the help of generating station.

In this case of oscillation, it was observed that while desynchronizing the hydro-unit from the grid, the voltage plots of PMU located near the particular generating unit were observing the oscillation. The oscillation observed was of very low-frequency nature, i.e. 0.066 Hz, and the oscillations were present in all three phases of voltage in the nearest PMU as shown in Fig. 8.25. Since these oscillations were occurring repeatedly while one of the units was being desynchronized at the particular hydro-power plant, details from that generating station were sought. On investigation by the generator, it was found that improper pulse from PLC during the stop sequence was causing variation in reactive power thereby causing oscillations in voltage.

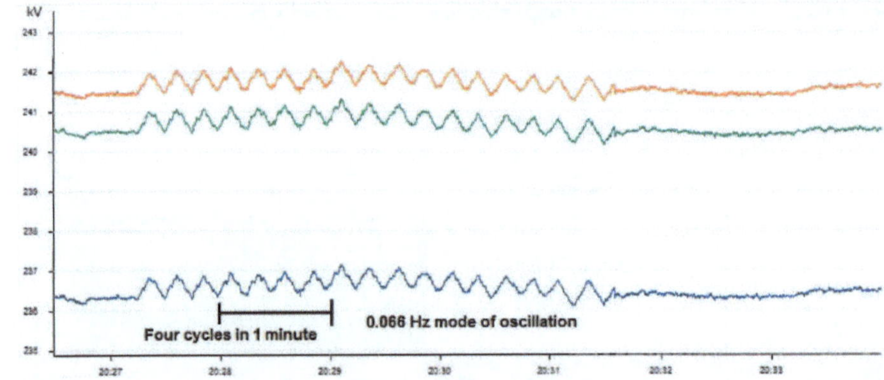

Fig. 8.25 Oscillation in the three-phase voltage of the node which is located nearest to hydro-power plant from PMU device

The generating station informed that when the following given conditions are met during stop sequence for any unit, then it automatically gets desynchronized from grid

1. Active power less than 9 MW
2. Reactive power within +3 MVAr

During automatic stop sequence execution, the PLC has been programmed to reduce the active and reactive power of the generating unit which is to be desynchronized. The PLC will give a pulse at regular intervals to reduce the active and reactive power to governor and excitation, respectively. It was observed that oscillation phenomenon was occurring when the stop command was executed for MVAR value beyond +3 MVAR (second condition not satisfied). It was observed that the automatic PLC sequence and measurement system were responding with some time gap of around 4–5 s delay. Due to this, AVR got several increase/decrease MVAR command which in turn led to large variations in MVAR (around +20 MVAR max) as shown in Fig. 8.26 from generator data acquisition system. The oscillatory frequency observed was also matching, i.e. 4 cycles/min = 0.066 Hz, as shown in Fig. 8.26.

Such large variation in MVAR was observed in the grid as oscillation in voltage, as captured by nearby PMU plot. The PLC pulse logic was modified accordingly, and the modification helped in eliminating the oscillations during the desynchronization as shown in Fig. 8.27.

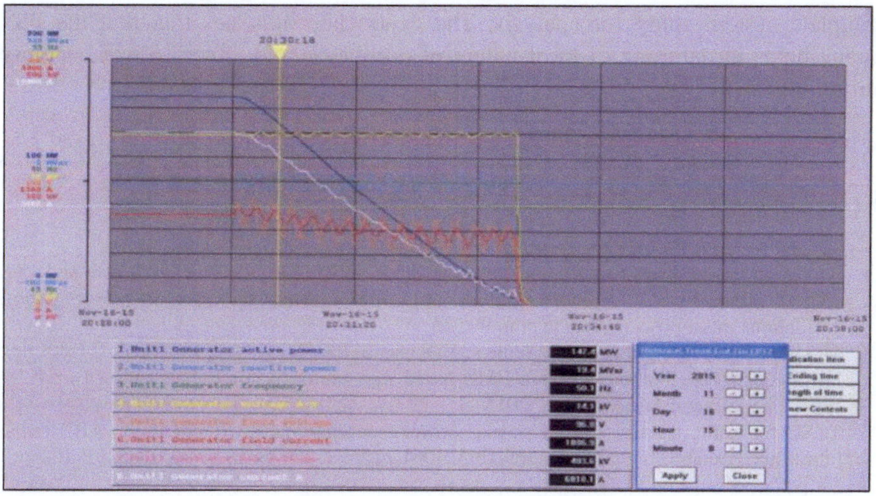

Fig. 8.26 Oscillations observed during desynchronization of unit as observed from generator SCADA (before modification in PLC pulse logic)

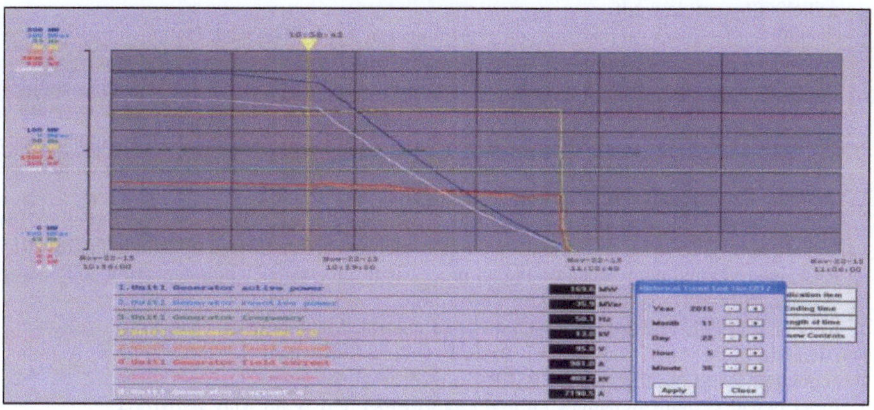

Fig. 8.27 Smooth desynchronization of unit (post-modification in PLC pulse logic)

8.3 Conclusion

In this chapter, various cases of small-signal stability have been discussed to familiarize the readers with the low-frequency oscillations occurring in Indian power system and how Synchrophasor measurements have helped operator in their observability and their mitigation after analysing the data. Based on the analysis, it can be seen that many of the local oscillations can be brought to conclusion; however, the same is not possible for inter- or intra-area mode as these require complete system study for analysis. The experience indicates that that the PSS tuning helps in damping of local nature of oscillation and affects system stability during transient.

References

1. Report on power system oscillations experienced in Indian Grid on 9th, 10th, 11th and 12th August 2014. POSOCO, India. Online: https://posoco.in/download/report-power-system-oscillations-india-posoco-final/?wpdmdl=535
2. Mukhopadhyay S, Agrawal VK, Soonee SK, Agarwal PK, Anumasula R, Kumar C Detecting low frequency oscillations through PMU-based measurements for Indian national grid. In: IEEE PES General Meeting, July 2015
3. TCSC for stable transmission of surplus power from eastern to western India, ABB India. https://library.e.abb.com/public/1e36fc1bf524064ec125787800258583/A02-018520E_Raipur_LR.pdf
4. Nayak RN, Vikas S, Sehgal YK, Subir S, Manju G (2006) Testing of TCSC damping controller installed in a tie line interconnecting two large areas—a case study. In: CIGRE 2006 session
5. The biggest FACTS project worldwide—the FSC and TCSC Purnea Gorakhpur project in India was put into commercial operation in 2006, Siemens. http://www.ptd.siemens.de/artikel0606_tcsc_low.pdf
6. Operational feedback on transmission constraints Jan 2014, POSOCO. Online: https://posoco.in/download/nldc_operational-feedback_january-2014/?wpdmdl=7215
7. Operational feedback on transmission constraints Apr 2014, POSOCO. Online: https://posoco.in/download/nldc_operational_feedback_april-2014/?wpdmdl=7216
8. Pentayya P, Pandey V, Anumasula R, Kumar C, Patil S, Chitturi S (2014) Spontaneous oscillations and modal resonance in Indian grid: a case study. In: Power systems conference (NPSC), 2014 Eighteenth National, pp 1–6, 18–20 Dec 2014
9. Mukhopadhyay P, Seshadri P, Gangrekar C, Kumar C, Rai AB, Reddy NM (Feb 2015) WAMS and synchrophasor experience during synchronization of large grids in India. In: 2015 IEEE international conference on signal processing, informatics, communication and energy systems (SPICES), pp 1–5, 19–21
10. Kundur P (1994) Power system stability and control. McGraw-Hill, New York
11. Alvarado F, DeMarco C, Dobson I, Sauer P, Greene S, Engdahl H, Zhang J (1999) Avoiding and suppressing oscillations. In: PSERC project final report
12. EPRI power systems dynamics tutorial (2009) EPRI. CA, Palo Alto, p 1016042
13. Wang XF, Yonghua S, Malcolm I (07 June 2010)Modern Power Systems Analysis, Springer Science & Business Media
14. Grigsby LL (25 Apr 2012) Power system stability and control, CRC Press
15. Bijaya KP, Ajith A, Swagatam D Computational intelligence in power engineering. Springer Science & Business Media, 20 Sept 2010
16. Debasish M, Abhijit C, Aparajita S Power system small signal stability analysis and control. Academic Press, 28 Apr 2014
17. Analysis and control of power system oscillations, CIGRE, Final report by the Task force 07 of Advisory Group 01 of Study Committee 38, Dec 1996
18. Robust control in power systems. (ed) Bikas P, Balarko C Springer, US
19. Pentayya P, Rajkumar A, Kumar C, Chakrabarty R, Patil S, Chitturi S (Dec. 2014) Modal resonance in power system-a case study, in Power Electronics, Drives and Energy Systems (PEDES), 2014 IEEE International Conference on, vol., no., pp. 1–6, 16–19 Dec. 2014

Chapter 9
Experiences of Oscillation Detection and Mitigation in Grid Operations at PEAK Reliability

Hongming Zhang

9.1 Introduction

The Western Interconnection Synchrophasor Program (WISP) began with a collaborative project to improve reliability of the Bulk Electric System (BES) in the Western Interconnection.

In 2010, the Western Electricity Coordinating Council (WECC) received $53.9 million in funding from the US Department of Energy (DOE) for WISP. The funding, awarded under the American Recovery and Reinvestment Act's Smart Grid Investment Grant initiative, matched dollars committed by nine WISP partners to extend and deploy synchrophasor technologies within their western electrical systems. The total funding for WISP was $107.8 million.

Between 2010 and March 2014, WISP installed more than 480 new or upgraded phasor measurement units (PMUs) in the Western Interconnection. PMUs deliver high-speed synchronized measurements using a secure communication network and software in order to better manage the power grid. Under the WISP project, WECC Reliability Coordinator (RC) installed many state-of-the-art synchrophasor applications and tools, such as

- OSI soft PI for PMU Historian
- V&R Energy ROSE Real-Time Voltage Stability Analysis (RT-VSA) software
- Alstom PhasorPoint platform integrated with Montana Tech developed Modal Analysis Software-MAS 1.0

H. Zhang (✉)
PEAK Reliability, Loveland, CO, USA
e-mail: hzhang@peakrc.com

S. Nuthalapati (ed.), *Power System Grid Operation Using Synchrophasor Technology*, Power Electronics and Power Systems,
https://doi.org/10.1007/978-3-319-89378-5_9

- Alstom Grid Stability Assessment (GSA) application to transfer synchrophasor application results to EMS
- Alstom OpenPDC and e-Terravision (replaced by other products in 2016).

The new synchrophasor infrastructures and advanced applications enabled RC playing a leadership role in promoting new technologies for visualizing power system disturbances, and minimizing the risk of these disturbances from evolving into a major, system-wide event on the Western BES.

Since WISP's phase-1 completion in March 2014, RC continued to improve the quality and use of the synchrophasor data it receives. Peak Reliability applied for an additional grant from the DOE for this next phase of WISP, subsequent to the bifurcation of WECC into a Regional Entity (WECC) and a Regional Coordinator (Peak Reliability). The new DOE grant project is named Peak Reliability Synchrophasor Program (PRSP), which was widely considered WISP 2.0. The PRSP project was budgeted at $12.4 million. It consists of nine partner entities: BPA, CAISO, Idaho Power, NV Energy, PacifiCorp, PG&E, Salt River Project, Southern California Edison, and Peak Reliability.

Through the PRSP project, Peak Reliability plays an important role in the program by housing the PMU registry, MAS 2.0 software enhancement, historical data archives, PMU data accuracy and event data export tools, disturbance reports, linear state estimation, inter-area oscillation mode meters/forced oscillation detection, as well as the wide-area visualization tools.

Peak Reliability currently receives inbound 4000 PMU signals of 412 PMU from 18 WECC utilities. Most of the PMU are down-sampled sent out to SCADA. Hundreds of down-sampled PMU voltage phasors are enabled in EMS state estimation solution.

The oscillatory characteristics of power systems due to cyclic load variations were first uncovered in mid-1960s [1]. Since the 1990s, BPA has been one of pioneering utilities in development and implementation of system modal oscillation detection technology using high-resolution PMU signals in collaboration with Montana Tech, University of Wyoming, and PNNL et al. research institutions [2–4]. In the last decade, forced oscillations in power systems are caused by external sources such as cyclic loads and control failures in generation power plant. The identification of these forced oscillations has been challenging for the industry. Because of increase in PMU visibility, there is significant work being done in detecting and locating forced oscillations [5–9]. Thanks to advancement of PMU-based oscillation detection technology over the last decades, now more utilities are moving forward to implement real-time oscillation detection applications in control room settings.

In the following sections, we will focus on Peak Reliability experience in implementation of inter-area mode meters, forced oscillation detection, and source locating.

9.2 Overview of Montana Tech MAS Tool

Peak Reliability currently utilizes the GE-Alstom PhasorPoint integrating MAS 1.0 for real-time system oscillation monitoring in Production. The MAS tool, including Oscillation Detection Module (ODM) and Mode Meter Module (MMM), was developed by Dr. Dan Trundnowski from Montana Tech University and Dr. John Pierre from the University of Wyoming. Peak Reliability MAS engine inherited most key features of BPA custom MAS tool that was developed and improved over the last decades [2, 3].

9.2.1 Mode Meter Functionality

The mode meter is designed to accurately track a known system mode. This purpose is different than detecting and estimating all system modal activity. The research focus for this type of tool is accuracy, and therefore, much effort has been expended in assessing and refining the most appropriate numerical and signal processing method for a known system mode under different stimulus conditions. For example, a numerical method that provides excellent accuracy while the system is being excited from a known external input may not represent the best algorithm when "probing" is turned off. In another example, an algorithm that performs well under low damping conditions may not provide the best accuracy under high damping conditions. The mode meter incorporates an expert system that selects the numerical algorithm delivering the best accuracy for the current system conditions.

The mode meter delivers a "mode estimate" as its result. The primary components of the mode estimate are estimates of the frequency, damping, energy, and shape of the system mode. Frequency and damping estimates can be used in conjunction with one another to assess system "stress". Operating rules can be put in place that, when used in conjunction with system topology information, can direct system operators to reconfigure flows. Below is a list of Western inter-area oscillation modes widely recognized.

Peak Reliability successfully integrated MAS 1.0 into GE-Alstom PhasorPoint (PP) product for real-time mode meter validation and monitoring in Prod (NS modes A and B) and Test (All known modes) (Fig. 9.1).

Oscillation Detector Functionality: The oscillation detector is designed to rapidly detect oscillation energy in input signals. The research focus is speed, and therefore, much effort has been expended in assessing the fastest filtering algorithms that will accurately compute oscillatory energy. The oscillation detector is intended to be used for broad system coverage. Most often, the oscillation detector will be used operationally to detect forced oscillations. Forced oscillations are more common, in normal power system operations, than oscillations resulting from undamped natural modes. Most forced oscillations are not a threat to bulk grid stability, but early detection of forced oscillations can help system operators/

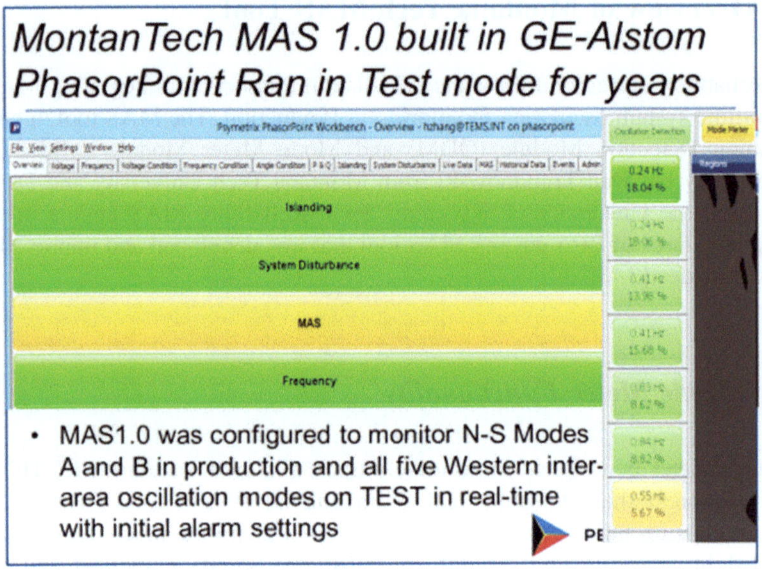

Fig. 9.1 GE-Alstom PhasorPoint/MAS displays

dispatchers to assess which of their generation assets may be experiencing pre-failure characteristics. For example, if an operator can be alerted of a forced oscillation at a given power plant, they may be able to start thinking about how they will reconfigure the system if that power plant trips in the next few minutes.

The primary output of the oscillation detector is oscillation energy in four broad frequency bands. The four broadbands help simplify the output for system operators, and the four bands can often be used to quickly find the source of a forced oscillation. The oscillation detector also provides, in its "Spectrum" output, finer frequency resolution of oscillation energy. We expect that the "Spectrum" output will be most often used by engineers in offline analysis, while the four broadbands described below will be most often used in operations.

- Band 1, with a passband of 0.01–0.15 Hz, monitors very slow oscillations that typically involve speed-governor controllers. The response time is 200 s or less.
- Band 2, with a passband of 0.15–1.0 Hz, is tuned to oscillations typically observed in the electromechanical oscillation range. The response time is 12 s or less. Most of known inter-area oscillation modes are within Band 2, including
- Band 3, with a passband of 1.0–5.0 Hz, is typically associated with local electromechanical modes and generator controls. The response time is 6 s or less.
- Band 4, with a passband of 5.0 Hz to Nyquist, may be associated with torsional dynamics of a generator, for example, or may be related to voltage or other relatively high-speed controllers.

Early versions of the Band 1 filter implemented in BPA showed false alarms for ramp conditions on the Pacific HVDC Intertie (PDCI). A new version of the Band 1 filter has been designed to reject ramps to avoid the false alarms.

MAS Analytics

The oscillation detection (ODM) algorithm is an "RMS Energy Filter" as derived and described in [2] and is shown in Fig. 9.2. This extends a classic energy detector [3] to multiple frequency bands. A PMU-derived "Input" signal (e.g., active power or voltage magnitude) is formed and passed thru a bandpass (BP) filter that focuses on the desired bandwidth for oscillation detection. After BP filtering, the signal is then squared, passed thru an "Averaging Filter"," and then square-rooted. The goal of the "Averaging Filter" is to estimate the mean of the squared signal and is matched to the BP filter. The resulting output signal will be an estimate of the RMS of the input signal in the bandwidth of the BP filter. A similar approach was used at BPA in the late 1990s for event detection.

The RMS energy filter approach has several distinct advantages as an oscillation detection algorithm [2]. Firstly, the BP filter can be designed to focus on a specific desired frequency range. Secondly, it provides useful engineering units for the output, that is, the total RMS content in the signal. Though it is often convenient to use units of peak-to-peak value of the oscillation energy when visually examining time-domain waveforms during an oscillation event, the RMS value is a more accurate and consistent measure. Lastly, it is not dependent on the oscillation having a single frequency; many oscillations have multiple frequencies (e.g., harmonics).

Four different RMS energy filters are implemented for the operation control application. The response time is the max time the RMS energy filter takes to respond to an oscillation.

MAS 2.0 Enhancements: With PRSP funding support, Dr. Dan Trundnowski and Dr. Matt Donnelly upgraded the MAS from version 1 to version 2 with many software improvements. The highlights are:

(1) An inverse time alarm (ITA) module is added. The ITA returns two Boolean outputs, "Alarm" and "Alert". The ITA is user-configurable to accept a single frequency band from a single OD as its input. For example, four ITA analytics must be configured for each OD if the user wishes to alarm on all four frequency bands of an OD. The parameters of the inverse time curve follow standard industry practice and are user-configurable.
Figure 9.3 contains a depiction of the inverse time curve capability. For any frequency band, the ITA will post an ALERT if oscillation energy exceeds the

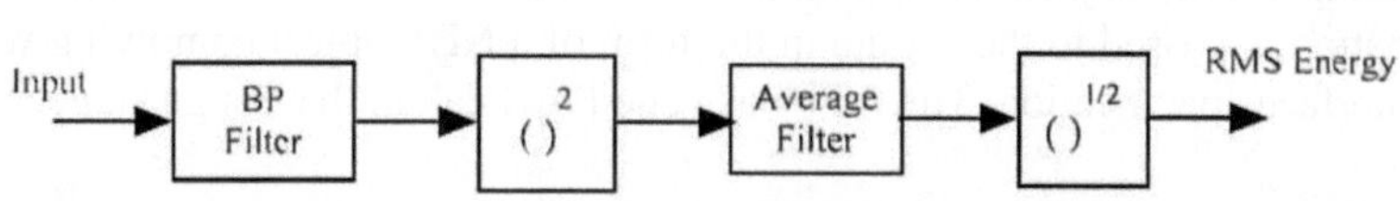

Fig. 9.2 RMS energy filter in oscillation detection application

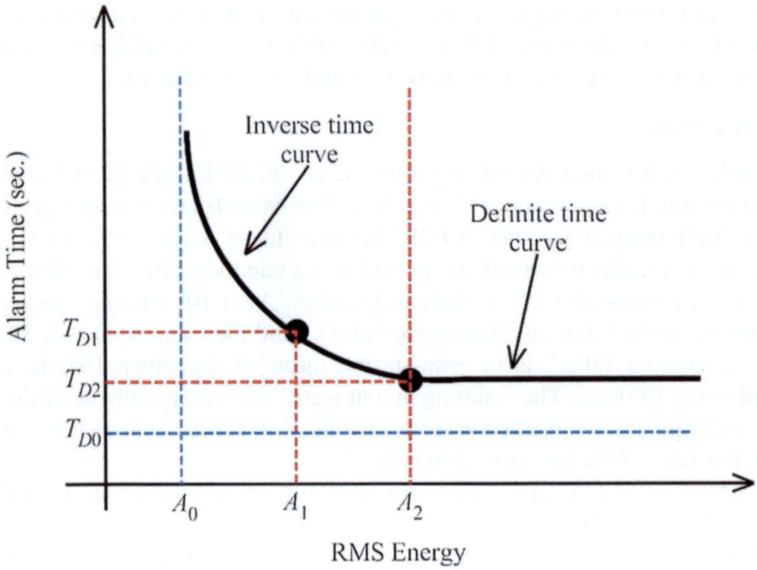

Fig. 9.3 RMS energy

A0 threshold and remains above the A0 threshold for T_{D0} s. If the oscillation energy and time duration in the band acting as the input to the ITA cross to above and to the right of the curve, the ITA will post an ALARM. It can be seen by examining the curve that the ITA will enter the ALARM state quickly if there is high energy in the band acting as the input to the ITA and more slowly if there is low energy.

(2) The mode meter has been significantly enhanced in MAS 2.0 by allowing for multiple input signals. In previous versions, a single instance of a mode meter could only accept one input. Users wishing to track a specific system oscillatory mode through multiple inputs needed to establish multiple mode meters. The problem with this approach comes in trying to determine which of the several result sets is most accurate for given system conditions. With multiple inputs, MAS 2.0 incorporates a decision structure that automatically excludes invalid data points. If one PMU stops reporting, MAS 2.0 is still able to provide accurate results from the other valid inputs.

(3) A transparent and public API was developed to facilitate incorporating the MAS 2.0 engine into online environments. Two significant features of the API are the logging framework described above as well as clearly defined.NET configuration objects and configuration validation. The configuration information is passed to the engine in the form of a.NET object complying with an interface specification. The API was vetted by Peak in the design stage and has

proven to be very flexible. The API has been successfully integrated into a commercial streaming synchrophasor service as well as into two separate offline analysis environments.

(4) A different numeric engine for MAS 2.0 was chosen. The MAS 1.0 software used an "unmanaged" numeric library. The use of "unmanaged" code in V1 resulted in one instance of a memory leak and required the design team to use a third-party interface library. The MAS 2.0 software uses 100% "managed" code which should provide users with a greater degree of confidence with respect to memory management.

(5) A set of tools making use of the MAS 2.0 engine was developed in MATLAB for offline analysis. The offline tools parse COMTRADE data files, provide the user with a sufficient degree of plotting and presentation capability, and are designed to facilitate automation. For example, the user can save configurations and run the stored configurations repetitively on any number of COMTRADE data sets. The offline tools use the same API approved by PEAK for the online engine. Therefore, the offline tool will provide results identical to the online version.

(6) A new derived signal type "Angle" as the raw angle of a synchrophasor input was created.

(7) A list of multiple derived signals can be used as inputs to a mode meter. The design team added logic that discards input signals with unusable content and uses the remaining input signals for multi-input mode meter processing. The goal is to provide a slightly more accurate answer, but more importantly to allow users to configure a mode meter with multiple inputs such that the mode meter automatically drops bad inputs. The alternative with v1 was to configure multiple independent mode meters.

(8) There is an additional "Spectrum" output for each band of the OD. The "Spectrum" output returns a scaled FFT of the energy content of the input signal within the band of interest. For example, the "Spectrum" output for Band 2 returns a spectrum from 0.15 to 1.0 Hz. The width of the frequency bins for the FFT is user-configurable. The benefit of the Spectrum output is better resolution of the primary frequency contributing to the oscillation energy in a given band. This feature adds lots of data to the result stream. It is anticipated that users will only retrieve the Spectrum output when the OD is in an alarm condition. It is anticipated that the "Spectrum" output will be most often used by engineers in offline analysis, while the four broadbands will be most often used in operations.

(9) The Band 1 filter was redesigned to reject ramps typical of routine operational re-dispatching of generating resources. Previously, the Band 1 filter would show significant energy content during a ramp and this was causing some false positives.

(10) The Band 4 filter was redesigned to extend to the Nyquist rate of the underlying derived signal input. Previously, the Band 4 filter had a cutoff at 15 Hz. This was appropriate for 30 sps PMU inputs but it unnecessarily limited the

usefulness of the Band 4 output when PMUs with 60 sps or 120 sps capabilities were used.

Peak has started to validate MAS 2.0 for offline system oscillation event analysis. The integration of MAS 2.0 into real-time application platform is under evaluation.

9.3 Peak's Experience with MAS Mode Meters (MMM)

BPA uses the same Montana Tech MAS tool as Peak in Real-Time, and a comparison of the real-time results was conducted to make sure Peak's PhasorPoint (PP) MAS MMM tool function as expected.

Results Comparison—Peak versus BPA

Several key dates of MAS MMM results are compared in the following sections, including PDCI trip event, brake test, and some normal days.

PDCI Trip Event on April 28, 2015:

On April 28, 2015, around 6:06 a.m. PDT, Pacific DC Intertie (PDCI) tripped. Both Peak's and BPA's MAS tool were running during that time. The NS mode A and B frequency and damping are compared and shown in Figs. 9.4 and 9.5.

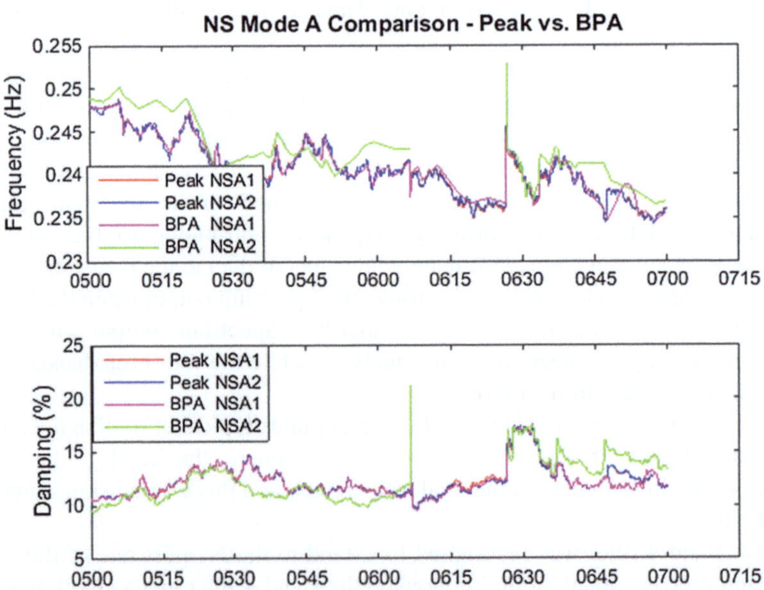

Fig. 9.4 PDCI trip event MAS Mode Meter NS mode A results comparison—peak versus BPA

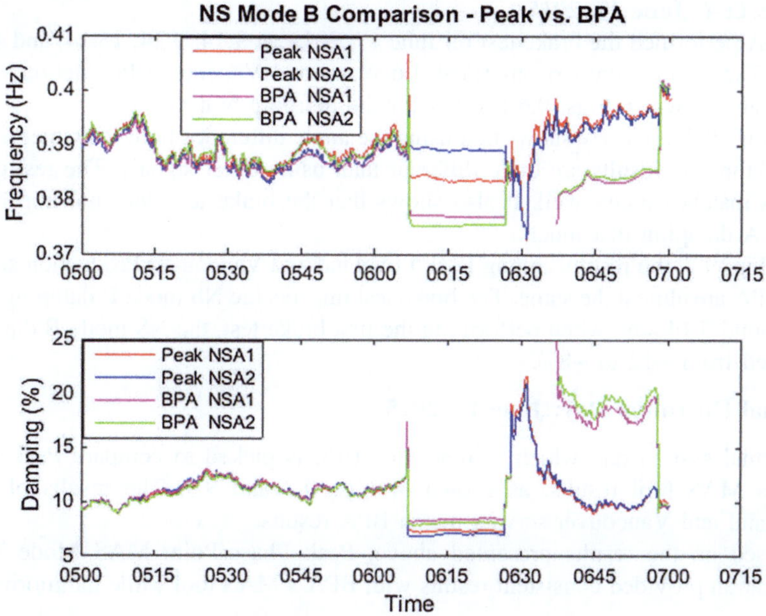

Fig. 9.5 PDCI trip event MAS mode meter NS mode B results comparison—peak versus BPA

As mentioned in the previous section, there is a difference in results of the second NS mode A, the second signals used for BPA is the angle difference between Malin and Chief Joseph, instead of Custer and Captain Jack. Chief Joseph PMU measurements were not configured in Peak's PP yet during the test period. The results match pretty well through pre-disturbance, during the disturbance, and post-disturbance.

By comparison of the damping variations during the event, PDCI trip event impacts NS mode B more than NS mode A. In Fig. 9.5, there is a clear difference in both the frequency and damping results during the disturbance and post-disturbance.

Since the signals used (Big Eddy and John Day) are really close to the DC intertie, during the disturbance, the signals are impacted by the disturbance a lot and thus may not provide actual estimation of modal frequency and damping. For the PDCI event, the fault is really close to Big Eddy and John Day, the PMU locations. If the PMU signals are used to estimate the mode, the results could be biased.

Post-disturbance situation, the results between Peak and BPA are quite different. Prior to performing modal analysis using MAS tool, BPA has a bad data check. It clearly shows in both frequency and damping results which BPA's MAS tool stopped calculating for a while and restarted. Around 7:00 a.m., Peak and BPA's results match again.

Brake Test: June 17, 2015:

BPA performed the brake test on June 17, 2015, at 9:14, 9:24, 15:14, and 15:24 PDT. The results comparisons (Peak Loveland and Vancouver Production versus BPA) are shown in Figs. 9.6 and 9.7 for the morning brake test.

In Fig. 9.5, it clearly shows that using the angle difference between Chief Joseph and Malin, the results are quite different than using other signals. The rest of the results matches pretty well. It also shows that the brake test does not impact NS mode A damping that much.

In Fig. 9.7, the results among Peak Loveland and Vancouver Production servers and BPA are almost the same. The brake test impacts the NS mode B damping a lot. At around 9:14 a.m. when performing the first brake test, the NS mode B damping dropped from ~12 to ~8%.

Normal Operation Day: June 16, 2015:

A normal system day which is June 16, 2015, is picked to compare Peak's and BPA's MAS tool results, as shown in Figs. 9.8 and 9.9. The results of Peak Loveland and Vancouver servers match BPA results.

Based on the results presented above, Peak PhasorPoint MAS Mode Meter application provided consistent results with BPA's MAS tool while monitoring NS

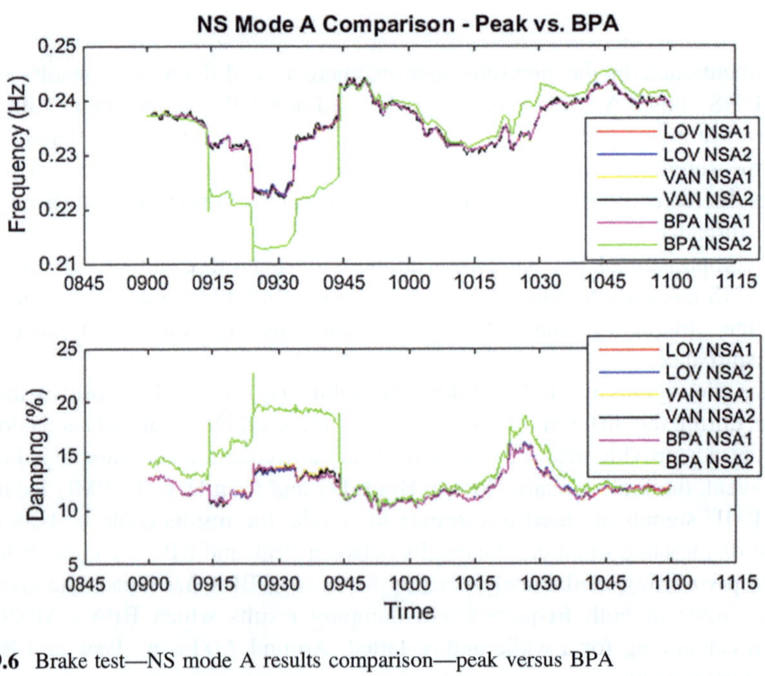

Fig. 9.6 Brake test—NS mode A results comparison—peak versus BPA

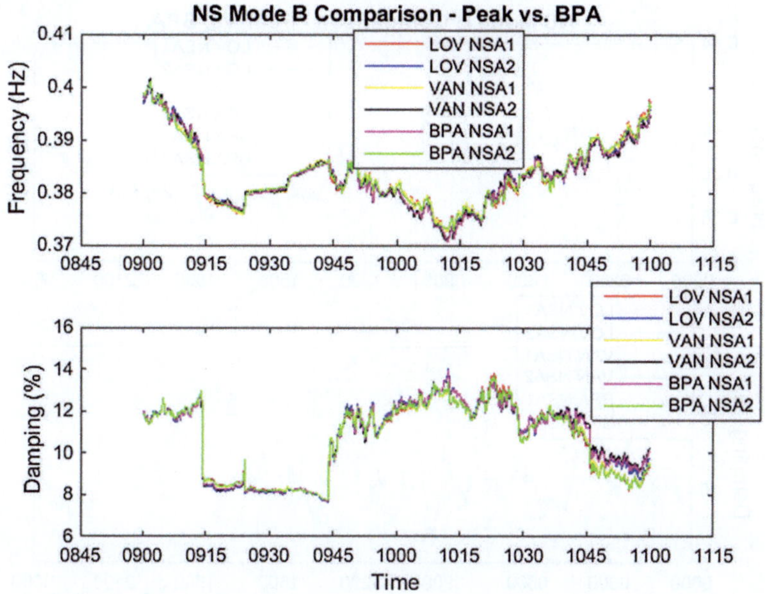

Fig. 9.7 Brake test—NS mode B results comparison—Peak versus BPA

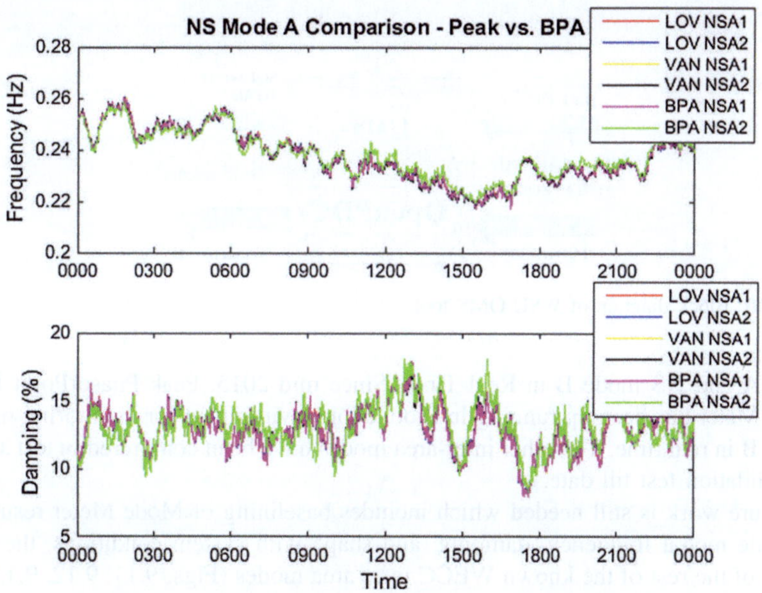

Fig. 9.8 June 16, 2015 NS mode A results comparison—Peak versus BPA

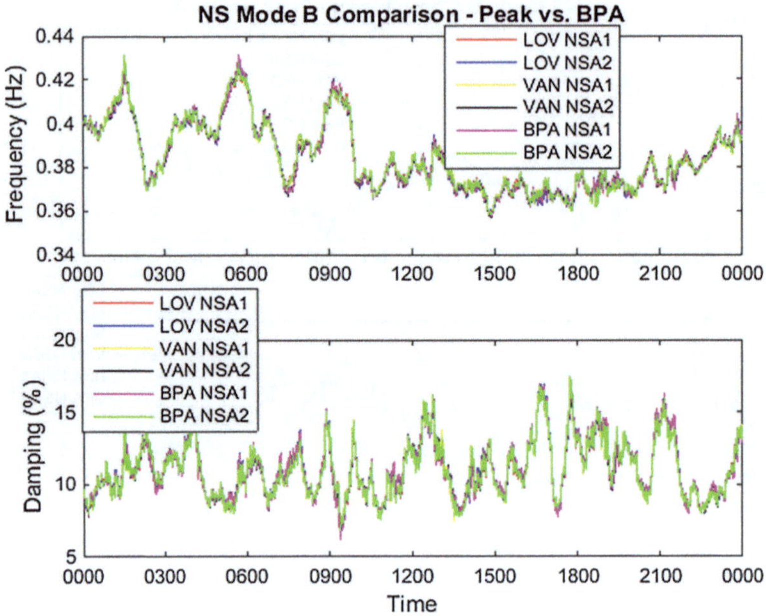

Fig. 9.9 June 16, 2015 NS mode B results comparison—Peak versus BPA

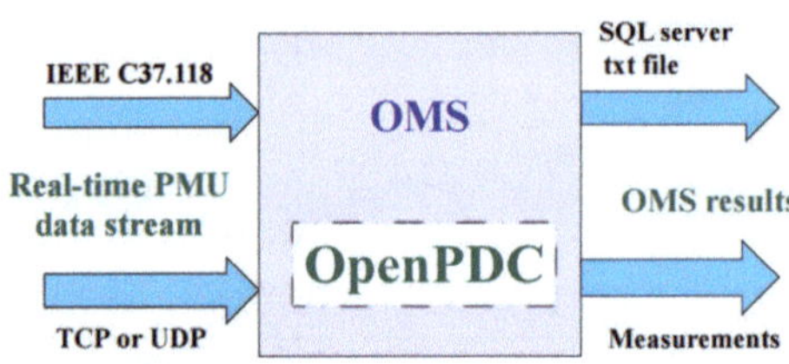

Fig. 9.10 Block diagram of WSU OMS tool

mode A and NS mode B in Real-Time. Since mid-2015, Peak PhasorPoint MAS Mode Meter has been up running in Production environment for monitoring modes A and B in real time. The other inter-area modes have been configured in test server for validation test till date.

Future work is still needed which includes baselining of Mode Meter results to correlate modal frequency, damping, and shape with system conditions, the validation of the rest of the known WECC inter-area modes (Figs. 9.11, 9.12, 9.13 and 9.14).

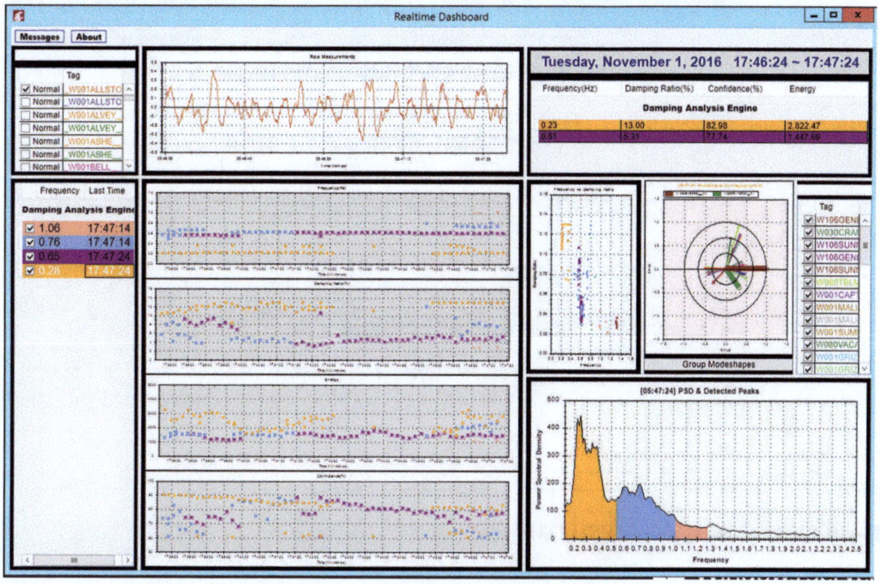

Fig. 9.11 User interface of WSU OMS Tool—Real-time dashboard

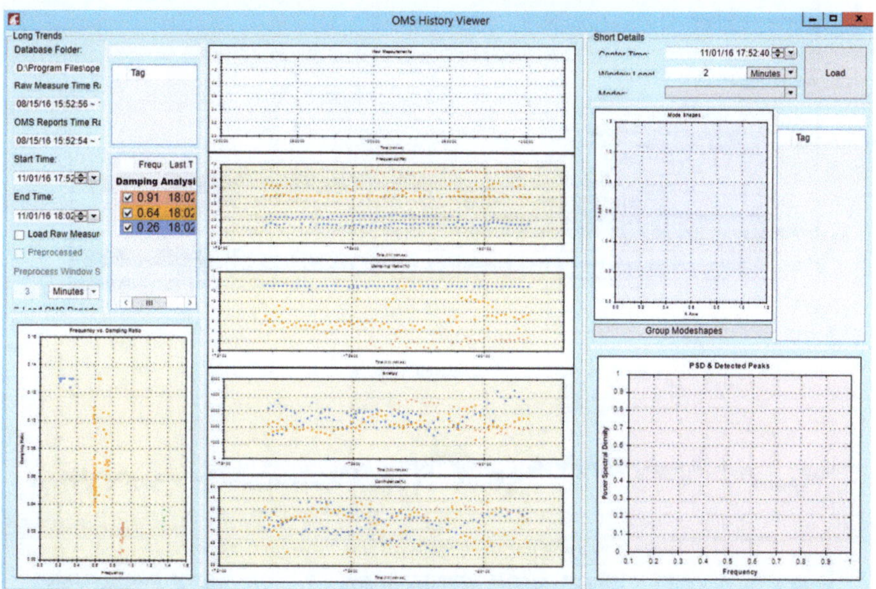

Fig. 9.12 User interface of WSU OMS tool—history viewer

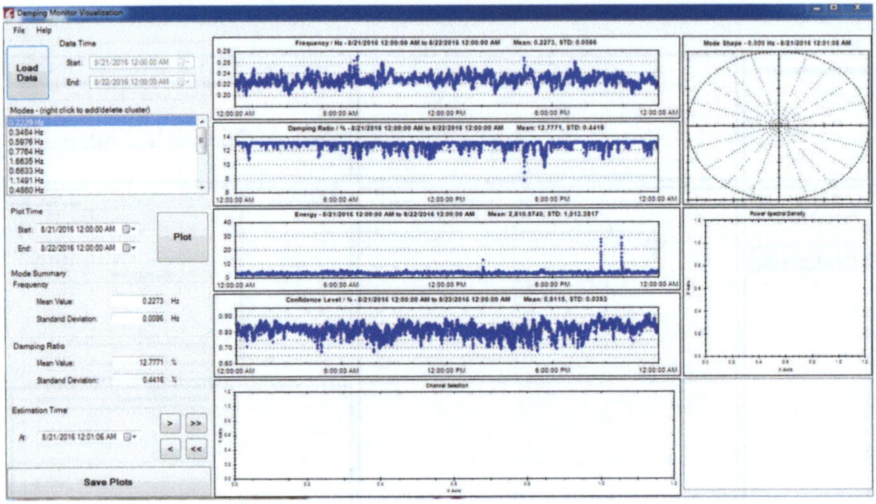

Fig. 9.13 User interface of WSU OMS tool

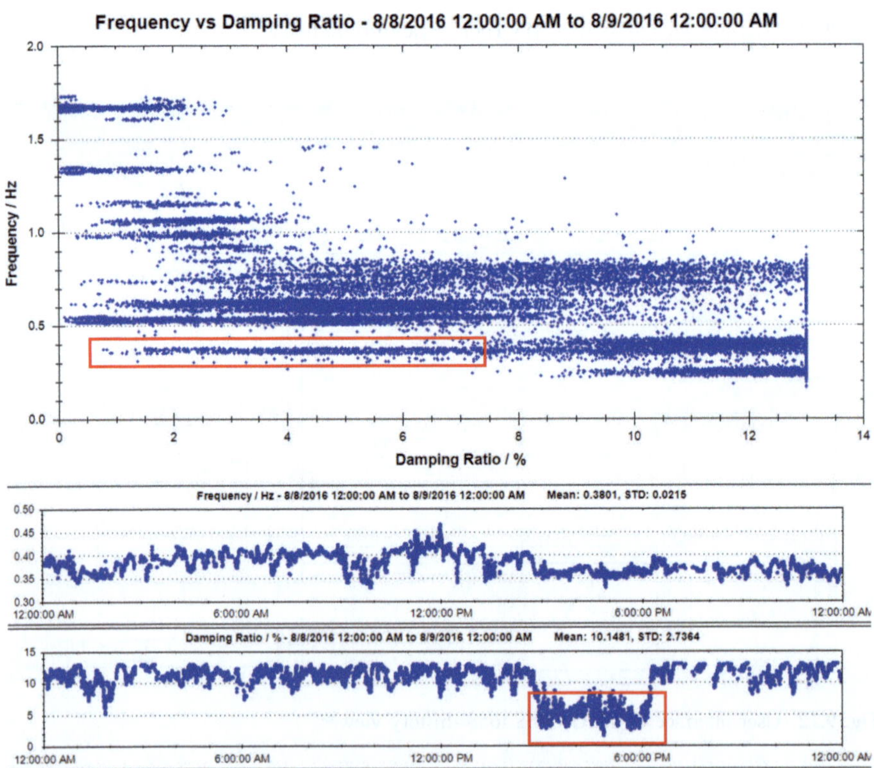

Fig. 9.14 User interface of WSU OMS tool

9.4 Forced Oscillation Detection and Analysis at PEAK

Forced oscillations occurred in WECC system. Usually they are caused by abnormalities in unit excitation and governor systems. They were characterized by

- Very low damping (near zero)
- High-energy and unknown modes
- A clear peak in power spectral density.

BPA was the first Western utility using MAS tool-oscillation detection application to locate a large number of oscillation events in real time since the tool was implemented in 2013. Most of the events were local forced oscillations due to equipment issues or bad operating point summarized below [2, 3]:

- **Wind Power Plant Oscillation (in bands 3 and 4)** due to the problematic wind turbine voltage controller.
- **Hydro Power Plant Rough Zone Oscillation (in band 2)**: Hydro turbines are designed for most efficient operation at nominal head and flow. Hydrodynamic instability (rough zone operation) occurs at partial load, typically 25–60% of rated generator power. Multiple sustained oscillations were detected at the hydro power plants, while the hydro units were continuously operated in a rough zone. Hydro power plant rough zone oscillation is very undesirable because mechanical forces cause premature wear of equipment or even catastrophic damage.
- **Hydro plant oscillation due to interactions between power system stabilizers (PSSs)** and under-excitation limiters (UELs) (in bands 3 and 4): The oscillation occurred in 2015. That year was a low hydro generation year in the Pacific Northwest, as a result, transmission voltages were high due to lighter transmission loading, and fewer generators were available to absorb the increased reactive power. The oscillation problem was resolved successfully after BPA technical staff re-tuned the UEL gains later.
- **Large Power Plant Active Power Oscillation** (in band 1) due to issues with an interface between a plant controller and a generating unit's governor.
- **Central Oregon Plant Oscillation** (in band 3) as a result of the plant control system being supplied an erroneous measurement from its local meter. The wires terminating at the meter had been poorly crimped and generated enough heat to start a fire. The plant operator powered down the unit in response to the fire, and the sustained oscillation stopped 5 min later.
- **Generator oscillations** because of transmission outages (in band 2): Those oscillation events usually clear in 5–10 min, mainly by taking oscillating generators offline.
- **PDCI high energy oscillation** (mainly in bands 3 and 4) due to equipment failure at converter stations.

From the above oscillation events detected by BPA, forced oscillation is usually caused by local generators operating abnormally. Unlike system oscillation modes that are usually well-damped, forced oscillation shows up as a sustained oscillation and can damage power plant equipment. It will be present in the grid until the underlying problem is corrected.

Through DOE CERTS project [10], Peak first adopted Washington State University (WSU) offline Oscillation Monitoring System (OMS) software for oscillation event analysis in 2015. Then, we installed WSU online OMS tool at Peak engineering laboratory and kept it running since May 2016. The objective of WSU online OMS tool is to provide real-time oscillation monitoring results in the system. OMS tool is a group of custom action adapters in OpenPDC. OMS results include oscillation frequency, damping ratio, modal energy, confidence level, and mode shapes.

Along with that, WSU also provides two C sharp.net-based user interface applications which allow us to visualize real-time and historical OMS results. Real-time visualize tool allows us to monitor power system oscillation stability in real time to enhance control room situation awareness. And the historical visualization tool enables us to review oscillation events and perform post-event analysis. Peak EMS Network Applications team has been testing the software and expanding use cases within engineering laboratory environment [11].

WSU OMS Tool Framework

WSU online OMS tool has three OpenPDC custom action adapters. They are FFDD, EVENT, and HISTORIAN. The input of OMS is real-time streaming PMU data in c37.118 format. OMS gets the data through OpenPDC. The output of OMS (oscillation frequency, damping ratio, modal energy, confidence level and mode shapes) is saved as OpenPDC measurements which allows us to publish them directly into PI data system just as what we did with raw PMU measurements. The results can also be saved as csv files in local drive.

The FFDD adapter is the engine for damping monitor based on Fast Frequency-Domain Decomposition algorithm. This is the one that outputs continuous results for every 10 s. The EVENT adapter would be switched in if it detects an oscillation event. The type of results from EVENT adapter is the same as in FFDD adapter. HISTORIAN adapter saves all FFDD, and EVENT adapter results into OpenPDC Historian.

WSU OMS Tool Highlights

Comparing to Montana Tech MAS tool, the main benefits of WSU OMS tool are:

(1) It uses a much shorter time window (1–5 min) comparing to 20–30 min time window used by MAS Mode Meter. Shorter time window means the engine could react and detect much faster to oscillatory system changes.

(2) Unlike Mode Meter which requires user to specify input signals and frequency bands for each mode, WSU tools simply uses all available PMU signals in OpenPDC to analyze. In this way, it can capture more oscillation modes than Mode Meter, especially for forced oscillation modes.

(3) It automatically detects bad PMU signals and excludes them from oscillation analysis. It estimates the confidence level for each detected oscillation mode, e.g., higher confidence level, more creditable oscillation detection results.

(4) OMS results can be directly saved into PI data system which is easier for baselining studies.

(5) WSU OMS tool is self-reliant, e.g., unnecessary to be integrated into another vendor platform such as GE-Alstom PhasorPoint.

Since WSU OMS tool and Montana Tech MAS tool use different algorithms, different inputs, and different setting parameters, it is not very meaningful to compare their damping results directly. But, overall, results from both engines meet each other in a closed range for most cases.

WSU OMS Tool UI Features

The WSU OMS tool Real-Time Dashboard allows users to monitor system oscillation in real time. It provides trend charts for oscillation frequency, damping ratio, modal energy, and confidence level for each mode it detected (up to 4 modes at one time). Users can detect poorly damped modes in real time by using this tool.

WSU OMS tool saves oscillation results in two forms. The first one saves results directly into OpenPDC historian, and the other one creates daily csv files in the local drive. It has provided two user interface API to view each kind of historical results. The OMS History Viewer pulls results directly from OpenPDC historian and plots them into trend charts for user to review.

There is another visualization tool which could load csv files to perform analysis review.

In the example below, we can see NS mode B was poorly damped for about 3 h. By using this tool, we can better learn the behaviors of these system dominant modes.

WSU FFDD Tool for Forced Oscillation Detection

An oscillation event (denoted Event 1) occurred on January 27, 2015, in western interconnection power system. It has been detected during offline studies using Fast Frequency-Domain Decomposition (FFDD) algorithm [13].

As shown in Fig. 9.15a, the oscillation started at approximately 11:00 a.m. and lasted till about 11:40 a.m. according to FFDD analysis of the WECC PMU data. Analysis window length for FFDD was 180 s which was updated every 10 s in a moving window formulation [8]. The mode frequency was at around 1.12 Hz with a low damping ratio below 1%. Figure 9.15b shows the average mode shape for the 1.12 Hz oscillation of Fig. 9.15a. The mode shape pointed to an area with low PMU coverage consisting of more than 50 generation substations. The presence of the forced oscillation can be clearly seen in the line current measurement from one of the PMUs in Fig. 9.15c. Although the oscillation magnitude is small, the FFDD algorithm [13] can detect it well as shown in the summary plot Fig. 9.15a.

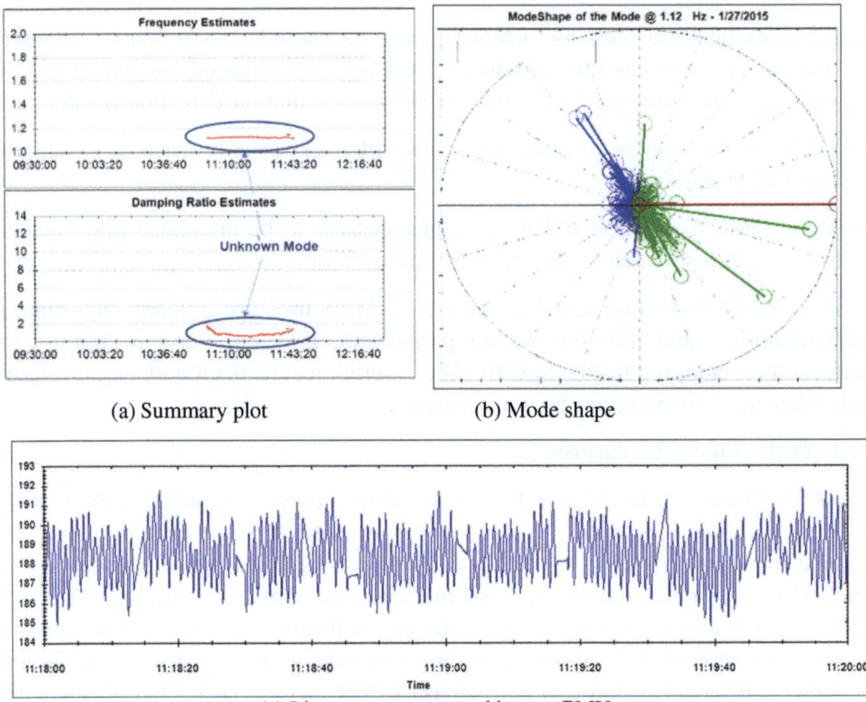

(a) Summary plot (b) Mode shape

(c) Line current measured by one PMU

Fig. 9.15 Detection of the forced oscillation using FFDD

In order to identify the specific source of the forced oscillation, three-hour SCADA data (9:30 a.m. to 12:30 p.m.) recorded from over 2000 generators in western interconnection power system was used for analysis. The discussion in this paper assumes the oscillation source to be a generator, while similar analysis can be applied for load sources as well.

Because of the vast amount of SCADA data, even with the indication from mode shape results, it took several weeks to identify the oscillation source through the manual search process. Subsequently, the oscillation event motivated Peak to work with WSU team for development of efficient methods for locating the source of forced oscillations automatically from among thousands of SCADA signals.

Use of SCADA Data for Source Location

Both WSU OMS and Montana Tech MAS tools have allowed engineers to detect oscillation issues that may have previously gone undetected. Although such an oscillation can be flagged and its mode shape can indicate the general vicinity of its source, low penetration of synchrophasors means that a specific generator or load that is the root cause of an oscillation cannot easily be pinpointed. Fortunately, SCADA serves as a much more readily available telemetered source of data if only at a relatively low sampling rate of 1 sample every 1–10 s. According to Nyquist criterion, this low sampling rate would be too slow for analyzing typical forced oscillations that occur in the frequency range of 1–5 Hz. However, since the SCADA measurements are not time-synchronized, the asynchronous sampling does preserve the oscillation amplitude to a great extent which can be exploited.

Through the DOE CERTS project, Peak collaborated with WSU to design and employ an online forced oscillation detector and source locator. Real-time WSU damping monitor results (oscillation frequency, damping ratio, and confidence level) are used to detect a forced oscillation. When a forced oscillation of significant energy is detected in the PMU data, the source locator would then pull the corresponding MW and MVAR SCADA data for all the generators in the Western Interconnection. Two algorithms are then used to find the source of oscillation by analyzing the generator SCADA data. For multiple recent oscillation events, the new developed methods were successful in correct identification of the oscillation source which was confirmed in each case by discussion with respective generation plant owners.

Unlike the high sampling rate PMU data, the SCADA system only provides data from generators and substations every 1–10 s. Measurement and collection of SCADA data are not time-synchronized. Therefore, SCADA data has been first interpolated like a synchronized stream of data with one sample every 10 s. The sampling rate then becomes 0.1 Hz which suggests that the corresponding Nyquist frequency is 0.05 Hz. It is impossible to analyze a forced oscillation at 1.12 Hz of Event 1 with such low sampling rate data using any existing signal processing technique.

Another challenge in using SCADA data is that the signals are recorded with different resolutions (tenth of MW or MVAR for some channels, and one MW or MVAR for some channels). Moreover, the sensitivities of the signals vary a lot in different areas of the system. Therefore, some of the signals can be very noisy, whereas others may remain unchanged for long periods.

Two methods that can handle all these practical issues are proposed in this section. Purely from the very large number of generators in a large power system, it is very difficult to identify the oscillation source effectively while avoiding false alarms. For this purpose, the two slightly different methods can be used to reinforce the ranking of possible source locations.

The start time and end time of the forced oscillation need to be known from a PMU-based oscillation detection engine such as by using FFDD of [13] before applying these two methods to the SCADA data. The time window in between the start and end times of oscillation detection will be referred to as the *oscillation window* for the rest of the paper. Non-oscillation time period is accordingly referred to as the *ambient window*. The oscillation window of Event 1 in Sect. 9.2 is from 11:00 a.m. to 11:40 a.m.

In order to properly distinguish from a generator which shows oscillating characteristics within the oscillation window versus another generator that has these characteristics throughout the entire data set, an initial ambient window is needed for baselining. Both methods will compute the ranking index $K \in \Re^n$ for all the generator outputs, where n is the number of channels.

A. Pattern Mining Algorithm (PMA)

To begin with, the data goes through a pre-screening process (denoted the data sanity check) to ensure that the generator is in use and there is some minimum variation within the output for the set. The channel whose maximal output is less than 10 MW or MVAR, and the one whose maximal difference for the entire data set is less than 1 MW or MVAR is ignored.

Next, a 25-point median filter is applied for detrending. The absolute values of the differences between the raw measurements and the filtered data, denoted the detrended data, can be used as a measure of the oscillation activity as seen in the SCADA signal. When an oscillation occurs, the amplitude of the differences during the oscillation window should be relatively higher than the amplitudes during the ambient window. Accordingly, a threshold is needed in order to rule out small differences. The threshold (denoted the 3σ threshold) is set to be three times the standard deviation of the detrended data in the ambient window.

This method then counts the number of the high-amplitude peaks in the raw measurements whose detrended values are outside the 3σ threshold in the oscillation window, denoted NUM_{osc}, and in the ambient window, denoted NUM_{amb}. In this context, the amplitude of the peaks is ignored per se. The ranking index of each channel is then formulated as,

$$K_{PMA\_i} = \frac{NUM_{osc\_i}}{Length_{osc}} - \frac{NUM_{amb\_i}}{Length_{amb}}, \quad i = 1,\ 2,\ \ldots,\ n, \tag{9.1}$$

where $Length_{osc}$ and $Length_{amb}$ represent the lengths (total number of samples) of the oscillation and the ambient windows, respectively (Table 9.1).

In order to determine relative ranking between the generators, the main steps of the pattern mining algorithm are summarized as below.

(1) Input SCADA data of generators and the oscillation event time as detected by FFDD using PMU data.
(2) Data sanity check.
(3) Apply the median filter and subtract the median filtered data from the raw data for detrending.

Table 9.1 Western inter-area oscillation modes

Western inter-area modes	Frequency Hz (normally)
North–South mode A	0.23
North–South mode B or Alberta mode	0.4
Montana mode	0.8
British Columbia mode	0.6
East–West mode A (first observed in 2013)	0.5
East–West mode B (found recently)	0.7

(4) Calculate the absolute values of the differences between the raw measurements and the filtered data.
(5) Reject the channel if the maximal absolute value of the differences is less than 1 MW or MVAR.
(6) Count NUM_{osc_i} and NUM_{amb_i}.
(7) Compute the ranking index K_{PMA_i} based on (1).
(8) Apply step 2–7 for the rest of channels.
(9) Select top 3 (user-configurable) channels based on the ranking index.
(10) Inspect the MW outputs of the possible oscillation sources for manual verification.

Event 1 from Sect. 9.2 is analyzed using the pattern mining algorithm. The three highest ranked possible sources are listed in Table 9.2, and Fig. 9.16 shows their MW outputs. In Fig. 9.16, the actual MW plots of each generator are presented on the left side of each subplot. The right side of the subplots shows the values of the detrended data. The red horizontal line in the right subplot of Fig. 9.16 depicts the 3σ threshold for this data set. The inflection points above this 3σ threshold are marked in red circle, and they contribute to the counts *NUM_amb* and *NUM_osc* in the ambient and oscillation windows, respectively.

The pattern mining algorithm selects generators 1085, 1088, and 1087 as the potential candidates for the oscillation source according to Table 9.2. Observable oscillations can be seen from all three MW outputs during the oscillation window. However, generator 1085 has a ranking index two times that of the indexes for the other two. Indeed, higher-amplitude oscillation activity can be seen in generator 1085 MW output in Fig. 9.16a compared to the MW outputs in Fig. 9.16b, c.

B. Maximal Variance Ratio Algorithm (MVRA)

The same data sanity check as in the pattern mining algorithm is first applied. The data is then detrended using a third-order bandpass filter. For the 0.1 Hz sampling rate (10 s SCADA update rate), the corner frequencies are set to be 0.005–0.035 Hz for the bandpass filter. Then, the MW (or MVAR depending on the nature of the oscillation) output of the generator causing the oscillation is expected to show sustained oscillation (like in Fig. 9.15) with the highest "amplitude" among all such signals.

Two key factors are considered when calculating the ranking index K_{MVRA} in this approach. One is the number of times the data values cross their mean value within the oscillation window N_{osc}, which indicates how much the MW data is showing sustained oscillations. The other one is the average standard deviation of the SCADA signal, which is a measure of the oscillation amplitude.

In order to accommodate the slow sampling rate of the SCADA data, we suggest estimating the oscillation amplitude by taking an average of standard deviations from multiple moving windows. That is, let us first compute the standard deviation σ_1 of a defined analysis window, say 30 samples (5 min). And then move the analysis window along the time axis with a fix step, say 6 samples (1 min). Next, calculate the standard deviation σ_2 of the new window. Keep moving the analysis

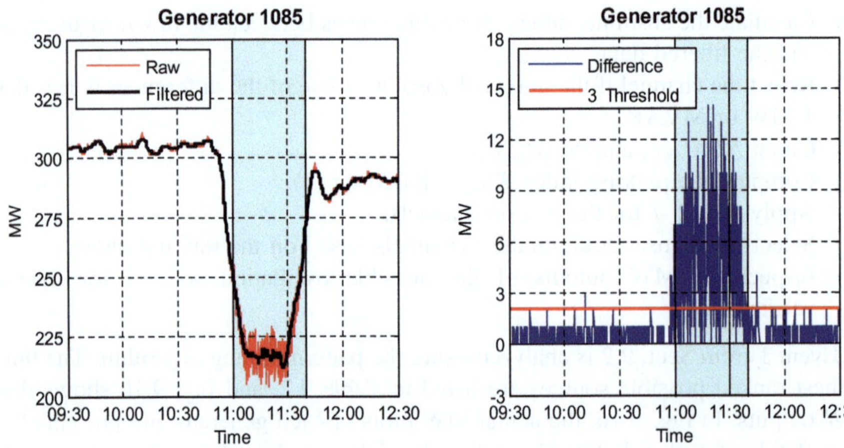

(a) Rank 1

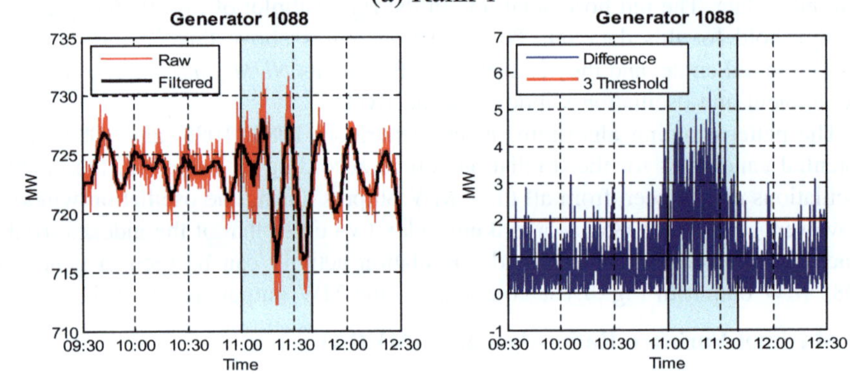

(b) Rank 2

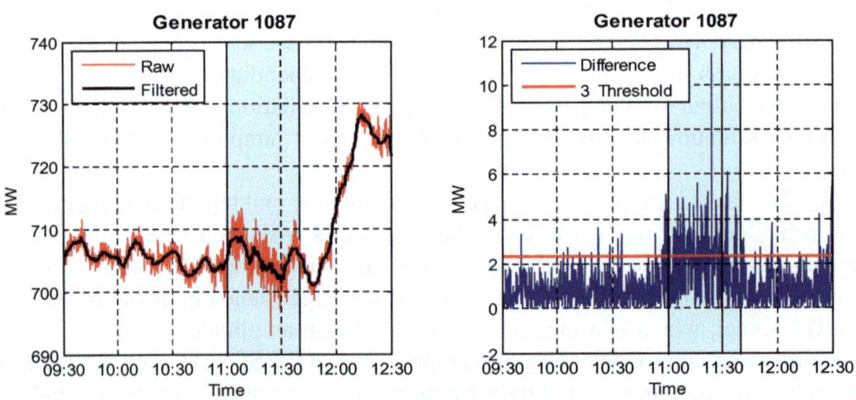

(c) Rank 3

Fig. 9.16 MW outputs of the three highest ranked generators using PMA window and computing the standard deviation σ_i (i = 3, 4,…) until the end of the data.

The initial ambient window is set to be the first 20 min of the data set (9:30 a.m. to 9:50 a.m.). The moving standard deviations are calculated over both the initial ambient window and the oscillation window, and the averages of the moving standard deviations for the two windows are denoted as STD_{amb} and STD_{osc}, respectively.

This moving window approach can be used to extend the algorithm toward online implementation in the future. STD_{amb} can easily be estimated in online framework from routine ambient SCADA data that is available all the time. Then, once a sustained oscillation is detected by a PMU-based oscillation detection algorithm such as FFDD in [13], if the oscillations persist long enough (say longer than 5 min), the corresponding SCADA data during the oscillation time period can be used to estimate STD_{osc} from SCADA data.

The ranking index for each signal is defined as

$$K_{MVRA\_i} = N_{osc\_i} \frac{STD_{osc\_i}}{STD_{amb\_i}}, \quad i = 1,\ 2,\ \ldots,\ n. \tag{9.2}$$

It is noted that K_{MVRA_i} will become very large when STD_{amb_i} is too small. This can happen for the generators which were off during the window and for SCADA signals with low resolution or low sensitivity. Such signals that remained mostly unchanged during the initial ambient window will be excluded with Filter 1. For example, Fig. 9.17a shows a generator MW output whose STD_{amb_i} is zero, so that its K_{MVRA_i} in (2) is infinity.

The second check is for the moving variance STD_{osc_i} of the signal during the oscillation window. Low value of STD_{osc_i} suggests there was mainly ambient activity (not oscillations) in the MW data during the oscillation window. Such generators are not candidates to be oscillation sources and can be omitted from further analysis. An example of such a signal is shown in Fig. 9.17b, which will be ruled out by Filter 2.

As a next example, the signal in Fig. 9.17c has a high STD_{osc_i} value because of many spikes in the oscillation window even though the oscillation activity amplitude is relatively small. In our experience, forced oscillations tend to be sustained with high amplitude over the entire oscillation window (like in Fig. 9.16a) for the potential oscillation sources. Therefore, generator outputs like in Fig. 9.17c with many spikes will be ruled out from the analysis with Filter 4. This example in Fig. 9.17) is from another oscillation event which will be introduced in Sect. 9.4.

Filter 4 rejects the channel whose crossover number N_{osc} is too small. For example, the generator output in Fig. 9.17d ramped up and down during the oscillation window. Because of the MW ramp, it has a high moving standard deviation STD_{osc_i}. However, the generator can also be excluded because it did not show much oscillations during the time window of interest.

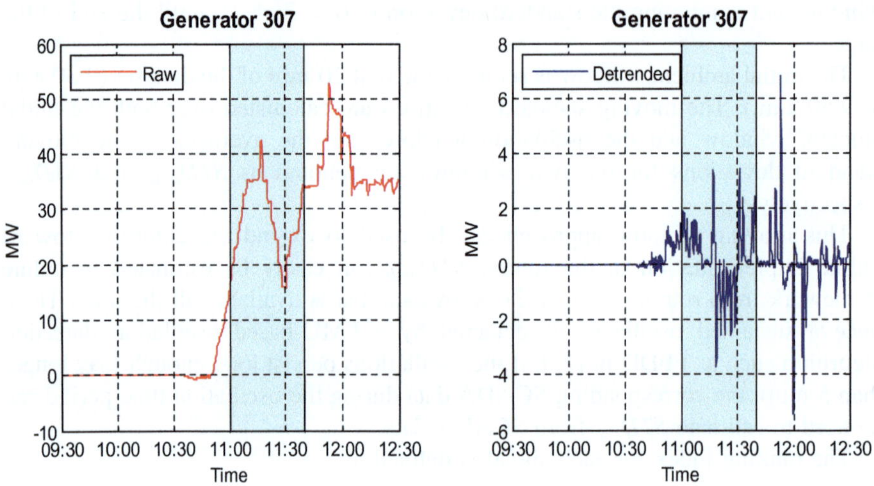

(a) Filter 1

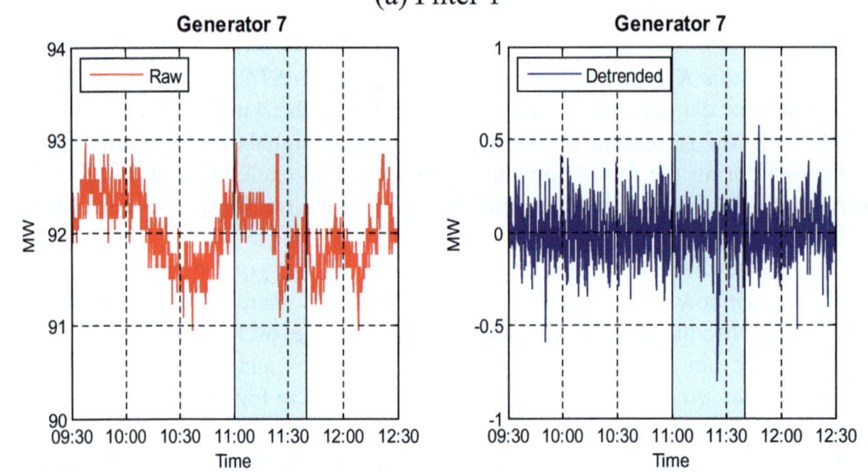

(b) Filter 2

Fig. 9.17 Example of signals ruled out by different filters in MVRA

Detrending using a bandpass filter noted in the beginning of Sect. 9.3, and a clipping limiter have been applied in order to remove the slow trends of the signals and to reduce the effect of sudden data spikes, respectively.

The main steps of the maximal variance ratio algorithm are summarized below.

(1) Input SCADA data of generators and the oscillation event time from FFDD analysis of PMU data.
(2) Data sanity check.

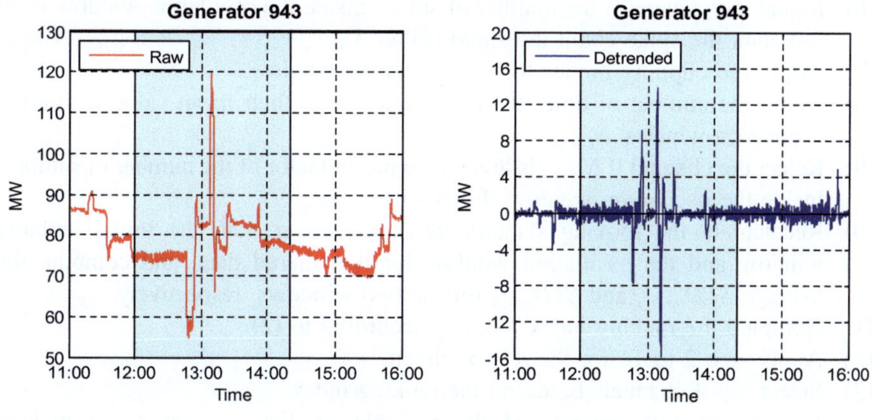

(c) Filter 3

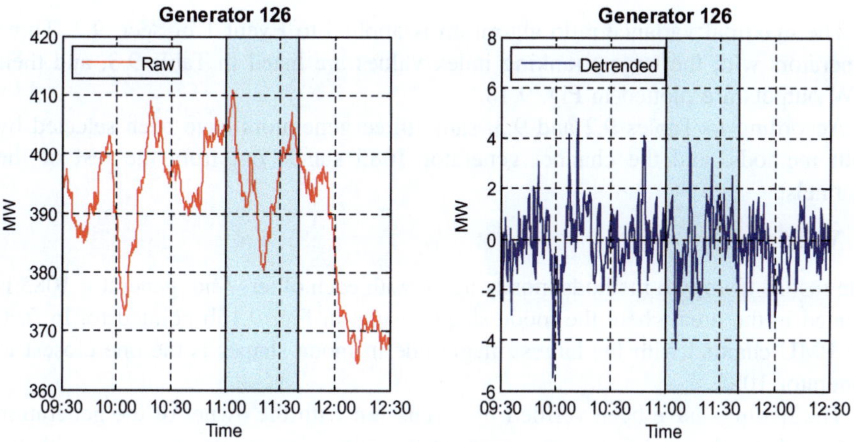

(d) Filter 4

Fig. 9.17 (continued)

(3) Calculate the average of the moving standard deviations for the initial ambient window. Reject the channel if the maximal difference of the data during the window is less than a preset multiple of the average standard deviation (Filter 1).

(4) Calculate the average of the moving standard deviations for the oscillation window. Reject the channel if this average standard deviation is less than the minimum oscillation threshold (Filter 2).

(5) Detrend using the bandpass filter.

(6) Reject the channel if the number of spikes inside the oscillation window is no less than the spike count threshold (Filter 3).
(7) Apply the clipping limiter for the oscillation window.
(8) Count the number of times the data values cross their mean value within the oscillation window N_{osc_i}.
(9) Reject the channel if N_{osc_i} is less than a preset factor of the number of samples inside the oscillation window (Filter 4).
(10) Recalculate the moving standard deviations over both the initial ambient window and the oscillation window for the filtered data, and compute the average STD_{amb_i} and STD_{osc_i} for the two windows, respectively.
11) Compute the ranking index K_{MVRA_i} according to (2).
(12) Apply step 2 to 11 for the rest of channels.
(13) Select top 3 channels based on the ranking index.
(14) Inspect the MW outputs of the possible oscillation sources for manual verification.

The maximal variance ratio algorithm is applied to Event 1 in Sect. 9.2. Three generators with the largest ranking index values are listed in Table 9.3, and their MW outputs are plotted in Fig. 9.18.

According to Tables 9.2 and 9.3, same three generators have been selected by both methods, and the channel generator 1085 stands out from the rest of the channels.

C. Validation

The results from two methods mostly agree with each other. And, generator 1085 is located in the area where the mode shape results in Fig. 9.15b pointed to. In fact, the PMU channel with the largest magnitude in mode shapes is the one closest to generator 1085.

The findings have been verified by discussion with the owner of the generation station. A mechanical failure occurred on the particular generation unit we identified, which caused the forced oscillation in the system. The second- and third-ranked generators 1087 and 1088 are two other units in the same generation plant, and they were responding to the forced oscillation in unit 1085. Therefore, they likely had the next highest amplitudes after unit 1085. In conclusion, *the ranking by the two methods PMA and MVRA has correctly identified the source of the forced oscillation from over 2000 generators in the system using SCADA data.*

Other Oscillation Events

The proposed methods have been applied to two other oscillation events found in western interconnection power system for validation.

D. Event 2 on January 28, 2015

The same oscillation discussed in Event 1 on the next day which serves as the second validation case. Estimation results of PMU data using FFDD [13] are provided in Fig. 9.19. The oscillation returned from around 12:00 p.m. to 2:20 p.m.

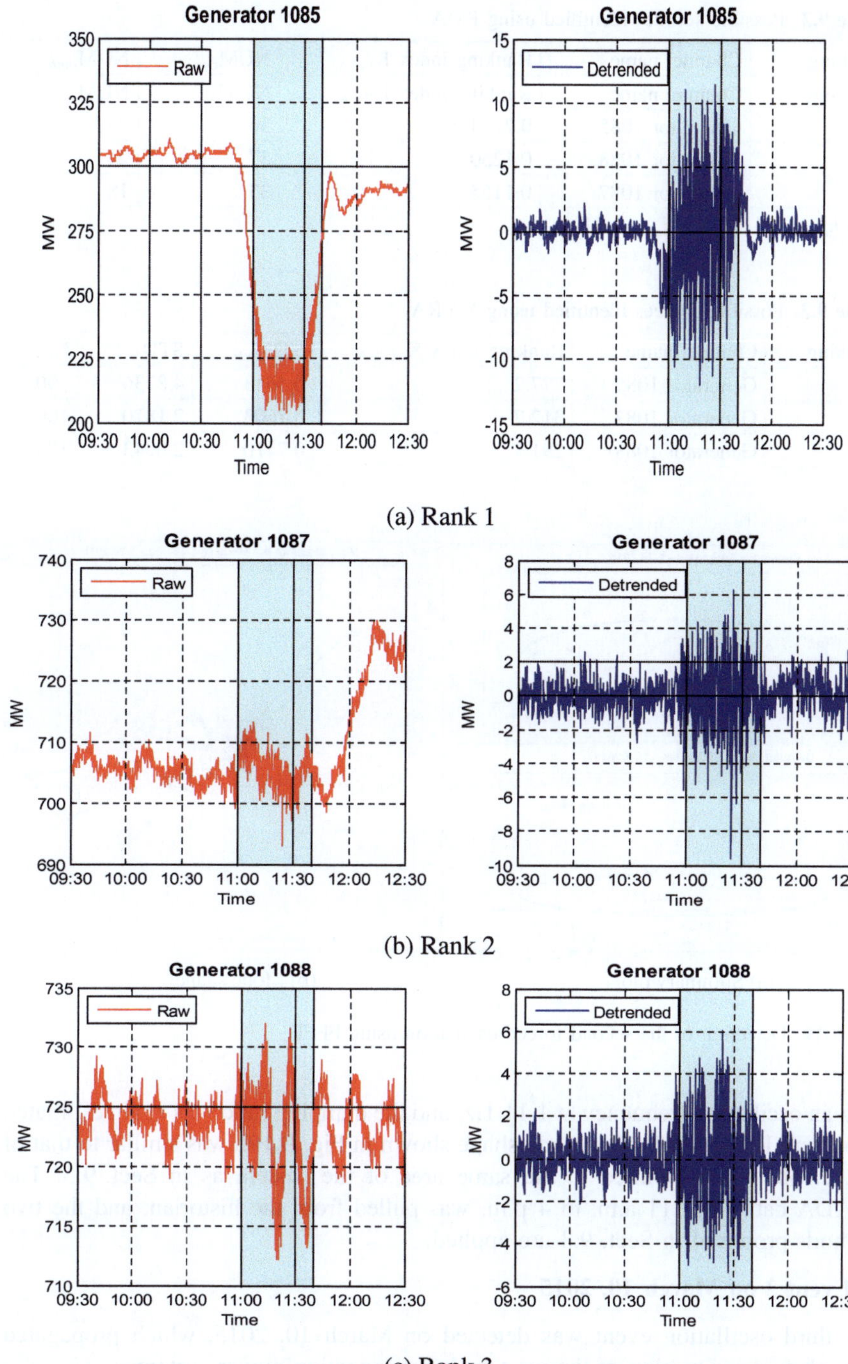

Fig. 9.18 MW outputs of the three highest ranked generators using MVRA

Table 9.2 Possible sources identified using PMA

Ranking	Channel name	Ranking index K_{PMA}	NUM_{osc}	NUM_{amb}
Ranking	Channel name	Ranking index K_{PMA}	NUM_{osc}	NUM_{amb}
1	Generator 1085	0.2264	56	5
2	Generator 1088	0.1250	37	24
3	Generator 1087	0.1155	33	18

$Length_{osc} = 241$, $Length_{amb} = 840$

Table 9.3 Possible sources identified using MVRA

Ranking	Channel name	Ranking index K_{MVRA}	STD_{amb}	STD_{osc}	N_{osc}
1	Generator 1085	777.2	0.5643	4.8736	90
2	Generator 1087	312.7	0.7603	2.1810	109
3	Generator 1088	291.1	0.7410	2.0943	103

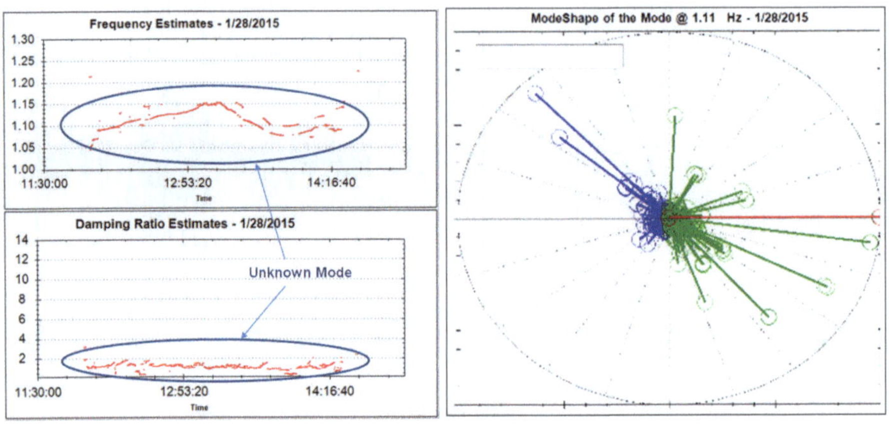

(a) Summary plot (b) Mode shape

Fig. 9.19 Detection of the second forced oscillation using FFFD

with an oscillation frequency at 1.11 Hz, and the damping ratio was again estimated to be very low at 1.0%. The mode shape shown in Fig. 9.19b was similar to that of Fig. 9.15b, and it pointed to the same area of the system as in Sect. 9.3. The SCADA data from 11 a.m. to 4 p.m. was pulled from the historian, and the two methods proposed in Sect. 9.3 are applied.

E. Event 3 on March 10, 2015

The third oscillation event was detected on March 10, 2015, which propagated through a major portion of the western interconnection power system.

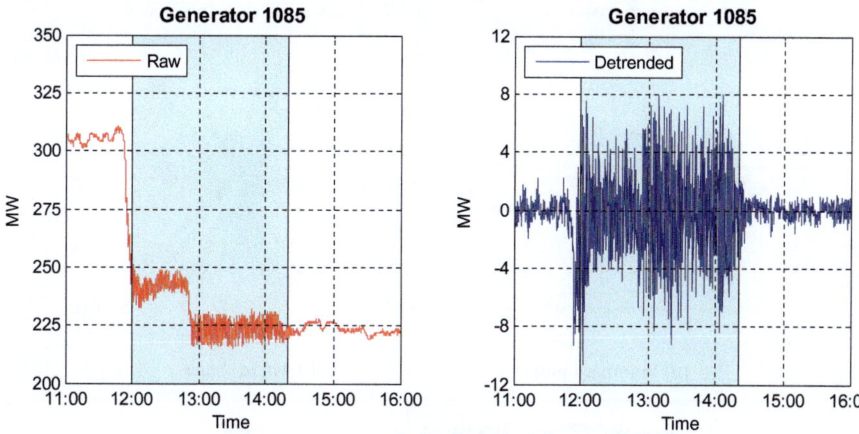

(a) Rank 1

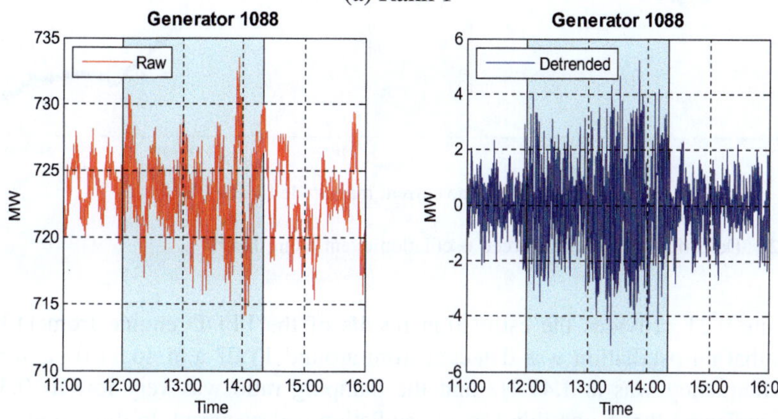

(b) Rank 2

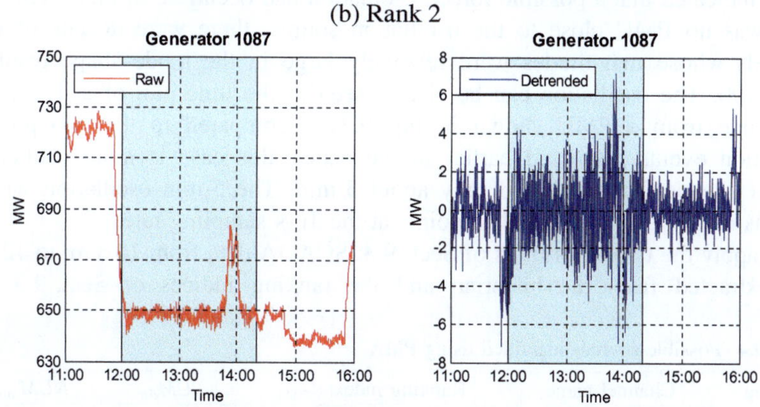

(c) Rank 3

Fig. 9.20 MW outputs of the three highest ranked generators using MVRA

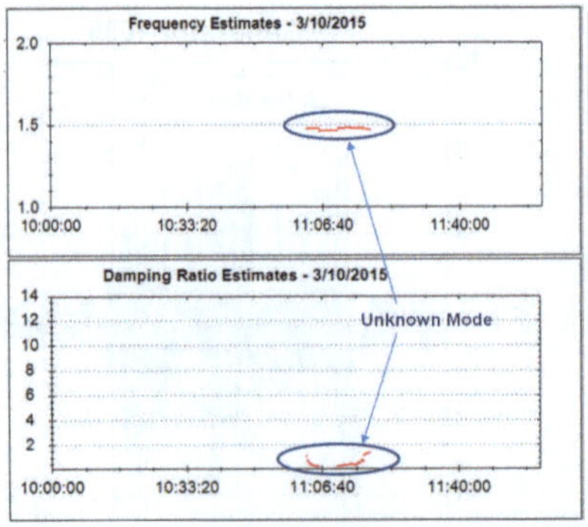

(a) Summary plot

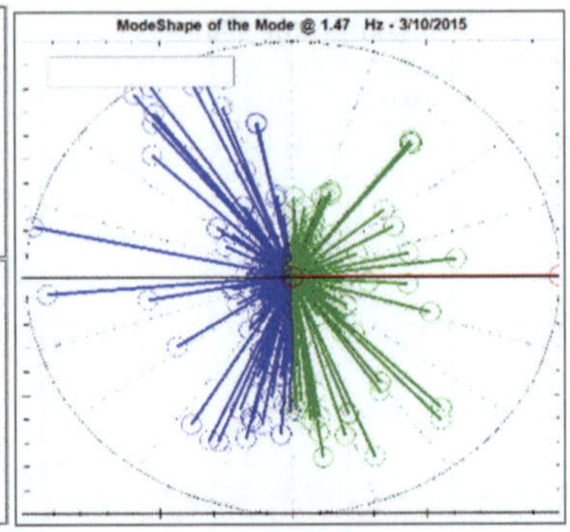

(b) Mode shape

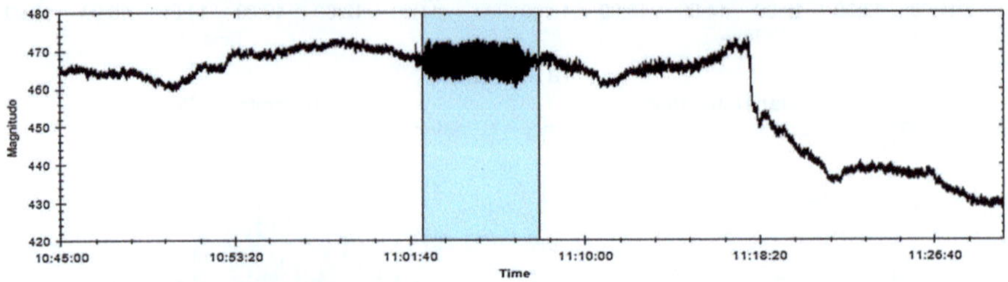

(c) Time-plot of a line current magnitude from a PMU

Fig. 9.21 Detection of the third forced oscillation event using FFDD

Figure 9.21 provides the estimation results of the FFDD engine from [13]. It shows that an oscillation was detected from around 11:02 a.m. to 11:07 a.m. The mode frequency was at 1.47 Hz and the damping ratio was very low as 0.34%, which indicated that a possible forced oscillation had occurred. In this event, since there was no PMU close to the oscillation source, there were dozens of PMU channels whose magnitudes were relatively large in the mode shape results in Fig. 9.19b. The oscillation can be clearly seen in the time plot of a line current magnitude from a PMU shown in Fig. 9.21c. Compared to the two previous oscillation events discussed earlier in the paper, this case is more challenging because the oscillations lasted only about 5 min. The 5-min oscillation window consists of only 30 SCADA data points at the 10-s sampling rate.

To apply the two algorithms of Sect. 9.3, SCADA data from 10 a.m. to 12 p.m. was extracted from the historian and the ranking indices of Sect. 9.3 were

Table 9.4 Possible sources identified using PMA

Ranking	Channel name	Ranking index K_{PMA}	NUM_{osc}	NUM_{amb}
1	Generator 1215	0.2903	9	0
2	Generator 2278	0.2186	7	5
3	Generator 945	0.1410	5	14

$Length_{osc} = 31$, $Length_{amb} = 690$

Table 9.5 Possible sources identified using MVRA

Ranking	Channel name	Ranking index K_{MVRA}	STD_{amb}	STD_{osc}	N_{osc}
1	Generator 1215	1031.1	0.1142	9.8159	12
2	Generator 2278	113.1	0.1397	1.5795	10
3	Generator 2280	56.3	0.7653	3.9141	11

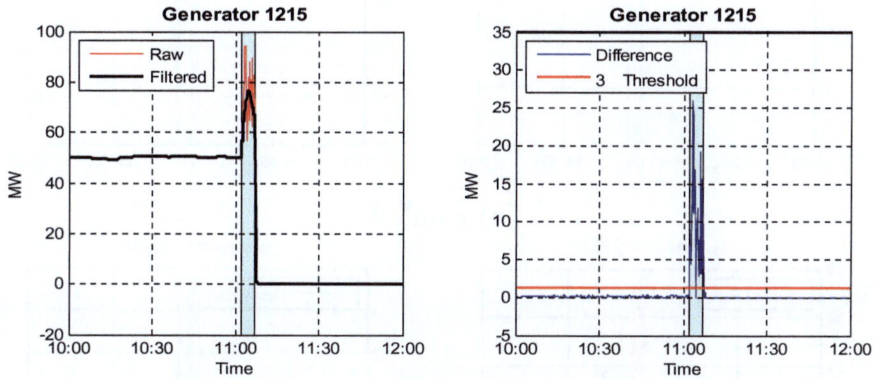

(a) Rank 1

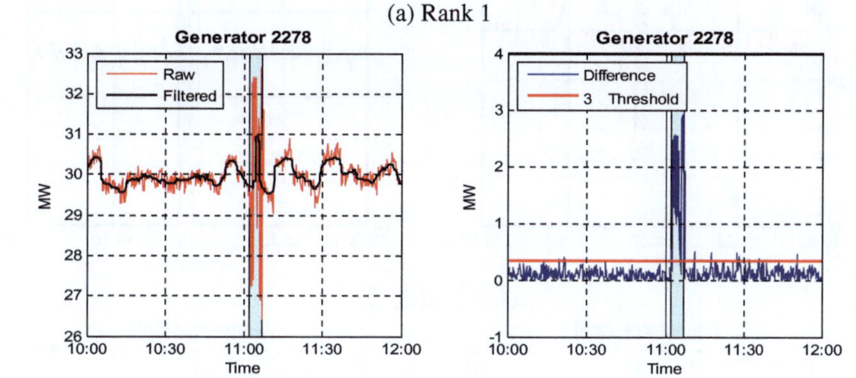

(b) Rank 2

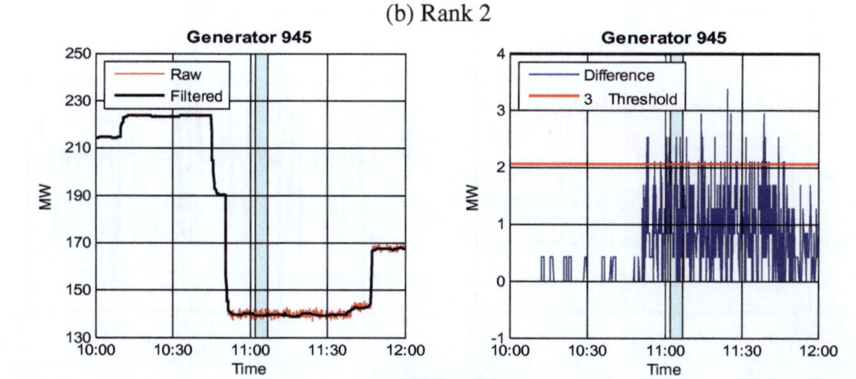

(c) Rank 3

Fig. 9.22 MW outputs of the three highest ranked generators using PMA

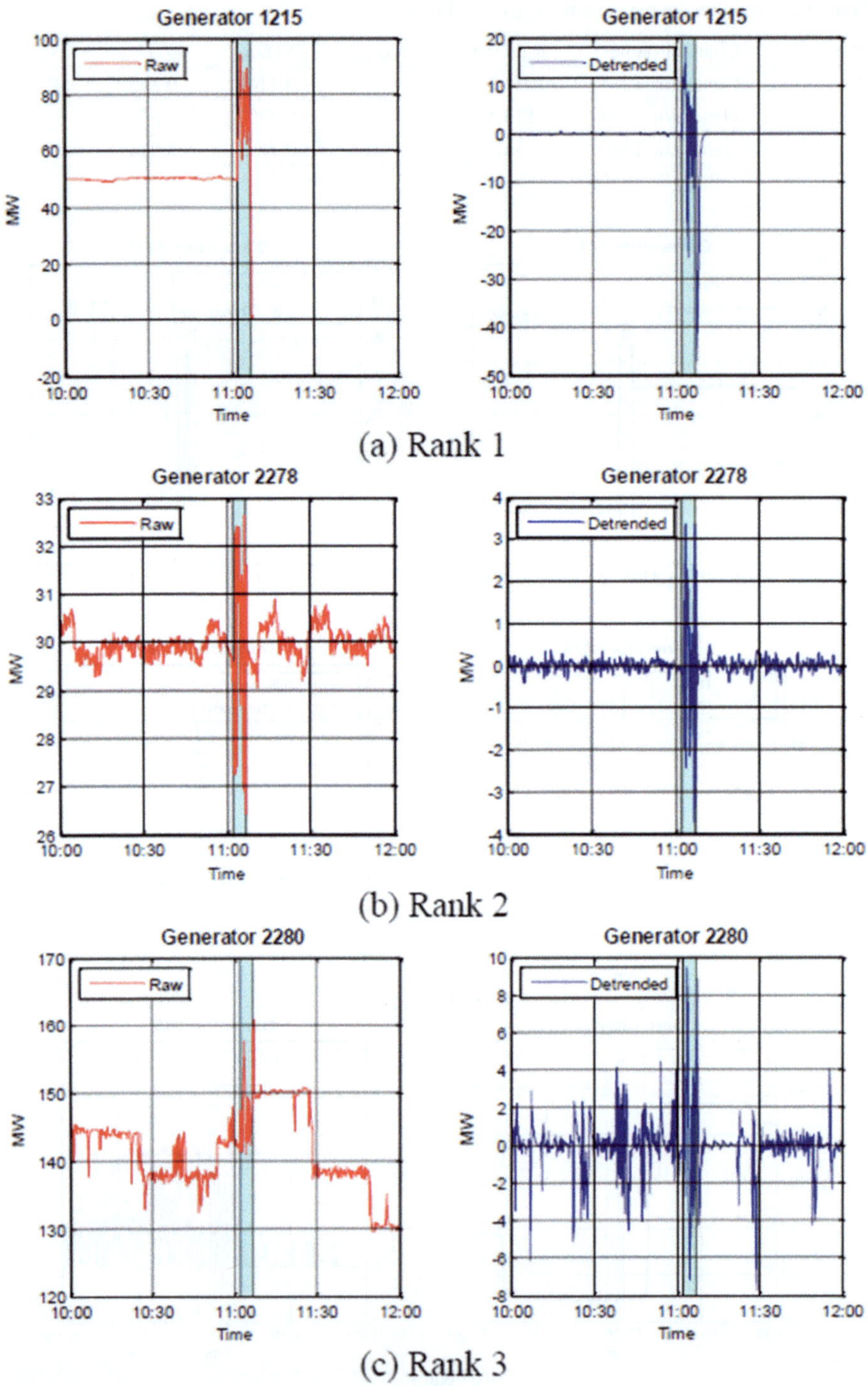

Fig. 9.23 MW outputs of the three highest ranked generators using MVRA

Table 9.6 Computational time (s)

Event no.	1	2	3
PMA	1.6845	2.8922	1.2313
MVRA	1.7032	3.8294	0.8529

estimated. The three highest ranked possible oscillation sources identified using PMA and MVRA are summarized in Table 9.4 and Table 9.5, respectively. Their SCADA generation outputs are shown in Figs. 9.22 and 9.23.

Generator 1215 has been identified as the most likely cause of the forced oscillation by both methods. The generator owner has confirmed that a control problem occurred during the time period of the forced oscillation which validates the findings from the two proposed methods. It appears that the third-ranked generators in both methods look a little suspect, and they may be unrelated to the oscillation event at generator 1215. However, both methods are able to correctly identify the oscillation source, generator 1215 in Figs. 9.22a and 9.23a. They also correctly point to the next highest ranked unit, namely generator 2278, which is clearly reacting to the same oscillation event as shown in Figs. 9.22b and 9.23b.

F. Computational Times

The computational times for source locations of all three events using both methods are summarized in Table 9.6. They are reported from tests using MATLAB 8.1 on a workstation with Intel Xeon CPU E5-2650 v2 @ 2.60 GHz and Windows 7 operating system. All the times in Table 9.6 are well below the SCADA update rate of 10 s which indicates that the two methods can be extended for online implementation in the future.

Wide-Area Oscillation Resonance Event on September 5, 2015

On September 05, 2015, Peak real-time MAS Engine (Montana Tech) detected very low damping on the NS 'B' mode under normal operation conditions:

- No topology changes (quick)
- No gen/load/tie-line flows drop (slow)

When further analyzing the real-time Mode Meter results for NS mode B, a low damping about 2% for a period of ~ 30 min was identified as shown in Fig. 9.24.

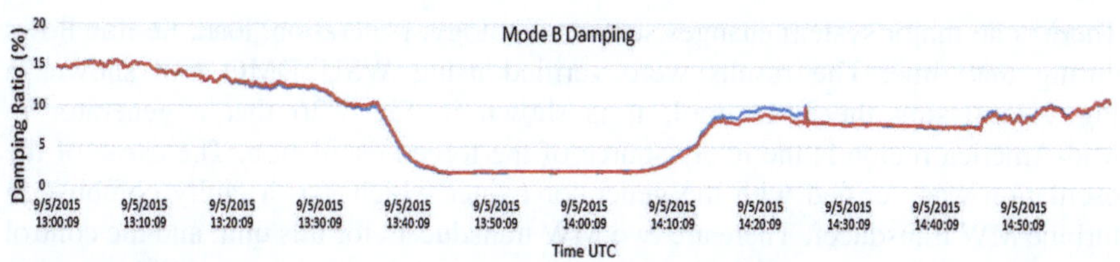

Fig. 9.24 September 05, 2015 event—real-time Mode Meter results

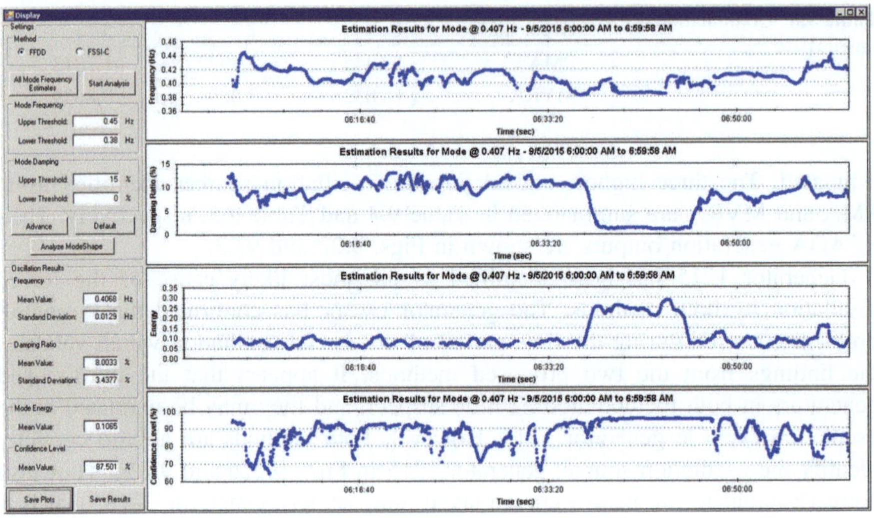

Fig. 9.25 FFDD analysis of September 05, 2015

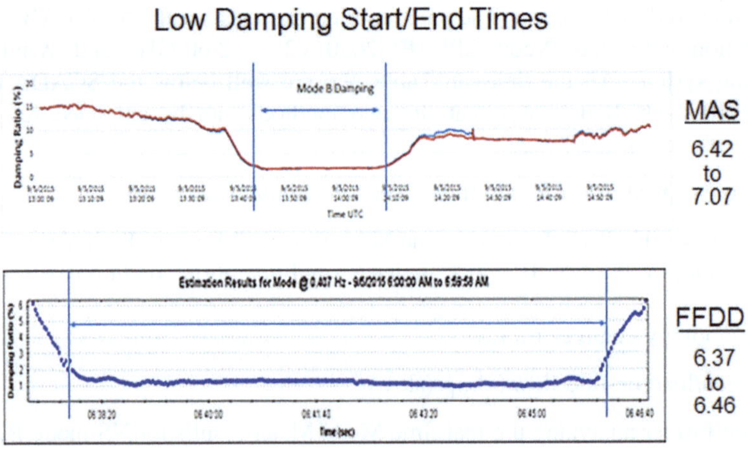

Fig. 9.26 September 05, 2015 event—DMO versus MAS results

There is no major system changes such as topology, generation, load, tie-line flows during that time. The results were verified using WSU DMO tool shown in Fig. 9.25. Using the PMA tool, it is shown in Fig. 9.26 that a generator in Mid-America region is the likely source of the forced oscillation. The cause of the oscillation was verified with the generator owner which was a faulty combustion turbine MW transducer. There are two MW transducers for this unit, and the control logic uses a select logic of the two values to control the output to a MW set point.

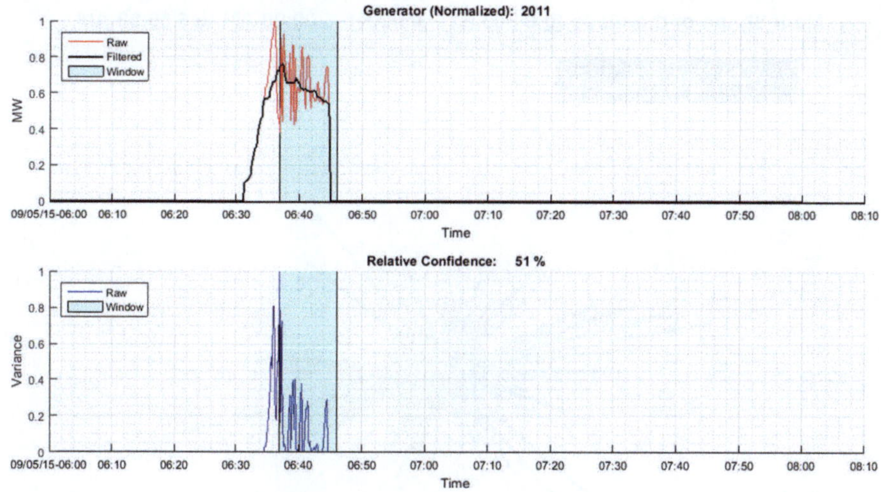

Fig. 9.27 September 05, 2015 event—PMA results

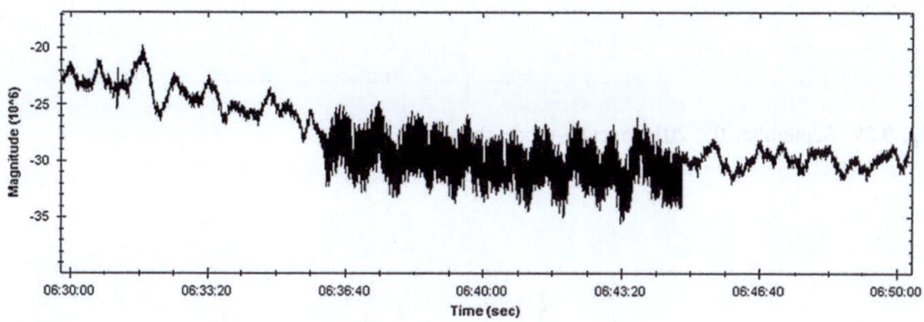

Fig. 9.28 September 05, 2015 PMU validation of the oscillation source

Of course if the measured output is incorrect or oscillating, then the control system attempts to correct to the set point and there will be real output swings.

Peak engineer ran PMA and MVRA algorithm to identify the source unit of the forced oscillation.

PMU validation for MW flow on a line nearby confirms the actual oscillation start/end times are 6.35–6.45. WSU FFDD tool detects the oscillation start/end times are 6.37 and 6.46, which is very close to what is found from PMU signal.

The mode shape plot in Fig. 9.29 clearly shows the source of the oscillation.

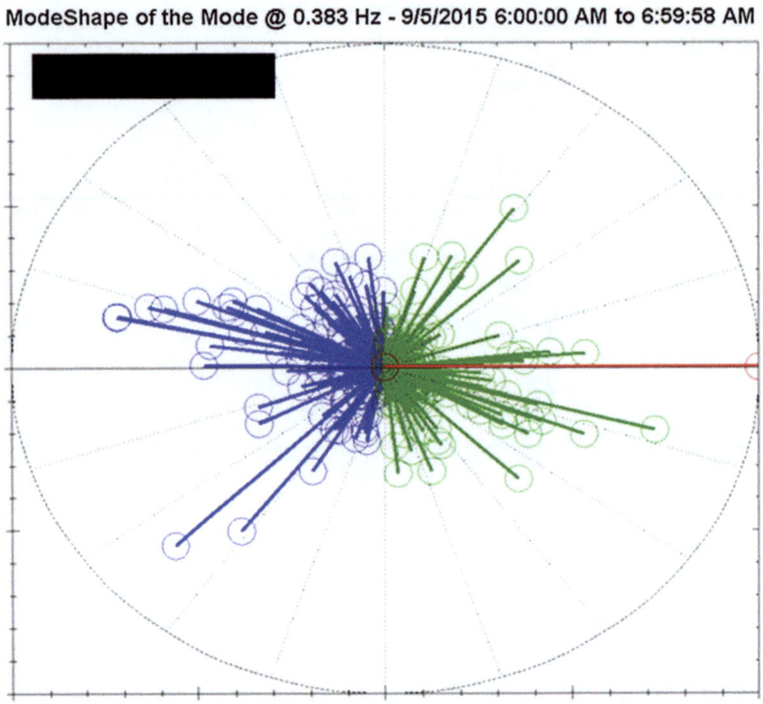

Fig. 9.29 September 05, 2015 event—mode shape

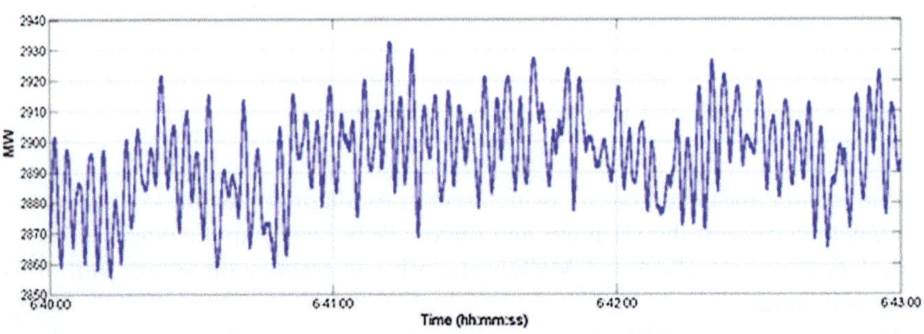

Fig. 9.30 September 05, 2015 event—Intertie MW flow

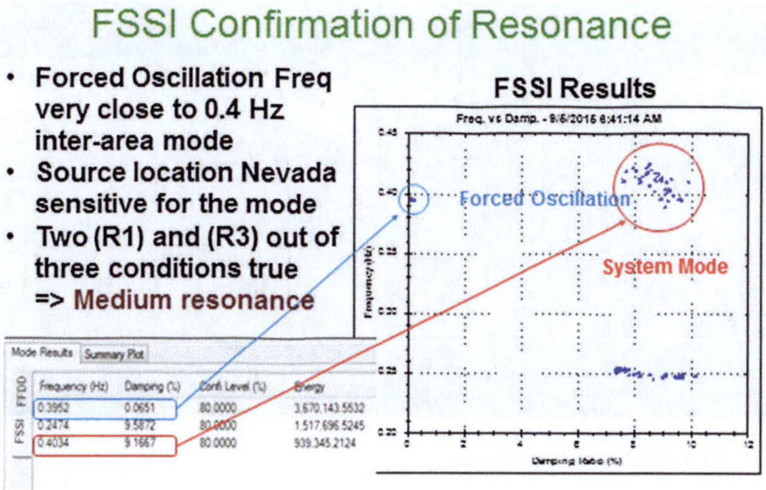

Fig. 9.31 September 05, 2015 event analysis results with FSSI

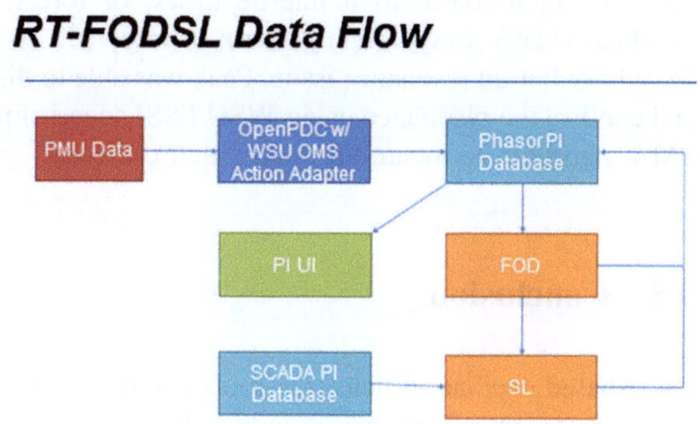

Fig. 9.32 RT-FODSL data flow diagram

The significance of this event is the forced oscillation (~ 6 MW) resonance with the 0.4 Hz NS mode B which showed up on a major intertie in the Western Interconnection for ~ 40 MW as shown in Fig. 9.30.

In this event, MAS tool and WSU FFDD tools gave false alarms of low damping on NS mode B. In a frequency-domain simulation, an oscillation that appears at a frequency that is very close to that of another mode, it can "trick" the program into thinking that there is low damping on the inter-area mode. Control room need to know the difference between a "false alarm" due to a forced oscillation and an

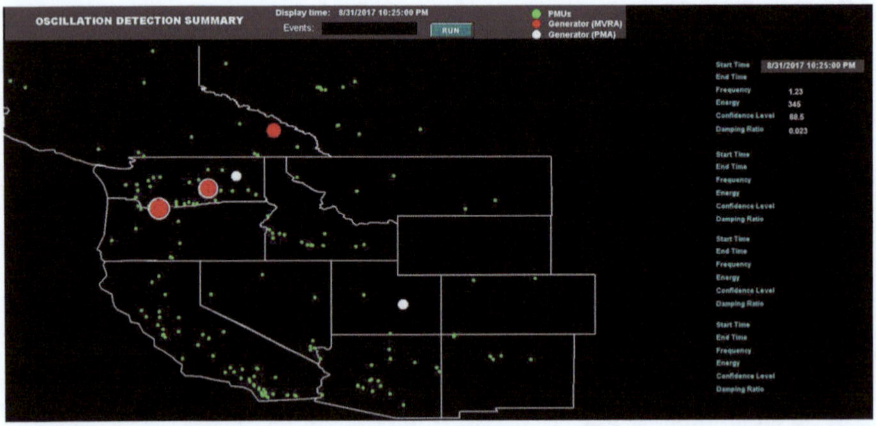

Fig. 9.33 RT-FODSL data flow diagram

actual low damping event is a problem [8, 10]. Then, an operator could take actions where none is needed. Actual low damping on inter-area modes is usually caused by either light loads, high intertie flows, or forced outages, which need to be checked when a low damping alarm is received. To deal with the complexity of forced oscillation resonance issue, Peak was able to distinguish two different modes and confirm the resonance using WSU FSSI successfully [14]. Meanwhile, we used PMA algorithm to locate the source unit correctly.

9.5 Conclusion

PMU-based oscillation detection tools can detect occurrence of forced oscillations in power systems. They can provide a clear indication of the time window when the oscillations were present and related information such as frequency, damping ratio, and the mode shape. Although mode shape results can point to a specific area of the system that might cause the problem, they are often not conclusive enough. Further investigation is needed in order to locate the particular source of the oscillations.

In this section, SCADA data is used for automatic source location of forced oscillations for the first time. It is not straightforward to use SCADA data by itself to detect the oscillations due to its slow sampling rate and high noise level. However, this paper shows that SCADA data becomes extremely useful for source location when combined with oscillation monitoring results from PMU data.

Ongoing Oscillation Detection Work

In later 2016 Peak started a pilot project in collaboration with WSU team. The project aimed to use WSU damping monitor results to detect forced oscillation and then pull all generator's MW/MVAR SCADA data around the oscillation time to determine the source of the forced oscillation. PI database is used to provide both WSU damping monitor results and SCADA data for the engine. And eventually, the results will be sent back to the PI database and be ready to be presented in a geographic wide-area visualization tool. By such forced oscillation detection tool, RC can contact the generator owner to report and confirm the issue and develop a resolution if the oscillation causes any reliability issue.

This new tool targets to be deployed in engineering laboratory by mid-2017 for daily monitoring by Network Applications engineers. The architecture of the integrated real-time forced outage detection and source locating (RT-FODSL) is described below.

A GIS-based overview to present in Fig. 9.33 to include two major executable components:

(1) Forced Oscillation Detector

- Start/end time, oscillation frequency, damping ratio, oscillation energy, confidence level
- Average normalized mode shape, length = magnitude, east = 0 degree angle

(2) Oscillation Source Locator

- PMA result (red), MVRA result (white)
- Area = ranking index

A few recent oscillation events were detected successfully by RT-FODSL (initial version). Peak team continually works on the tool development and tuning under WSU's support. Currently, the tool is still under development and fine-tuning in Peak engineering laboratory. Ultimate goal is to deploy the tool in control room for system oscillation monitoring and source locating [12].

Acknowledgements The author would like to thank Prof. V. Venkatasubramanian (Mani), Prof. Trudnowski, Dan, Mr. James O'Brien, Mr. Ran Xu, Dr. Jiawei "Alex" Ning, Dr. Tianying "Lily" Wu, Dr. Haoyu "Harry" Yuan, Mr. Mark Bowl, Mr. Andrew Esselman et al., who have contributed to the work described in this chapter. I would also like to thank the project teams of WISP, PRSP, and CERTS who supported our effort on system oscillation detection and source locating development work.

We thank US Department of Energy for supporting this research through funding from the Consortium for Electric Reliability Technology Solutions.

References

1. Van Ness J (1966) Response of large power systems to cyclic load variations. IEEE Trans Power App Syst PAS-85(7), 723–727
2. Kosterev D, Burns J, Leitschuh N, Anasis J, Donahoo A, Trudnowski D, Donnelly M (2016) Implementation and operating experience with oscillation detection application at bonneville power administration. in CIGRE US national committee 2016 grid of the future symposium

3. Donnelly M, Trudnowski D, Colwell J, Pierre J, Dosiek L (2015) RMS-energy filter design for real-time oscillation detection. in Proceedings of the 2015 IEEE PES general meeting, July 2015, http://ieeeexplore.ieee.org/stamp/stamp.jsp?arnumber=7286192
4. Hauer JF, Vakili F (1990) An oscillation detector used in BPA power system disturbance monitor. IEEE Trans Power Syst 5(1):74–79
5. Ma J, Zhang P, Fu H, Bo B, Dong Z (2010) Application of Phasor measurement unit on locating disturbance source for low-frequency oscillation. IEEE Trans Smart Grid 1(3):340–346
6. Zhou N (2016) A cross-coherence method for detecting oscillations. IEEE Trans Power Syst 31(1):623–631
7. Follum J, Pierre JW (2016) Detection of periodic forced oscillations in power systems. IEEE Trans Power Syst 31(3):2423–2433
8. Nezam SA, Venkatasubramanian V (2016) Inter-area resonance in power systems from forced oscillations. IEEE Trans Power Syst 31(1), 378–386
9. Kim K et al (1997) Methods for calculating oscillations in large power systems. IEEE Trans. Power Syst. 12(4):1639–1648
10. Wide-area oscillation resonance event in the western interconnection on September 5 2015. https://www.naspi.org/naspi/sites/default/files/2017–03/02_Mani_NASPI_WSU_Peak_Resonance_20170322.pdf
11. Real-time forced oscillation detection and source location in the western interconnection. https://www.naspi.org/naspi/sites/default/files/2017-03/01_Ning_Peak_Reliability_WSU_Real-time_foced_OD_20170322.pdf
12. Peak reliability synchrophasor technology implementation roadmap and current progress. https://www.naspi.org/naspi/sites/default/files/2017-03/01_pr_zhang_synchrophasor_roadmap_20161019.pdf
13. Source location of forced oscillations using synchrophasor and SCADA data. in Proceedings of the 50th Hawaii international conference on system sciences 2017. https://scholarspace.manoa.hawaii.edu/bitstream/10125/41541/1/paper0392.pdf
14. Resonant effect of forced oscillation to inter-area mode from a system event. https://scholarspace.manoa.hawaii.edu/bitstream/10125/41541/1/paper0392.pdf

Chapter 10
Online Oscillations Management at ISO New England

Slava Maslennikov and Eugene Litvinov

10.1 Introduction

Installation of Phasor Measurement Unit (PMU) infrastructure in ISO New England (ISO-NE) power system in 2012 has enabled monitoring of the actual dynamic behavior of the system. Multiple instances of oscillations in the frequency range from 0.05 to 2 Hz with significant MW magnitude and lasting from few seconds to hours have been observed. Majority of these oscillations can be classified as "forced oscillations" and caused mainly by the failure of equipment, failure or inappropriate settings of control systems or unplanned operating conditions. Forced oscillations represent the threat to the power system security and need to be mitigated. The critical step in the successful mitigation is finding the source of oscillations which is typically a specific generator or power plant. The chapter describes the practically observed oscillations in the ISO-NE power system, forced oscillation mitigation, and the Oscillation Source Locating (OSL) application which is an online tool for the robust estimation of the source of sustained oscillations.

10.2 Practically Observed Sustained Oscillations in Power Systems

Modern interconnected power systems are very complex electromechanical objects with inherent dynamic properties exhibiting oscillatory behavior. Oscillations are always present in power systems and ideal steady-state conditions practically do not

S. Maslennikov (✉) · E. Litvinov
ISO New England Inc, Holyoke, MA 01040, USA
e-mail: smaslennikov@iso-ne.com

S. Nuthalapati (ed.), *Power System Grid Operation Using Synchrophasor Technology*, Power Electronics and Power Systems,
https://doi.org/10.1007/978-3-319-89378-5_10

exist. From the practical standpoint, the oscillations could be categorized as (a) natural/ambient, (b) forced, and (c) transient.

Natural/ambient oscillations are the inherent response of a power system to constantly occurring small disturbances caused by the changing load, generation, and tripping of transmission elements. Characteristics of natural oscillations could be estimated by using conventional modal analysis based on eigenvalues and eigenvectors of the state matrix of the linearized dynamic model of a power system. They typically consist of frequency, damping and mode shape. Characteristics of natural oscillations reflect the dynamic response of a power system to a "small" disturbance when the behavior of system variables can be described by the linearized model and not significantly impacted by the system nonlinearity. Natural oscillations are typically characterized as electromechanical and torsional oscillations. Electromechanical oscillations reflect the interaction of rotating masses of generators' rotors over the network with frequencies ranging from 0.2–0.25 to 2–4 Hz. Electromechanical oscillations are further categorized into terms of spreading in the system as system wide, inter-area, and local.

Torsional oscillations are in the frequency range from about 5–60 Hz, and they are caused by the resonance conditions between highly compensated transmission and the mechanical modes of the rotating mass of generators' rotors or by the interaction of control systems.

Very low-frequency natural oscillations below 0.2–0.25 Hz can also exist, mainly in the small-size power systems. They are caused mainly by interaction of governor controls.

With the proper design of the power system and all controllers, properly functioning equipment and restricting operating conditions based on the accepted security criteria, all natural oscillations are well damped and do not present any threat to the system.

Forced oscillations are the oscillations caused by the periodic injection of energy into the power system via various mechanisms, including unexpected failure of the equipment and control systems, unplanned operating conditions, unexpected and unplanned interaction of the control systems, etc. Forced oscillations occur unexpectedly; typically, they have very low damping and can last from second to hours and days until the conditions periodically injecting energy in the system exist. The frequency of the forced oscillations depends on the frequency of the forced signal and can exist in the entire range of natural oscillations. Non-sinusoidal periodic injection of energy by a forced signal creates not a single frequency oscillation but a spectrum of harmonics. Magnitude of forced oscillations could be very significant, especially in the resonance conditions, when the frequency of the forced signal equals or close to the frequency of the natural mode.

Transient oscillations are the natural oscillations with elevated magnitudes excited by a large disturbance like tripping of a generator or a transmission line. Because of its large magnitude, the system nonlinearity can impact the characteristics of oscillations, especially at the beginning of the transient after the disturbance. As soon as the magnitude decays, the oscillations could be characterized by the damping and frequency derived from the analysis of linearized model of the system.

All the above described oscillations, except torsional, could be monitored by PMUs. Maximal frequency of the oscillations monitored by PMUs depends on the sampling rate and theoretically equals to the Nyquist frequency which is the half of the frequency of the sampling rate [1]. For example, the maximal monitored frequency is 15 Hz for a sampling rate of 30 fps. In actual implementation, the maximal frequency is lower than the theoretical one due to the filtering inside of a PMU and is not higher than 8–10 Hz for 30 fps.

Poorly damped natural and forced oscillations together can be classified as sustained oscillations. Sustained oscillations, especially with a significant magnitude, potentially represent a threat to the power system and should be mitigated.

Installation of the PMU infrastructure in the ISO-NE power system in 2012 has allowed monitoring actual power system dynamics. PMUs are installed at about 40 locations and provide full observability of the 345 kV backbone network. The PhasorPoint application by GE is used for an automatic detection and characterization of oscillations and generating alarms if the magnitude and/or damping exceed predefined thresholds.

Monitoring of the oscillations during 5 years by running above PMU infrastructure allows making the following conclusions:

- Natural oscillations are well damped practically all the time, and their ambient magnitude is low and does not impose significant risk for the power system security. This observation means that the power system in general is well designed and operated within security limits in terms of dynamic properties and electromechanical oscillations have typically sufficient damping.
- Multiple instances of the forced oscillations in the frequency range from 0.03 to 1.8 Hz with the magnitude from few MW up to 100 MW peak-to-peak have been detected. The oscillations have lasted from several seconds to several hours.
- Forced oscillations can suddenly emerge and often disappear without any obvious for the Control Room operator reasons. Offline analysis combining different sources of information is typically needed to investigate the reasons for oscillations.
- Forced oscillations can be localized to a limited area or spread system wide. Mode shape of some oscillations based on PMU data from ISO-NE footprint only indicates that majority of the ISO-NE generators oscillate in phase, suggesting that available PMU data provide only partial observability of a system-wide mode spread in the Eastern Interconnection (Fig. 10.1).

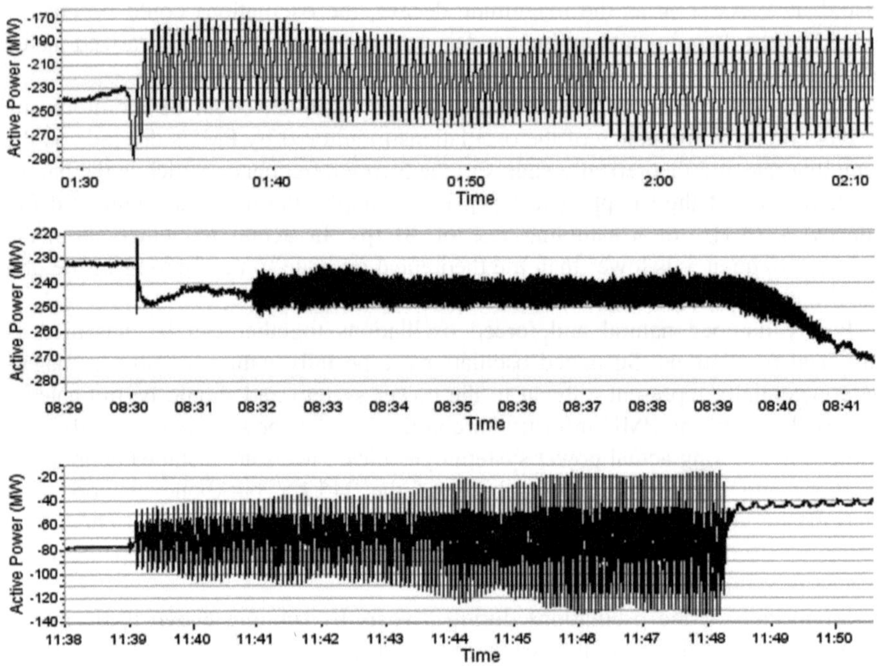

Fig. 10.1 Examples of forced oscillation at 0.04, 1.7 Hz, and multi-frequency process

10.3 Mitigation of Oscillations

Sustained oscillations, especially with significant magnitude, regardless of being forced or natural, represent a threat to the power system as potentially causing uncontrolled cascading outages and undesirable mechanical vibration in its components. Mechanical vibration increases the probability of equipment failure, reduces the lifespan of equipment, and results in the increased maintenance requirements. Mechanical vibration caused by sustained oscillations mainly contributes to the reduction of the equipment lifetime but also could be a triggering factor for catastrophic failures.

Catastrophic event at the Sayano–Shushenskaya hydroelectric power station in 2009 in Siberia [2] could be considered as an example of the consequences caused by the vibration in the turbine. One of the units, prone to vibrations at a specific narrow MW output range, was selected for AGC and was forced to operate in the prohibited MW range for a prolonged time. Resulting mechanical vibrations eventually caused a catastrophic failure of the rotor and flooding of the turbine hall and engine room. Seventy-five people were killed during the accident, and the entire 6400 MW power plant was out of service for several years (Fig. 10.2).

Sustained oscillations should be mitigated. Mitigation measures differ depending on the nature and the magnitude of oscillations. Systematic approach to mitigating

Fig. 10.2 Turbine hall of the Sayano–Shushenskaya power station before and after the accident in 2009 [2]

natural oscillations is based on the proper design and tuning of the control systems and preventing the system from operating under the undesirable conditions in terms of insufficient damping of natural oscillations. This task is mainly addressed at the planning stage. In operations, the sustained natural oscillations can occur if the system enters into unexpected operating conditions not studied for damping. The likely scenario here could be the need to transfer large MW amount over a weak network. Reduction of such MW transfer could be used as an operational mitigation measure.

Another likely scenario for sustained natural oscillations could be unplanned or wrong settings in the governor, excitation systems, or power plant-level control. Mitigation measures require finding the wrong settings and fixing them.

Because of the nature of forced oscillations, the mitigation measures cannot be designed/applied at the planning stage and should be implemented in operations only. The most efficient way to mitigate forced oscillations is to locate the source and eliminate the reason causing periodic injection into the system. Operationally, that often means identifying the system component causing forced oscillations and disconnecting it from the network.

The critical step for efficient mitigation of sustained oscillations is the identification of the system component causing forced or poorly damped natural oscillations. Overwhelming majority of the detected sustained oscillations based on the five-year history of oscillations monitoring at ISO-NE indicate that the source of sustained oscillations are generators with their control systems or power plant control systems. Cycling load, HVDC systems, and FACTS devices could theoretically be the source of forced oscillations, but none of such actual cases has been reported so far. So, practically important is identifying the source down to a power plant level or to a specific generator within the plant.

As soon as the source of oscillations is located, the mitigation measures could include the following steps:

- If oscillations do not impose an immediate reliability risk, ISO's personnel can communicate the suspect power plant offline in an attempt to find the source within the power plant and fix it.
- If oscillations impose a potential reliability risk and situation allows, the control room operator can communicate the suspected-source power plant online in attempt to find the source within the power plant and fix it. The fix can include the reduction of MW output, shutting down the suspected generator or other measures, as deemed necessary, at the power plant level and coordinated with ISO.
- If oscillations impose a real reliability risk, the control room operator can reduce MW output or request to shut down the suspect generator or the entire power plant.

In bulk interconnected power systems consisting of multiple control areas, the sustained oscillations could be observed in multiple control areas, while the source is located only in one of them. That, for example, can happen when the frequency of forced oscillations is close to the frequency of a natural inter-area mode. In this situation, the operator in each area needs to be able to determine whether the source is located within his control area or outside. The above-listed mitigation measures are applicable only within the control area. For the source located outside, the control room operator can communicate the findings to the neighboring control areas in an attempt to locate the source.

10.4 Locating the Source of Sustained Oscillations

Locating the source of sustained oscillations is the first key step of the mitigation process. Locating the source for the Control Room operator means answering the following questions. Is the source located in my control area or outside? If outside, which of my neighboring control areas is the suspect? If inside, which specific substation and specific generator is the suspect? Correctly answering these questions allows selecting a proper set of remedial actions. In other words, locating the source means providing actionable information to the Control Room operator. Locating of the source down to a specific device level is not considered here.

Locating the source should be measurement-based. Model-based approach can be used in some situations; however, it cannot be considered as reliable and, particularly for online application, when the identification error and time could be critical. Majority of sustained oscillations have a forced nature with unknown a priori the nature and location of a forced signal and that prevents the efficient use of the model-based methods.

Sustained oscillations have multiple features and characteristics such as natural, forced, local, widespread, resonance conditions, multiple simultaneous existing sources, harmonics. These characteristics can impact the efficiency of the source locating methods. Practically meaningful online method should satisfy the following requirements:

- A rigorous theoretical foundation;
- Be universally efficient for a broad variety of practically feasible situations and do not produce false alarms;
- Able to provide useful information even with the partial system observability;
- Robustness in the presence of missing/noisy measurements;
- Measurement-based without use of the system model.

10.4.1 Methods for Locating the Source of Oscillations

The oscillation source location problem has been a topic of multiple publications during the last two decades and particularly since 2010, following widespread installations of PMUs. Reference [3] provides a survey of the publications. The proposed methods can be split into four major categories:

- Traveling wave-based methods utilize the principle of the electromechanical wave propagation [4] to locate the oscillation source by detecting the abnormal wave speed among all monitored transmission lines. The method assumes the accurate detection of the arrival time of the oscillations at different locations and the availability of the accurate wave speed map with significant observability of the network by PMU [5]. It is shown in [6] that the earliest detected peak of the oscillation always first appears at the generator with forced oscillation and that the oscillation phase of the location far away from the disturbance is lagging that of the bus close to the disturbance [7].
- Damping torque-based methods assume the ability to calculate a damping torque coefficient for generators in multi-machine system, and the generator with a negative damping torque coefficient is identified as the oscillation source. These methods depend on the extension of the damping torque concept from a single-machine-infinite-bus system to a multi-machine system. Several extensions have been proposed [8, 9] including a technique to calculate a damping coefficient of each generator for each dominant mode [10].
- Mode shape estimation-based methods analyze the relative magnitude and phases of the dominant modes of oscillations throughout the system. Mathematically, the method estimates the right eigenvector of the state matrix of the linearized system model. Leading generator in the leading group in terms of the mode shape phases is assumed to be the source of oscillations. This property was discovered in a two-machine system and intuitively extended to a multi-machine system [11]. It was shown in [12] that the mode shape method can provide wrong estimation of the source. Also, the estimated mode shape

under resonance conditions could be different compared to the one from eigenanalysis [13].

- Energy-based methods utilize the Transient Energy Function (TEF), which is defined as the sum of potential and kinetic energies, as an application of Lyapunov function in power system stability analysis. It was shown that, under certain assumptions, the dissipation component of the TEF calculated for generators for a specific mode of oscillations is consistent with the damping torque. Generator producing the dissipating energy is the source. Calculation of the dissipating energy flow in network elements allows to locate the source [14–17].

Other source locating methods include the following:

- Equivalent circuit-based method adopts the phasor concept for the deviation of the power and frequency and builds an "equivalent circuit" representation for the power system [18]. The source in the "circuit" is identified as the source of oscillations.
- Hybrid simulation-based method utilizes a hybrid model-measurement by replacing a part of variables with measurements. The source of oscillation is located in the area with the largest difference between simulated and metered quantities [19].
- Artificial intelligence (AI)-based method [20] proposes to train a decision tree offline for a variety of operating conditions by using the Characteristic ELLipsoid (CELL) and to use the decision tree online with PMU measurements for the identification of the source.
- A statistical signature of a mechanism creating the oscillations is proposed in [21] as a useful information to locate the source.
- Graph-theoretic-based method [22] proposes to construct the input–output transfer matrix and calculate the array of nominal localizations keys. PMU measurements are used for calculating of estimated localization keys and the comparison with the nominal ones.
- Generalized linear model-based method [23, 24] uses PMU measurements to fit a generalized linear model for the system. The fitted model may help identifying variables that have significant effect on the mode damping, which might be related to the oscillation source.

Each of the proposed methods has advantages and disadvantages and can be successfully used only for some situations. Unfortunately, none of the methods satisfies the set of the requirements for practically meaningful online implementation. So far, none of the methods have been demonstrated as being a universal and reliable practical tool, except for the dissipating energy flow method, applicable for a broad range of possible actual situations in power systems.

10.4.2 Magnitude of Oscillations as an Indicator of the Source Location

Relative magnitude of forced oscillations observed in multiple locations of the power system depends mainly on two factors: (a) closeness of the frequency of forced signal to the frequency of natural mode and (b) damping of that natural mode [25]. Forced oscillations that occur in the power systems are generally locally contained and could reach significant magnitude only in the limited area. That is a typical case when the frequency of the forced signal is far away from the frequencies of natural modes. In this situation, the location with maximal magnitude of oscillations is a good indicator of the source.

However, if the frequency of forced signal equals or close to the frequency of the inter-area mode, the forced signal can "excite" the mode so that the oscillations propagate through the system. This situation is referred to as the resonance conditions. In the resonance conditions, and particularly for a poorly damped inter-area mode, the magnitude of the oscillation near the source could be much smaller than the magnitude in the places far away from the source. A relatively small source can create oscillations with much more significant magnitudes hundreds and thousands miles away. As an example, on November 29, 2005, a steam extractor valve failed at one of the power plants in Alberta, Canada, leading to the 20 MW forced oscillations in the active power output of the plant. The frequency of the forced signal was 0.27 Hz, creating near-resonance conditions with 0.26 Hz inter-area mode of the Western interconnection. As a result, 200 MW oscillations were observed 1100 miles away at the California Oregon Intertie interface [26].

In the case of non-resonance conditions when the frequency of the forced signal deviates significantly from the frequency of natural modes, the magnitude of oscillations can be used as an indicator of the source location. That inspires an idea of locating the source when the frequencies of natural modes are known and the frequency of the observed oscillations is not equal to any of natural modes. Unfortunately, the frequencies of natural modes fluctuate with the change of operating conditions and particularly in the marked-driven environment or at high penetration of renewables where the power flow pattern can change significantly following fuel prices and wind/solar volatility. As an example, Fig. 10.3 shows the variability of frequencies of inter-area modes estimated by the PhasorPoint application at ISO-NE every 5 s within 24 h.

The histogram is not a combination of sharp peaks and shallow valleys which would be expected at the stable frequencies of oscillatory modes. Frequency of the same mode significantly fluctuates over time. Accurate estimation and robust tracking of frequencies of natural modes online with the noisy measurements is difficult by itself. Oscillation spectrum is dynamic, and frequencies can change within hours and minutes. Thus, the use of magnitudes of oscillations as an indicator of the source of oscillations in online environment is unreliable.

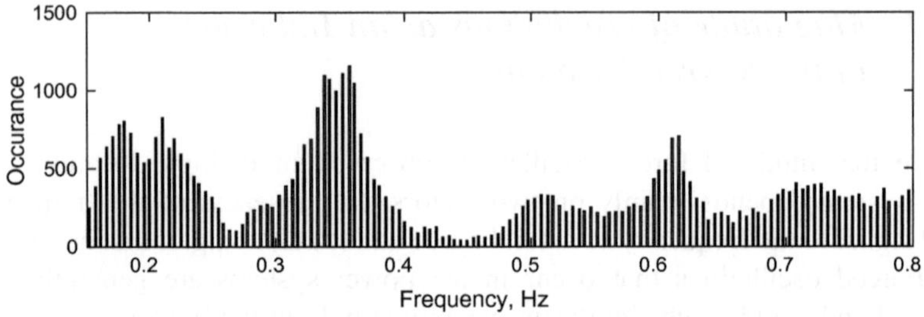

Fig. 10.3 Histogram of oscillation frequencies within 0.15–0.8 Hz range estimated in ISO-NE every 5 s during 24 h

10.4.3 The Dissipating Energy Flow Method

Several proposed in the literature oscillation source locating methods were analyzed at ISO-NE and the energy-based method [14] was selected as the candidate for practical implementation. The method is based on the primary attribute of oscillations, i.e., energy, and thus expected to be more robust in multiple possible situations, while other methods based on other attributes of oscillations experience difficulties.

10.4.3.1 Original Energy-Based Method

An energy-based method calculates the flow of dissipating transient energy in the network and demonstrates that the calculated energy is equivalent to the energy dissipated by a damping torque. The flow of dissipating energy in a branch ij is expressed by using bus voltage angles or bus frequencies as follows [14]:

$$\begin{aligned} W_{ij}^{D} &= \int (\Delta P_{ij} d\Delta\theta_{ij} + \Delta Q_{ij} d(\Delta \ln V_i)) \\ &= \int (2\pi \Delta P_{ij} \Delta f_i dt + \Delta Q_{ij} d(\Delta \ln V_i)) \end{aligned} \tag{10.1}$$

where ΔP_{ij} and ΔQ_{ij} are the deviations from the steady-state values of the active and reactive power flow in branch ij; Δθi and Δfi are the deviations from the steady-state values of the bus voltage angle and frequency at bus i; Vi is the bus voltage magnitude. ΔlnVi = lnVi – lnVi,s, where Vi,s is the steady-state voltage magnitude.

The formula (10.1) was derived by using the assumption of the lossless network and constant power load model.

The flow of dissipating energy by (10.1) can be considered as a regular power flow through network in terms of the source–sink relation and allows the tracing of the source of sustained oscillations. The source of transient energy for sustained oscillations is the source of oscillations. It was demonstrated in [14] that the method also works at varying amplitude of oscillations and with realistic model of generators by extracting the information about the dissipating energy from the slope of W_{ij}^{D}curve rather than from its instantaneous values. It was also demonstrated in [15] that in the linear single machine system, the rate of dissipated energy calculated by (10.1) for a specific mode is approximately proportional to the real part of the eigenvalue of that mode. That makes it possible, under above assumptions, the utilization of the dissipated energy approach for the estimation of the damping contribution of each individual generator or any other system element into damping of a specific mode of oscillations.

Standard PMU measurements of bus frequency, voltage, and current phasors are sufficient for the estimation of the dissipating energy flow by (10.1). Note that power quantities, i.e., P and Q, are calculated from the voltage and current.

10.4.3.2 Challenges in Actual Power Systems

The underlying assumptions on the lossless network, system linearity, and constant power load model do not hold for actual bulk power systems, and the impact of these assumptions needs to be evaluated.

The most challenging factor to apply the method proposed in [14] is the need to find the deviations from the steady-state values for all quantities used in (10.1). Ideal steady-state conditions in actual power systems practically do not exist, so the de-trending process is required in order to estimate steady-state quantities. Reasonably accurately de-trending can be done only for a short period of time which is often too short to reliably trace the flow of dissipating energy. A particular difficulty is the de-trending of PMU bus voltage angle measurements. Due to the nature of PMU measurements, the absolute value of an angle may change by several hundred degrees over 10–30 s at off-nominal frequency, while the deviation of the angle from its steady-state value, which is required in (10.1), is typically less than 1°. Steady-state values from traditional state estimation (SE) cannot be reliably used here due to time misalignment, accuracy of SE results, and absence of bus frequency values.

10.4.3.3 Modification of the Method for Use with Actual PMU Data

The calculation of deviations for all variables in (10.1) can be effectively done without the need of de-trending and estimation of the steady-state quantities by filtering modes of interest from PMU data in the frequency domain. Filtered mode for any variable has zero steady-state value. Then, dissipating energy flow is calculated for each mode separately.

Formula (10.1) cannot be directly used for filtered signals because lnVi is defined only for positive values of voltage, but a filtered signal has both positive and negative values with a zero mean, which makes calculation of lnVi impossible. To address this problem, the term d(ΔlnVi) is replaced by the approximately equivalent d(ΔVi)/Vi* accounting for a single oscillatory mode only [17]

$$
\begin{aligned}
W_{ij}^{D} &\approx \int (\Delta P_{ij} d\Delta\theta_{ij} + \Delta Q_{ij}\frac{d(\Delta V_i)}{V_i^*}) \\
&= \int (2\pi \Delta P_{ij}\Delta f_i dt + \Delta Q_{ij}\frac{d(\Delta V_i)}{V_i^*})
\end{aligned}
\tag{10.2}
$$

where Vi* = Ṽi + ΔVi and Ṽi is the average voltage in the studied period. For discrete PMU signals, a discrete-time version of (10.2) has the following form:

$$
W_{ij,t+1}^{D} = W_{ij,t}^{D} + 2\pi \Delta P_{ij,t}\Delta f_{i,t} \cdot t_s + \Delta Q_{ij,t}\frac{\Delta V_{i,t+1} - \Delta V_{i,t-1}}{2V_{i,t}^*}
\tag{10.3}
$$

Quantities with Δ in (10.2) and (10.3) are the filtered components for a mode; t_s is time interval between PMU samples; index t reflects the time instant. Integration time limits in (10.2) are determined from the transient when sustained oscillations have significant magnitude larger than noise. Duration of the time interval should preferably contain 20–40 periods of oscillations. Smaller interval containing as few as 4–6 periods can also work as well but with less robustness.

Reasonably good filtering of the mode of interest with frequency f_s can be done by using a 4th–6th-order Butterworth band pass filter with the pass frequencies fp = (1 ± ε) · fs, where ε = 0.05; cutoff frequencies fc = (1 ± 2ε) fs; 1 dB of ripple allowed and 10–15 dB attenuation at both sides of the pass band.

Calculation of the dissipating energy flow is sensitive to the phase angle among all quantities in (10.2) and (10.3). That is why the accurate calculation by (10.2) and (10.3) requires preserving the phase angle relation among all quantities during the band-pass filtering. Zero-phase distortion is achieved by applying filtering in both forward and reverse directions for all signals. Filtering typically creates significant deviations of the filtered signals at the edges of the interval [27]. Keeping 30–50% of the middle part of filtered signal only works best in (10.3).

Calculated by (10.3), energy W_{ij}^D in each branch is either increasing (positive) or decreasing (negative) over time. Negative means the dissipating energy flows into the bus from the source. Positive means the energy flows from the source bus to system elements dissipating transient energy.

Dissipating energy W_{ij}^D is a function of time, and it can be simplified for the purpose of oscillation source location by making a linear approximation for each branch:

$$W_{ij}^{D}(t) = DE_{ij} \cdot t + b_{ij}, \tag{10.4}$$

where DE is a slope or rate of change of dissipating energy and b is a constant.

DE is constant for sustained oscillations with constant magnitude, while it is not constant for the oscillations with varying magnitude; however, the ratio of DE values for different branches remains almost constant. Based on this observation, it is most convenient using the normalized value:

$$DE_{ij}^{*} = DE_{ij} / \max_{i,j}\{|DE_{ij}|\} \tag{10.5}$$

The sign of DE_{ij} indicates the direction where the energy is coming from. The source of oscillations is located in the system elements producing the dissipating energy.

Calculation of DE coefficient in multiple network elements allows to trace the source of sustained oscillations. Figure 10.4 illustrates the general rule for interpretation of DE results. An arrow indicates the DE calculated from PMU measurements at a specific location. Direction of arrow indicates the sign of DE, and the size indicates the value of DE. The largest DE values are observed in the branches surrounding the source of oscillations. Figure 10.4a illustrates the interpretation of DE for PMU measurement at the bus i. Figure 10.4b illustrates the DE calculated for multiple branches monitored by the same PMU.

The described above modification of the energy-based method with the PMU processing is called the dissipating energy flow (DEF) method [17]. The DEF method provides a practical way for calculating the DE coefficients for the network elements monitored by PMU. The major difference of the DEF method from the original energy-based methods is PMU signal processing which is a key step in converting the energy-based method into a robust automated methodology for the use with actual PMU data. Proposed PMU processing also creates an opportunity using the DEF method in a new type of PMU application: a decentralized online estimation of the contribution of a generator into the damping of a specific oscillation mode. Some details are described below in Sect. 10.8.

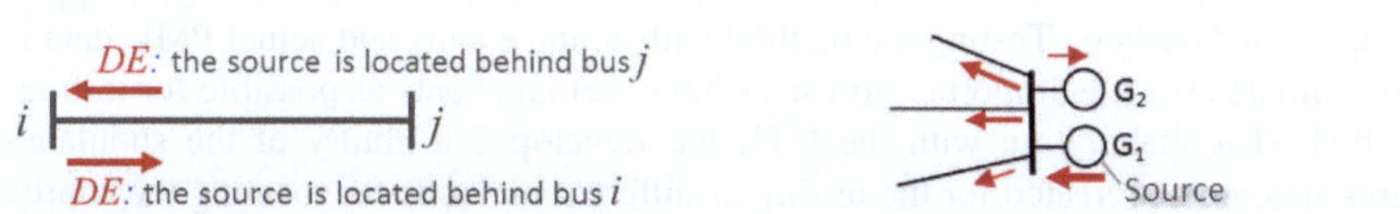

(a) Single branch monitored by PMU (b) Multiple branches monitored by PMU

Fig. 10.4 Interpretation of the DE coefficients

10.4.4 Testing the Oscillation Source Locating Methods

The method suitable for practical implementation should satisfy the set of requirements described above. Practically observed and potentially feasible sustained oscillations have multiple features, and practically meaningful method should be able to identify the source for entire variety of situations or at least for practically significant part of sustained oscillations. The features of poorly damped natural oscillations include

- Local or inter-area modes or their combination;
- Presence of a single or multiple sources of negative damping.

Important features of the forced oscillations include

- Forced signal not equal to the frequency of natural modes;
- Forced signal resonating with a single local or inter-area mode;
- Non-harmonic periodic forced signal creating spectra of oscillations which can be non-resonating or resonating with local and inter-area modes simultaneously;
- Presence of a single or multiple sources of forced signals with the same or different frequencies.

Sustained oscillations with significant magnitude regardless of their natural or forced nature represent the threat for the power system, and the practically meaningful method should be able to locate the source of oscillations for any sustained oscillations.

The ultimate test for the method should be the use of PMU data from an actual event when the source of oscillation is known with a high confidence level. At the same time, it is difficult to expect that the set of actual events for all above situations is available. The practical approach for rigorous testing is to use a combination of simulated and actual cases. The simulated approach allows for a flexibility to synthesize any desirable situation where the output of time-domain simulation is treated as PMU data. The use of simulated cases should be considered as a qualification test only to verify the general usefulness of the method in a variety of possible situations. Simulated PMU data can be considered as idealized and lacking some real-life PMU features such as noise, systematic errors, trending, dropped samples, and outliers. Testing the method with actual events and actual PMU data is the ultimate test. It is important to use as many actual events as possible for testing.

ISO-NE collaborating with the UTK has developed a library of the simulated cases specifically created for the testing of different methods for locating the source of the sustained oscillations [12]. Test cases are based on a simplified Western Electricity Coordination Council (WECC) 179-bus, 29-generator system model [28] and contain nine cases of poor damped natural modes and twelve cases of forced oscillations. Test cases are representative of electromechanical oscillations which are observed and could be observed in actual systems including: (a) undamped local and inter-area modes and their combinations, (b) single and multiple sources of oscillations, (c) forced periodic and harmonic injection signals creating resonance and near resonance with local and inter-area modes. All cases in [12] have network with losses and constant power load model.

The library of the simulated test cases is publicly accessible via Internet [29] and contains a complete description of all cases with the sets of simulated PMU measurements for voltages in all buses, currents in all branches, rotor angles, and speeds for all generators. The library contains also a complete set of input files for TSAT software [30], results of modal analysis, and the sensitivity of real parts of eigenvalues for damping coefficients of all generators. This information can be used for the creation of new cases if needed.

A publicly available collection of real events of sustained oscillations with corresponding PMU measurements which can be used for testing of the source locating methods does not exist today. The IEEE task force on Oscillation Source Location has been created in 2016 [31]. One of the goals of this task force is creating publicly available case library of actual cases of sustained oscillations and make it available at the task force Web site.

10.5 Testing the Dissipation Energy Flow Method

10.5.1 Simulated Cases

The test case library was used for testing of the DEF method. Table 10.1 shows the DEF method test results with cases of poorly damped natural oscillations. The method was applied to the transient from 5 to 40 s. Negative damping in [12] is created by setting the negative value of the damping coefficient D of a specific generator. Contribution of a generator with classical model into negative damping, shown in column 5, is evaluated by using the following linear approximation $\Delta\alpha = \partial\alpha/\partial D \cdot D$, where $\partial\alpha/\partial D$ is the sensitivity of the real part of the eigenvalue w.r.t. the parameter D. DE value in the last column is normalized per the largest DE value of the generator branches. The column lists only positive values greater than 0.01 indicating the sources of energy. The DEF method reliably identifies the main source of oscillations for all poorly damped oscillations and oscillations with negative damping. In ND6, ND7, ND8 cases, where multiple sources impact the same mode, DE value qualitatively reflects the contribution of individual generator into negative damping. Generators 6 and 30 in ND3 case have slightly positive DE while not being the sources of negative damping (Table 10.1).

Table 10.2 contains results for the cases of forced oscillations. Forced oscillations in [12] were created by adding a periodic forced signal into the excitation system of specific generator(s). In case F6-2, the forced signal is applied to generator 79 and has rectangular shape creating spectra of odd harmonics with the lowest frequency of 0.2 Hz. In other cases, the forced signal has a sinusoidal shape at the specific frequency listed in column 3. Forced signal in cases F1–F3 has a

Table 10.1 Results for simulated natural oscillations

Case	Natural modes		Actual source		DE results	
	Frequency (Hz)	Damping (%)	Gen	$\Delta\alpha$ p.u.	Detected source	DE
ND1	1.41	0.01	45	1.0	45	1.0
ND2	0.37	0.02	65	1.0	65	1.0
ND3	0.46	2.22	11	1.0	11	1.0
			–	–	6	0.05
			–	–	30	0.03
	0.70	1.15	11	1.0	11	1.0
ND4	0.46	0.68	11	1.0	11	1.0
	0.70	−0.58	11	1.0	11	1.0
	1.63	0.54	11	1.0	11	1.0
ND5	0.46	0.69	11	1.0	11	1.0
			–	–	112	0.02
	0.70	−0.19	11	1.0	11	1.0
ND6	1.41	−0.93	45	1.0	45	1.0
			159	0.45	159	0.38[a]
ND7	1.41	−0.4	45	0.55	45	0.55
			159	1.0	159	1.0
ND8	1.27	−1.06	36	1.0	36	1.0
			45	0.10	45	0.27
	1.41	−0.22	36	0.01	36	0.02
			45	1.0	45	1.0
ND9	0.46	−0.86	11	1.0	11	1.0
	0.69	−1.81	11	1.0	11	1.0

[a]Time interval of transient from 34 to 40 s was used. For the time interval from 5 to 40 s, generator 159 is not detected as the source

frequency equal to the frequency of certain natural modes, thus creating resonances. Frequencies of the forced signal in cases F4-2 and F5-2 are respectively in between of two close frequencies of natural local and inter-area modes. Case F7-1 has two simultaneous sources at different generators creating resonance conditions with two different natural modes. Case F7-2 has two sources at different generators but creating resonance conditions with the same natural mode. Results for some of the tested cases from [12] are not shown in the table because they represent small variations of the F4–F6 cases. Figure 10.5 shows the power spectral density of oscillations excited in the F6-2 case. Figure 10.6 shows W_D of all 29 generators as a function of time for the first four harmonics on the forced signal. Red curve representing the generator 79 clearly indicates this generator as the source for all harmonics.

The DEF method reliably identifies the source of oscillations for all cases.

Table 10.2 Results for simulated forced oscillations

Case	Injected signal			DE results	
	Type	Frequency (Hz)	Actual source	Detected source	DE
F1	sin	0.86	4	4	1.0
F2	sin	0.86	79	79	1.0
F3	sin	0.37	77	77	1.0
F4-2	sin	0.85	79	79	1.0
F5-2	sin	0.46	79	79	1.0
F6-2	Periodic rectangular creating odd harmonics	0.2	79	79	1.0
			–	77	0.02
		0.6	79	79	1.0
		1.0	79	79	1.0
		1.4	79	79	1.0
F7-1	sin	0.65	79	79	1.0
		0.43	118	118	1.0
F7-2	sin	0.43	70	70	0.14
		0.43	118	118	1.0

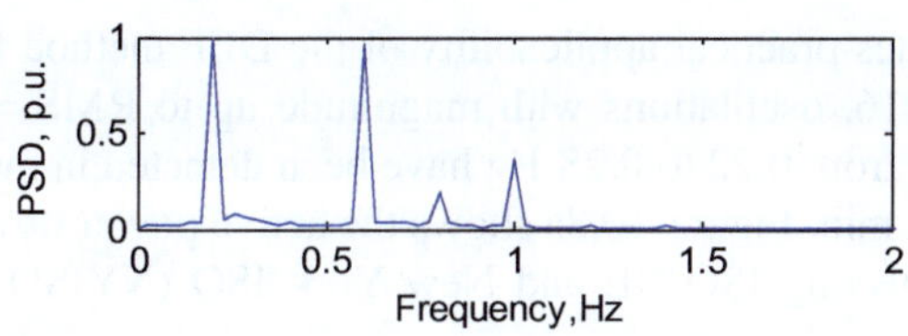

Fig. 10.5 Power spectral density of the system oscillations in F6-2 case

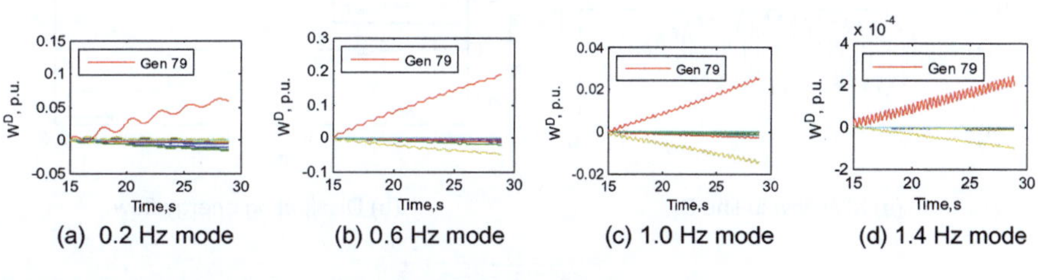

Fig. 10.6 Dissipating energy flow of all 29 generators in F6-2 case

10.5.2 Actual Events in ISO-NE System

The DEF method was tested with more than 40 instances of actual events for oscillation frequencies in the range from 0.05 to 1.7 Hz where the actual source of oscillation was known with a high confidence level based on operational information, analysis of multiple data sources, and communication with power plants. The set of PMU data includes frequency and voltage measurements from 44 locations and currents from 252 branches (lines and transformers) of the 345 and 115 kV network. Each location has more than one branch monitored by PMU. The DEF method has demonstrated high efficiency for all tested cases in identifying the source of forced oscillations. Calculated dissipating energy flow in transmission elements also demonstrates high consistency with the expected direction of flow from the source to sinks. Below are some examples of actual events.

April 5, 2013 event. Rapidly growing in magnitude oscillations of 0.12 Hz were observed in a significant part of the system. The magnitude of oscillations reached up to 100 MW peak-to-peak in some transmission lines. Oscillations gradually disappeared in 3 min. Figure 10.7a shows MW flow in one of the lines.

The DEF method has clearly identified one power plant as a source of the oscillations, and Fig. 10.7c illustrates the DE flow in the 345 kV network around that power plant. Figure 10.7b shows the flow of dissipating energy over time at the connection point of the power plant to the network. Note that the energy coming from the generator practically equals to the sum of energy flowing over lines A and B and provides a reasonable nodal balance:

$$DE_{Gen} + DE_A + DE_B = 1.000 - 0.636 - 0.397 = -0.032$$

That demonstrates practical applicability of the DEF method for actual systems.

On June 17, 2016, oscillations with magnitude up to RMS = 11 MW and frequency fluctuating from 0.22 to 0.28 Hz have been detected in multiple locations of ISO-NE lasting 45 min. Figure 10.8a shows the active power flow in one of the two 345 kV lines connecting ISO-NE and New York ISO (NYISO) during the initial

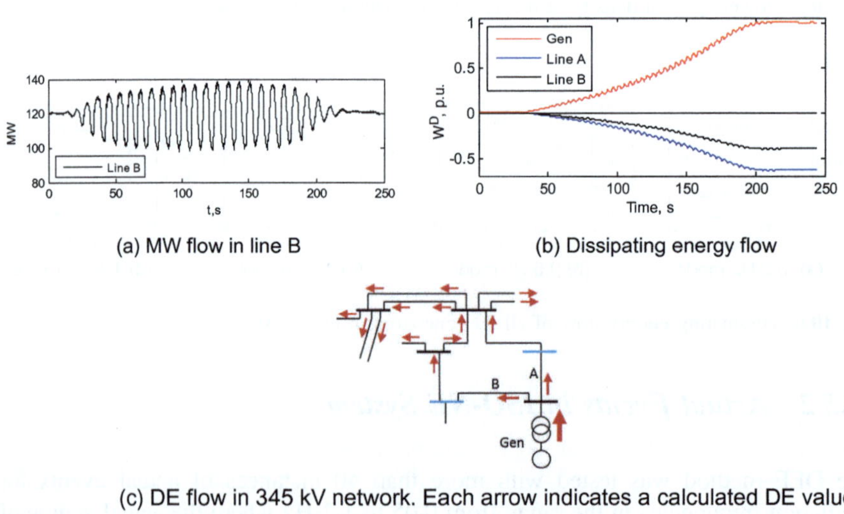

Fig. 10.7 Oscillation source locating for April 5, 2013 event

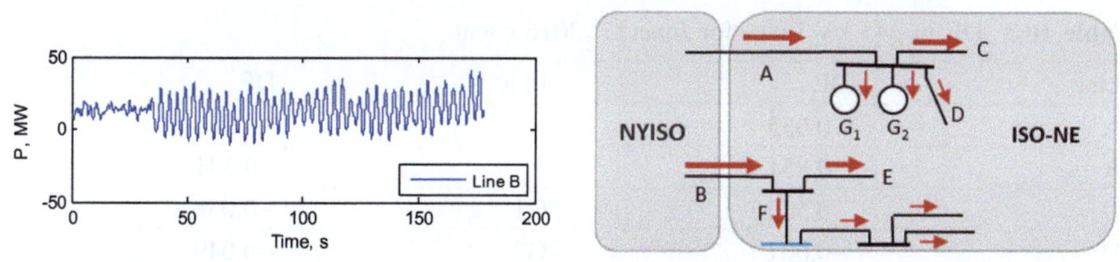

(a) MW flow in 345kV line B

(b) DE flow in the 345 kV network

Fig. 10.8 Locating the source of oscillations for June 17, 2016 event

stage of the event. The mode shape showed that the majority of ISO-NE generators were oscillating in phase, suggesting that ISO-NE PMUs allow the observation of only a part of the system-wide event. Control room operators in this situation need to know whether the source of oscillations is located inside or outside of the control area in order to take proper course or actions. Figure 10.8b shows the DE values for the 345 kV part of the network connecting ISO-NE and NYISO, and Table 10.3 contains these DE values. DE flow clearly indicates that the source of oscillations was located outside ISO-NE.

On June 30, 2016, three generators G1, G2, and G3 at one of the power plants were operating in the same conditions producing 237 MW each. A failure suddenly occurred at the excitation system of G3 resulted in 1.3 Hz oscillations with RMS magnitude of 18 MW and consequential tripping of this unit. Figure 10.9a shows power plant terminal voltage. Excitation failure at G3 started at $t = 5$ s, and the unit was disconnected at $t = 24$ s. All three units were monitored by PMU, and Fig. 10.9b illustrates the results of the DEF method. Unit G3 is reliably identified as the source of oscillations producing energy, while other two units were contributing into damping by dissipating energy. Note that the G3 has the highest DE flow in the network. This example illustrates the ability of the DEF method to identify a specific unit as the source from multiple units within a power plant even for very short oscillatory event lasting only 8 s.

10.5.3 Actual Events in WECC

The DEF method was also tested with two events where the widespread forced oscillations with significant MW magnitude were observed in the WECC system on March 10 (1.48 Hz event) and on November 17, 2015 (1.17 Hz event), respectively. PMU from 55 locations metering flows in 271 transmission elements were kindly provided by the Bonneville Power Administration (BPA) for testing. Mapping of the PMU data with the system topology was not known during the test. Results of the DEF method in the form of the DE largest values for 20 transmission elements were sent back to BPA. As an example, Table 10.4 shows some of the DE

Table 10.3 DE in 345 kV Lines for June 17, 2016 event

Line	DE	Line	DE
A	0.635	E	–0.514
B	0.851	F	–0.341
C	–0.525	G1	–0.058
D	–0.007	G2	–0.049

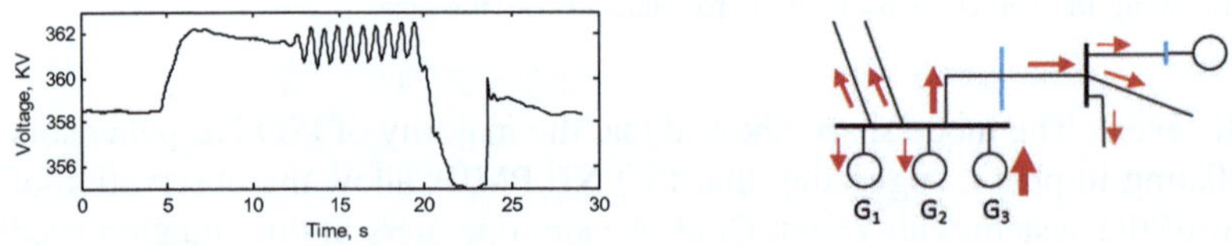

(a) Power plant terminal voltage during event (b) DE flow in 345 kV lines around suspect power plant

Fig. 10.9 Oscillation source locating for June 30, 2016 event

Table 10.4 DE of transmission lines for November 17, 2015 event

Line	L1	L2	L3	L4	…	L20
DE	–1.0	0.88	–0.86	0.84	…	–0.43

values for the November 17, 2015 event. BPA personnel have mapped DE values to the system topology and confirmed that DE values correctly trace the source of forced oscillation in both cases.

10.5.4 Impact of the Energy-Based Method Assumptions

The underlying assumptions of the energy-based method on the lossless network and constant power load model are applied to the DEF method as well. Actual power systems have losses, and the characteristics of loads deviate from the constant power model. Comparison of the results for lossless and network with losses shows that the losses have very little impact on DE values. The sum of DE values from the both sides of a transmission line or transformer with losses is not always equal to zero and could be positive or negative depending on the location of a transmission element. Per the energy-based method, a transmission line with negative sum of DEs consumes transient energy; a line with positive sum of DEs produces transient energy and thus could be formally classified as the source of oscillations. Such a classification would be wrong because a passive transmission line cannot be a source of oscillations. An artifact of the energy-based method that a passive transmission line can generate or produce transient energy is the factor complicating the location of the oscillation source by looking at the DE flow in the network.

Similar issue exists for the loads modeled differently than a constant power. Per the energy-based method, a load with constant resistive component has DE value not always equal to zero but can be positive or negative depending on the location in the network. The load with a positive DE produces transient energy and formally can be classified as the source of oscillations. Again, that is an artifact of the energy-based method and it could be observed for a load modeled as a constant impedance, ZIP, or constant current [17, 32, 33]. Numerical impact of the constant resistance load model on DE values could be much higher than the impact of neglecting network losses. DE values of constant resistance loads can be comparable with the DE value form a generator—actual source of oscillations. That fact complicates the interpretation of the DE flow in the network and locating the source of oscillations.

The good news is that the DE values of generators are much less impacted by the constant resistance loads or network losses compared to DE values in transmission lines. Importantly, the DE values of generators only can still be used to identify the generator—source of oscillations. Theoretically, a cycling load, HVDC, or FACTS can be a source of forced oscillations. Actual monitoring of power system dynamics shows that the sources of oscillations for practically all detected cases were generators. That indicates that the practically important for the source locating method is the ability to identify the generator—source of oscillation.

Analysis of the simulated cases shows that the negative impact of the constant resistance load model is more severe for load models with larger percentage of constant resistance. Actual characteristics of the loads in the presence of sustained oscillations are often unknown, and the severity of the negative impact of load modeling on DE values needs to be evaluated based on actual events. Application of the DEF method for more than 40 actual events in the ISO-NE system with the high confidence knowledge of the source of oscillations indicates that the potentially negative impact of network losses and unknown characteristics of the load do not prevent from successfully identifying the generator—source of oscillations by the DEF method. The DE flow in the network in actual events also does not significantly deviates from the ones which would be expected in an ideal situation of lossless network and constant power load characteristics.

Difficulties in interpretation of DE results related to network losses and unknown load characteristics are not critical, and the DEF method can still be effectively used in vast majority of practical situations when the source of oscillations is a generator. In testing, the DEF method has demonstrated universal and high efficiency for locating the source of oscillations for a variety of actual and simulated events. In our opinion, the DEF method far over performs any other available methods for practical needs and is a good candidate for practical implementation in power systems.

10.6 Installation of PMUs for Locating the Source of Forced Oscillations

The resolution of the source location depends on the system observability by PMUs. For the majority of practically observed oscillations with large magnitude, the most probable sources of oscillations are individual generators or control systems at a plant level. Considering that fact, the most efficient installation of PMUs for locating the source power plant is at the point of radial interconnection of that power plant to the grid. Availability of PMU measurements at a radial connection of individual generator to the grid allows increasing the location resolution to an individual generator within the power plant. The more generators are monitored by PMUs the higher the chances for an accurate identification of the source of oscillations.

Availability of PMU measurements in the major transmission lines between control areas allows to calculate the DE flow in these lines and thus to identify the area—source of oscillations. So, the general recommendation for PMU installation targeting the oscillation source detection is to install PMU at the points of interconnection of every generator or power plant and at major transmission lines between control areas. Additionally, monitoring HVDC and FACTS by PMUs allows to track these devices as a source of oscillations.

10.7 Online Oscillation Management at ISO-NE

The DEF method was implemented in the Oscillation Source Locating (OSL) application developed at ISO-NE, which is a part of the online oscillation management illustrated in Fig. 10.10. The objective of the online oscillation management is

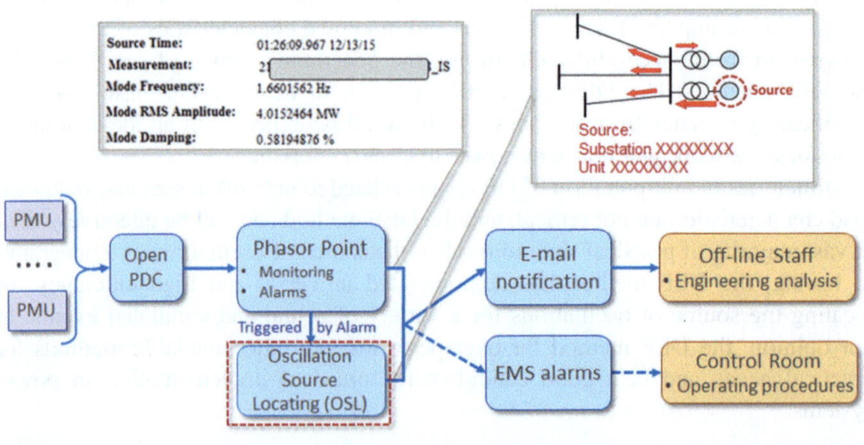

Fig. 10.10 Data flow and functional diagram of the online oscillation management concept at ISO-NE

to detect dangerous sustained oscillation, locate the source, and provide results of the analysis to the engineering staff for offline analysis and to the Control Room operator.

PMU measurements are constantly analyzed in the PhasorPoint application. PhasorPoint analyzes the spectrum of oscillations every 5 s at multiple system locations and estimates the frequency, amplitude and damping for every dominant mode in several pre-defined frequency bands covering the entire range of electro-mechanical oscillations from 0.05 to 4 Hz. Should the combination of any dominant mode's damping and magnitude exceed a corresponding pre-defined threshold for an alert or alarm, PhasorPoint generates an alert or alarm.

Every oscillatory alarm triggers the OSL application, which calculates the DE flow in the network and identifies a suspect source of oscillations. Combined information on the characteristics of the detected oscillations together with the information of the suspect source of oscillations is sent by e-mail to the pre-defined list of the recipients.

The same information is sent to the EMS Alarm system which is the main alarming tool for the Control Room operator. It does not make sense to send an alarm to the operator indicating that the system experiences dangerous oscillations without simultaneously providing the instruction on how to mitigate the threat. Information on the suspect source of oscillations combined with the exact instructions on the remedial actions applicable at different situations from operating procedures represents the set of actionable information for the operator (Fig. 10.11).

Robustness of the online OSL application depends on the efficiency of automatic selection of the time interval for the analysis. PhasorPoint produces an alarm with some delay after the detection of oscillations, and the duration of the delay can fluctuate depending on the parameters of oscillations. The oscillations also can be

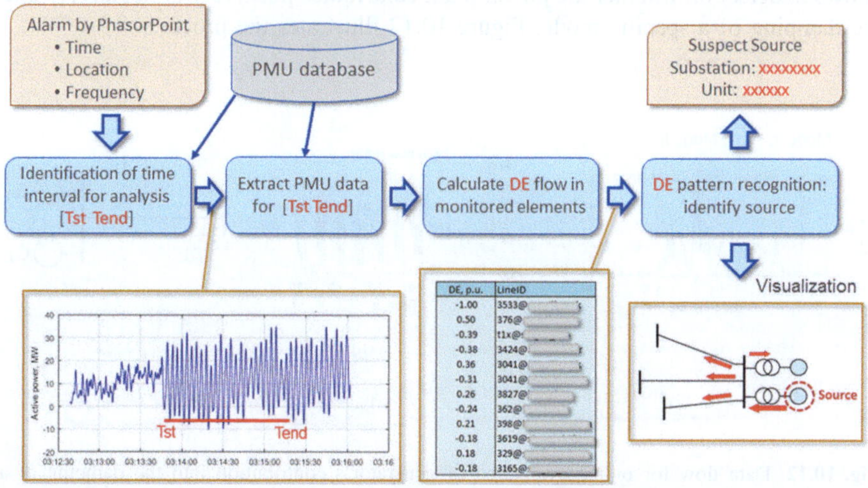

Fig. 10.11 Data flow diagram of online version of the OSL application

triggered by the switching in the system, creating a significant transient. Oscillations can be also suddenly stopped by a tripping event. All these factors can negatively impact the efficiency of the DEF method if the studied period is improperly selected. Additionally, some instances of bad PMU data can falsely create an oscillatory alarm and PMU data itself can have outliers and dropped periods. Automatic selection of the time interval for study in the OSL application is designed to minimize the negative impact of the above issues.

Regular output of the DEF method provides the DE coefficients in a tabular form for all network elements monitored by PMU. Tabular form is inconvenient for analysis, and more efficient way is to visualize the DE flow on the oneline diagram. Additionally, the OSL application includes a pattern recognition module which converts DE flow into short information indicating a specific area, substation, or specific generator as the source of oscillations.

The oscillation monitoring and oscillation source locating system works automatically, 24/7, and ensures that all dangerous oscillations are detected and characterized and corresponding information is communicated to the Control room operator and designated engineering personnel.

10.8 Online Monitoring of the Generator Damping Contribution

The DEF method opens an opportunity for a new type of PMU application—decentralized online estimation of generators' contribution into damping of a specific oscillatory mode. Application of the DEF method by using measurements from only one PMU installed at the Point Of Interconnection (POI) of a power plant allows to detect on whether the power plant contributes positively or negatively into the damping of a specific mode. Figure 10.12 illustrates the process.

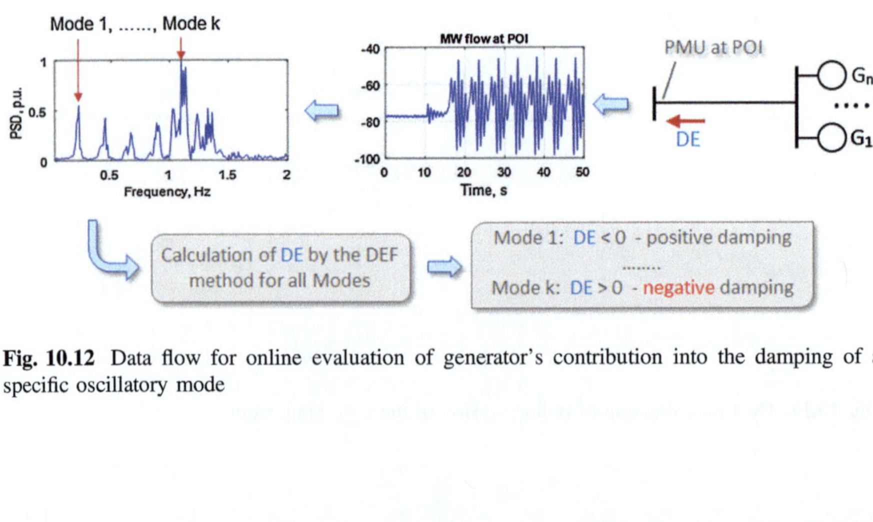

Fig. 10.12 Data flow for online evaluation of generator's contribution into the damping of a specific oscillatory mode

PMU measurements are converted into the frequency domain for identification of the frequencies of the dominant oscillatory modes. The DEF method can be applied to any of these modes and can provide corresponding DE values. Positive DE means a power plant negatively impacts the mode damping. Negative DE means a power plant contributes into the mode damping. The same analysis can be applied to an individual generator if PMU is metering the POI of an individual generator.

The process could be triggered by an oscillatory event or even can work at ambient conditions continuously. It was demonstrated in [34] that the transient energy of a generator estimated at ambient conditions could be potentially used for quantitative estimation of the generator's contribution into the mode damping. The accuracy of such estimation for real power systems needs further investigation. However, qualitatively distinguishing of whether a generator contributes positively or negatively into the mode damping is feasible.

New PMU application could be useful for an integrated monitoring of the health of a power plant or individual generator in terms of the damping contribution. At a normal situation, a generator should be positively contributing to the dominant modes damping. Detection of the negative damping is an indicator of the issues in the equipment or in the control systems.

10.9 Conclusions

The objective of the oscillation management is an online detection and characterization of sustained oscillations based on PMU measurements and efficient mitigation of dangerous oscillations. At ISO-NE, the detection and characterization of oscillations is provided by using PMU measurements and the PhasorPoint application. PhasorPoint generates alerts and alarms if the magnitude and damping of detected oscillations exceed pre-defined threshold values. Oscillation Source Locating application automatically processes every alarm and estimates the source location. Information on oscillation parameters and the estimated source location is sent to the designated personnel.

This information combined with operating procedure describing specific mitigation measures is an actionable information for the Control Room Operator for efficient mitigation of dangerous oscillations online. The same information can be used offline in coordinated activities of ISO and the personnel of the suspect-source power plant to identify the exact reasons causing oscillations and for developing actions preventing the occurrence of such oscillations in the future.

References

1. Black HS (1953) Modulation theory. New York-London, 363pp
2. https://en.wikipedia.org/wiki/2009_Sayano%E2%80%93Shushenskaya_power_station_accident
3. Wang B, Sun K (2016) Location methods of oscillation sources in power systems: a survey. J Modern Power Syst Clean Energy. https://doi.org/10.1007/s40565-016-0216-5
4. Thorp JS, Seyler CE (1998) Phadke AG Electromechanical wave propagation in large electric power systems. IEEE Trans Circuits Syst I, Fundam Theory Appl 45(6):614–622
5. Markham PN, Liu Y Electromechanical speed map development using FNET/GridEye frequency measurements. In: IEEE PES general meeting | conference and exposition, national Harbor, MD, USA, 27–31 Jul 2014
6. Hu N, Li X, Yang Y et al (2013) Research for space-time variational features of frequency for power system forced oscillations and disturbance source location. J Sichuan Univ (Eng Sci Edn) 45(6):135–142
7. Geng T, Zhang Z, Xiang L et al (2015) A locating and splitting scheme for disturbance source of forced power oscillation based on the propagation characteristic. Power Syst Protect Control 43(6):98–103
8. Shaltout AA, Feilat EA (1992) Damping and synchronizing torque computation in multimachine power systems. IEEE Trans Power Syst 7(1):280–286
9. Zheng X (2000) The complex torque coefficient approach's applicability analysis and its realization by time domain simulation. Proc CSEE 20(6):1–4
10. Li Y, Huang Y, Liu J et al (2015) Power system oscillation source location based on damping toruqe analysis. Power Syst Protect Control 43(14):84–91
11. Ashwal NA, Wilson D, Parashar M (2014) Identifying sources of oscillations using wide area measurements. In: Grid of the future symposium, CIGRE USNC, Houston, TX, 19–21 Oct 2014
12. Maslennikov S, Wang B, Zhang Q, Ma F et al (2016) A test cases library for methods locating the sources of sustained oscillations. In: 2016 IEEE PES general meeting, Boston, MA, 17–21 July 2016
13. Myers R, Trudnowski D (2013) Effects of forced oscillations on spectral-based mode-shape estimation. IEEE PES GM, Vancouver, BC, Canada, pp 21–25
14. Chen L, Min Y, Hu W (2013) An energy-based method for location of power system oscillation source. IEEE Trans Power Syst 28(2):828–836
15. Chen L, Min Y, Chen Y, Hu W (2014) Evaluation of generator damping using oscillation energy dissipation and the connection with modal analysis. IEEE Trans Power Syst 29(3): 1393–1402
16. Li Y, Shen C, Liu F (2013) A methodology for power system oscillation analysis based on energy structure. Autom Electr Power Syst 37(13):49–56. https://doi.org/10.7500/aeps 201211185
17. Maslennikov S, Wang B, Litvinov E (2017) Dissipating energy flow method for locating the source of sustained oscillations. Electr Power Energy Syst (88):55–62
18. Yang YQ, Liu TQ, Li XY et al (2015) Discussion on energy mechanism of negative damping and oscillation source locating method based on equivalent circuit method. Autom Electr Power Syst 39(10):63–68. https://doi.org/10.7500/aeps20140409012
19. Ma J, Zhang P, Fu H et al (2010) Application of phasor measurement unit on locating disturbance source for low-frequency oscillation. IEEE Trans Smart Grid 1(3):340–346
20. Jiang C, Liu J, Liu Y et al (2015) Online forced oscillation detection and identification based on wide area measurement system and CELL theory. Electr Power Autom Equip 35(2): 125–132
21. Wang X, Turitsyn K (2016) Data-driven diagnostics of mechanism and source of sustained oscillations. IEEE Trans Power Syst PP(99):1–11. https://doi.org/10.1109/tpwrs.2015.2489656

22. Nudell T, Chakrabrtty A (2015) Graph-theoretic methods for measurement-based input localization in large networked dynamic systems. IEEE Trans Autom Control 60(8): 2114–2128
23. McNabb P, Bochkina N, Bialek J (2010) Oscillation source location in power systems using logic regression. In: IEEE PES ISGT Europe, Gothenburg, Sweden, 11–13 Oct 2010
24. McNabb P, Bochkina N, Wilson D, Bialek J (2010) Oscillation source location using wavelet transforms and generalized linear models. In: IEEE PES transmission and distribution conference and exposition, New Orleans, LA, USA, 19–22 Apr 2010
25. Sarmadi SAN, Venkatasubramanian V (2015) Inter-area resonance in power systems from forced oscillations. IEEE Trans Power Syst PP(99):1–9
26. Reliability Guideline (2017) Forced oscillation monitoring and mitigation. North American Electric Reliability Corporation
27. Gustafsson F (1996) Determining the initial states in forward-backward filtering. IEEE Trans Signal Process 44(4):988–992
28. Electric Power Research Institute "DC Multi-Infeed Study", EPRI TR-TR-104586s, Projects 2675-04,-05, Final Report (1994)
29. http://web.eecs.utk.edu/~kaisun/Oscillation/
30. http://www.powertechlabs.com/software-modeling/dynamic-security-assessment-software/# serv193
31. http://web.eecs.utk.edu/~kaisun/TF/index.html
32. Duan Q, Zhu Y, Zhou H, Ma J (2012) Impact of load model on disturbance sources locating based on energy function. Shaanxi Electr Power 8:16–19
33. Yang Y, Liu T, Li X et al (2013) An analysis to the concentric relaxation phenomenon of power system forced oscillations. J Sichuan Univ (Eng Sci Edn) 45(4):163–170
34. Chen L, Sum M, Min Y, Xu X, Xi J, Li Y (2017) Online monitoring of generator damping using dissipation energy flow computed from ambient data, IET Digital Library, http://ietdl.org/t/DzHJd. February 10, 2017

Chapter 11
Operational Use of Synchrophasor Technology for Power System Oscillations Monitoring at California ISO

Jim Hiebert, Aftab Alam, Veera Raju Vinnakota and Dede Subakti

11.1 Introduction

The phenomenon and concerns around power system oscillations [1] have been discussed in the power industry and academia around the world for decades, and solutions for monitoring and mitigation are being actively researched, and while some of them are being implemented. Recent advances in synchrophasor technology [2, 3] provided an opportunity to take a fresh look at possible solutions for detecting, monitoring and controlling power system oscillations. Reference [4] provides a broad overview of the sources, characteristics, and analyzes of natural and forced oscillatory behaviors in power systems.

Western Electricity Coordination Committee's (WECC) Joint Synchronized Information Subcommittee have identified and verified the existence of various modes of inter-area power oscillations in Western interconnection of USA [4]. System disturbances or other factors can further excite these modes which can be observed as oscillatory behavior on the system. Awareness of the existence of certain modes of inter-area oscillations due to phasor technology provides an opportunity to monitor them which otherwise may get unnoticed. Awareness provides a further possibility of assessing the potential risk of such modes in their gird for their specific operating conditions. Wide-scale use of synchrophasor data with appropriate tools provides an opportunity to allow real-time monitoring of modes.

From the authors' perspective, another important consideration in favor of using synchrophasor data is in the market-based grid operation that has been increasingly becoming a normal operating practice of the grid around the world. However, the markets have put pressure on operators to operate the grid closer to the limits, warranting the need to monitor closely for stability concerns. PMUs provide an

J. Hiebert · A. Alam · V. R. Vinnakota · D. Subakti (✉)
California ISO, Folsom, CA, USA
e-mail: dsubakti@caiso.com

S. Nuthalapati (ed.), *Power System Grid Operation Using Synchrophasor Technology*, Power Electronics and Power Systems,
https://doi.org/10.1007/978-3-319-89378-5_11

opportunity to closely monitor them providing an opportunity of economic benefits from markets perspective by operating closer to the limits, determined using synchrophasors once the confidence on the measurements and tools to monitor the limits is obtained.

Forced oscillations on the bulk electric system (BES) may originate from any component connected to the BES, and these can come from a variety of sources, such traditional and variable energy resources, loads, plants, and grid control systems. Synchrophasor data provides the opportunity to effectively monitor forced oscillations using real power, voltage and frequency signals [5]. Determining the source of forced oscillations is particularly challenging. However, monitoring the magnitude or energy of the oscillation is an important indicator of the source.

In addition, there is general interest to monitor high-frequency local oscillations taking benefit of the PMUs currently installed in the grid providing the phasors data and to alert the operating engineers of the participating transmission owners (PTOs). If sustained undamped inter-area small-signal oscillations are noticed, the regional reliability coordinator could be informed as a preventative step and to analyze the source of oscillation to work with the utilities involved with the oscillations. The effort is to identify the existence of oscillations which involves CAISO and whether it requires investigations.

Dispatchers in North America are primarily governed by operating procedures. An alarm is expected out of any serious anomaly in the system operation upon which dispatcher can take corrective action. Questions are raised upon alerts due to certain oscillations in control room, such as the ones that can impact grid stability, where response from the operator is not good enough to take any meaningful control action to prevent a potential disaster.

Alarms could be set-up just for monitoring purposes as the first source of detection of the problem to pass it on to regional entities or to engineers for further analysis. However, as technology matures and usage of tools to monitor the potential dangerous events improves, remedial action schemes involving phasors could be part of the solution.

At CAISO, monitoring the oscillations is in its initial stage where operating experience is being gained to identify alarms as described above. Some of these monitoring experiences with specific examples at CAISO are shared in this chapter.

Section 11.2 provides an overview of architecture for gathering synchrophasor data for the purpose of monitoring the oscillations, including a few other applications that are being evolved [6].

Section 11.3 provides details of three events as examples of oscillations monitoring at CAISO.

11.2 Synchrophasor Data Gathering Architecture at Caiso

A phasor measurement unit (PMU) measures voltage and current signals and estimates a time-synchronized phasor representation (magnitude and phase angle) of these electrical quantities. These voltage and current phasors are referred to as synchrophasors. The synchrophasor phase angle is explained as the following:

A PMU estimates synchrophasor phase angle based on the nominal system frequency synchronized to UTC (global positioning system (GPS)). The PMU estimates the sinusoidal component of the AC waveform from a voltage or current input. Using a time input, usually from a GPS source, it constructs a synchronized reference cosine waveform at the nominal system frequency (60 Hz) such that positive peak is at a UTC second rollover. The synchrophasor phase angle is the phase difference between these signals at the given reporting time.

The CAISO receives synchrophasor data from multiple sources throughout the western US.

Referring to the schematic diagram of CAISO oscillations detection and monitoring system shown in Fig. 11.1, measurements coming from each substation are collected in the PMUs then transmitted to a phasor data concentrator (PDC) device. Each of the PMUs' time is synchronized with the GPS satellite time using a GPS

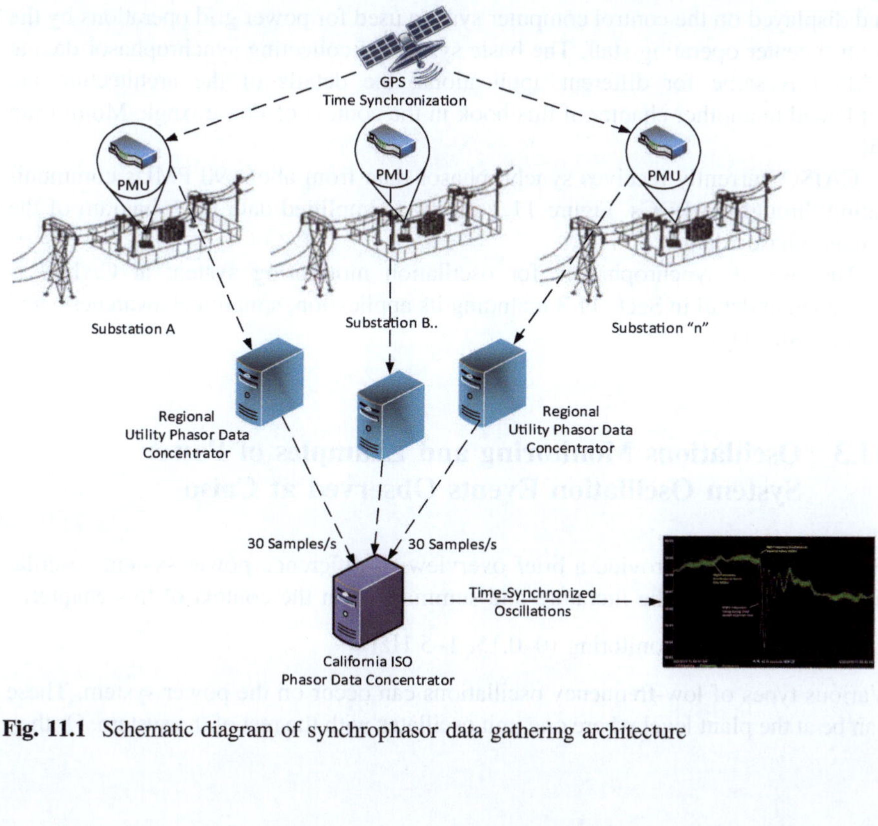

Fig. 11.1 Schematic diagram of synchrophasor data gathering architecture

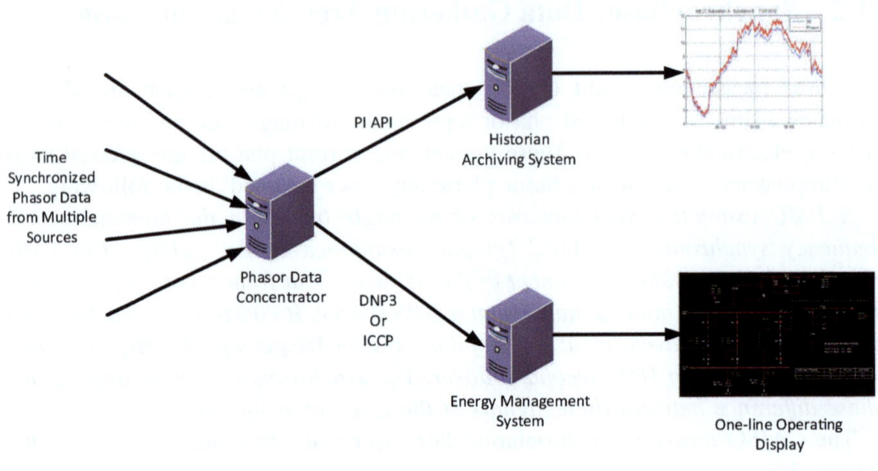

Fig. 11.2 Flow diagram of CAISO synchrophasor data

receiver as the communication interface. Time-synchronized stream of data from the utility (regional) PDC is transmitted to redundant PDCs located at CAISO—sometimes referred to as "Super PDCs." The data are archived in a PI historian system. Down-sampled signals are fed to the energy management system (EMS) and displayed on the control computer system used for power grid operations by the control center operating staff. The basic system of collecting synchrophasor data at CAISO is same for different applications; the details of the architecture are explained in another chapter of this book in the context of Phase Angle Monitoring [6].

CAISO currently receives synchrophasor data from about 90 PMUs communicating through 14 PDCs. Figure 11.2 shows a simplified data flow diagram of the synchrophasor data.

The use of synchrophasors for oscillation monitoring system at CAISO is explained in detail in Sect. 11.3 including its application, situational awareness, and operator displays.

11.3 Oscillations Monitoring and Examples of Power System Oscillation Events Observed at Caiso

Following sections provide a brief overview of difference power systems oscillations gathered from the literature and summarized in the context of this chapter.

Local Oscillations Monitoring (0–0.15, 1–5 Hz)

Various types of low-frequency oscillations can occur on the power system. These can be at the plant level where one unit oscillates with the rest of the system. Or they could exist between multiple plants at the same geographic area. These generally occur during heavy loading and by generator controls; oscillations are monitored and mitigated by the plant operators. These may not always appear on the transmission level voltage, frequency, and power flow signals.

In addition, oscillations in 1–5 Hz frequency range can be caused by local generator controls such as excitation control systems or DC circuits. oscillations observed on the bulk power system may be caused by some of these local oscillations.

Inter-area Oscillations Frequency Oscillation Monitoring (0.15–1 Hz)

Inter-area oscillations by and large, tend to be across larger geographic region or larger span of the electrical grid. These are generally characterized by coherent parts of the system oscillating against each other. Inter-area oscillations are predominant in systems with relatively weak interconnections [3]. Low-frequency oscillations. These are generally monitored by using synchrophasor data from transmission level signals of voltage, frequency, and power signals and utilizing the signals in mode meter applications to calculate the energy and damping of the modes.

The mode meters generally monitor the energy and damping of modes with mode frequencies in the range of 0.2–1 Hz. One of the well-known modes in the Western interconnection is the 0.26 Hz North–South A mode popularly referred to as the North–South mode with Alberta as a key participant in the 0.26 Hz system mode. This mode has been known to cause oscillations on key power transfer interfaces between Alberta and California. Some of the other known modes [7] are a 0.34 North–South mode B, a 0.45 East–West mode, a 0.6 Hz Alberta mode and a 0.8 Hz Montana mode.

Validation of observed concerns in real-time monitoring can be performed by conducting ring-down analysis of synchrophasor data. In addition small-signal stability analysis packages also provide another way to validate the existence of the modes and perform sensitivity analysis for various topology and operating conditions.

Higher-Frequency Oscillation Monitoring (5–50 Hz)

High-frequency oscillations can be torsional in nature caused by resonance between a conventional or variable energy resource and highly compensated transmission lines (generally referred to as sub-synchronous resonance). Various methods have been proposed to utilize synchrophasor information to monitor sub-synchronous resonance.

Control systems for variable energy resources have also known to cause high-frequency forced oscillations on the system. The oscillations may be caused due to a variety of reasons such as wind turbine controls, plant-level controls, and other types of controls. When PMUs are located near or at a generating unit, synchrophasor data from the PMU can be used to monitor any high-frequency oscillation at this generator and also provide the trigger for any necessary mitigating actions by the operators.

From among the above classification of oscillations, the following sections describe three events of local oscillations monitoring at CAISO.

11.3.1 Voltage Oscillations at 500 KV Station March, 2016

The engineering team at the California ISO observed a local oscillation event during the annual testing of the Chief Joseph Brake insertion. High speed trends of frequency and voltage from several PMUs were being monitored for the brake insertion test when a "super-imposed" oscillation of higher frequency was recorded and is shown in Fig. 11.3. It was obvious by studying the trend results that a local oscillation event was occurring at the same time as the brake insertion event (Fig. 11.4).

Further analysis of this event showed that the higher-frequency oscillation was the strongest (or highest energy) at a southern California substation PMU and appeared to be driven by local generating unit's controls. The California ISO operators and engineers were able to correct the oscillation by working directly with the generating unit operators. The immediate correction required a reduction in unit output with longer-term corrections involving controller tuning.

Oscillations of this type should be dealt with as soon as they are detected. At the time of this event, the California ISO operators did not have sufficient situational

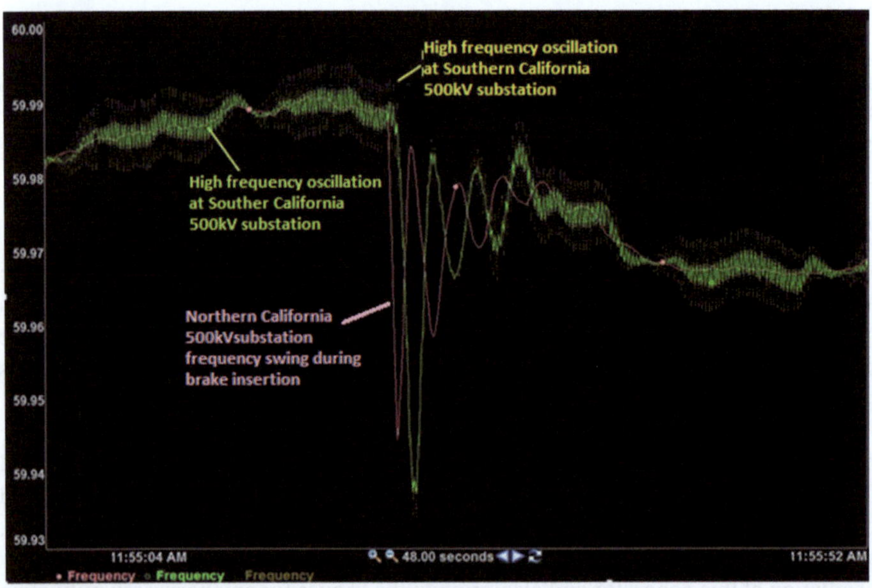

Fig. 11.3 High-frequency local oscillations super-imposed over lower-frequency inter-area oscillations

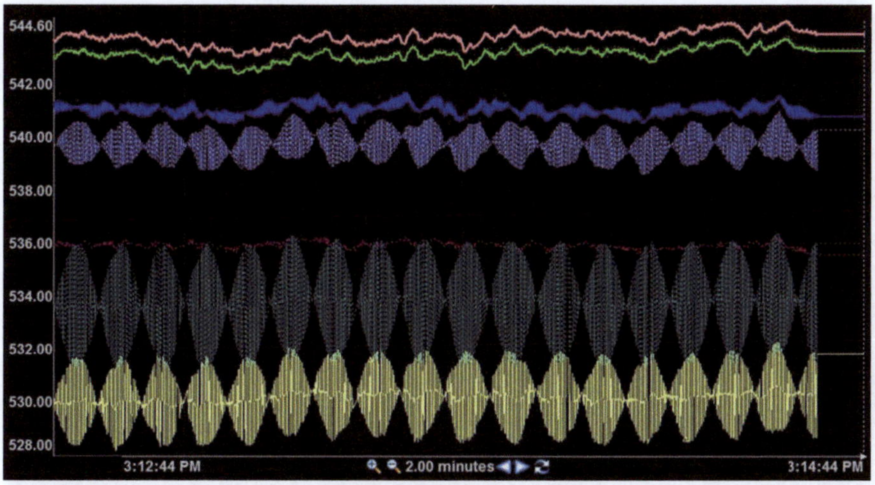

Fig. 11.4 Relative amplitudes of oscillations at different PMU locations

awareness or alerting to warn them of this oscillation and the potential for further destabilization. It is the opinion of the authors that good oscillation detection and alerting methods are necessary using multiple PMUs to expose and provide the location of these local oscillations. A typical solution would include both a situational awareness display as well as alarms in the energy management system in the control room. Care must be taken in the design of situational awareness displays to provide the most important information (such as location) as quickly and as simply as possible without cluttering the normal operating displays in order to minimize time delays for taking appropriate control actions. The text describing the alarm should contain simple and concise information to lead the operator to the solution quickly. In most cases, the company operations engineers will have to be involved in the solution. The company should provide sufficient training to increase the operators' knowledge of oscillations including documented procedures to mitigate the problem from the known incidences in the past.

11.3.2 High-Frequency Solar Plant Local Oscillation in SCE Area Near Devers on April 25, 2014

CAISO Engineers analyzed an event in April, 2014 captured by several PMUs in the southwest area. The event was discovered during routine examination of PMU data in the Southern California region.

Analysis of the PMU results indicated a 6.85 Hz oscillation as shown in Fig. 11.5. The oscillations were detected by local PMUs but were also slightly observable by distant PMUs 800 miles from the source.

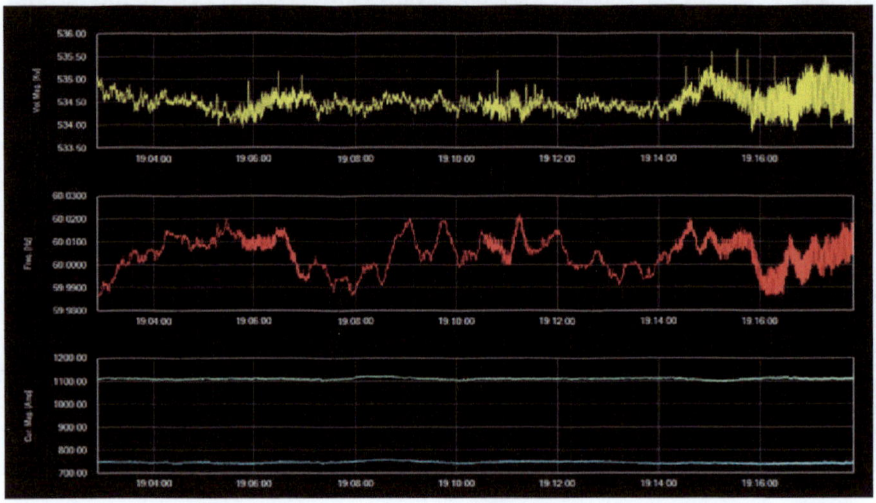

Fig. 11.5 High-frequency solar plant-based local oscillation event

Another view of the same event with frequency on the horizontal scale is shown in Fig. 11.6. The correction of the forced oscillation required California ISO Engineers working with the local utility as well as the solar plant operators to reduce the oscillations by reducing the plant output to a level that caused the oscillations to diminish.

11.3.3 Local Oscillations at Pacific DC Intertie (PDCI) Station in Oregon in Jan, 2008—Operating Logs

In January 2008 California ISO, operators in the Folsom control room (Northern California) were alerted to a 3–4 Hz frequency oscillation from PMU measurements in Southern California. The oscillations appeared to have the greatest energy as measured by the PMU at a Southern California Substation and progressively lower energy as measured by adjacent substation PMUs (Fig. 11.7). These oscillations were not readily visible on EMS SCADA systems in the area due to the slow time sampling of SCADA measurements. In addition, these oscillations did not appear on the locally measured high-speed frequency displays in the Folsom control room due to the fact that Folsom is far away from the source of the event in Oregon. The southern California utility control room noted both voltage and frequency oscillations on their analog trend recorders but no oscillations on their energy management system digital trends.

At the same time, a utility in Oregon located in the Pacific Northwest reported separation of several high-voltage transformer banks at an Oregon substation near

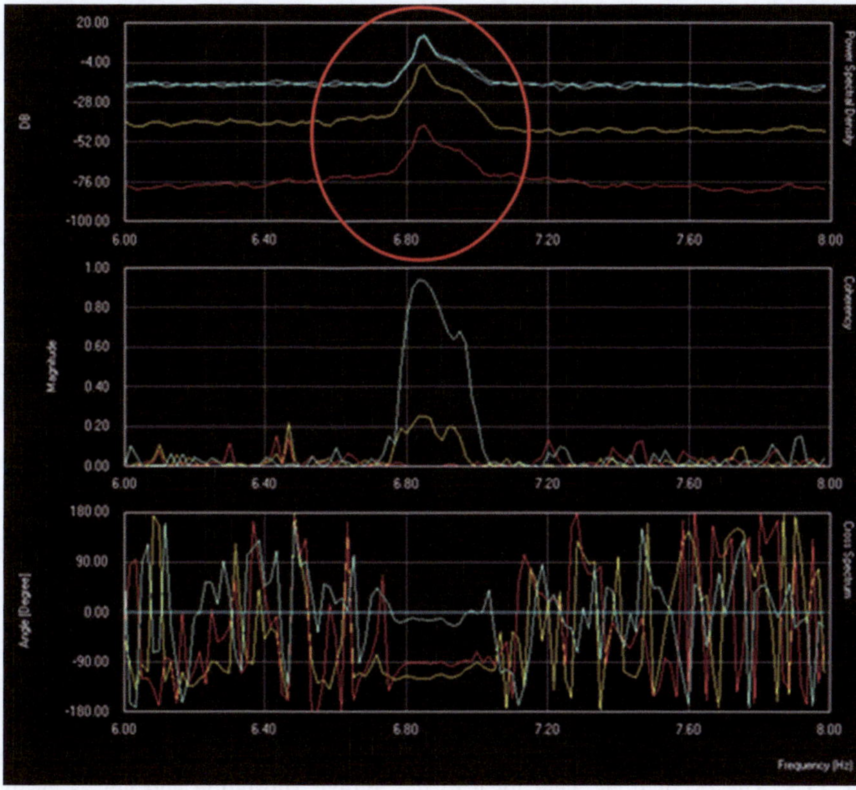

Fig. 11.6 Another view of solar plant-based local oscillation event

the DC circuit due to poor weather conditions. The California ISO Reliability Coordinator (Folsom) confirmed initially the oscillations were local to the southern California substation and may have been related to the event in Oregon due to the fact that Oregon substation was located near the DC circuit substation and the Oregon and southern California substations form the two ends of the Dc circuit. The Pacific Northwest utility also reported tap changer equipment failure at the Oregon substation which would have contributed to control problems on the DC circuit.

Subsequent reduction of the DC circuit flow and eventual removal from service reduced the oscillations and allowed the utilities to fix the equipment failures.

An analysis of the PDCI event demonstrated the difference between normal EMS digital trend readings and real-time PMU trend readings. Most of this event was not visible from the EMS displays in the control rooms. This was due to the fact that PMU results arrive at 30 samples/s (minimum) while typical EMS readings occur as one sample every 2–4 s. The result was these oscillations were not detected by the

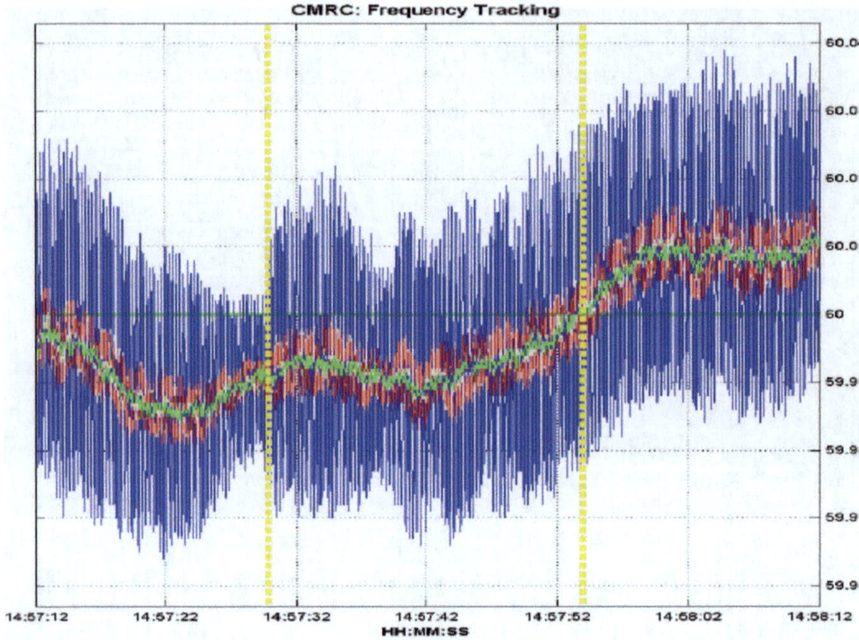

Fig. 11.7 Local oscillations at a DC circuit station in Oregon

normal control room displays. A relative comparison of the graph is shown in Fig. 11.8. The oscillations were local to the Sylmar substation and not as easily detected on high-speed frequency recorders located farther away.

The authors have concluded that steps should be taken to alert engineers and operators of under-damped local oscillations detected by any PMU located within all major substations. The alert should filter out normal, low-energy oscillations that may occur from time to time and alarm on substantial high-energy local oscillations that may be a precursor to larger grid problems.

Bringing high-speed synchrophasor data into real-time operator displays has greatly improved the awareness of control room operators to dynamic control problems. For example, Fig. 11.8 demonstrates the dramatic difference between a typical SCADA data trend and a synchrophasor trend. The synchrophasor trend shows the detail that is "hidden" on the typical SCADA trend. An event may not be noticed on the normal SCADA trends, while significant detail of the event is seen on the synchrophasor trend. This kind of situational awareness made it possible for the operator in the PDCI event to identify the affected area of grid control and focus on the resolution very quickly.

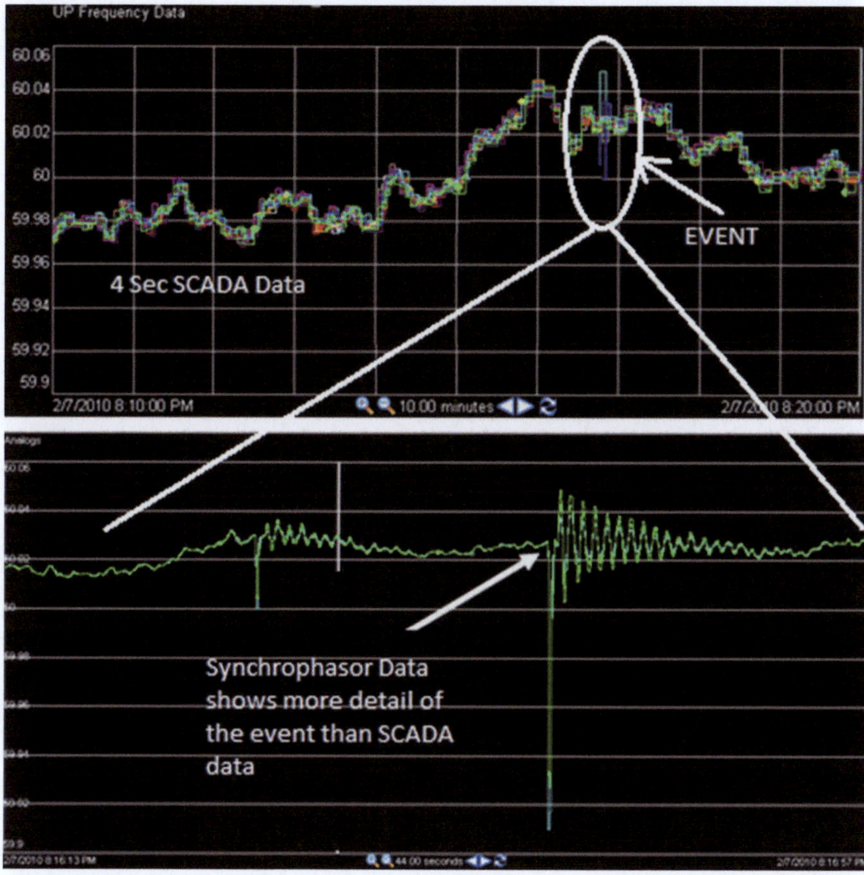

Fig. 11.8 Example of SCADA data versus synchrophasor data of a local oscillation

11.4 Potential Future Work

There is a general interest to monitor inter-area and high-frequency local oscillations to utilize synchrophasor information from installed PMUs and develop alerting mechanism to alert the operators for real-time monitoring and mitigation and for operational engineers to work with the utilities in performing engineering analysis.

Development of operating procedures will generally call for operators to monitor damping and energy levels of known modes and report any observances of sustained undamped inter-area oscillations to the impacted entities and the reliability coordinator as a preventative step. In addition, operating procedures will call for operators to work with the utilities to analyze the source of oscillations when higher-frequency oscillations are observed. Also, transmission operators and the reliability coordinator would need to determine the extent of the observed forced oscillations to see if the oscillation is more localized or widespread as mitigation actions can differ depending on the extent of the oscillations.

If the source of oscillations is determined to be a component connected to the bulk power system, further investigation will likely need to be done by the transmission operator and the reliability coordinator. For example, if the cause of oscillation is linked to a particular plant, further analysis may be required utilizing local SCADA or other data collected by the generator operator to determine and identify root-causes.

The well-known western interconnection oscillations of 1996 could have been mitigated with proper alarming for the inter-area oscillations. Dispatchers in North America are primarily governed by operating procedures. An alarm is expected out of any serious anomaly in the system operation upon which dispatcher can take corrective action. Continuous training and education of operators about oscillations and guidelines on coordination for mitigation actions is as important as developing monitoring and alarming tools to detect oscillations and prevent harmful impacts to the bulk power system and its components.

Acknowledgements The authors acknowledge California ISO for encouraging and supporting the work and in supporting dissemination of the information. The authors acknowledge the discussions and interactions with Electric Power Group (EPG) professionals during the past several years of synchrophasor-related projects at CAISO and for the interesting ideas on visualization of the data.

References

1. Book on Power System Stability and Control by Dr. Prabha Kundur, McGraw-Hill Inc publication
2. NASPI Synchrophasor Technology Fact sheet, Oct 2014
3. Power system oscillatory behaviors: sources, characteristics, and analyses document reference 2017 Version 1.0 NASPI-2017-TR-003 PNNL-26375 prepared by North American Synchrophasor Initiative (NASPI)
4. Using synchrophasor data for, oscillation detection. NASPI control room solutions task team paper, Sept 2017
5. Reliability Guideline—Forced Oscillation Monitoring and Mitigation—Technical Reference Document prepared by NERC, June 2017
6. Hiebert J, Subakti D, Raju Vinnakota V, Alam A Operational use of synchrophasor technology for wide-area power system phase angle monitoring. In: Power system grid operation using synchro-phasor technology, Springer
7. WECC JSIS Modes of Inter-Area Oscillations-2013-12-REV1.1

Chapter 12
Operational Use of Synchrophasor Technology for Wide-Area Power System Phase Angle Monitoring at California ISO

Jim Hiebert, Dede Subakti, Veera Raju Vinnakota and Aftab Alam

12.1 Introduction and Chapter Overview

The application described in this chapter is based on the following two concepts in utilizing the new exciting technology:

1. The real benefit of using synchrophasors in real time or off-line (which are captured several times faster than SCADA measurements) should be uniquely distinct from the near real-time information provided by SCADA (updated every 1–2 s) or by other energy management system (EMS) applications such as State Estimator, Contingency Analysis (20 s or slower), which use SCADA measurement and or real-time network model.
2. Control room operators prefer to have an alarm primarily if mitigation of the alarm is possible by following operating procedures.

Considering the above background, the application of phase angle monitoring using synchrophasor technology is implemented at California ISO (CAISO) and is described in this chapter.

Phase angle difference between bus voltage phasors, as measured by PMUs on the bulk power system, is a measure of the active power transfer from one substation or area to another substation or area [1]. Large measured angle differences correlate to a large amount of power being transferred; thus, it can be used as an indication for potential system stress and stability. An angle difference within a predetermined limit is acceptable but needs to be monitored closely for early warnings. An increasing phase angle difference can lead to a serious problem when the deviation gets large enough to cause instability, either pre- or post-contingency.

J. Hiebert · D. Subakti (✉) · V. R. Vinnakota · A. Alam
California ISO, 250 Outcropping Way, Folsom, CA 95630, USA
e-mail: dsubakti@caiso.com

S. Nuthalapati (ed.), *Power System Grid Operation Using Synchrophasor Technology*, Power Electronics and Power Systems,
https://doi.org/10.1007/978-3-319-89378-5_12

This chapter provides a concise view, as most of the information related to the chapter topic is covered in Reference [1]. However, the details of implementing and using the phase angle monitoring system at the CAISO are included and are explained in this chapter.

12.2 Basic Principles of Phase Angle Monitoring and Implementation Overview at CAISO

Synchrophasor Phase Angle [1]

A phasor measurement unit (PMU) measures voltage and current signals and estimates a time-synchronized phasor representation (magnitude and phase angle) of these electrical quantities. These voltage and current phasors are referred to as synchrophasors. The synchrophasor phase angle is explained as the following:

A PMU estimates synchrophasor phase angle based on the nominal system frequency synchronized to UTC (global positioning system (GPS)). The PMU estimates the sinusoidal component of the AC waveform from a voltage or current input. Using a time input, usually from a GPS source, it constructs a synchronized reference cosine waveform at the nominal system frequency (60 Hz) such that positive peak is at a UTC second rollover. The synchrophasor phase angle is the phase difference between these signals at the given reporting time.

Implementation Overview at CAISO

The CAISO receives synchrophasor data from multiple sources throughout the Western USA. The phase angle difference between two substations is calculated from raw time-synchronized phasor data sourced from PMUs located at each substation.

Referring to the schematic diagram of CAISO phase angle monitoring system shown in Fig. 12.1, time-synchronized measurements coming from each substation are collected in the PMUs then transmitted to a phasor data concentrator (PDC) device. Each of the PMUs' time is synchronized with the GPS satellite time using a GPS receiver as the communication module located in each of the PMUs to communicate with the GPS satellite(s). The GPS satellites are synchronized with an atomic clock maintained on pre-identified ground station(s) as part of the GPS satellite system.

A time-synchronized stream of data from the utility (regional) PDC is transmitted to redundant PDCs located at CAISO—sometimes referred to as "Super PDCs." The data is archived in a PI historian system. Down-sampled signals are fed to the energy management system (EMS) and displayed on the control computer system used for power grid operations by the control center operating staff.

The synchrophasor data streams arriving at the CAISO PDC are combined, time-aligned, and consumed by various applications. CAISO receives synchrophasor data from about 90 PMUs communicating through 14 PDCs. Figure 12.2 shows

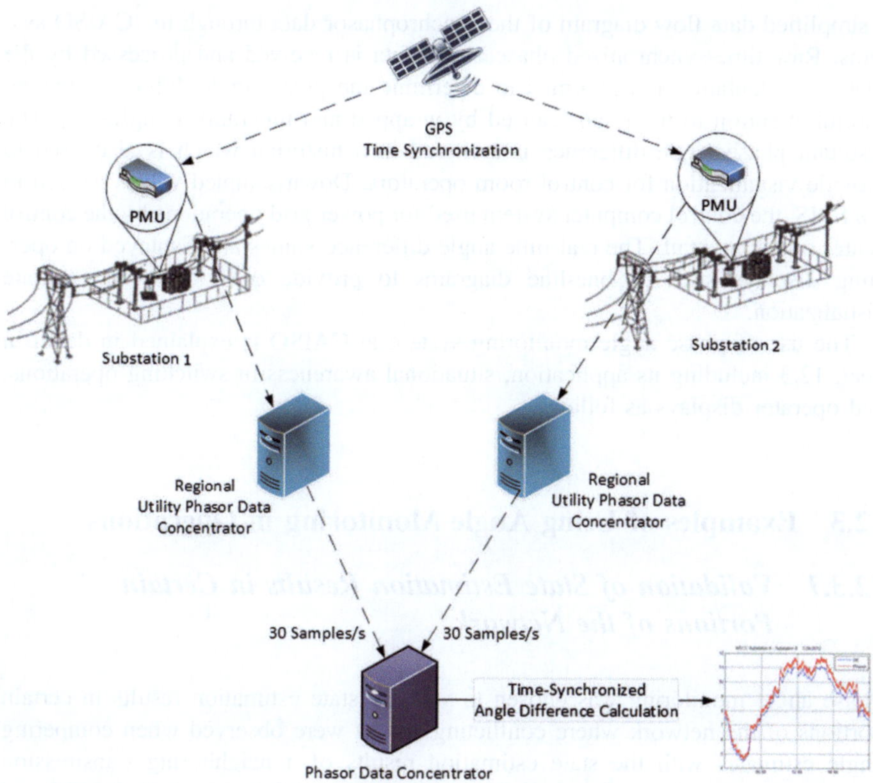

Fig. 12.1 Schematic diagram of phase angle monitoring system

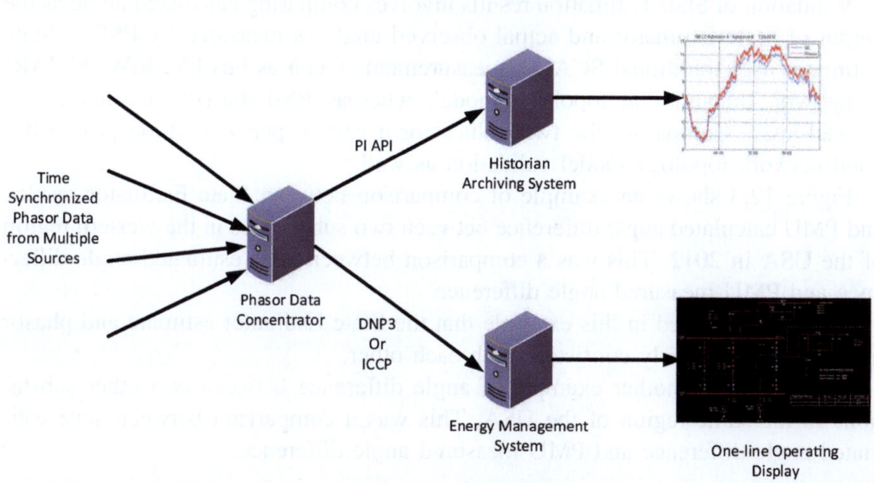

Fig. 12.2 Data flow diagram of CAISO phase angle monitoring system

a simplified data flow diagram of the synchrophasor data through the CAISO systems. Raw time-synchronized phase angle data is received and processed by the PDC. A calculation is performed to determine the phase angle difference paying careful attention to the issues caused by wrapped and unwrapped angles [2]. The resultant phase angle difference is archived in a historian which is also used to provide visualization for control room operators. Down-sampled values are fed to the EMS, the control computer system used for power grid operations by the control center operating staff. The real-time angle difference values are displayed on operating displays such as one-line diagrams to provide operators with accurate visualization.

The use of phase angle monitoring system at CAISO is explained in detail in Sect. 12.3 including its application, situational awareness in switching operations, and operator displays as follows.

12.3 Examples of Using Angle Monitoring in Operations

12.3.1 Validation of State Estimation Results in Certain Portions of the Network

Phase angle monitoring was chosen to validate state estimation results in certain portions of the network where conflicting results were observed when comparing angle estimates with the state estimation results of a neighboring transmission operator. Synchrophasor data being a common source of data for both CAISO and neighboring transmission operators, real-time phasor data was used as an approach to validate the state estimation results.

Validation of State Estimation results involves comparing calculated angle as the output of State Estimator and actual observed angle as measured by PMUs. State Estimator uses traditional SCADA measurements (such as bus kV, MW, MVAR) along with transmission topology model, whereas PMU-based measurement is "model-less." Comparing the two values for a certain period of time provided a good network topology model validation as well.

Figure 12.3 shows an example of comparison between State Estimator results, and PMU calculated angle difference between two substations in the western region of the USA in 2012. This was a comparison between state estimated angle difference and PMU measured angle difference.

It can be observed in this example that the State Estimator estimate and phasor measurement are fairly consistent with each other.

Figure 12.4 is another example of angle difference between two other substations in the same region of the USA. This was a comparison between state estimated angle difference and PMU measured angle difference.

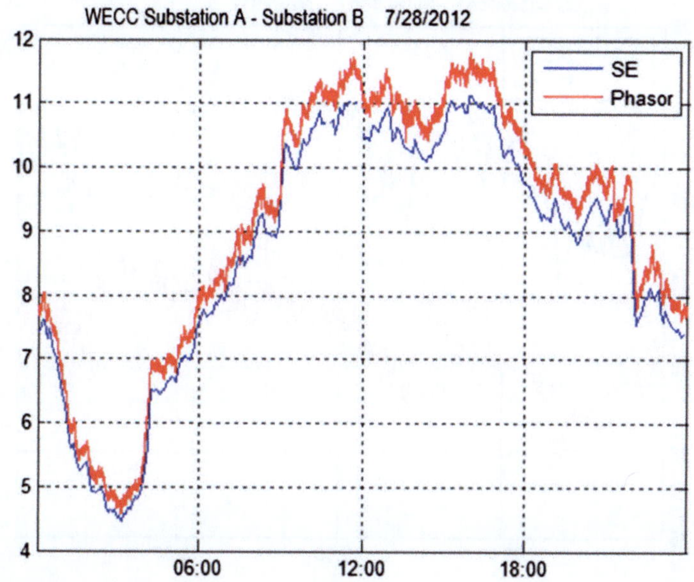

Fig. 12.3 24 h tracking of phase angle difference computed by SE and by synchrophasor measurements—example 1

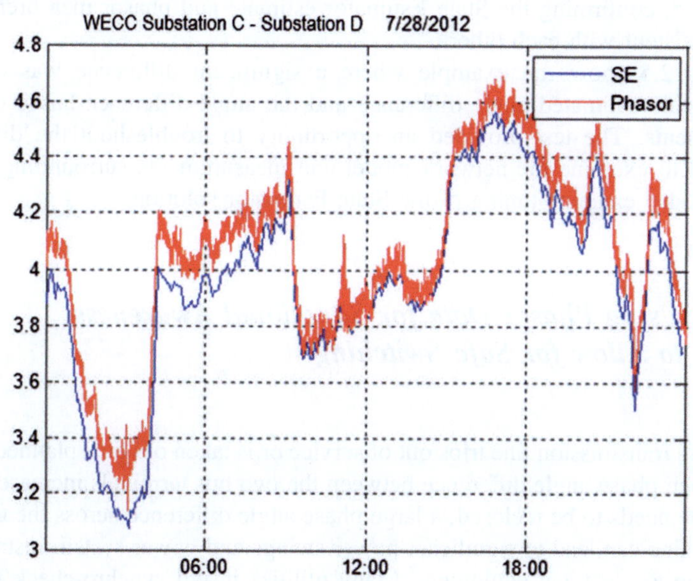

Fig. 12.4 24 h tracking of phase angle difference computed by state estimator and by synchrophasor measurements—example 2

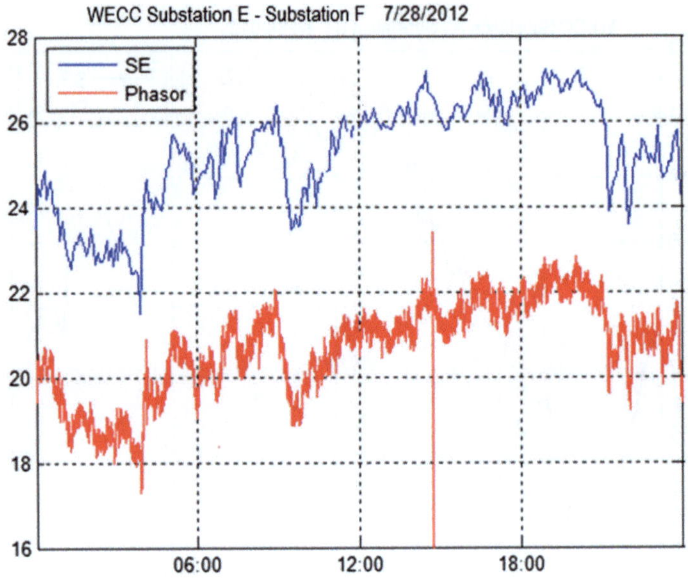

Fig. 12.5 24 h tracking of phase angle difference computed by SE and by synchrophasor measurements—example 3

It can be observed that the angle differences are very small, monitored over a 24 h period, confirming the State Estimator estimate and phasor measurement are fairly consistent with each other.

Figure 12.5 shows an example where a significant difference was observed between state estimated angle difference and the angle difference based on PMU measurements. The test provided an opportunity to troubleshoot the difference, which led to examine the network model and measurements surrounding the stations and also examine tuning of the State Estimator solution.

12.3.2 Using Phasor Data for Situational Awareness to Allow for Safe Switching

- When a transmission line trips out of service or is taken out on a planned outage, the open phase angle difference between the two bus terminals increases. When the line needs to be reclosed, a large phase angle difference across the terminals of the line can lead to significant power swings and power system instability or damage to electrical equipment. Many utilities install synchro-check relays to prevent line closing at an unacceptable phase angle difference. In other words, utilities expect that the system operator would adjust the system and reduce the

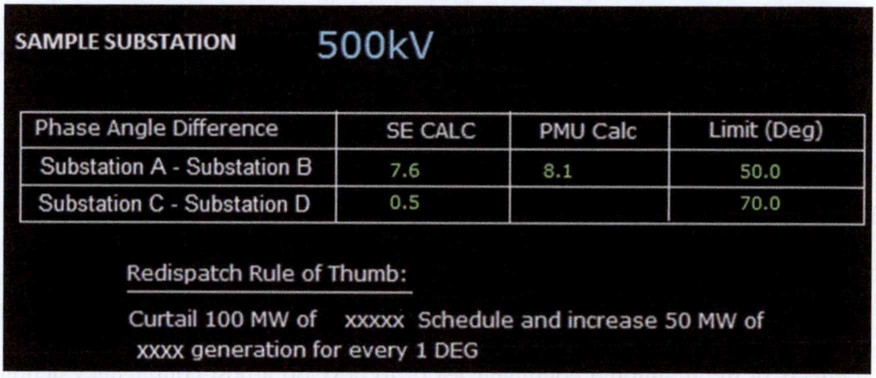

Fig. 12.6 Sample EMS display monitoring angle difference

phase angle differences between the terminal voltages to below a predetermined threshold as set by the synchro-check relay.

- Currently, the CAISO uses State Estimator calculated values and PMU calculated values to measure the open angle difference. The State Estimator calculated values are updated (every 30 s). PMU calculated values are updated much more frequently.
- Both State Estimator solution and PMU calculated angle pairs provide situational awareness for the system operator and provide them immediate feedback on whether it is safe to restore the transmission line or if further system re-dispatch is needed prior reclosing the line. This information is crucial in time-sensitive decision-making processes. For example, knowing how large the open angle is, an operator will not waste time in attempting to reclose the transmission line if they already know that the phase angle is larger than what the synch-check relay in the station would allow.
- Figure 12.6 shows a simple EMS display that is used by CAISO operators to monitor the phase angle difference between buses in two stations.

12.3.3 Operator Displays

At the CAISO, system operators currently use pre-contingency angle differences computed by using State Estimator solutions and PMU data. The operator is also provided with the post-contingency angle differences calculated using the real-time contingency analysis (RTCA) tool. These are available to the operators in the EMS display and allow operators to know these things:

1. Referring to Fig. 12.6, the operators look at the existing State Estimator and PMU-based phase angle difference across the terminals of a line and what is the synchro-check relay setting for the corresponding line. This information is

displayed to the operators on substation one-lines where the synch-check relay resides allowing operators to make an informed decision on when it is safe to close the breakers at the terminals of a line that is currently out. In addition, operators are also provided approved mitigation steps in the same one-line next to the phase angle differences allowing operators to take quick actions in the event a line on outage is to be brought back into service.
2. Operators would also know what the RTCA-based phase angle difference would be following a credible contingency and how the post-contingency phase angle difference would compare to the synch-check relay setting for the corresponding line. This would provide operators with an indication of system stress following the loss of a line and if any pre-contingency actions are necessary to mitigate the effects of the monitored contingencies or to develop a post-contingency action plan necessary to reduce post-contingency angle below synchro-check relay limit that would be needed to reclose the transmission line.

12.4 Potential Future Work

Static or dynamic path limits are conventionally based on MW flow. Using phase angle difference may be an additional indicator that provides system operators with situational awareness of system stress, particularly for stressed outage conditions where MW flow changes minimally but phase angle increases significantly due to path separation or other outages.

Baselining typical angle differences have been considered. However, from the authors' experience, the approach can involve extensive off-line studies supported by real-time data. Limitations of the approach would be that the values of typical angle differences may not be the same for different topology conditions. This requires investigation if the approach is predictable and cost effective for the result it provides.

Using PMU data for island detections is being looked at as another potential application of synchrophasors.

Another area of exploration is using PMUs for calculating balancing area frequency response by accurately measuring tie-line MW changes upon loss of a resource.

Acknowledgements The authors wish to acknowledge the EMS team of CAISO in general and, Sirajul Chowdhury in particular who provided the displays, including the design, for showing the results. The authors also acknowledge California ISO for encouraging and supporting the work and in supporting dissemination of the information.

References

1. Phase Angle Monitoring: Industry Experience Following the 2011 Pacific Southwest Outage Recommendation 27—Technical Reference Document prepared by NERC, June 2016
2. North American SynchroPhasor Initiative—Phase Angle Calculations: Considerations and Use Cases (2016)

Chapter 13
Synchrophasor-Based Linear State Estimation Techniques and Applications

Kevin D. Jones and Alireza Rouhani

13.1 Introduction to Synchrophasor-Based State Estimation Analytics

If you have ever worked in or around the control room of an electric transmission organization, then state estimation is a term that probably rolls off your tongue. Your power system state estimator is central to nearly every analytic activity that goes on in a control room, and its importance is unquestioned. While its direct purpose also rolls off the tongue—"The state estimator gives me my base case."—it takes a deeper understanding of its history and methodology to understand its *fundamental role* as well as the role of modern estimation techniques such as synchrophasor-only state estimation.

As previously alluded, the common understanding of the state estimator is that it simply takes raw measurements from the field and provides the network applications with a base case that can be used for N − 1 contingency analysis and perhaps, depending on the organization, more advanced applications that inform on the stability and economics of real-time operation. This is certainly not disputed. However, it is a superficial narrative. To understand this, we can recall that during the early years of SCADA, raw measurements were used to populate a power flow model that operators would attempt to solve and compare with previously studied, off-line, planning cases. Cleverly enough, this was called the "operators power flow" [1]. Convergence was highly unreliable. Rather than enumerating the drivers for poor convergence one by one, it is more useful to acknowledge that many of the issues that plagued the operators load flow arose from poor data quality.

K. D. Jones (✉) · A. Rouhani
Dominion Energy Virginia, 2400 Grayland Ave, Richmond, VA 23220, USA
e-mail: kevin.d.jones@dominionenergy.com

A. Rouhani
e-mail: alireza.rouhani@dominionenergy.com

S. Nuthalapati (ed.), *Power System Grid Operation Using Synchrophasor Technology*, Power Electronics and Power Systems,
https://doi.org/10.1007/978-3-319-89378-5_13

Fortunately, a man by the name of Fred Schweppe was quick (even by today's standards of technology development) to propose a solution [2–4]. Schweppe put forward that rather than using the data gathered from the system directly in our calculations we should first process the data using as many measurements as possible (even if we were measuring the same points twice) and everything we know about the system (in the form of a model) and statistically combine these two descriptive blocks of information into a "clean" solution that represents the current operating point of the system. This preprocessing step fed into the operator's power flow and drastically improved the convergence rate because it improved the quality of the data provided to the power flow algorithm. This technology is called *power system state estimation*.

What does this story tell us is missing from our correct yet incomplete assertion that "The state estimator gives me my base case"? First, we can observe that the purpose of state estimation is not solely to provide base cases. Rather, it is a mechanism for improving the quality of the data used in downstream analytics. Second, comparing a measured and an estimated quantity provides a way for engineers to get feedback about their analytics pipeline. Statistical tools such as bad data detection and identification are not just useful feedback loops for state estimators but are great feedback loops for engineers to fix data quality problems at their upstream source. In summary, state estimators improve the quality of the data for downstream analytics in two ways: directly, by transforming the data, and indirectly, by providing feedback about the data and its sources.

Since its inception, the field of power system state estimation has grown substantially within the research community and the traditional WLS power system state estimator is now the cornerstone of modern control rooms. However, the variety of industry implementations of these technologies is limited, particularly in the USA, when compared to the vast amounts of methodologies and techniques available in the research community. This cannot and will not last. The reason for this is due in no small part to the rising tide of Phasor Measurement Units (PMUs) and the infamous synchrophasor measurements they produce. Recalling that the role of state estimation is to process raw measurements to improve data quality, we can observe that this shift in the type and volume of field measurements (from PMUs!) begs for an industrial adoption of these techniques that mirrors the variety in the research community. Instead of talking about *THE* state estimator in the singular sense, we will equip our analytics pipeline both inside and outside of the control room with arrays of estimators and similar techniques whose fundamental purposes will be to improve the data quality both directly and indirectly and whose specifications and performance will be dictated by the downstream analytics that they serve.

The purpose of this chapter is to illuminate several of the contemporary estimation techniques that are either in use or will be in use across the industry in the very near term. The scope of these estimation techniques is limited to synchrophasor measurements but many varieties are presented—all of which are centered around practicality of technological implementation. We begin with a perspective of the deeper role that state estimation techniques play in the analytics pipeline as well as an introduction to network observability and topology processing in the context of a completely synchrophasor-based measurement set. This is followed by a presentation of both a linear weighted least square (WLS) and a robust linear least absolute value (LAV) power system state estimator in both a positive sequence and a three-phase formulation. From there, we will examine a distributed approach to robust dynamic state estimation followed by the joining of these technologies—demonstrating the nature of data flow in the analytics pipeline.

13.1.1 The Service that State Estimators Provide to the Analytics Pipeline

Above it was established that the role of state estimators in the analytics pipeline was to improve data quality both directly and indirectly. Direct improvements come from transformation of data, and indirect improvements come from internal feedback loops that eliminate bad data from the calculation and external feedback loops that inform engineers about data quality problems. This section outlines in more detail the key roles played by synchrophasor-only state estimators in the analytics pipeline. Specifically, these state estimators:

1. Improve synchrophasor correctness and completeness per channel,
2. Extend network observability beyond direct measurement (adds channels),
3. Increase signal availability for noncritical measurements,
4. Establish synchrophasor data in the context of the network model,
5. Estimate key parameters that cannot physically be measured (dynamic state estimation), and
6. Improve the confidence level in data used by downstream analytics.

Therefore, in the most general sense, these estimation techniques provide data conditioning services to certain classes of synchrophasor analytics.

13.1.2 Considerations Regarding Network Observability with Synchrophasors

The determination of what signals can and cannot be estimated for a power system is a question of importance both for the engineer–mathematician and for the engineers that chose when and how to deploy a state estimator. If a signal can be estimated, it is said to be *observable*. Network observability is a subdomain of power system state estimation which deserves a chapter to itself. However, for the sake of brevity, this section describes in the simplest terms possible how observability is defined when formulating a network state estimation problem using only synchrophasors.

The scope of the state estimation problem (the state) is defined not as every node voltage in the network but rather as every node voltage in the observable network. This is an important note because it affects how engineers perceive the cost of entrance to synchrophasor-only state estimation. Many perceive that full network observability is a requirement to make use of a linear state estimator but this simply is not the case. It is true that the usefulness of a state estimator is proportional to the measurement footprint but it is not an all-or-nothing value proposition.

Now that we understand that the observable node voltages define the scope of the problem formulation we can establish a set of rules for characterizing the observability of each node. All nodes in the power systems network at any given point in time can be assigned to one of the following three categories: (1) directly observed, (2) indirectly observed, and (3) not observed or unobserved. The criteria for each of these three categories are as follows:

1. A *directly observed* node is a node which is directly measured by a voltage phasor. In a positive sequence formulation, only a single voltage phasor is required. In a three-phase formulation, each of the phases A, B, and C is required. Pseudo-measurements cannot provide this status to a node.
2. An *indirectly observed* node is a node which is not directly measured by a voltage phasor and therefore can be estimated from knowledge of the model and other voltage and current measurements or pseudo-measurements.
3. An *unobserved node* is one which meets none of the above criteria.

The only interesting category that requires further description is that of the indirectly observed nodes. Nodes can achieve this status in several ways. These include:

1. If the node in question is adjacent to another node that is either directly or indirectly observed, then the node in question is indirectly observed. For two nodes to be adjacent, they must be connected through a negligible impedance like a circuit breaker or a bus bar.
2. If the node in question is separated from another node by an unbroken series impedance (no taps) that is directly or indirectly observed and the unbroken series impedance is measured by at least one current phasor, then the node in question is also indirectly observed.

For example, consider the sample multi-station nodal topology shown in Fig. 13.1. Imagine a line relay–PMU placed at Station 2 that measures both the voltage phasor at Node B_2 and the current flow phasor originating at B_2 directed toward D_1. Let us assign observability statuses to each node. Table 13.1 shows the observability status of each node in the example if all breaker statuses are known and closed.

Now consider the same measurement placement but with an additional line relay–PMU at Node D_1 and a transformer relay–PMU at Node C_1 (Table 13.2).

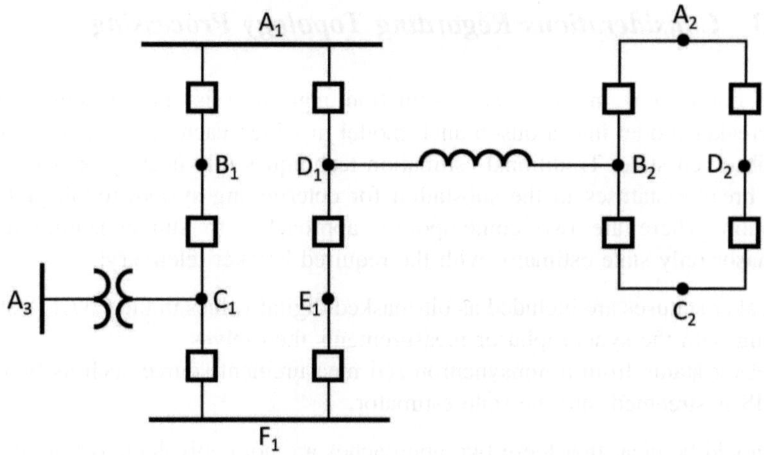

Fig. 13.1 Sample multi-station nodal topology

Table 13.1 Observability example 1

Node	Observability status
A_1	Indirectly observed
B_1	Indirectly observed
C_1	Indirectly observed
D_1	Indirectly observed
E_1	Indirectly observed
F_1	Indirectly observed
A_2	Indirectly observed
B_2	Directly observed
C_2	Indirectly observed
D_2	Indirectly observed
A_3	Unobserved

Table 13.2 Observability example 2

Node	Observability status
A_1	Indirectly observed
B_1	Indirectly observed
C_1	Directly observed
D_1	Directly observed
E_1	Indirectly observed
F_1	Indirectly observed
A_2	Indirectly observed
B_2	Directly observed
C_2	Indirectly observed
D_2	Indirectly observed
A_3	Indirectly observed

13.1.3 Considerations Regarding Topology Processing

A key component in any state estimation algorithm is the reduction of the node-breaker model into a bus–branch model in which each bus represents a separate observed state. Traditional estimation techniques rely heavily on knowledge of the breaker statuses in the substation for determining system topology before estimation. There are two contemporary approaches to supplementing a synchrophasor-only state estimator with the required breaker telemetry:

1. Breaker statuses are included as bit-masked digital values in the C37.118 stream along with the synchrophasor measurements themselves.
2. Breaker status from a nonsynchronized measurement source such as SCADA/EMS is streamed into the state estimator.

It should be clear that these two approaches are not equivalent. When possible, breaker statuses should be brought in through a time-synchronized measurement source such as the C37.118 stream. This is the only way to perform near-perfect topology processing. The downside of this approach is that few organizations have had the foresight to configure their PMUs and PMU-enabled devices to include these values in their output. For those that have, there may still be missing statuses. It is because of this that the second approach exists at all. However, this second approach is nothing short of a temporary work around. While nonsynchronized SCADA breaker statuses can help provide accurate topological information for a large percentage of computation frames, it lags the time-synchronized values and therefore the topological accuracy is compromised during switching conditions. From a data perspective, these are high-valued periods of time and their absence should be weighed heavily against blue-sky conditions.

As an example of breaker status availability on nodal observability, consider the same measurement placement in Example 2 from the previous section but assume that no breaker telemetry is available to discern connectivity (Table 13.3).

Table 13.3 Observability example 3 (no breaker status)

Node	Observability status
A_1	Unobserved
B_1	Unobserved
C_1	Directly observed
D_1	Directly observed
E_1	Unobserved
F_1	Unobserved
A_2	Unobserved
B_2	Directly observed
C_2	Unobserved
D_2	Unobserved
A_3	Indirectly observed

13.2 Static Weighted Least Squares Linear State Estimation Techniques

This section describes several contemporary static linear state estimation techniques using synchrophasor measurements. It includes a weighted least squares (WLS) positive sequence estimator and a WLS three-phase estimator.

13.2.1 Positive Sequence Linear WLS State Estimation

The introduction to linear state estimation starts with the two-port π-model [1].

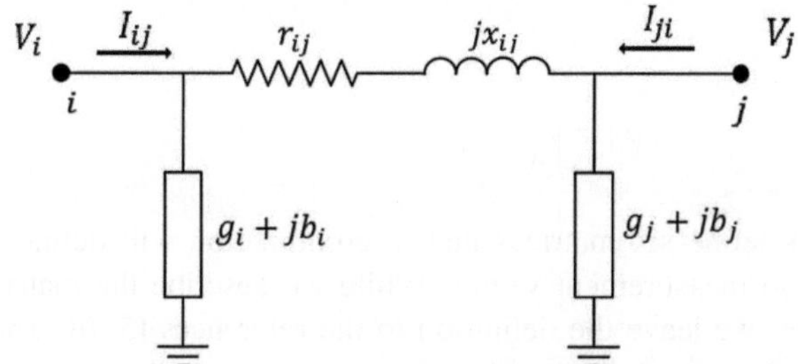

With voltage phasor measurements at either end of the π-section and current phasors looking into either end of the pi section, then the state of this system is simply the following complex-valued vector:

$$[x] = \begin{bmatrix} V_i \\ V_j \end{bmatrix}$$

And the error-free measurement vector (also complex-valued) is as follows:

$$[z] = \begin{bmatrix} V_i \\ V_j \\ I_{ij} \\ I_{ji} \end{bmatrix}$$

Then, the network structure and network parameters of the two-port π-model are used to transform the state vector into the measurement vector.

$$\begin{bmatrix} V_i \\ V_j \\ I_{ij} \\ I_{ji} \end{bmatrix} = \begin{bmatrix} 1 & 0 \\ 0 & 1 \\ y_{ij} + y_{i0} & -y_{ij} \\ -y_{ij} & y_{ij} + y_{j0} \end{bmatrix} \begin{bmatrix} V_i \\ V_j \end{bmatrix}$$

This demonstrates the simplicity of the relationships between the phasor values of the network and the phasor values of the state. However, we need a bit more sophistication to better define the linear state estimation problem. The continuation of this exercise defines the measurement vector as a vertical concatenation of an enumerated column vector of voltage phasor measurements, an enumerated column vector of current phasor flow measurements, and an enumerated column vector of current injection measurements.

$$[\boldsymbol{z}] = \begin{bmatrix} \boldsymbol{V} \\ \boldsymbol{I}_{flow} \\ \boldsymbol{I}_{inj} \end{bmatrix}$$

We will assume the measurements are imperfect and define the state vector with a different nomenclature.

$$[\boldsymbol{x}] = \begin{bmatrix} x_i \\ x_j \end{bmatrix}$$

From here, we need to define six matrices that in combination will define the transformation from state to measurement vector. While we describe the matrices and the role in this chapter, we leave the definition to the references [5, 6]. These matrices are:

1. $[A]$—*Current Flow Measurement–Bus Incidence Matrix*—relates the position of current flow measurements to the "buses" (states) which they originate from and point toward.
2. $[II]$—*Voltage Measurement–Bus Incidence Matrix*—taller than it is wide due to measurement redundancy, it relates the position of voltage measurements to the "buses" (states) which they directly observe.
3. $[Y]$—*Series Admittance Matrix*—a square diagonal matrix where the diagonals are the admittances corresponding to the lines that are being measured.
4. $[Y_s]$—*Line Shunt Admittance Matrix*—relates the location of each current measurement with the shunt admittance of the line that it is measuring.
5. $[Y_{sh}]$—*Shunt Device Admittance Matrix*—serves the same purpose as the series admittance matrix but for current injections from shunt devices. This matrix has only been defined in [6].
6. $[W]$—*Covariance Matrix*—defines the error models for each synchrophasor measurement.

Together, these matrices can be assembled in the following way to relate the measurement vector and the state where $[e]$ is the error in the synchrophasor measurements.

$$[z] = \begin{bmatrix} \boldsymbol{V} \\ \boldsymbol{I_{flow}} \\ \boldsymbol{I_{inj}} \end{bmatrix} = \begin{bmatrix} \boldsymbol{II} \\ \boldsymbol{YA} + \boldsymbol{Y_s} \\ \boldsymbol{Y_{sh}} \end{bmatrix} [x] + [e] = [\boldsymbol{B}][x] + [e]$$

Weighing each measurement equally (a least squares solution), the solution of the over-defined set is as follows:

$$[x] = \left[\left(\boldsymbol{B^T B} \right)^{-1} \boldsymbol{B^T} \right] [z] = [\boldsymbol{H}][z]$$

To assign different weights to each measurement (a *weighted* least squares), the covariance matrix must enter the equation.

$$[x] = \left[\left(\boldsymbol{B^T W^{-1} B} \right)^{-1} \boldsymbol{B^T W^{-1}} \right] [z] = [\boldsymbol{H_e}][z]$$

13.2.2 Three-Phase Linear WLS State Estimation

The three-phase linear WLS estimator has a formulation which is nearly identical in structure to the positive sequence formulation and only requires each element of each matrix to become either a 3 × 3 block matrix or a 3 × 1 column vector itself. There only a few rules which dictate a reformulation in three phase.

1. Each measurement element in the measurement vector becomes a 3 × 1 column vector which is a vertical concatenation of the A, B, and C phase measurements, respectively.

$$z_i \overset{yields}{\rightarrow} \begin{bmatrix} z_{ia} \\ z_{ib} \\ z_{ic} \end{bmatrix}$$

2. Each state element in the state vector becomes a 3 × 1 column vector which is a vertical concatenation of the A, B, and C phase states, respectively.

$$x_i \overset{yields}{\rightarrow} \begin{bmatrix} x_{ia} \\ x_{ib} \\ x_{ic} \end{bmatrix}$$

3. Each element in both incidence matrices becomes a 3 × 3 block diagonal of itself.

$$1 \overset{yields}{\rightarrow} \begin{bmatrix} 1 & 0 & 0 \\ 0 & 1 & 0 \\ 0 & 0 & 1 \end{bmatrix} \; and \, 0 \overset{yields}{\rightarrow} \begin{bmatrix} 0 & 0 & 0 \\ 0 & 0 & 0 \\ 0 & 0 & 0 \end{bmatrix}$$

4. (And most importantly) Each element in the admittance matrices becomes a 3 × 3 block matrix that represents the three-phase impedance values considering different phase impedances and mutual coupling.

$$y_{ij} \overset{yields}{\rightarrow} \begin{bmatrix} y_{ij,aa} & y_{ij,ab} & y_{ij,ac} \\ y_{ij,ba} & y_{ij,bb} & y_{ij,bc} \\ y_{ij,ca} & y_{ij,cb} & y_{ij,cc} \end{bmatrix}$$

Obviously, the computational burden of the three-phase linear state estimation carries an approximate threefold increase compared to the positive sequence estimator.

13.2.3 Implementation History

The first three-phase linear state estimator was installed at Dominion Energy, Virginia, in July 2012 as part of a Department of Energy (DOE) Technology Demonstration Project entitled "Three-Phase Linear State Estimator and Its Applications" [7, 8]. In August 2013, the technology was demonstrated before an industry audience at Dominion facilities and released open source [6] to promote technology transfer.

As seen above, the mathematical difference between a three-phase linear state estimator and a positive sequence linear state estimator is marginal. The same is true for the software implementation of these technologies. While the original driver was to create a three-phase linear state estimator, the libraries that defined the three-phase linear state estimator also included the logic for a positive sequence version of the same algorithm. From the user's perspective, to switch between these two is a matter of modeling and configuration, not of technology and software.

Following the three-phase estimator demonstration, the positive sequence was exercised in a demonstration for the Lawrence Berkeley National Laboratories as part of the CERTS Data Conditioning and Validation project [9]. While the code base is still maintained to support three-phase estimation, all derivative demonstrations as well as all derivative technologies [10] have utilized only positive sequence.

Further improvements to the open source code base have been achieved in the three-year span since the 2013 and 2014 demonstrations. In addition to many improvements through the work done in [10], the technology served a central role in a recent DOE FOA project targeting the commercialization of synchrophasor applications [11]. These include four primary contributions:

1. Available native integration with the openECA [12] platform,
2. Encapsulation as a standard installation package for easy deployment [13],
3. Automation of much of the labor-intensive work associated with modeling and measurement mapping [6], and
4. Addition of topology estimation features which enable better topology processing in the absence of breaker telemetry using only phasor measurements [6, 13].

The above contributions were demonstrated through real-time simulation and with real-time field data at Dominion Energy, in Glen Allen, VA, on November 8, 2017.

13.3 Dynamic State Estimation Techniques

The positive sequence and three-phase linear state estimators described above, like the most prominent industrial implementations, are static state estimators. Estimators of this type do not consider the dynamic state variables of the power system. Perhaps the main reason that most of the earlier state estimation approaches were focused on the estimation of static variables of the system was related to the slow rate of available measurements provided by SCADA.

However, the invention and proliferation of Phasor Measurement Units (PMUs) eliminated the limitation imposed by the legacy measurement collection system. The availability of measured voltage and current phasors by PMUs at rates commensurate with system dynamics made it possible for researchers and engineers to investigate implementation of dynamic state estimation for electric power systems.

Accurate dynamic models are necessary for dynamic security assessment of power systems. Dynamic state estimators (DSEs) can be used in real time to estimate or identify the dynamic state variables and unknown parameters associated with the dynamic models of the power system and its components. Thus, just as how static state estimators enable a static security assessment, dynamic state estimation can be used to conduct dynamic security assessment of power systems. Dynamic estimators can be used to procure these models by tracking the state variables and parameters associated with the models in real time providing the most recent estimates for the system and machine dynamic variables.

13.3.1 Robust Linear Phasor-Assisted Dynamic State Estimation

However, static and dynamic state estimations are not necessarily independent of one another. In fact, dynamic state estimation can be considered as an extension of static state estimation through the provision of high fidelity initial conditions from the static state estimator. Inthis section, a Two-Stage Robust Dynamic State Estimation (TRODSE) [14] technique which takes advantage of a robust linear phasor estimator and an efficient dynamic state estimator will be presented. Implementation of TRODSE on the power system allows us not only to estimate the static state variables of the power systems such as the voltage phasors of the buses, but also to estimate the dynamic state variables and unknown parameters associated with the nonlinear dynamic models such as synchronous generators and loads in real time.

The first stage of the TRODSE is the Least Absolute Value (LAV) linear state estimator. It has been shown [15] that LAV-based linear state estimators provide a more robust static state estimation technique compared to the widely used WLS-based linear state estimator like those presented earlier in the chapter. Robustness in this case is indicative of the ability of the state estimator to provide unbiased results in the presence of bad data. In nonrobust estimators such as the WLS, even the presence of a single bad measurement can result in a biased solution. The second stage of the TRODSE, the Unscented Kalman Filter (UKF)-based dynamic state estimator will be used. The primary advantage of using UKF dynamic state estimators as opposed to the widely used Extended Kalman Filter (EKF)-based DSEs is that there is no need to linearize the nonlinear dynamic equations associated with the synchronous generators or loads to obtain the propagated mean and covariance associated with the state vectors through the nonlinear functions [16]. By using UKF-based DSEs, the propagated mean and covariance of a state vector through the nonlinear functions are obtained by transforming the deterministic vectors called *sigma points* through the nonlinear dynamic equations. The following subsection presents the compact form of the LAV and UKF.

13.3.2 Least Absolute Value Linear State Estimator

The LAV estimator minimizes the L1 norm of the measurement residuals. The following optimization problem is solved by LAV [15]:

$$\textit{minimize} \sum_{i=1}^{m} |r_i|$$

$$\textit{subject to } Z - H\widehat{X} = r$$

where

m is the number of measurements,
H is the measurement Jacobian matrix as defined in the previous section,
$\widehat{X}$ is the estimated state vector made up of the voltage phasors of each observable node,
Z is the measurement vector,
r is the measurement residual vector.

The LAV state estimation problem can be expressed as an equivalent linear programming problem as follows:

$$\min c^T y$$

$$\textit{subject to } My = Z$$

$$y \geq 0$$

where

$$c^T = [\, Z_n \quad O_m \,]$$

$$y = [\, X_a \quad X_b \quad U \quad V \,]^T$$

$$M = [\, H \quad -H \quad I \quad -I \,]$$

Z_n is the $1 \times 2\,n$ vector including zeros. O_m is the $1 \times 2\,m$ vector consisting of ones. X_a and X_b are $1 \times n$, and U and V are $1 \times m$ nonnegative vectors where:

$$X = X_a^T - X_b^T \textit{ and } r = U^T - V^T$$

Given the close relationship between the WLS and the LAV estimators, the technology hurdles to implement a LAV estimator as part of existing solutions such as in [6–13] would be minimal. The challenge would only include small modifications to the source code and configuration as well as the addition of a library dependency for linear programming.

The following section outlines the algorithm associated with the UKF.

13.3.3 UKF-Based Dynamic State Estimation

The UKF uses the unscented transformation [16] for solving nonlinear problems by considering system dynamics and measurement equations as follows:

$$x_{k+1} = f(x_k, k) + w_k$$

$$z_k = h(x_k, k) + v_k$$

where

$x \in \mathfrak{R}^n$	is a discrete state vector,
$z \in \mathfrak{R}^m$	is a discrete measurement vector,
$w_k \sim N(0, Q_k)$	is Gaussian process noise at time, k,
$v_k \sim N(0, R_k)$	is Gaussian measurement noise at time, k,
Q_k and R_k	are covariance matrices of w_k and v_k, respectively.

The algorithm of the UKF based on $2n$ sigma points is expressed as follows: First, UKF is initialized:

$$\hat{x}_0^+ = E(x_0),\ P_0^+ = E\left[\left(x_0 - \hat{x}_0^+\right)\left(x_0 - \hat{x}_0^+\right)^T\right]$$

Next, time updates are performed with the calculation of sigma points:

$$\hat{x}_{k-1}^{(i)} = \hat{x}_{k-1}^+ + x_*^{(i)} \qquad i = 1, \ldots, 2n$$

$$x_*^{(i)} = \left(\sqrt{nP_{k-1}^+}\right)_i^T \qquad i = 1, \ldots, n$$

$$x_*^{(n+i)} = \left(\sqrt{nP_{k-1}^+}\right)_i^T \qquad i = 1, \ldots, n$$

Note that $\left(\sqrt{nP}\right)_i$ is the ith row of $\sqrt{nP}$.

$$\hat{x}_k^{(i)} = f\left(\hat{x}_{k-1}^{(i)}, k-1\right), \quad \hat{x}_k^- = \frac{1}{2n}\sum_{i=1}^{2n} \hat{x}_k^{(i)}$$

$$P_k^- = \frac{1}{2n}\sum_{i=1}^{2n}\left(\hat{x}_k^{(i)} - \hat{x}_k^-\right)\left(\hat{x}_k^{(i)} - \hat{x}_k^-\right)^T + Q_{k-1}$$

Then, the measurement update is as follows with the calculation of new sigma points:

$$\hat{x}_k^{(i)} = \hat{x}_k^- + x_*^{(i)} \quad i = 1, \ldots, 2n$$

$$x_*^{(i)} = \left(\sqrt{nP_k^+}\right)_i^T \quad i = 1, \ldots, n$$

$$x_*^{(n+i)} = -\left(\sqrt{nP_k^+}\right)_i^T \quad i = 1, \ldots, n$$

Subsequently,

$$\widehat{z}_k^{(i)} = h\left(\widehat{x}_k^{(i)}, k\right), \widehat{z}_k = \frac{1}{2n}\sum_{i=1}^{2n} \widehat{z}_k^{(i)}$$

$$P_z = \frac{1}{2n}\sum_{i=1}^{2n} \left(\widehat{z}_k^{(i)} - \widehat{z}_k\right)\left(\widehat{z}_k^{(i)} - \widehat{z}_k\right)^T + R_k$$

$$P_{xz} = \frac{1}{2n}\sum_{i=1}^{2n} \left(\widehat{x}_k^{(i)} - \widehat{x}_k\right)\left(\widehat{z}_k^{(i)} - \widehat{z}_k\right)^T + R_k$$

$$K_k = P_{xz}P_z^{-1}$$

$$\widehat{x}_k^+ = \widehat{x}_k^- + K_k(z_k - \widehat{z}_k)$$

$$P_k^+ = P_k^- - K_k P_z K_k^T$$

13.3.4 Implementation of TRODSE

For implementation of TRODSE, the first step is to identify the zone or zones for which a linear state estimator will be implemented [14]. Zones may be chosen based on geographical areas or voltage levels or any other criteria of interest to the utility operator. It is assumed that a robust linear state estimator (in particular, an LAV estimator) is implemented for each zone that estimates the voltage phasors of all buses inside the zone $\left(\widehat{V}\, and\, \widehat{\theta}\right)$ using synchrophasor measurements. Estimated active and reactive power injected by the generators $\left(\widehat{P}\, and\, \widehat{Q}\right)$ can also be obtained based on the estimated voltage phasors.

The LAV estimator described herein not only provides more accurate estimates of bus voltages and power injections but also eliminates the need to have a PMU at each generator terminal as long as the existing set of PMUs render the entire power system observable. Note that this is an important advantage over the measurement-based approach since most generators may not be monitored by PMUs at their terminals. Furthermore, even when all units have installed PMUs, measurements from these PMUs are still subject to data quality issues resulting from measurement error or communication system failures. In these cases, the proposed linear estimator will remain robust and yield unbiased state estimates. Figure 13.2 shows the block diagram associated with the implementation of the TRODSE.

As an example, TRODSE will implement on the NPCC test system with 140 buses and 48 generators. The performance of the TRODSE in the presence of bad data will be compared to the conventional dynamic state estimation approaches

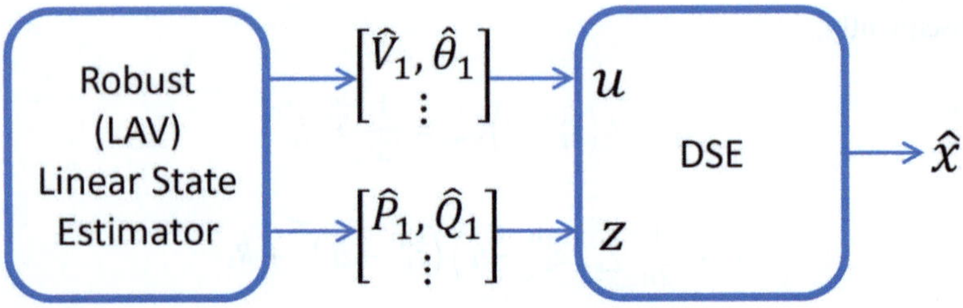

Fig. 13.2 Implementation of TRODSE

where measurements are directly used by the DSE. For this example, the following scenario is considered:

- Simulation scenario involves three line switching events occurring at $t = 1.6$ and 15 s, respectively.
- Starting at $t = 15$ s, the voltage magnitude measurement at the generator terminal is replaced by half of its true value (simulating a scaling error) for a duration of 5 s. Generator terminal voltage (magnitude and phase) measurements have white noise added with zero mean and 0.005 p.u. standard deviation for the entire simulation period.
- All other measurements have white noise added with zero mean and 0.001 p.u. standard deviation for the entire simulation period.

Figures 13.3 and 13.4 show the estimation results for the rotor angle of one of the generators in the test system using both TRODSE and the conventional dynamic state estimation approach, respectively. Note that in these figures the dashed lines indicate the estimated results and the blue solid lines indicate the true trajectories.

As evident from Figs. 13.3 and 13.4, the estimated results for the rotor angle of the monitored synchronous generator remain accurate during the occurrence of the bad data thanks to the usage of the LAV as a robust linear phasor estimator in the algorithm. However, when measurements are directly used, as in the case of the conventional dynamic state estimator, the estimated results for the rotor angle diverge from the corresponding true trajectory.

Fig. 13.3 Actual and estimated rotor angle for a synchronous generator in NPCC test system–using TRODSE

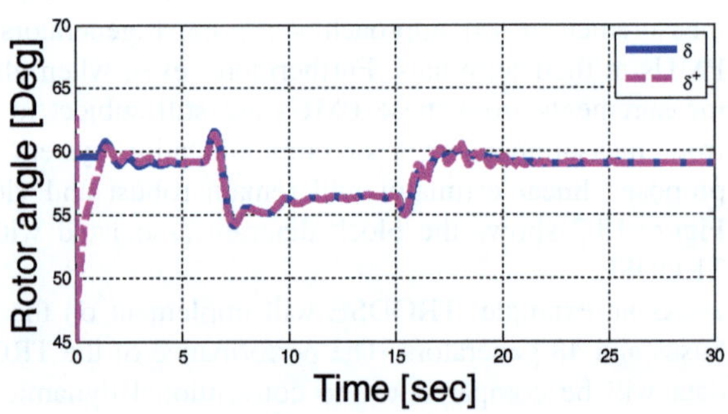

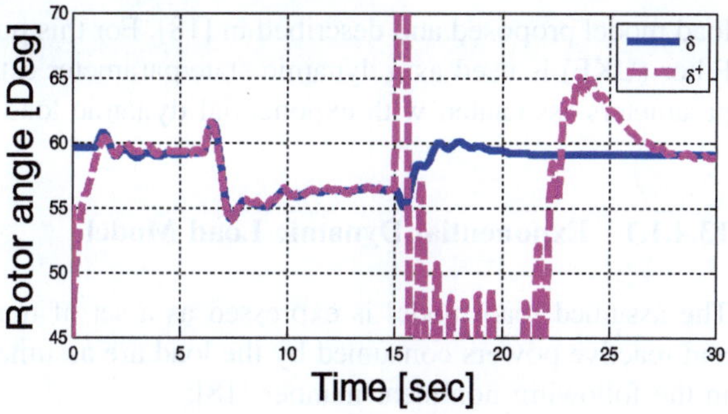

Fig. 13.4 Actual and estimated rotor angle for a synchronous generator in NPCC test system–conventional approach

13.4 Applications

While the previously presented state estimator formulations exercise new and better measurement footprints, they are still reminiscent in scope to that of traditional estimation techniques. This section extends the estimation paradigm into both a sampling of applications that can depend on the output of one or more of the above analytics and additional analytics that utilize estimation techniques on their own to create new and useful analytic results.

13.4.1 Load Modeling

Obtaining more accurate load models that properly reflect dynamic behavior of loads under various disturbances is a challenge of great importance to a variety of analytics. Various online and off-line applications which rely on dynamic simulation studies require detailed and accurate load models. Use of models that fail to accurately capture the dynamic behavior of loads may lead to inconsistent results for dynamic stability and voltage collapse studies [17]. A load model is a mathematical representation related to the measured voltage and/or frequency at a bus, and the real and reactive power consumed by the load. Therefore, load modeling is considered as a system identification problem.

Load models can be broadly classified as either static or dynamic [17]. A static load model does not depend on time, and therefore it relates the active and reactive power at a given time to the voltage and/or frequency at the same instant of time. Static load model is suitable to represent static load components such as resistive loads and light bulbs. While they are also used to approximate the dynamic load components, their accuracy is not sufficient for all analytics. On the other hand, a dynamic load model describes the load behavior as a function of time and therefore provides a much more accurate tool for dynamic simulations.

This chapter considers one of the most widely accepted dynamic load models and aims to identify and track its parameters online. This is the exponential dynamic load model proposed and described in [18]. For this purpose, an Unscented Kalman Filter (UKF) is used as a dynamic state/parameter estimator to track the unknown parameters associated with exponential dynamic load model.

13.4.1.1 Exponential Dynamic Load Model

The assumed load model is expressed as a set of nonlinear equations, where real and reactive powers consumed by the load are assumed to be related to the voltage in the following nonlinear manner [18]:

$$T_P \frac{dP_r}{dt} + P_r = P_0 \left(\frac{V}{V_0}\right)^{\alpha_s} - P_0 \left(\frac{V}{V_0}\right)^{\alpha_t}$$

$$P_l = P_r + P_0 \left(\frac{V}{V_0}\right)^{\alpha_t}$$

where

V_0 and P_0	are the voltage and power consumption before a voltage change,
P_r	is the active power recovery,
P_l	is the total active power response,
T_p	is the active load recovery time constant,
α_t	is the transient active load–voltage dependence coefficient,
α_s	is the steady state active load–voltage dependence coefficient.

Similar equations are also valid for reactive power.

This formulation focuses mainly on the real-time estimation of the unknown parameters associated with the load model based on measurements. It also includes the development of a real-time dynamic load model which represents the behavior of the monitored load with an acceptable accuracy. This is accomplished by using an UKF which is implemented as a dynamic parameter estimator for the unknown parameters and state variables of the assumed load model. For the real power model, the unknown parameters are α_s, α_t *and* T_p and the state variable is P_r.

13.4.1.2 Proposed Approach

State dynamics and the measurement equations from the previous section can be developed for the above considered exponential dynamic load model by discretizing the equations using the second-order Runge–Kutta method, a numerically stable discretization method.

$$P_r(k+1) = P_r(k) + \frac{A+B}{2}$$

$$A = \frac{\Delta t}{T_p(k)}\left(-P_r(k) + P_0\left(\frac{V(k)}{V_0}\right)^{\alpha_s(k)} - P_0\left(\frac{V(k)}{V_0}\right)^{\alpha_t(k)}\right)$$

$$B = \frac{\Delta t}{T_p(k)}\left(-(P_r(k)+A) + P_0\left(\frac{V(k)}{V_0}\right)^{\alpha_s(k)} - P_0\left(\frac{V(k)}{V_0}\right)^{\alpha_t(k)}\right)$$

$$\alpha_s(k+1) = \alpha_s(k)$$

$$\alpha_t(k+1) = \alpha_t(k)$$

$$T_p(k+1) = T_p(k)$$

$$P_l(k) = P_r(k) + P_0\left(\frac{V(k)}{V_0}\right)^{\alpha_s(k)}$$

Note that the discretized equations associated with the reactive part can be obtained in a similar way.

In order to validate the dynamic estimator performance, a set of measurements is created by dynamic simulations on a system with known load model parameters. UKF is then implemented and used to estimate the augmented state vector which includes the state variable and unknown parameters associated with the exponential dynamic load model:

$$x_k = [\ P_r(k) \quad \alpha_s(k) \quad \alpha_t(k) \quad T_p(k)\ ]^T$$

where the measurement is assumed to be:

$$z_k = P_l(k) + v_k$$

13.4.1.3 Implementation of the Proposed Approach with Historical Field Measurements

The proposed approach is evaluated based on historical synchrophasor data from the field, where synchronized voltage and power measurements are acquired every 6 s for a utility distribution feeder. The total duration of the recordings is 1440 min or 24 h. Figure 13.5 shows that the voltage of the feeder, *V*, increases during this period (from almost 115–116.5 kV).

UKF is used to estimate the unknown parameters of the assumed exponential dynamic load model. Figures 13.6, 13.7, 13.8, 13.9 show the estimated values of the assumed exponential load model for the study duration.

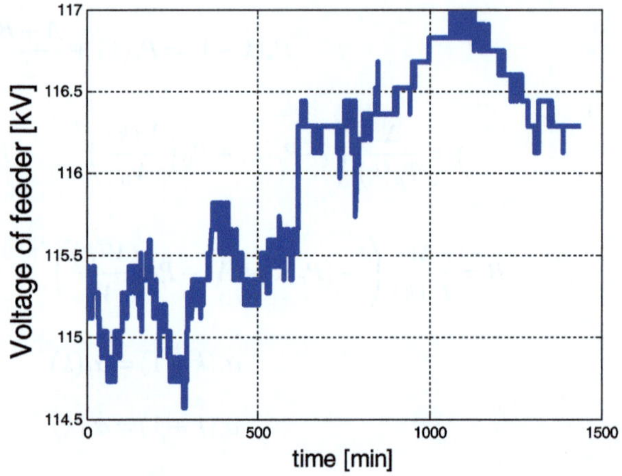

Fig. 13.5 Voltage of the distribution feeder (V)

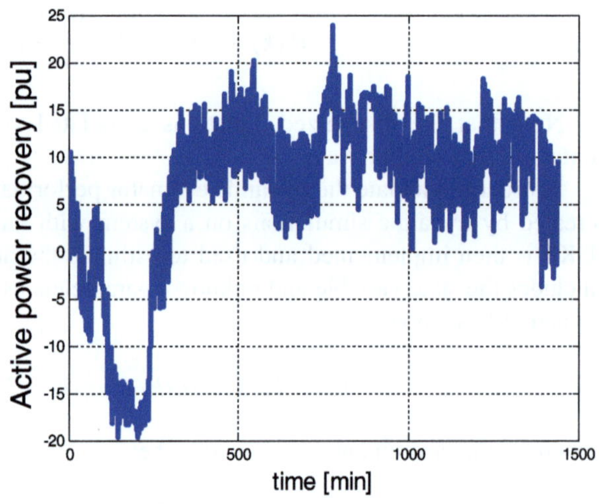

Fig. 13.6 Estimated values for P_r

Note that the true values are not known, so it is not possible to comment on the accuracy of the parameters; however, given the long duration of the study time (1440 min or 24 h), one can observe UKF's tracking of parameters as they gradually change in time.

While the actual parameters of the model are unknown and cannot be directly measured, total active and reactive power are available as measured values. Thus, performance of the UKF in tracking the total active power can be evaluated by observing the measured and predicted total active power as shown in Fig. 13.10.

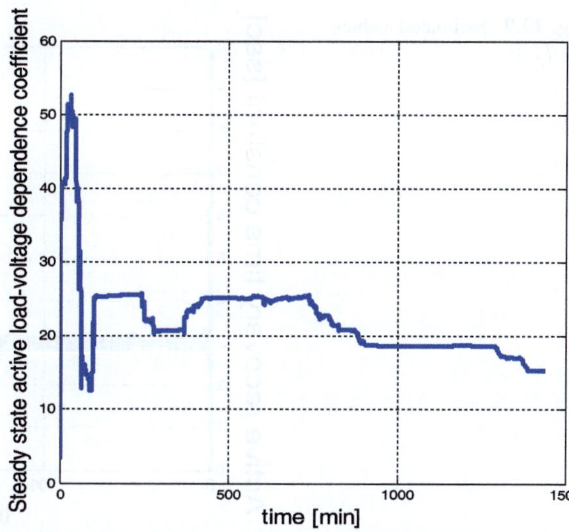

Fig. 13.7 Estimated values for α_s

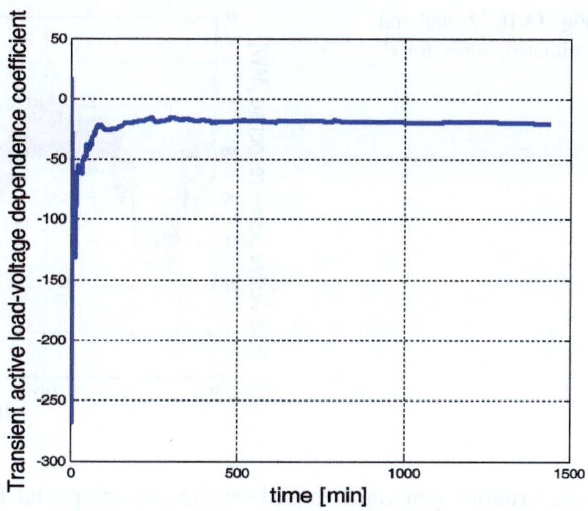

Fig. 13.8 Estimated values for α_t

13.4.2 Failure Detection of the Synchronous Generator Excitation Systems

A common problem that involves generator control systems is the failure of exciters because of diode failures [19]. It is important to detect such failures in a timely fashion in order to avoid risk of exciter and voltage regulator damage. While there are direct methods to detect exciter diode failures, they are typically expensive to implement. A popular indirect method to detect diode failure is via the exciter field current. In this section, a practical solution for detection of exciter failures in

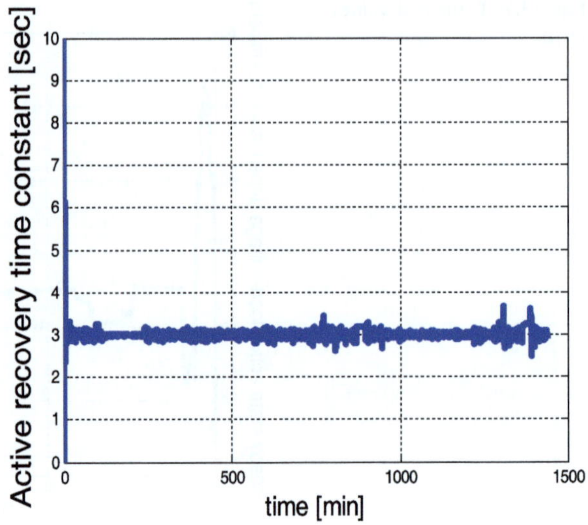

Fig. 13.9 Estimated values for T_p

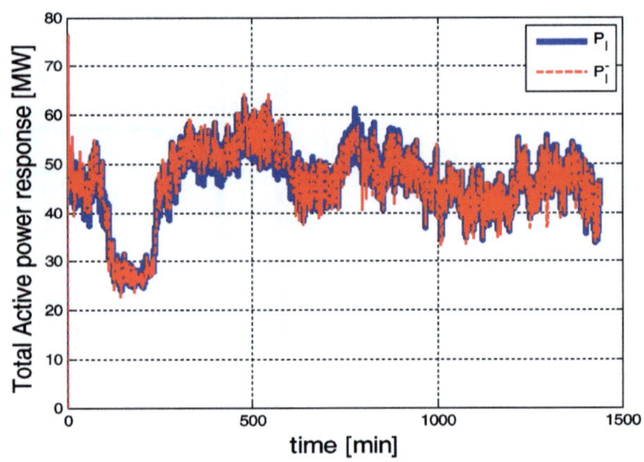

Fig. 13.10 Actual and estimated values for P_l

synchronous generators based on the use of special DSEs will be described. The approach is motivated by considering the case where at some point during the daily operation of a synchronous generator, the dynamic model related to the exciter may abruptly change. This could be due to an exciter rotator diode failure [19]. Thus, the proposed dynamic state estimation approach in this section will not only remain robust against exciter failure but also detect its occurrence in a timely fashion. Hence, the system operator can be promptly informed about the type and location of the exciter failure without requiring costly local investigations on the performance of the specific excitation system. This section will describe the multiple model estimation technique which will be used to identify the true model among several alternative models of a given generator and its associated control systems.

13.4.2.1 Multiple Model Estimation Technique

Let the set, p, given by $\{p_1, p_2, \ldots, p_N\}$ represent the set of possible values which can be assumed by a given system model [19]. Considering N different models corresponding to the parameters $\{p_1, p_2, \ldots, p_N\}$ and a given measurement z_k at time step k, the probability of p being equal to p_j can be computed as:

$$\Pr\left(p_j|z_k\right)$$

This probability is expected to converge to a negligible value during the simulations when an invalid model parameter is used and it will converge to a value close to 1.0 for the valid parameter. The probability given above can be computed and updated using the following algorithm:

1. Initialize the probability of each parameter set before any measurements are obtained:

$$\Pr\left(p_j|z_0\right) \quad j=1, \ldots, N$$

2. For each assumed parameter set, run a DSE and obtain $\hat{x}^{-}_{k,j}$ and $P^{-}_{k,j}(j=1, \ldots, N)$ from each DSE.
3. For each parameter set, approximate the pdf of z_k given p_j as follows:

$$pdf(z_k|p_j) \approx \frac{\exp(-r_k^T P_{z,j}^{-1} r_k/2)}{(2\pi)^{m/2}|P_{z,j}|^{1/2}}$$

where $P_{z,j}$ is as defined in a previous section, $r_k = z_k - \hat{z}_{k,j}$ and m is the total number of the measurements.

4. Estimate the probability that $p = p_j$ as follows:

$$\Pr(p_j|z_k) = \frac{pdf(z_k|p_j)\Pr(p_j|z_{k-1})}{\sum_{i=1}^{N} pdf(z_k|p_i)\Pr(p_i|z_{k-1})}$$

13.4.2.2 Proposed Approach for the Detection of Exciter Failure in Dynamic State Estimation

Under the normal operating conditions of a synchronous generator and associated excitation system, it is desired to estimate the state variables of both synchronous generator and the exciter. However, when a problem such as the exciter rotator diode failure occurs, the assumed dynamics of the exciter will not be valid anymore and therefore using the inaccurate dynamic model of the exciter can significantly degrade the accuracy of the estimated results of the state variables provided by the DSE. The following approach is proposed in order to circumvent this problem. Note that this approach not only remains robust against the occurrence of the exciter failure but also properly detects the occurrence of this failure. The steps of the proposed approach are listed below [19]:

- Use an UKF for the dynamic state estimation of the synchronous generator and associated exciter.
- Use an Iterated Unscented Kalman Filter (IUKF) for the estimation of the augmented state vector associated with synchronous generator assuming that excitation voltage is an unknown input.
- Use multiple model estimation technique for the detection of exciter failure. This will be accomplished by executing both UKF and IUKF in parallel for tracking the states of the synchronous generator and associated exciter. When the exciter failure occurs, the first DSE (UKF) will perform poorly due to the inaccurate dynamic model used for the exciter whereas the performance of the second DSE (IUKF) will not be affected. By using the multiple model estimation technique, the calculated probabilities corresponding to the models used by the UKF and IUKF will converge to 0 and 1, respectively, following the occurrence of the exciter failure.

The proposed dynamic state estimation approach is tested using the two-axis model of synchronous generator with the following assumption:

- At time = 10 s, the value of the exciter amplifier gain (K_a) is changed from 50 to 30 (pu). This change is intentionally ignored, and the old gain value is continued to be used in the dynamic model of the exciter. Dynamic state estimation is carried out using the UKF as described in previous section.

Figure 13.11 shows the estimated rotor angle by the UKF following the change in the exciter amplifier gain. As evident from this figure, the estimated rotor angle diverges from the true trajectory due to the exciter model error. On the other hand, IUKF continues to provide accurate estimates for the rotor angle (Fig. 13.12) since the performance of the IUKF remains independent of the assumed exciter model.

Note that during a real exciter failure scenario the true trajectory will not be known. Therefore, in order to detect the exciter failure it is necessary to evaluate the performance of both estimators. As mentioned before, the multiple estimation technique can detect the occurrence of the exciter failure. For example, Fig. 13.13 shows that the probabilities for the UKF and IUKF converge to 0 and 1, respectively, following the exciter failure.

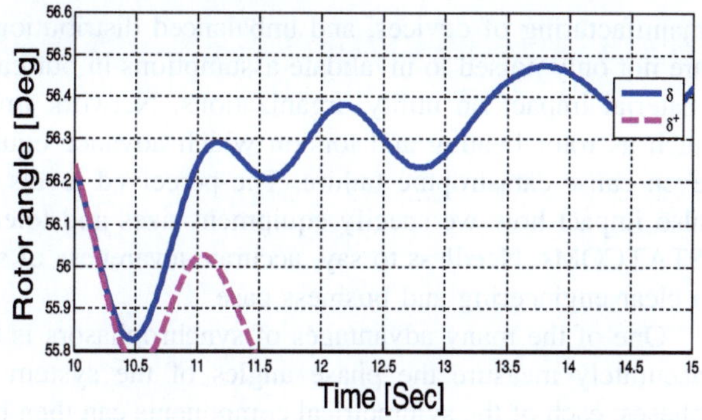

Fig. 13.11 Estimated and true trajectories of the rotor angle-UKF is used as DSE

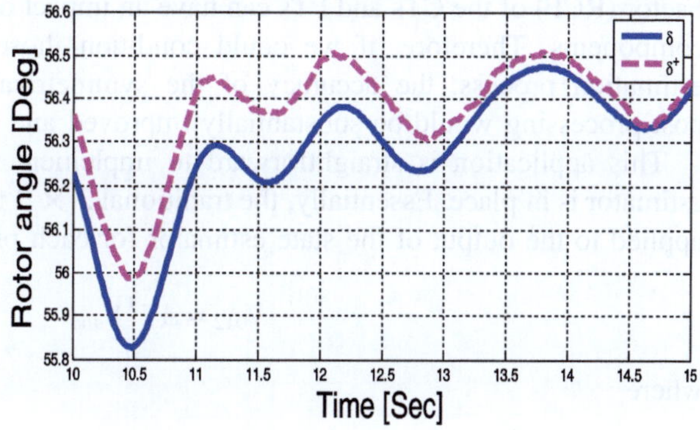

Fig. 13.12 Estimated and true trajectories of the rotor angle-IUKF is used as DSE

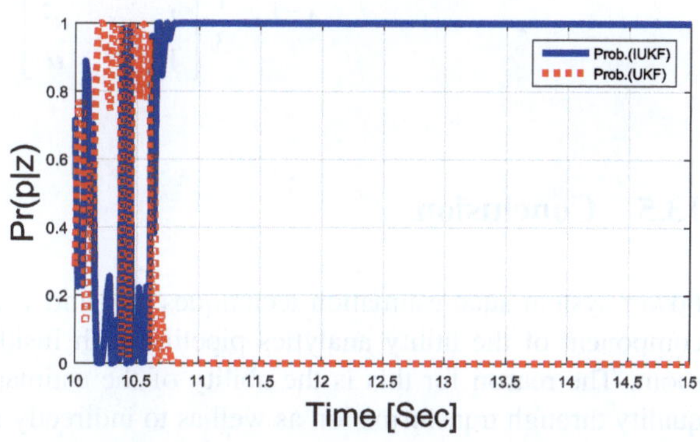

Fig. 13.13 Result of the multiple estimation technique for UKF and IUKF after the occurrence of the exciter failure

13.4.3 Post-estimation Symmetrical Component Computation

The assumption of a balanced three-phase power system is an important one. However, it is flawed. Imbalances between the phases arise from multiple agents including, but not limited to, poor or absent line transposition, slight variations in manufacturing of devices, and unbalanced distribution loading. These imbalances are not only poised to invalidate assumptions in our calculations but can have real, material impact on utility organizations. Network imbalances can lead to issues such as rotor heating and torsion which advance equipment degradation and can even cause catastrophic failure. The perceived levels of symmetrical components also impact how we specify equipment sizes and tolerances such as in SVCs and STATCOMs. Needless to say, accurate awareness of symmetrical components has a clear engineering and business case.

One of the many advantages of synchrophasors is their ability to precisely and accurately measure the phase angles of the system. When measuring all three phases, each of the symmetrical components can then be computed from the phasor values. However, errors in the measurements, particularly the Ratio Correction Factor (RCF) of the CTs and PTs can have an impact on the calculated symmetrical components. Therefore, if we could condition those measurements through an estimation process, the accuracy of the symmetrical components computed in post-processing would be substantially improved and more trustworthy.

This application is straightforward to implement if a three-phase linear state estimator is in place. Essentially, the traditional 3 × 3 transformation matrix will be applied to the output of the state estimator for each phase group.

$$V_{012} = A^{-1} V_{abc}$$

where

$$A^{-1} = \frac{1}{3} \begin{bmatrix} 1 & 1 & 1 \\ 1 & \alpha & \alpha^2 \\ 1 & \alpha^2 & \alpha \end{bmatrix}$$

13.5 Conclusion

Power system state estimation techniques are, and will continue to be, an integral component of the utility analytics pipeline both inside and outside of the control room. The reason for this is the ability of the estimators to directly improve data quality through transformation as well as to indirectly improve data quality through feedback loops that inform algorithms and engineers about the presence of bad data so that steps can be taken to mitigate it. This is no less true as synchrophasor measurements become an increasingly voluminous subset of utility data.

This chapter presented a WLS linear state estimator in both positive sequence and three phase as well as an extension of this formulation that utilizes LAV for robustness. Furthermore, a robust dynamic state estimator that utilizes a UKF was presented as the second stage to the static LAV linear state estimator. In combination, the two form the TRODSE. Following this, several applications that employ estimation techniques and/or estimation results were presented including dynamic load modeling, exciter failure detection, and symmetrical component calculation.

References

1. Phadke AG, Thorp JS (2008) Synchronized phasor measurements and their application, springer science + business media
2. Schweppe FC, Wildes J (1970) Power system static-state estimation, part I: exact model. In: IEEE transactions on power apparatus and systems, vol PAS-89, pp 120–125
3. Schweppe FC, Rom DB (1970) Power system static-state estimation, part II: approximate model. In: IEEE transactions on power apparatus and systems, vol PAS-89, pp 125–130
4. Schweppe FC (1970) Power system static-state estimation, part iii: implementation. In: IEEE transactions on power apparatus and systems, vol PAS-89, pp 130–135
5. Jones KD (2011) Three phase linear state estimation with phasor measurements. Master's thesis, Virginia Tech
6. Jones KD (2017) Synchrophasor analytics—LSE core libraries. https://www.github.com/kdjones/lse
7. Phadke AG, Thorp JS, Centeno V (2013) Synchrophasor based tracking three-phase state estimator and its applications, DE-OE0000118 Final Report, 31 Aug 2013
8. Jones KD, Thorp JS, Gardner RM (2013) Three-phase linear state estimation using phasor measurements. In: IEEE power and energy society general meeting (PES)
9. Jones KD, Pal A, Thorp JS (2015) Methodology for performing synchrophasor data conditioning and validation. IEEE Trans Power Syst 30(3):1121–1130
10. eLSE (2017) Electric power group. http://www.electricpowergroup.com/else.html
11. [Reference to DOE FOA 970]
12. Grid Protection Alliance (2017) The open and extensible control and analytics platform (2017). https://www.github.com/GridProtectionAlliance/openeca
13. Jones KD (2017) The open source linear state estimator (2017). https://www.github.com/kdjones/openlse
14. Rouhani A, Abur A (2016) Linear phasor assisted dynamic state estimator. IEEE Trans Smart Grid, early access, Mar 2016
15. Gol M, Abur A (2014) LAV based robust state estimation for systems measured by PMUs. IEEE Trans Smart Grid 5(4):1808–1814
16. Simon D (2006) Optimal state estimation: Kalman, H infinity and nonlinear approaches. Wiley, New Jersey
17. Rouhani A, Abur A (2015) Real-time dynamic parameter estimation for an exponential dynamic load model. IEEE Trans Smart Grid 7(3):1530–1536
18. Karlsson D, Hill D (1994) Modeling and identification of nonlinear dynamic loads in power systems. IEEE Trans Power Syst 9(1)
19. Rouhani A, Abur A (2016) A robust dynamic state estimator against exciter failures. In: Proceedings of the IEEE north American power symposium, Denver, USA, 18–20 Sept 2016

Chapter 14
Implementation of Synchrophasor-Based Linear State Estimator for Real-Time Operations

Lin Zhang, Heng Chen, Anthony Faris, Megan Vutsinas, Tim Bradberry, Evan Phillips and Josh Bui

14.1 Introduction

An electric power system is a network of electrical components deployed to supply, transfer, and use electric power. The complex cyber-physical system can be broadly divided into generation, transmission, and distribution. The generation system supplies the electricity to transmission network. The transmission system carries the power from the generating centers to the load centers. Phasor Measurement Units (PMUs) are usually deployed at transmission system; so in this chapter, we will only focus on the transmission system. Energy Management System (EMS) is used by utilities for real-time operation: monitor, dispatch, and control the transmission network. As one of the key functions of EMS, State Estimator (SE) plays a critical role in providing the best estimate of the system state for downstream advanced applications, such as power flow, contingency analysis, optimal power flow, dynamic stabilities, etc. SE derives the voltage phasor and current phasor values iteratively based on system network model, voltage magnitude, power injection, and power flow measurements. The iterative solution often faces issues like failing to convergence and time-consuming. PMUs provide phasor measurements in a high-sampling rate in real time, which can be used to form a linear SE. Using voltage and current phasors rather than power flow data makes SE process linear.

L. Zhang (✉) · H. Chen
Electric Power Group, 251 S Lake Ave, Suite 300, Pasadena, CA, USA
e-mail: zhang@electricpowergroup.com

A. Faris
Bonneville Power Administration, 905 NE 11th Ave, Portland, OR, USA

M. Vutsinas · T. Bradberry · E. Phillips
Duke Energy, 550 S Tryon St, Charlotte, NC, USA

J. Bui
Southern California Edison, 2244 Walnut Grove Ave, Rosemead, CA, USA

S. Nuthalapati (ed.), *Power System Grid Operation Using Synchrophasor Technology*, Power Electronics and Power Systems,
https://doi.org/10.1007/978-3-319-89378-5_14

For a set of linear measurement functions, the Weighted Least-Square (WLS) method has a non-iterative and closed form solution. Hence, it becomes feasible that the so-called Linear State Estimator (LSE) can solve as fast as PMU data rate, e.g., 30/60 frames per second, and capture the system behavior in response to transient events. In the USA, several pilot projects have been created with the objective to develop the LSE algorithm for practical use at control centers.

In this chapter, we provide an overview of the practical LSE use cases for real-time operation. The benefits of LSE for real-time operation have been identified after the successful deployment of LSE application. The LSE can address PMU data quality issues by validating and conditioning PMU data against network model. LSE operating with only synchrophasor measurements can provide independent and reliable situational awareness in parallel with EMS. In addition, due to the high expense of PMU device, commission, and communication bandwidth, the PMU coverage is very limited in some utilities. LSE can be employed to achieve pseudo-observability and extend the PMU coverage by creating virtual PMUs at neighboring substations, such that a better wide-area situational awareness can be achieved cost-effectively.

14.2 Synchrophasor-Based Linear State Estimator: Theory

Before diving into synchrophasor-based LSE, it is necessary to first introduce the existing traditional state estimator which has been deployed and in operational use for several decades.

14.2.1 Traditional State Estimation Algorithm

Traditional power system state estimation is one of the key EMS applications in existing control centers. The purpose of this traditional SE is to obtain the best estimate of the real-time model of the power system. For SE to run, it needs input from a static database and a real-time database. The static data base of the EMS provides network configuration data, such as line impedance, device connectivity within the substation, and branch connectivity between substations. The real-time database is made up of (SCADA), both analog and digital measurements. Analog measurements typically comprise voltage magnitudes, real and reactive power injections, and power flow. Digital measurements comprise breaker and switch status. Supervisory Control and Data Acquisition (SCADA) is a complex monitor and control system consisting of a central host or master, known as master terminal unit (MTU); several field data gathering and control units or remotes are known as remote terminal units (RTUs); and a collection of standard and/or custom software used to monitor and control remotely located field data elements.

The static state of a power system can be described by the vector of complex voltage. The state estimation is to calculate the states from inherent network parameters and an adequate set of voltage and power flow measurements. The formulation of state estimation problem can be described as follows,

$$\arg\min \boldsymbol{\sigma}^T \boldsymbol{W} \boldsymbol{\sigma}$$

$$s.t.\ \boldsymbol{Z} = \boldsymbol{h}(\boldsymbol{X}) + \boldsymbol{\sigma} \tag{14.1}$$

where $\boldsymbol{Z}$ is the vector of measurements consisting of voltage magnitude, power flow, and power injections; $\boldsymbol{X}$ is the vector of complex voltage; $\boldsymbol{\sigma}$ is the measurement error vector, $\boldsymbol{W}$ is the diagonal matrix of weighting factors of each measurement; $\boldsymbol{h}$ is a set of nonlinear measurement functions.

In transmission network, the number of measurements m is usually much greater than the number of state n. The redundancy r of the measurements opposed to state is defined as follows,

$$r = m/n \tag{14.2}$$

To solve this overdetermined system, the WLS method in regression analysis is the best estimate. The detailed formulation is given as follows.

Let us assume $i, j \in (1, \ldots, n)$. The measurement vector $\boldsymbol{Z}$ consists of:

Bus power injections: P_i, Q_i

Branch power flows: $P_{ij}, Q_{ij}, P_{ji}, Q_{ji}$

Voltage magnitude: V_i

The set of measurement Eq. (14.1) can then be written as,

$$P_i = V_i^2 G_{ii} + \sum_{j=1, j \neq i}^{n} V_i V_j Y_{ij} \cos(\theta_{ij} + \delta_j - \delta_i)$$

$$Q_i = -V_i^2 B_{ii} + \sum_{j=1, j \neq i}^{n} V_i V_j Y_{ij} \sin(\theta_{ij} + \delta_j - \delta_i)$$

$$P_{ij} = -V_i^2 G_{ij} + V_i V_j Y_{ij} \cos(\theta_{ij} + \delta_j - \delta_i)$$

$$Q_{ij} = V_i^2 B_{ij} - V_i V_j Y_{ij} \sin(\theta_{ij} + \delta_j - \delta_i)$$

$$V_i = V_i$$

If we assume that the number of injections P_i, Q_i is n, the number of power flow P_{ij}, Q_{ij} is b, and the number of V_i is n, then the redundancy can be

$$r = (3n + 4b)/2n \tag{14.3}$$

The classic WLS [1] method has been used to solve the typical nonlinear unconstraint SE problem. The detailed derivation of the solution can be found in many of optimization book, and hence will not be given in this book. Since the measurement function is nonlinear, the WLS solution is iterative.

14.2.2 Linear State Estimation Algorithm

A PMU is essentially a digital recorder with a synchronized time stamp provided by Global Positioning System (GPS). Time synchronization allows synchronized real-time measurements at multiple remote measurement points on the grid. A PMU can be a dedicated device, or the PMU function can be incorporated into a protective relay or Digital Fault Record (DFR). The PMU hardware device was first invented in the late 1980s by Dr. Arun G. Phadke and Dr. James S. Thorp at Virginia Tech [2, 3].

Synchrophasor Measurement Systems (SMS) have been expanded dramatically via the installation of numerous PMUs in a number of synchrophasor initiatives, e.g., Department of Energy Smart Grid Investment Grant (SGIG). Wide-area monitoring and situational awareness systems have been established at several Independent System Operators (ISOs) and Transmission System Operators (TSOs). Such systems can provide operators with great observability of system dynamic and operation margins, data analytics, and system event detection to assist maintaining power system stability.

With more and more PMUs installed, LSE utilizing PMU data becomes feasible. PMUs can directly measure the magnitude and angle of voltage and current. The measurements are voltage phasors and current phasors while the states are still the complex voltage phasor. The relationship between measurements and states can be linear under the condition when the phasors are in the form of a rectangular coordinate. The formulation of linear state estimation problem can be described as rewritten of Eq. (14.1),

$$\arg\min \overline{\boldsymbol{e}}^T \boldsymbol{W} \overline{\boldsymbol{e}}$$

$$s.t.\ \overline{\boldsymbol{Z}} = \overline{\boldsymbol{H}}\overline{\boldsymbol{x}} + \overline{\boldsymbol{e}} \tag{14.4}$$

where $\overline{\boldsymbol{Z}}$ is the complex vector of phasor measurements consisting of voltage and current; $\overline{\boldsymbol{X}}$ is the complex vector of bus voltage phasor; $\overline{\boldsymbol{e}}$ is the complex measurement error vector, $\boldsymbol{W}$ is the diagonal matrix of weighting factors of each measurement; $\overline{\boldsymbol{H}}$ is a set of complex admittance matrix. Note that $\overline{(.)}$ denotes a complex variable.

The WLS method is applied here again to solve the LSE problem. The detailed formulation is given as follows. Let us assume $i, j \in (1, n)$. The measurement vector $\overline{\boldsymbol{Z}}$ consists of:

Bus current flow injections: $\overline{I}_i$
Branch current flows: $\overline{I}_{ij}, \overline{I}_{ji}$
Voltage magnitude: $\overline{V}_i$

The classic Weighted Least-Square (WLS) method has been used to solve the linear unconstraint SE problem. Since the measurement equations are linear, the WLS solution has a closed form without iteration [4–6],

$$\overline{\boldsymbol{X}} = (\overline{\boldsymbol{H}}^T \boldsymbol{W} \overline{\boldsymbol{H}})^{-1} \overline{\boldsymbol{H}}^T \boldsymbol{W} \overline{\boldsymbol{Z}} \tag{14.5}$$

where the gain matrix is shown as

$$\overline{\boldsymbol{G}} = \overline{\boldsymbol{H}}^T \boldsymbol{W} \overline{\boldsymbol{H}} \tag{14.6}$$

Without the need of iteration, LSE can run at the sub-second cycle. PMU is able to stream phasor data measurements as fast as 30 frames per second. With microprocessor becomes much faster and more powerful than before, LSE should be able to run as fast as PMU data rate. As more and more PMUs are installed, a LSE utilizing only PMU data becomes feasible.

14.3 Synchrophasor-Based Linear State Estimator: Implementation

In this section, implementation and integration of LSE algorithm at the control center and the LSE application components are demonstrated in detail.

14.3.1 LSE Integration

Similar to traditional SE, an LSE also requires a static network model and real-time measurements including analogs and digitals. The static network model can be very much similar except it may be in different formats. Another major difference is that the PMUs usually are not able to cover the whole system, and hence, the LSE model has to be reduced to a subset of the full EMS model that is observed by the PMU measurements. The real-time analog measurements are different from a traditional SE, as the analog measurements are voltage and current phasors. As for digital measurements, e.g., breaker statuses, most of the existing PMUs do not report the digital status of circuit breakers; therefore, the breaker status has to be retrieved to the LSE in a different way. Here, we introduce that the breaker statuses

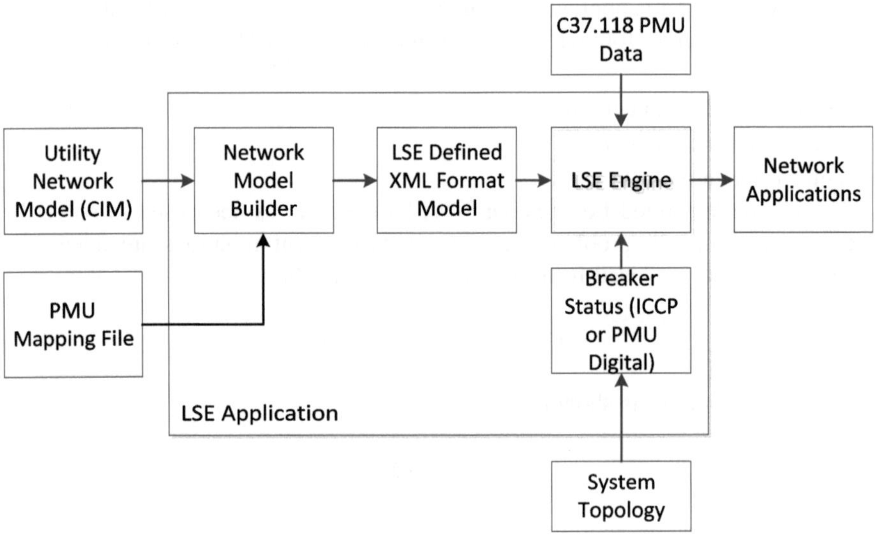

Fig. 14.1 LSE application context

can be obtained from the SCADA system through an Inter-Control Center Communications Protocol (ICCP) link [7–10]. Figure 14.1 shows the overall LSE application context.

14.3.1.1 Network Model Integration

The network model builder in Fig. 14.1 is a tool that generates the LSE customized Extensible Markup Language (XML) model from the EMS static system network model. Different utilities may use different format network model from their own vendors. Here, we introduce the so-called Common Information Model (CIM), which is an open standard that defines how the electrical equipment in an IT environment is represented as a common set of objects and relationships between them. The CIM format of EMS node-breaker model has been used to create the LSE XML format model. The choice was made because CIM network model is open to all users (not vendor specific).

14.3.1.2 PMU Data Mapping

The PMU signals need to be mapped to node voltages, the current of transformers, transmission lines, and shunt/reactor compensators as a mapping file. The network builder reads the mapping file and automatically maps signal to components into the LSE model. Note that the node voltage phasor has a 1 to 1 relationship with the node, whereas each line has current measurements at two ends. Hence, a current measurement signal needs to be properly mapped to the exact end of the line.

14.3.1.3 Real-Time Topology Update

A link which can transfer the real-time breaker statuses to LSE engine is required. Ideally, PMU digital channel is the best choice to fulfill this requirement. However, in reality, not many utilities actually activate this function of PMUs. Instead, ICCP is being specified by utility organizations to provide data exchange over wide-area networks (WANs) between utility control centers, utilities, power pools, regional control centers, and non-utility generators. ICCP is also an international standard: International Electro-technical Commission (IEC) Tele-control Application Service Element 2 (TASE.2). Hence, the LSE engine first looks to use the breaker status from PMU digitals. If not available, breaker status information from EMS will be used through ICCP.

14.3.2 LSE Application Components

The traditional state estimation function consists of four major program modules solved sequentially and the LSE is very similar in that respect. In this section, these four modules are described briefly.

14.3.2.1 Topology Process

The function of the network topology processor is to determine the present topology of the network from the telemetered status of circuit breakers. The LSE model extracted from the system static network model provides a detailed description of the network connectivity in terms of bus sections, circuit breakers, and switches. The LSE topology processor uses all these substation connectivity information and breaker/switch status to construct the network bus-branch model from the node-breaker model. A typical substation level description of a power system is shown in Fig. 14.2.

14.3.2.2 Real-Time Observability Analysis

The result of topology processor provides the present connectivity of the power system. When the topology of the system changes in real time by circuit breakers switching, a real-time observability checking is required to correlate the PMU measurement with the topology, so that the LSE algorithm is able to determine how

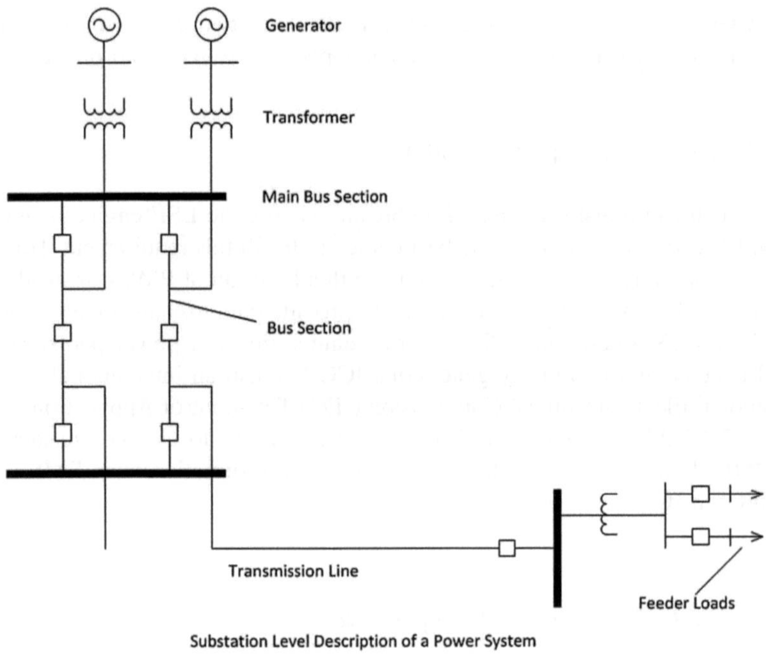

Fig. 14.2 A typical substation level description of a power system

many observable islands there are and how the buses are connected in each observable island.

The power system is algebraically observable with respect to the given measurements if the Jacobian matrix $\overline{\boldsymbol{H}}$ has the rank equals to a number of states. The numerical observability analysis looks conceptually simple, and by checking the rank of Jacobian matrix $\overline{\boldsymbol{H}}$, we can make sure the state estimation problem is algebraically observable or not. However, as a numerical algorithm, the round-off errors may make the process difficult if the diagonal entry of Jacobian matrix $\overline{\boldsymbol{H}}$ has zeros.

Unlike the numerical method, the topological method makes use of topology information from topology processor. The basic idea of topological observability method is to find an observable spanning tree, given a set of measurements and the topology of the power system. A spanning tree of the network graph is an Observable Spanning Tree (OST) if and only if it is possible to assign a current flow measurement to each one of the tree branches, such that no two branches are associated with the same measurement.

If it is not able to find an OST for the network, the topological observability analysis method has to find a maximal forest with multiple trees and each one of them in the forest is an observable spanning tree. Given the measurement set $\boldsymbol{M}$, the method constructs a fundamental spanning forest by assigning each line flow measurement to the corresponding branches and identifying the observable buses which are connected to form a spanning tree.

The topological method is straightforward and easy to implement in the program compared with the numerical method. Hence, in LSE program, the topological method is also being used to do the real-time online observability check.

14.3.2.3 LSE Matrix Formulation

The linear state estimation is essentially a well-defined linear optimization problem. Its result provides the best estimate of state variables of the observable network. The analog quantities measured by the meter system are sent to control center, where it has computer units to process these data through communication links. Here, in this case, PMU data are sent to control center. However, the measurements are not noisy free, so the LSE is needed to filter out those noises.

As the LSE algorithm introduced in previous sections, the measurement function becomes linear so the solution has no iterations. The measurement inputs are the bus voltage vector $\overline{\boldsymbol{V}}$, flow current vector $\overline{\boldsymbol{I}}_f$:

$$\overline{\boldsymbol{V}} = \begin{bmatrix} \overline{V}_1 \\ \overline{V}_2 \\ \vdots \\ \overline{V}_n \end{bmatrix} \tag{14.7}$$

$$\overline{\boldsymbol{I}}_f = \begin{bmatrix} \overline{I}_{flow,1} \\ \overline{I}_{flow,2} \\ \vdots \\ \overline{I}_{flow,i} \end{bmatrix} \tag{14.8}$$

where i is the index of flow measurement.

The measurement function can be formulated and written in a matrix format, as follows:

$$\overline{\boldsymbol{Z}} = \begin{bmatrix} \overline{\boldsymbol{V}} \\ \overline{\boldsymbol{I}}_f \end{bmatrix} = \overline{\boldsymbol{HV}} + \overline{\boldsymbol{e}} = \begin{bmatrix} \overline{\boldsymbol{I}} \\ \overline{\boldsymbol{Y}} \end{bmatrix} \overline{\boldsymbol{V}} + \overline{\boldsymbol{e}} \tag{14.9}$$

where $\overline{\boldsymbol{I}}$ is the identity matrix, $\overline{\boldsymbol{e}}$ is the measurement error vector, $\overline{\boldsymbol{Y}}$ is an admittance matrix and represents the relationship between two bus voltages and it can be obtained from network model. The π model is shown in Fig. 14.3, and R, X, and B are resistance, reactance, and line shunt susceptance, respectively. $\overline{\boldsymbol{V}}_1, \overline{\boldsymbol{V}}_2$ are the from-end and to-end complex bus voltages, whereas $\overline{\boldsymbol{I}}_{12}, \overline{\boldsymbol{I}}_{21}$ are from-end and to-end complex line currents. Hence, an example of line current measurement function is given in (14.10).

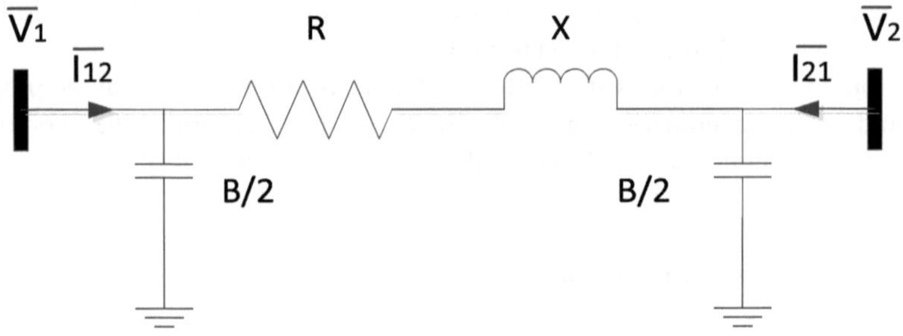

Fig. 14.3 A typical π model

$$\bar{I}_{12} = \bar{Y}_{12}(\bar{V}_1 - \bar{V}_2) + \bar{V}_1 * j\frac{B}{2} \tag{14.10}$$

14.3.2.4 Bad Data Detection and Identification

If the power system network model and acquired data measurement from PMUs are both accurate, there is a good reason to believe that linear state estimator can provide the best estimates using the weighted least-square algorithm. However, there is still a chance that a measurement is grossly erroneous with white Gaussian noise. One function of linear state estimator should be able to detect, identify, and remove those bad measurements, also known as bad data. Therefore, post-processing in the LSE algorithm includes a bad data detection and identification module which can filter out the bad PMU data to prevent it from contaminating the solution. The widely used method chi-squares test is applied to detect bad data, and the other method Largest Normalized Residual (LNR) is used to identify bad data.

Let us assume $\hat{e}$ the estimated measurement error,

$$\hat{e} = Z - Z = Z - HV = [I - HG^{-1}H^TW]e \tag{14.11}$$

Then, the mean of the estimated error is,

$$E(\hat{e}) = E(Z - Z) = E(Z - HV) = [I - HG^{-1}H^TW]E(e) = 0 \tag{14.12}$$

The expected value of ee^T can be easily obtained,

$$E(ee^T) = R \tag{14.13}$$

where $\hat{e}\hat{e}^T$ is shown as follows,

$$\hat{e}\hat{e}^T = (Z - Z)(Z - Z)^T = \left[I - HG^{-1}H^TR^{-1}\right]ee^T\left[I - R^{-1}HG^{-1}H^T\right]$$

Hence, the covariance matrix of the estimated error can be given,

$$\begin{aligned} E\left(\hat{e}\hat{e}^T\right) &= E\left(\left[I - HG^{-1}H^TR^{-1}\right]ee^T\left[I - R^{-1}HG^{-1}H^T\right]\right) \\ &= \left[I - HG^{-1}H^TR^{-1}\right]E\left(ee^T\right)\left[I - R^{-1}HG^{-1}H^T\right] = R - HG^{-1}H^T = R' \end{aligned} \tag{14.14}$$

We substitute estimated error e_i in the objective function of state estimation (14.1), then the weighted sum of square $\hat{f}$ is,

$$\hat{f} = \sum_{j=1}^{m} w_j \hat{e}_j^2 \tag{14.15}$$

where m is a number of measurement and the weighting factor w_j is set equal to $1/\sigma_j^2$.

Based on random process theory, $\hat{e}_j$ is a Gaussian random variable with zero mean, and then the weighted sum of square $\hat{f}$ is also a random variable which has a chi-square probability distribution, with the number of degrees of freedom $m - n$ (number of state variable),

$$E\left[\hat{f}\right] = \sum_{j=1}^{m} w_j \hat{e}_j^2 = m - n \tag{14.16}$$

The bad data detection procedure is described as follows,

(1) Solve the linear state estimator, and the estimated state vector $\hat{V}$ is obtained;
(2) Substitute estimates of states to the equation $\hat{Z} = H * \hat{V}$, hence the estimated error $\hat{e} = Z - \hat{Z}$;
(3) Evaluate the weighted sum of squares $\hat{f} = \sum_{j=1}^{m} w_j \hat{e}_j^2$;
(4) For the degrees of freedom $k = m - n$, and a specified probability α determines whether or not the value of $\hat{f}$ is less than the critical value corresponding to α. In practice, this means to check the inequality $\hat{f} < \chi_{k,\alpha}^2$
(5) If $\hat{f} > \chi_{k,\alpha}^2$, then bad data is detected; otherwise, no bad data.

The next major step is to identify the bad data and remove it from the measurements if there is bad data detected. The Largest Normalized Residue (LNR) method which uses the properties of the normalized residue is widely used in today's single bad data identification program. The identification procedure is shown below,

(a) Solve the LSE problem and then calculate the residue vector of measurement $\hat{e} = \boldsymbol{Z} - \hat{\boldsymbol{Z}}$;
(b) Calculate the covariance matrix of the estimated measurement error $\boldsymbol{R}' = \left(\boldsymbol{I} - \boldsymbol{H} * \boldsymbol{G}^{-1} * \boldsymbol{H}^T * \boldsymbol{R}^{-1}\right) * \boldsymbol{R}$;
(c) $R_{jj'}$ is the diagonal element of $\boldsymbol{R}'$, then the normalized residues: $r_i^N = \frac{r_i}{R_{jj'}}$;
(d) Find the largest normalized residue and compare it with pre-set threshold, if $r_i^N > threshold$, then remove it;
(e) Go to step 1 again, until all the $r_i^N < threshold$.

It should be pointed out that this function is even more useful during the commissioning of the LSE than during routine operation. Because newly installed PMUs and their communications may have many start-up issues, the LSE becomes a useful method of testing the accuracy and performance of the PMUs.

14.3.3 LSE Operation Procedure

Figure 14.4 shows the operation procedure of the LSE algorithm. The procedure of LSE operation is summarized in the following pseudo-codes.

Step ①: A LSE input interface is used to parse the LSE format network model, receiving and preprocessing PMU data and the breaker statuses. Some sanity check is applied in this step. For instance, the value of voltage and current magnitudes needs to be within a reasonable range in terms of the various corresponding voltage level. The particular PMU voltage measurement will be discarded if it fails to pass this sort of sanity check. The LSE model is read once when LSE starts, while PMU data and breaker statuses are updated continuously in real-time, 30 frames per second and in a timed interval, e.g., once per 2 s, respectively.
Step ②: Once the update of breaker status is received, the LSE program determines if there is any breaker status change. If so, the LSE program goes to step ③. If not, LSE program proceeds to identify if there is measurement state change once a new frame of PMU data is received. Measurement state refers to the validity and usability of the measurement determined by the sanity check in step ①. For instance, the measurement value changing from a valid and reasonable number to Not a Number (NaN) is considered as measurement state change. If measurement state changes, the LSE program goes to step ④. If there is no measurement state change, the LSE directly jumps to step ⑤. The rationale behind it is that the observability change of the network may be reflected in step ④ and LSE algorithm needs to reformulate the $\boldsymbol{H}$ and $\boldsymbol{W}$ in step ⑤.
Step ③: The network topology processor converts the system network node-breaker model to a bus-branch model. It uses telemetered status of circuit breakers to group the connected nodes into a bus, and therefore to determine the real-time system network topology.

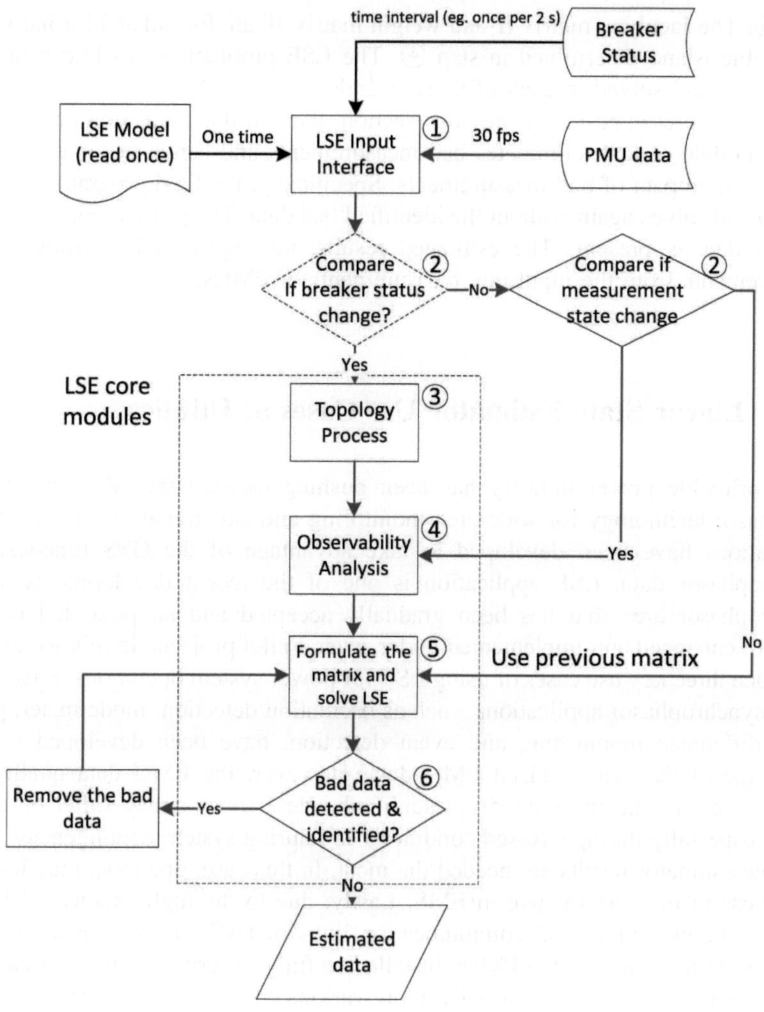

Fig. 14.4 Operation procedure of LSE

Step ④: The function of real-time observability analysis is to correlate the PMU measurements with the current system network topology determined in step ③. The unconnected portions of the network covered by PMU measurements are identified as individual observable islands. For instance, if two connected bus bars in a substation are split into separate buses, one may be observed by a direct PMU measurement (a PMU connected at the bus) or an indirect PMU measurement (PMU at the other end of the line). To the contrary, the other bus may not have a direct or indirect PMU measurement and hence becomes unobservable. Observability analysis is of the importance to keep the network model updated matching the measurements received from PMUs.

Step ⑤: The Jacobian matrix $\boldsymbol{H}$ and weight matrix $\boldsymbol{W}$ are formulated for individual observable island determined in step ④. The LSE problem in (14.1) can then be formulated and solved in each of those islands.
Step ⑥: As presented in the previous section, the bad data detection and identification module of LSE eliminates bad measurements and hence re-solves the LSE without the impact of bad measurements. Specifically, the LSE program returns to step ⑤ and solves again without the identified bad data. The process continues until no bad data is present. The estimated results are improved by removing bad measurements from the input raw measurements of PMUs.

14.4 Linear State Estimator Use Cases at Utilities

The worldwide power industry has been pushing forward the adoption of synchrophasor technology for wide-area monitoring and situational awareness. Many applications have been developed to take advantage of the GPS time-stamped synchrophasor data. LSE application is one of the recent developments in the synchrophasor area that has been gradually accepted and adopted. It has been adapted, enhanced and implemented under several pilot projects. In this section, we introduce three key use cases of using LSE for power system operations. First of all, many synchrophasor applications, such as oscillation detection, mode meter, phase angle difference monitoring, and event detection, have been developed to take advantage of the synchronized PMU data. However, the PMU data quality has always been a concern. Secondly, traditional state estimator may suffer on divergence, especially during stressed conditions and during system contingencies when the state estimator results are needed the most. In this case, operators may lose the awareness of the system state in EMS. Lastly, due to the high expense of installation, commissioning, and communication links of PMU devices in the field, it may not be feasible to have PMUs installed to fully observe a network in a short term. Furthermore, utilities normally apply various criteria to place PMUs in terms of their own objectives, such as protection, line transfer limit monitoring, and oscillation monitoring, etc. Hence, for a utility with limited PMU coverage, LSE can be used to expand synchrophasor measurement coverage, and so other synchrophasor applications can benefit from it.

The use cases of LSE addressing the above issues are summarized below,

1. Validate and condition the raw PMU data based on network model and redundant measurements
2. Provide synchrophasor-based LSE, independent situational awareness, and data analytics system to complement and backup EMS
3. Expand PMU coverage by creating virtual PMUs at neighboring substations.

The three use cases are individually demonstrated in sequence from Sects. 14.4.1–14.4.3 via practical utility case under pilot projects deployed at

Bonneville Power Administration (BPA), Duke Energy, and Southern California Edison (SCE). In Sect. 14.4.4, the real-time performance of LSE application is discussed in detail.

14.4.1 Synchrophasor Data Validation and Conditioning

PMUs have been widely installed in North America followed by the deployment of real-time synchrophasor data related applications. With this high exposure, data quality issues have become one of the top concerns of users at utilities and ISOs. However, without reliable PMU data, the trustworthiness of PMU-based applications is questionable. Moreover, the benefits brought by wide-area situational awareness systems and analytics applications are very limited without adequate PMU coverage.

Synchrophasor data is estimated by a PMU and sent by real-time communications to end users, often through several levels of real-time processing. Issues can arise in any of the processes including the transforming equipment, PMU hardware, communication framework, Phasor Data Concentrator (PDC), and the final applications. Errors and losses can occur at any stage.

PMU data quality can be generally categorized into four aspects:

- Availability: loss of data from one or several PMUs, or the loss of signals in a PMU.
- Corruption: data is corrupted by transmitting or receiving devices, in communication, or in processing.
- Timely delivery: inconsistent data rates and latencies. Excessive delay causes data loss; varying delays can cause loss through buffer overloads and application processing errors.
- Validity and accuracy: measurement errors such as phase errors created by bad timing, ratio and scaling errors, excessive noise in measurement, and frozen or repeated (stale) measurements.

Bad data quality impacts downstream applications. If the application results cannot be trusted, they cannot be used. As one of many synchrophasor pilot projects, the Synchrophasor Data Validation and Conditioning Application (SDVCA) project [11] was funded by the Western Electricity Coordinating Council (WECC) with the objective to develop and demonstrate algorithms to detect common PMU data quality issues and condition PMU data using LSE algorithm in real time. The LSE algorithm was implemented, tested, and deployed in BPA footprint in this project. Fifty-five PMUs have been installed at 37 substations in BPA transmission network. An off-line observability study has been conducted and the PMU observable area is extended to 65 substations by LSE, which consists of most of the 500 kV substations and a small portion of the 230 kV substations. The observable BPA system size is shown in Table 14.1. The LSE application is able to run in real

Table 14.1 Observable BPA system size

Elements	Number
PMU	55
Phasor measurements	220
Substations (w/PMU installed)	37
Substations (observed by LSE)	65
Lines	96
Line segments	126
Transformers	129
Nodes	3091
Breakers	849
Switches	2357
Series capacitors	18
Shunt capacitors	112

time (i.e., 60 frames per second) with all the 220 phasor measurements for these 65 substations. Note that some substations may have multiple observable buses due to multiple PMUs installed at different voltage levels.

14.4.1.1 Validation and Conditioning Historical Event Field PMU Data

A 1400 MW brake resistor was constructed at the BPA Chief Joseph substation in north central Washington. The brake resistor was designed for testing the BPA transmission system. Ten minutes of PMU data were archived for one Chief Joseph brake event and were fed into LSE application. The raw PMU data and LSE results for one of the 500 kV voltage signals are shown in Fig. 14.5. The LSE application is able to track the system behavior in response to the Chief Joseph brake event because it solves in real time as fast as PMU data rate (i.e., 60 frames per second). It is apparent that LSE runs fast enough to visualize the detail of the entire brake event. Note that it is not feasible in EMS/SCADA.

As seen, the brake event creates a disturbance to the system. The oscillation burst after the spike indicates the transient after the brake insertion. The LSE estimated voltage magnitude at the bus Chief Joseph tracks the raw value trajectory very well with generally less than 3 kV (less than 0.5%) difference. In the lower subplot of the angle difference between raw and estimated measurements, they match very well with a difference only less than 0.2°.

The LSE has also been tested with specific data quality issues, which were intentionally added to the raw PMU event data. Examples of these tests with a dropout and a constant offset data jump are shown in Fig. 14.6.

In time interval a, intentional dropouts of the raw PMU measurement are simulated by setting PMU data as dropout. Four and a half seconds of dropout is seen from roughly 6.2–10.7 s. The dropouts are filtered out by the sanity check in LSE

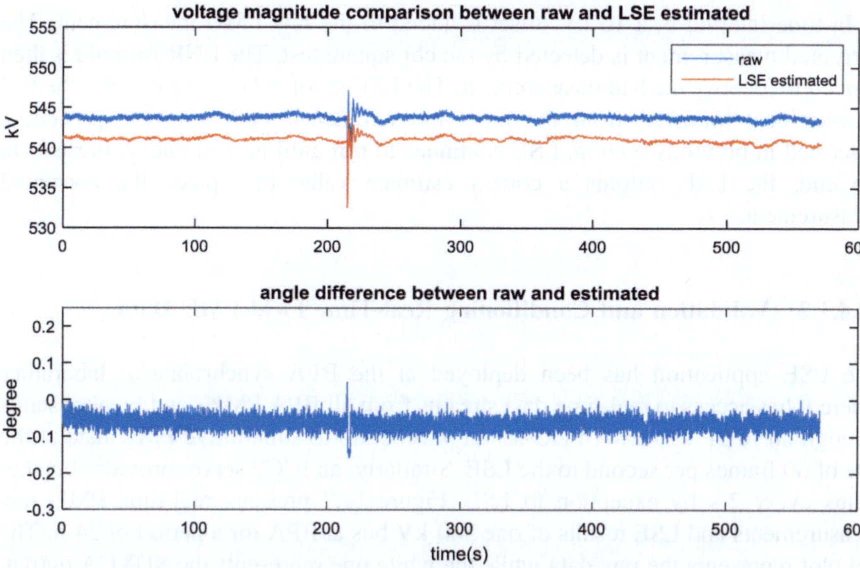

Fig. 14.5 Comparison between raw and LSE estimated—Chief Joseph brake event

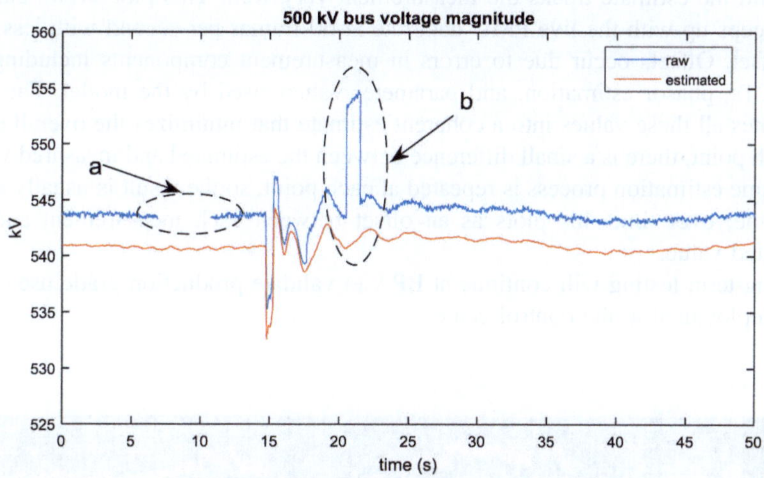

Fig. 14.6 Test scenarios of dropouts and bad data

input interface. When this particular measurement state changes from valid to dropout, it triggers LSE to conduct the real-time observability check, which concludes that this measurement can still be observed by other adjacent PMUs. Therefore, the LSE successfully bridges the data gap by replacing the missing values with estimated ones.

In time interval b, a 10 kV offset is added to the raw PMU measurement. The corrupted measurement is detected by the chi-square test. The LNR formula is then applied to identify the bad measurement. The LSE is solved over again after the bad measurement removed from the input measurements. Following the procedure described in previous section, LSE continues to run until no bad data is present. In the end, the LSE outputs a correct estimate value to replace the corrupted measurement.

14.4.1.2 Validation and Conditioning Real-Time Field PMU Data

The LSE application has been deployed at the BPA synchrophasor laboratory where it has access to real-time data streams from all BPA PMUs and breaker status through an ICCP server. A PDC server was set up to stream live PMU data at the rate of 60 frames per second to the LSE. Similarly, an ICCP server provides breaker status every 2 s by exception to LSE. Figure 14.7 presents real-time PMU raw measurements and LSE results of one 500 kV bus at BPA for a period of 24 h. The red plot represents the raw data while the white one represents the SDVCA output. Note that there are many changes of voltage magnitude due to switching events in this daily operation. The LSE can update the system topology accordingly in real time and the estimate tracks the measurement very well. This plot shows that the LSE keeps up with the live PMU data rate at 60 frames per second with less than 1% offset. Offsets occur due to errors in measurement components including the PTs, CTs, phasor estimation, and parameter values used by the model. The LSE combines all these values into a coherent estimate that minimizes the overall error. At each point, there is a small difference between the estimated and measured value. The same estimation process is repeated at each point, so the result is usually about the same; over time this plots as an offset between each measurement and the estimated value.

Long-term testing will continue at BPA to validate production-grade use of the LSE deployment at the control center.

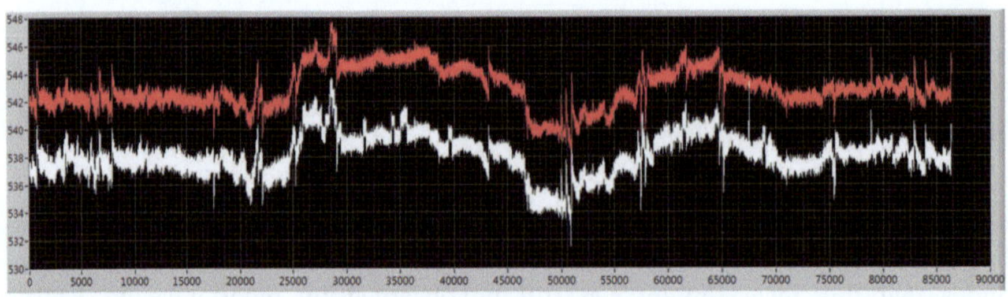

Fig. 14.7 24 h of real-time PMU raw measurements and LSE results (time axis in seconds)

14.4.2 *Independent Wide-Area Situational Awareness for Grid Resiliency and Synchrophasor Data Analytics*

Electric power grids rely on EMS for mission-critical functions of grid monitoring and control. When an EMS suffers an outage due to equipment failure, physical or cyber-attacks, grid operators lose situational awareness of the system. Given the role of EMS in control centers, the need for an alternative independent system is critical for grid resiliency so that operators have the ability to survive an EMS failure. It has to be pointed out that LSE can play a crucial role when EMS State Estimator fails to solve. In other words, the LSE can still provide situational awareness when operators need the situational awareness the most. Grid resiliency capability for the modern grid becomes more and more important for stakeholders—North American Electric Reliability Corporation (NERC), utilities, regulators and policy makers.

Duke Energy is one of the early adopters and has deployed the LSE to address the grid resiliency need. The LSE is currently deployed in the Quality Assurance (QA) environment for their bulk transmission system, which serves as an independent source for grid situational awareness, and provides assistance for operators to assess and diagnose current system conditions for proactive and necessary corrective actions. Long-term plan is to eventually move the LSE application from QA to production.

Based on an off-line observability study, the PMU observable area consists of most of the 500 kV substations and 230 kV substations in Duke Energy footprint. Note that unlike BPA, the observable system is not extended, thanks to Duke Energy's PMU placement strategy which was applied to have a full PMU coverage on their high-voltage network (230 kV and above). The observable Duke Energy system is listed in Table 14.2. PMUs are installed in 50 substations with 181

Table 14.2 Observable Duke energy system size

Elements	Number
PMU	132
Phasor measurements	181
Substations (w/PMU installed)	54
Substations (observed by LSE)	54
Lines	89
Line segments	94
Transformers	58
Nodes	2103
Breakers	620
Switches	2163
Series capacitors	0
Shunt capacitors	8

positive sequence voltage and current phasor measurements. The LSE is able to run in real time (in this case, it is 30 frames per second).

14.4.2.1 Wide-Area Situational Awareness

A wide-area situational awareness dashboard of the Duke Energy's subsystem includes a geospatial map of Duke Energy's footprint, corresponding network one-line diagram, system alarm view, and real-time frequency number. The geospatial map provides the physical substation locations and network connectivity. The system-level one-line diagram is developed as close as the one in EMS and it aims to provide operators a smooth transition from EMS visualization to this display. As shown in the lower left view of Fig. 14.8, each rectangle button icon represents a high-voltage substation within Duke Energy. Meanwhile, next to the rectangle button icons, the raw and LSE estimated voltage magnitudes at this substation are displayed and updated in real time. Active and reactive power flows calculated from the voltage and current phasors are also mapped to the corresponding transmission lines in the one-line diagram. Operators can monitor the critical buses and tie-lines of the high-voltage transmission network and obtain awareness of the current grid operating condition based on the LSE results. In the system alarm view, it shows the violations from synchrophasor applications, such as phase angle difference, voltage sensitivity, angle stability, oscillation analysis and detection, system islanding, generation trip, line trip.

In the dashboard, operators can drill down to any specific substation, where they can monitor the very detailed substation configuration—a node-breaker scheme. LSE estimated voltage magnitudes and angles are mapped to the corresponding nodes in the substation one-line diagram on the left side in Fig. 14.9. For a smooth

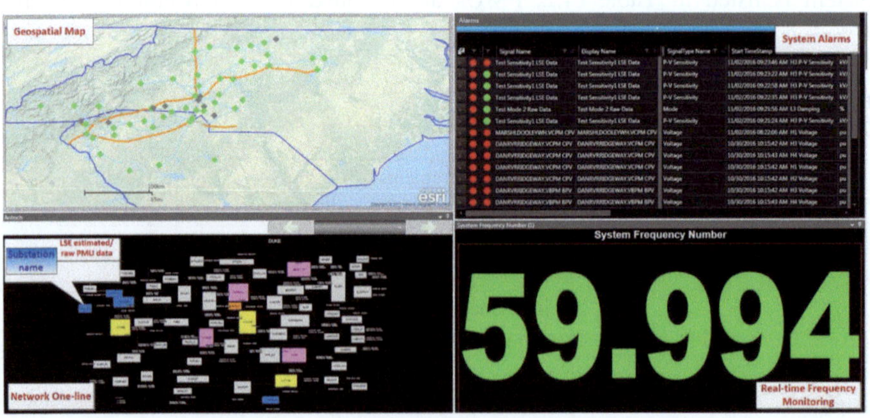

Fig. 14.8 Duke energy's independent grid situational awareness dashboard

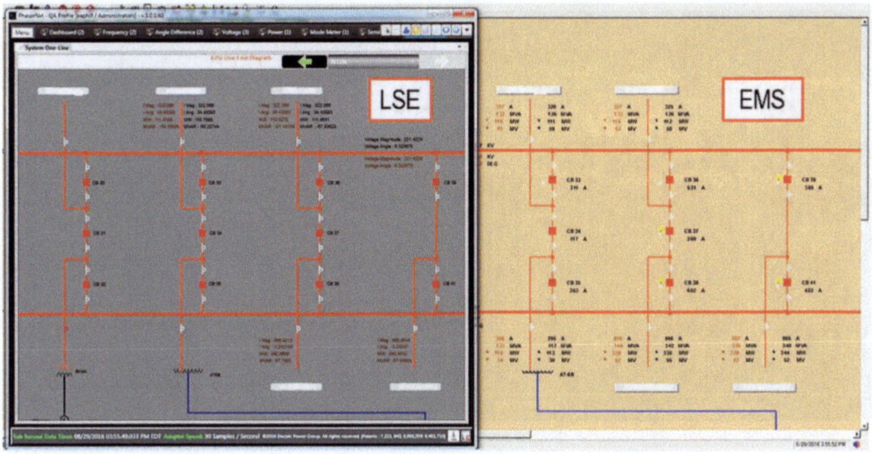

Fig. 14.9 Duke energy's substation one-line diagram—LSE versus EMS

transition and easy maintenance, the substation one-line diagrams are directly imported from Duke Energy's EMS, so that they are identical to those in EMS, as shown on the right side in Fig. 14.9. If EMS state estimator fails to converge due to equipment failure, physical or cyber-attacks, LSE results can benefit the operators by providing independent situational awareness of the system condition. LSE results can further be used as input to other synchrophasor-based grid analytical applications, such as phase angle difference and oscillation analysis and monitoring.

14.4.2.2 Phase Angle Difference for Grid Stress Monitoring

Phase angle difference serves as an indicator of grid stress. One use case is that a sudden rise in Phase angle difference implies an impedance increase mostly caused by the change of system topology, or an increase in power transfer between the two monitored buses. LSE can be used to resolve data quality issues and provide guaranteed phasor angle information so that the phase angle difference can be continuously monitored.

Here, we provide one example in Duke Energy system. Figure 14.10 shows the comparison of phase angle difference between the same two measured buses by using LSE results versus raw PMU data. Note that when the raw PMU data has dropouts, however, data from LSE is still good. It is because the LSE can estimate the bus voltage angle from adjacent substations.

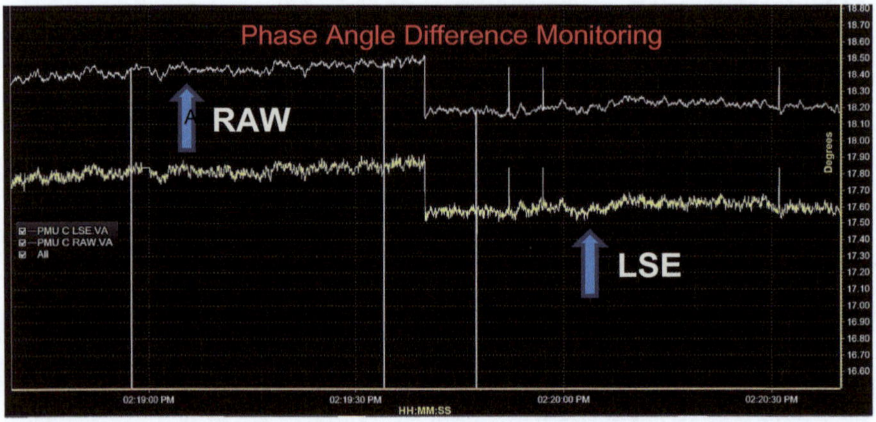

Fig. 14.10 Comparison of phase angle differences for raw PMU data and LSE data

14.4.2.3 Oscillation Analysis and Monitoring

Natural modes of oscillations in power system may become poor-damped or even un-damped under stressed or extreme system conditions. These oscillations, if not controlled, can lead to system breakup and separation. Improper operation or malfunction of control systems in power system can also cause abnormal system oscillations. In addition, such oscillations can also grow under stressed system conditions. It is critical to monitor and detect these sorts of oscillations in the power systems to ensure the reliability of grid operation. Oscillation analysis application is one of the key synchrophasor applications, which can also benefit from LSE validated and conditioned PMU data.

The LSE estimated measurements are also streamed into the oscillation analysis applications at Duke Energy. As a benefit of the conditioned PMU data from LSE, the oscillation analysis application is assured to have clean input data, which is much less sensitive to raw data drop out and invalid data; therefore, it is more robust to track oscillations in the system. Figure 14.11 below shows the comparison of oscillation analysis results using the same set of raw PMU data and LSE data, as shown in the upper plot. The lower plot gives oscillation analysis results, displaying calculated modal frequency, damping, and energy for a specific oscillation. One can observe that the oscillation results fail to update with an accurate result when dropouts are present in the raw PMU data. The dropouts are detected by the sanity check in LSE input interface. Through real-time observability check, this measurement can still be estimated by LSE using other electrically connected bus measurements. Eventually, the same oscillation analysis application using LSE results instead is not affected by the bad data. In other words, the one using LSE estimated measurements can provide continuous and reliable results.

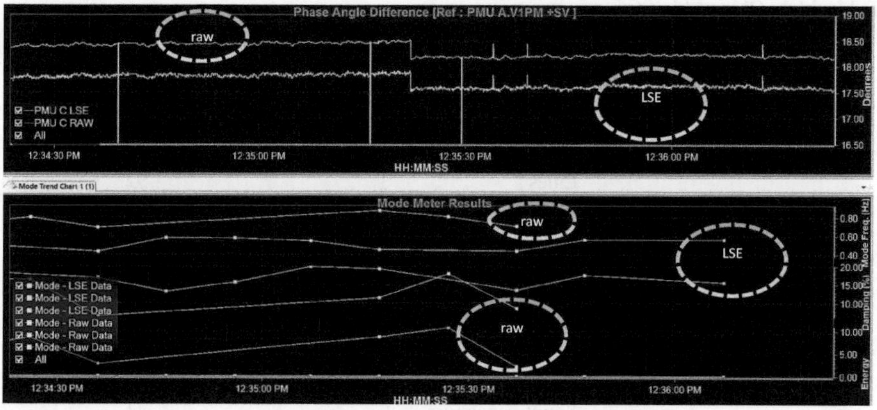

Fig. 14.11 Comparison of oscillation analysis for raw PMU data and LSE data

14.4.3 Extend Synchrophasor Measurement Coverage and Benefits for Potential Downstream Applications

Due to the high cost of installation and commissioning of PMU devices in existing substations, it may not be feasible to have PMUs installed to fully observe a network. Utilities typically use various criteria to place PMUs according to their own objectives, such as protection, transfer limit monitoring, and oscillation monitoring. For a utility with limited PMU coverage, LSE can be used to expand synchrophasor measurement coverage. The methodology is straightforward mathematics and LSE can extend the observability to neighboring substations using the measured voltages and line currents at a PMU-measured substation. A pilot project hosted by Southern California Edison (SCE) has been created to integrate and implement the LSE for expanded observability.

14.4.3.1 Methodology and Deployment at SCE

Based on the LSE algorithm, it is apparent that LSE can extend the observability to the neighboring substations using the measured voltages and line currents at a PMU-measured substation. SCE has 19 PMU deployments spreading across their transmission system network. With very limited observability provided to grid operators and the challenges to deploy more physical PMUs in the short term, it is difficult to realize values from existing synchrophasor infrastructure, especially for real-time operation and decision making based on synchrophasor data. No need to mention that, the PMU data quality is another concern as well.

Based on an off-line observability study for SCE system, the PMU observable area consists of few 500 kV substations and several 230 kV substations. The observable SCE system is listed in Table 14.3. Nineteen PMUs are installed in 13

Table 14.3 Observable SCE system size

Elements	Number
PMU	19
Phasor measurements	103
Substations (w/PMU installed)	13
Substations (observed by LSE)	52
Lines	92
Line segments	92
Transformers	154
Nodes	2775
Breakers	731
Switches	2004
Series capacitors	12
Shunt capacitors	41

substations with 103 positive sequence voltage and current phasor measurements, which are used by LSE to make additional 39 substations observable, as shown in Fig. 14.12. The LSE results have the same data rate as the PMU raw measurements, and the expanded synchrophasor data can be used in downstream synchrophasor applications.

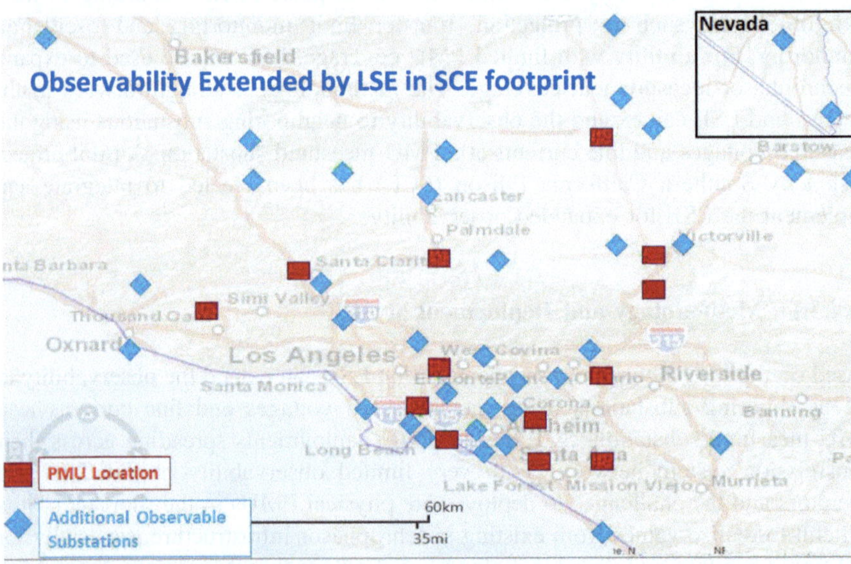

Fig. 14.12 LSE extends observability at SCE

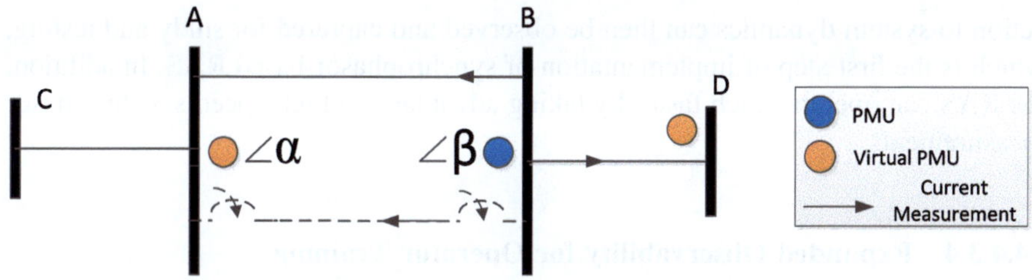

Fig. 14.13 LSE extends observability to neighboring substation

14.4.3.2 Expanded Observability for Line Closing

Synchrophasor is especially useful to assist line closing when the phase angle difference across a line is monitored by PMUs. In order to close the line breakers, the phase angle difference between the two ends of the line needs to be within the closing angle limit. LSE conditioned data is more advantageous for operator actions than raw PMU data. Furthermore, in the case where the PMU is installed on the line side when the line is de-energized, the PMU may not report useful angle measurement, whereas the LSE can estimate the bus voltage angle from adjacent substations through another line to substitute for the invalid raw PMU data. Given the LSE results, operators have reliable data to rely on to issue safe line switching action.

As seen in Fig. 14.13, the operators need to know the bus voltage angles α and β at respective bus A and B before closing the dashed line. LSE can create the virtual PMU at bus A through the other closed line under this circumstance. Hence, the operators are aware of the voltage angle difference at the two ends of the line in real time, which is very critical for closing a line. This can prevent the problems created by closing line with high-voltage angle difference from happening and make sure the closing process is successful.

14.4.3.3 Expanded Observability for Remedial Action Scheme (RAS) Testing

One other benefit of using the LSE extended synchrophasor measurements is to assist in developing and testing RAS. SCE currently has more than twenty of RAS implemented in the system. In order to deploy a RAS in operation, numerous simulations and test cases need to be performed. However, the grid response and RAS impact during transient events cannot be well studied due to the lack of high-resolution synchrophasor measurement, while LSE can supply more estimated PMU measurements. With the virtual PMUs created by LSE, the impact of RAS action to system dynamics can then be observed and captured for study and testing, which is the first step of implementation of synchrophasor-based RAS. In addition, the RAS can operate much faster by taking advantage of high-speed synchrophasor measurements.

14.4.3.4 Expanded Observability for Operator Training

One of the major challenges to adopt synchrophasor technology for real-time grid operation at control center is that operators are not familiar or used to the dynamic data provided by synchrophasor applications. Therefore, the training of how to make use of synchrophasor data and application becomes critical. However, with few PMUs in the system, it is not easy to provide operators with the complete system condition and training use case scenarios. With the extended synchrophasor measurements by LSE, system behavior after an event can be well observed and archived for replay to fulfill the operator training objective.

14.4.4 Performance Assessment

It is beneficial to provide some insights on the LSE performance in real time. The LSE performance testing is conducted on a 64-bit desktop PC with Intel i7-3770 3.4 GHz CPU and 16 GB RAM. Here, we use BPA system as an example. Table 14.4 shows the CPU time consumption for a complete run of LSE application with a single frame of BPA synchrophasor data. When breaker status or measurement state does not change, the execution time is 7 ms, which is less than the measurement stream interval (16.67 ms here for 60/s reporting), so there is no delay. When the measurement state changes, the execution time is 50 ms because the LSE needs to run real-time observability analysis. For that single frame, LSE lags behind $50 - 7 = 43$ ms. After that frame, the LSE will only take 7 ms to execute, so it takes around 5 frames (90 ms) for the estimates to catch up with the real-time data. The same logic applies to the case of breaker status change. For that single frame, LSE lags behind $150 - 7 = 143$ ms. After that frame, the LSE will need about 16 frames (250 ms) to catch up with the real-time data. Sufficient data buffering is included to prevent data loss during these interruptions.

Table 14.4 LSE CPU time consumption for a single frame

Scenario	Execution time (ms)
Breaker status change	150
Measurement state change	50
Neither of above	7

14.5 Conclusions

This chapter first gives a background of state estimator as a critical role for power system real-time operation. A theory of traditional state estimator is provided to demonstrate the inputs, outputs, and the algorithm. The LSE algorithm is then presented with different formulations with the synchrophasor measurements. The difference between LSE and traditional SE is briefly explained. Similarly to SE, LSE has four core modules, topology processor, real-time observability analysis, matrix formulation and WLS solution, and post-processing bad data detection and identification. We present the functionality of each module and the overall real-time operation procedure.

This chapter then presents the operational experience and business value of using synchrophasor-based LSE via actual examples from utilities in the USA. The LSE validates and conditions PMU data for grid operation, such as wide-area situational awareness, phase angle difference monitoring and oscillation analysis and detection. Historical event and live data case studies at BPA and Duke Energy systems demonstrate that the LSE can improve synchrophasor data quality and provide reliable and accurate state estimation solution for synchrophasor applications. The LSE can also expand synchrophasor measurement coverage by extending observability to neighboring substations. SCE's pilot project illustrates the effectiveness of using LSE to expand PMU observability and the benefit for downstream applications. The demonstration of the benefits and use cases from utilities' first-hand implementation and deployment experience is valuable to the researchers, engineers, and students. It is believed that the synchrophasor-based applications will provide greater observability and more dynamic insights to improve the grid stability in real-time grid operation with the deployments of LSE at control centers.

References

1. Abur A, Exposito AG (2004) Power system state estimation: theory and implementation. Marcel Dekker Inc
2. Phadke AG, Thorp JS (2008) Synchronized phasor measurements and their applications. Springer, New York, pp 150–163
3. Phadke AG, Thorp JS, Nuqui RF, Zhou M Recent developments in state estimation with phasor measurements. In: Proceedings of 2009 IEEE power systems conference and exposition conference, pp 1–7
4. Yang T, Sun H, Bose A (2011) Transition to a two-level linear state estimator—Part I: architecture. IEEE Trans Power Syst 26(1):46–53
5. Yang T, Sun H, Bose A (2011) Transition to a two-level linear state estimator—Part II: algorithm. IEEE Trans Power Syst 26(1):54–62
6. Zhang L, Bose A, Jampala A, Madani V, Giri J (2017) Design, testing, and implementation of a linear state estimator in a real power system. IEEE Trans Smart Grid 8(4):1782–1789

7. Zhang L, Chen H, Martin K, Faris A, Vutsinas M, Bradberry T, Phillips E, Abu-Jaradeh B, Bui J Successful deployment and operational experience of using linear state estimator in wide-area monitoring and situational awareness projects. IET Gener Transm Distrib. https://doi.org/10.1049/iet-gtd.2016.2028
8. Chen H, Zhang L, Mo J, Martin K (2016) Synchrophasor-based real-time state estimation and situational awareness system for power system operation. J Mod Power Syst Clean Energy 4(3):370–382
9. Zhang L, Chen H, Martin KE, Faris A Practical issues of implementation of linear state estimator in WECC. In: IEEE PES innovative smart grid technologies conferences (ISGT), Sept 2016. Minneapolis
10. Zhang L (2014) Validation, testing, and implementation of the linear state estimator in a real power system. PhD dissertation, School Electrical Engineering, Washington State University, Pullman
11. http://electricpowergroup.com/synchrophasor-data-validation.html

Chapter 15
Post-event Analysis in the ERCOT System Using Synchrophasor Data

Sidharth Rajagopalan, Patrick Gravois, Bill Blevins and Sarma Nuthalapati

15.1 Introduction

A valuable lesson from the August 14 2003 blackout [1] is the importance of having time-synchronised system data recorders. For the September 8th 2011 blackout event [2], PMU data proved valuable information in constructing the sequence of events and other post-event analysis in the Arizona-Southern California outages. With the industry citing the value of PMUs in post-event analysis ERCOT determined this use case as a key area of focus. Historically, ERCOT has experienced events in the past such severe cold weather events, storms and hurricanes such as RITA, IKE and HARVEY. Large system events take a long time to piece together and understand. Time synchronised measurements have the potential to cut down the time for analysis in events such as these. Additionally for smaller events which occur often on the grid, the benefit of being able to see the system performance and quickly determine if it was as expected is benefit to find and correct issues between large system events. Over the years, ERCOT has worked to improve its software and automate disturbance detection and automate event reports. During big system events, understanding and communicating the causes of such events are a matter of public trust. Better tools to identify how the grid responded have to be very important to explain when things happened and how they happened.

S. Rajagopalan · P. Gravois · B. Blevins (✉)
Electric Reliability Council of Texas (ERCOT), Taylor, TX 76574, USA
e-mail: bill.blevins@ercot.com

S. Rajagopalan
e-mail: Sidharth.Rajagopalan@ercot.com

S. Nuthalapati
Peak Reliability, Vancouver, WA 98662, USA

S. Nuthalapati (ed.), *Power System Grid Operation Using Synchrophasor Technology*, Power Electronics and Power Systems,
https://doi.org/10.1007/978-3-319-89378-5_15

This chapter presents experiences in the use of synchronised phasor measurement technology for post-event analysis in the Electric Reliability Council of Texas (ERCOT) interconnection, USA. The authors describe the types of events studied in such analyses, provide a general description of the parameters that are monitored and introduce some of the reporting that follows. Two case studies were provided from events on the ERCOT system—the post-event analysis of a direct loss of generation event and a compound event involving a fault, which, due to relay mis-operation, induced a loss of generation event.

15.2 Post-event Analysis

In developing a post-event Analysis process using PMUs, ERCOT considered the types of measurements that would be useful. ERCOT initially looked at only the real-time streamed PMU data. Measurements such as frequency, voltage current and phase angle were identified as key measurements. Ringdown analysis tools using offline software was improved over time such that PMU data could be merged with no-streamed time synchronised data. ERCOT focused on creating rules for generators above 20 MVA to be required to install PMUs [3]. This was in an effort to use post-event analysis in conjunction with model validation requirements from NERC. Gross errors in, for example, governor response could be measured and then corrected out of the event analysis. ERCOT expects to build on this analysis such that events can be used to validate dynamic models and improve the planning and operations of the ERCOT grid. Recent events like the South Australian outage also show that there is some difference in interpretation between vendors who build resources and grid operators with regard to ride-through. ERCOT has experienced events where some protection systems have tripped resources and those protection systems and not captured in current dynamic simulations. Post-Event Analysis is one way of finding those issues and addressing them.

ERCOT engineers perform post-event analysis using PMU data and create reports for any events in which system frequency drops below 59.9 Hz. ERCOT engineers may also create reports on other events that they deem significant, in order to analyse system performance and response to events such as loss of generation, system faults or oscillations. For generating the reports, engineers use Electric Power Group's Phasor Grid Dynamics Analyser (PGDA) to analyse four metrics for each event. The four metrics are system frequency, voltage magnitude swings, relative voltage angle swings and modes of oscillation.

15.2.1 System Frequency

PGDA contains a built-in analysis tool that allows for quick analysis of the system frequency response to an event. It calculates three points of frequency response labelled points "A", "B" and "C" (Fig. 15.1). Point "A" represents the interconnection frequency immediately before the disturbance, Point "B" represents the interconnection frequency at the point immediately after the frequency stabilises due to the governor action and before the contingent control area takes correction action and Point "C" represents the interconnection frequency at its maximum deviation due to the loss of rotating kinetic energy from the interconnection. The tool also calculates the maximum frequency swing due to loss of generation, as well as the time to stabilisation. ERCOT engineers also report the time between loss of generation and the lowest detected frequency.

For most events in the ERCOT interconnection, frequency calculated by the PMUs remains consistent throughout the entire system. Engineers typically see a slight variation between north and south PMUs while the frequency drops from Point "A" to Point "C". However, the minimum frequency recorded by the PMUs throughout the system stays within a ~0.002 Hz range across the system. The only events that have been found to diverge from this trend are when a large unit trips in the Rio Grande Valley region. For such events, engineers have seen PMUs in the Valley region record frequencies up to 0.05 Hz lower than the rest of the system.

Another event detail reported by engineers related to frequency is the PMU that records the first response in frequency. ERCOT engineers have found that the PMU that first shows a decrease in frequency is typically the PMU closest to the unit trip.

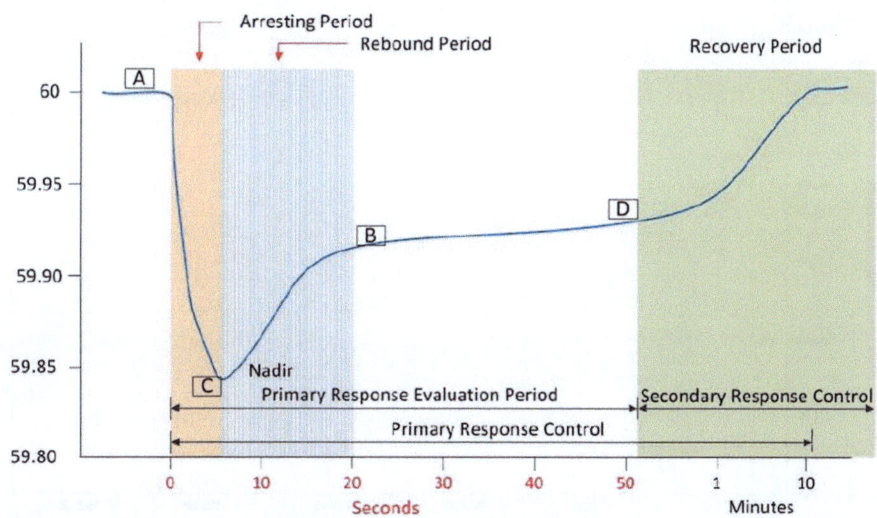

Fig. 15.1 Frequency response characteristics (*Source* NERC)

15.2.2 Voltage Magnitude Swings

ERCOT engineers also analyse system voltage recovery across the system, and determine the location with the largest swing in voltage magnitude. Unlike the first PMU to respond to frequency, the largest voltage magnitude swing does not necessarily occur at a location closest to the unit trip. Dips in voltage magnitude may occur due to both system faults and loss of generation. There have been several events over the past few years in which PMUs record a fault closely followed by a loss of generation. In such events, ERCOT engineers will determine the location with the largest voltage swing due to the fault, as well as the location with largest voltage swing due to loss of generation. Figure 15.2 shows a large voltage magnitude drop due to a fault, followed by a quick voltage recovery before the voltage once again drops due to loss of generation. For this event, engineers found that the PMU shown in Fig. 15.2 was closest to the unit trip. On the contrary, Fig. 15.3 shows a smaller dip in voltage due to the fault, and a larger dip due to the loss of generation.

In order to find the PMU that recorded the largest drop in voltage magnitude, for either a fault or loss of generation, ERCOT engineers use tools embedded within PGDA. The voltage magnitude for each available PMU is plotted, with the X-axis starting a few tenths of a second before the start of the event (Point "A" from frequency analysis above). All voltage magnitude trends are then detrended by first value by PGDA, which set the starting kV of each PMU to zero. Engineers may then zoom in to select the PMU(s) that show the largest voltage magnitude

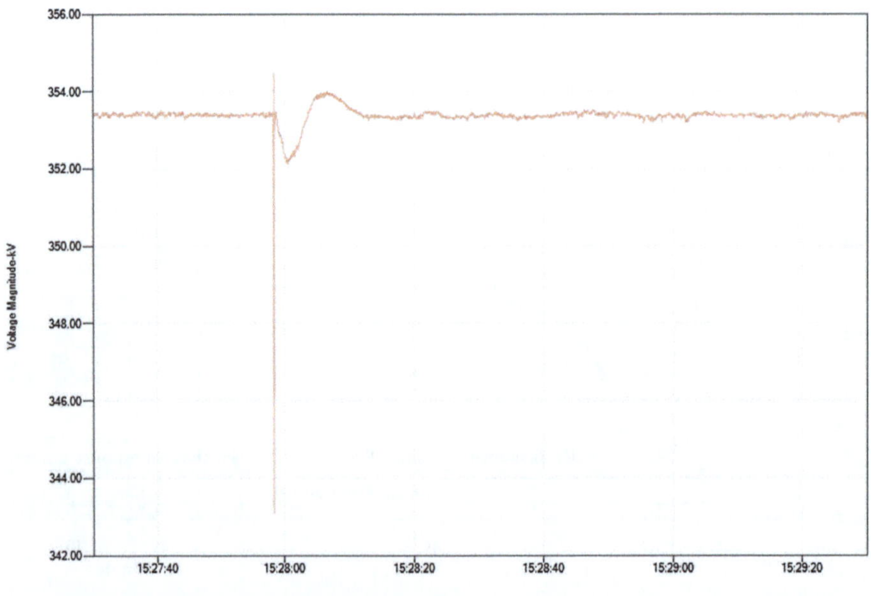

Fig. 15.2 PMU with largest voltage drop due to fault

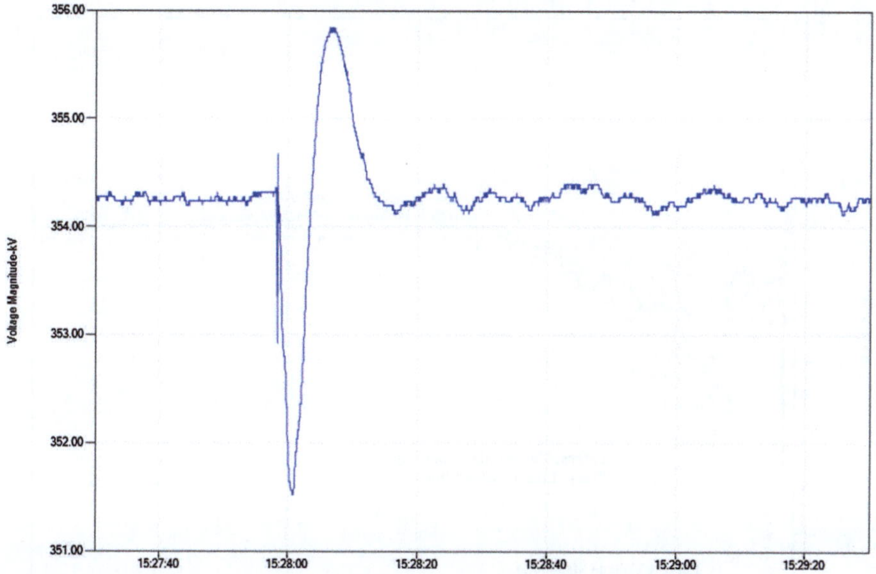

Fig. 15.3 PMU with largest voltage drop due to loss of generation

deviation for either a fault or loss of generation. Figure 15.4 shows a plot of all voltage magnitude signals detrended by first value for the above event.

15.2.3 Voltage Angle Swings

Voltage angle swings are analysed in a similar manner as voltage magnitude. ERCOT engineers study how voltage angles change after a large event to determine how units in different regions interact and stabilise. The process using PGDA to analyse voltage angle swings involves the same tools used to analyse voltage magnitude. All voltage angles are plotted with reference to a centralised PMU. The signals are once again detrended by first value, and the start time is set to slightly before Point "A" from the frequency response trend. An example of the resulting plot is shown in Fig. 15.5.

Figure 15.5 represents a typical response seen by ERCOT engineers after a large unit trip. The voltage angle in the South and South Central regions swing against the North, West and Panhandle regions. The largest voltage angle swing after a unit trip is mostly found from a PMU in the Rio Grande Valley in the south.

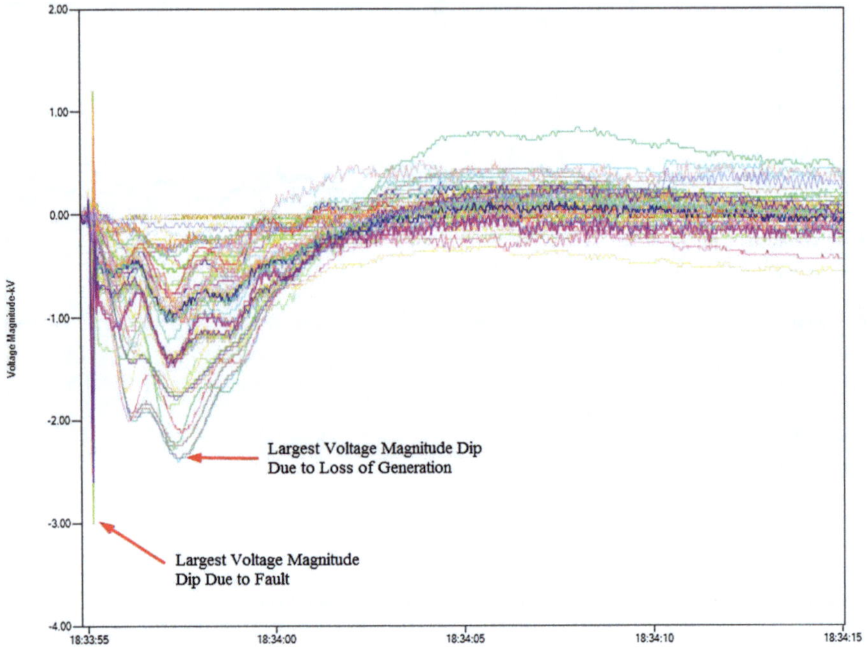

Fig. 15.4 All voltage magnitude signals detrended by first value

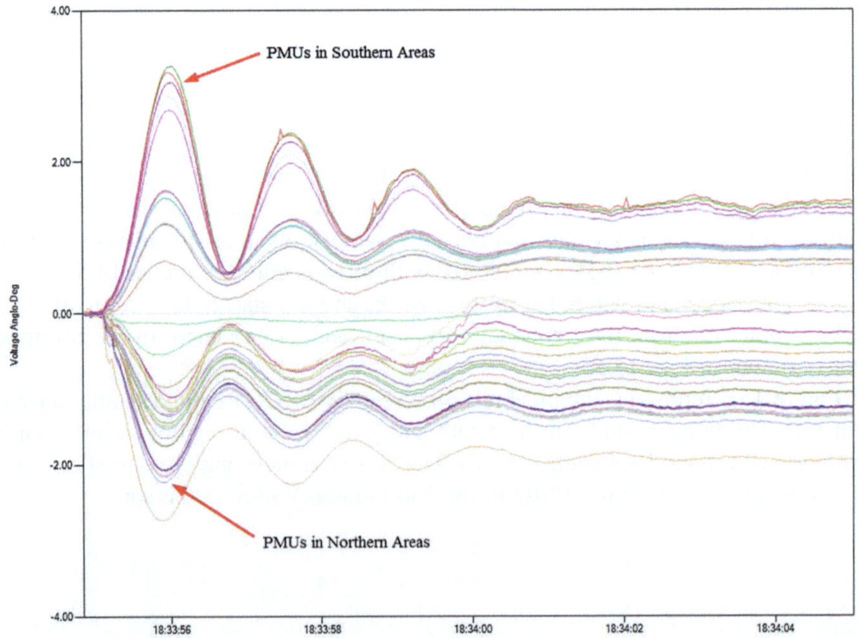

Fig. 15.5 Voltage angle trends w.r.t. Central PMU (detrended by first value at start of event)

15.2.4 Oscillation Modes

PGDA also includes a ringdown analysis tool that allows ERCOT engineers to identify any associated oscillation modes and their damping during a system event. As shown in Fig. 15.4, the system mode that is usually excited by system events is the North to South mode that ranges between 0.60 and 0.80 Hz. Occasionally, a second dominant system mode in the 0.9–1.0 Hz range is detected during an event.

To perform the ringdown analysis, ERCOT engineers will usually select the voltage angle signal with the largest angle swing referenced to a centralised PMU, as long as it shows a clear damped oscillation as shown in Fig. 15.5. The engineer will select a time frame that begins with the system event, and ends once the oscillation is damped down to its normal ambient fluctuation. The ringdown analysis tool will then calculate any modes found in the oscillation and construct a best fit mathematical representation of the oscillation using these modes. The engineer may them determine which modes are dominant using the magnitudes in the mode shape chart, as well as visually determining which modes have the biggest impact on matching the reconstructed signal to the original. An example of this process will be provided in Sect. 15.3.4.

15.3 Case Study 1—Loss of Generation Event

The following sections provide a real-life example of the PMU-based event analysis performed by ERCOT engineers on a loss of generation event that occurred on September 17, 2015. For this event, a fault was recorded by PMUs seconds before the loss of generation. Each section will provide a detailed process on how the engineer analysed the four metrics discussed in Sect. 15.2.

15.3.1 System Frequency

To begin, ERCOT engineers load five minutes of PMU data surrounding the event into PGDA. All frequency signals are plotted on a Time Series chart, and bad data is filtered out using embedded filtering tools in PGDA. PGDA's Event Analysis tool is then used to find an event in which frequency drops a minimum of 90 MHz. The tool then calculates the "A", "B" and "C" points described in the previous section. For this event, the average pre-event frequency value, Point "A", was 59.976 Hz. This is a rather low starting frequency, which at times is caused by a unit run back before the trip, causing generation to drop below the load demand. Point "C", or the Nadir Frequency, for this event was 59.833 Hz. Point "B" for this event was calculated to be 59.904 Hz, which is when frequency stabilises due to unit governor and AGC action. Figure 15.6 shows the frequency response of the

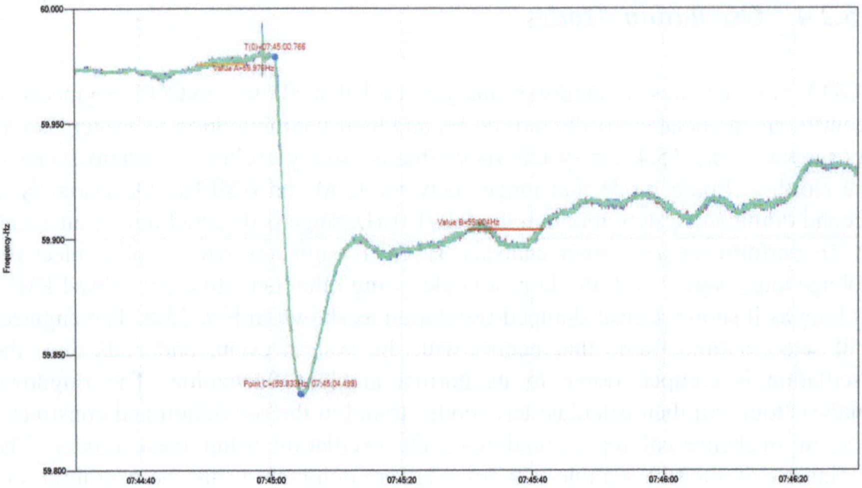

Fig. 15.6 System frequency Captured by the PMU with event analysis points

event with the corresponding Points. Take note of the frequency disturbance seconds before the drop in frequency.

ERCOT engineers also use PGDA to report other parameters associated with frequency response, such as largest frequency swing, first PMU to respond to drop in frequency, time taken to return to pre-event frequency or 60 Hz, time between Point "A" and Point "B" and time between Point "A" and Point "C". The largest frequency swing for this event was 0.143 Hz, which was recorded by multiple PMUs across the system. The first PMU to respond was the PMU closest to the unit that tripped, shown in Fig. 15.7. Frequency took 5 min and 33 s to return to its pre-event value, time between Point "A" and Point "B" was 26.1 s, and time between Point "A" and Point "C" was 3.6 s.

15.3.2 Voltage Magnitude Swings

As mentioned in Sect. 15.2.2, the largest swings in voltage magnitude are found by plotting all voltage signals, setting the start time slightly before the event, and detrending the plots by first value, effectively setting the starting points at zero. for this example however, start time was set roughly 5 s before Point "A" to include the fault. Figure 15.8 shows the resulting plot.

From Fig. 15.8, we clearly see a fault followed by a loss of generation event. ERCOT engineers determine and report the PMUs with the largest swings in voltage magnitude due to both the fault and loss of generation. For this event, the

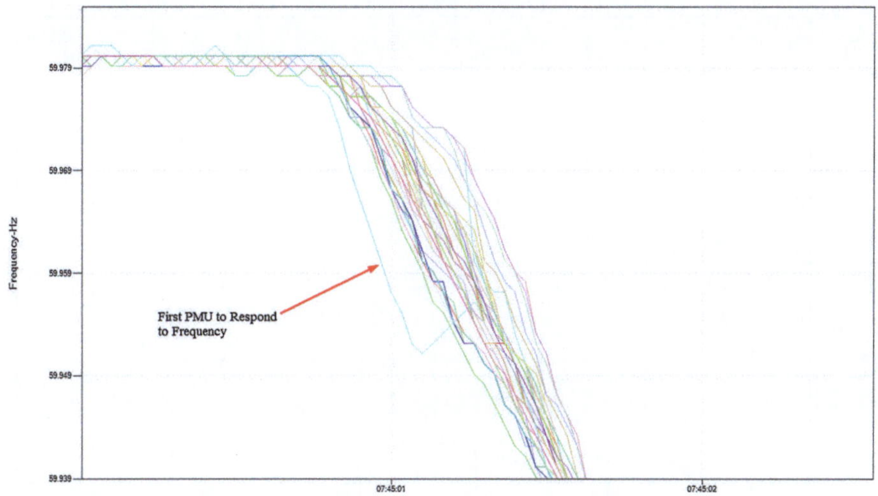

Fig. 15.7 First PMU to respond to frequency

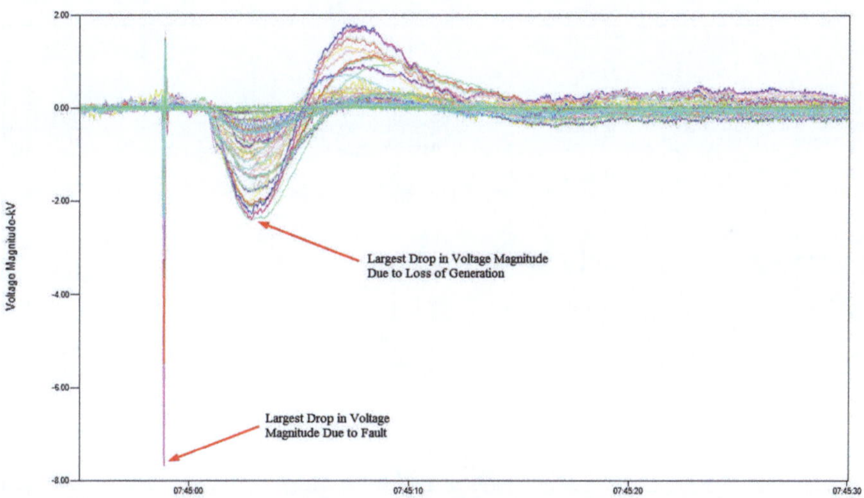

Fig. 15.8 Voltage magnitude signals (detrended by first value)

PMU that recorded the largest voltage swing due to the fault was determined to be the closest to the unit that tripped. From this information, engineers can deduce that the fault was likely near the loss of generation, and the two are likely related. For this event, ERCOT operator logs reported that a lightning arrestor failed at a nearby substation causing a breaker in the plant to open. PMU data confirms this assessment. Figure 15.9 and Fig. 15.10 show the largest voltage magnitude swings due to the fault (7.48 kV drop) and loss of generation (2.16 kV drop), respectively.

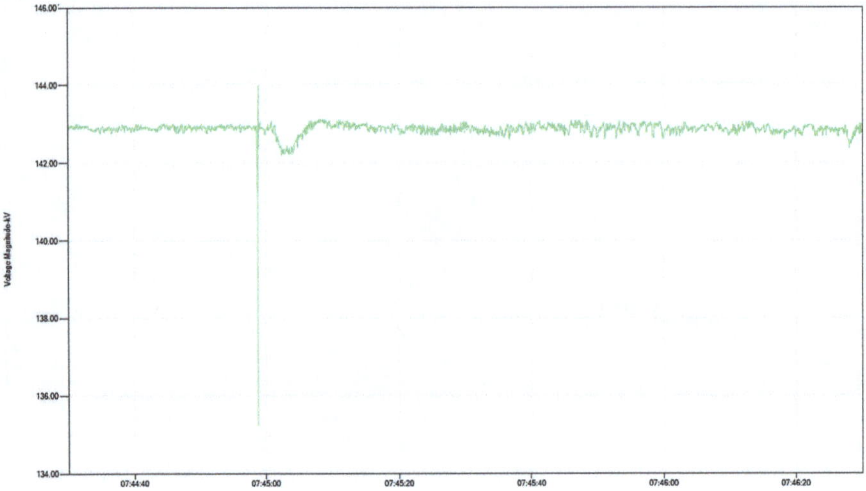

Fig. 15.9 PMU with largest voltage magnitude swing due to fault (138 kV nominal)

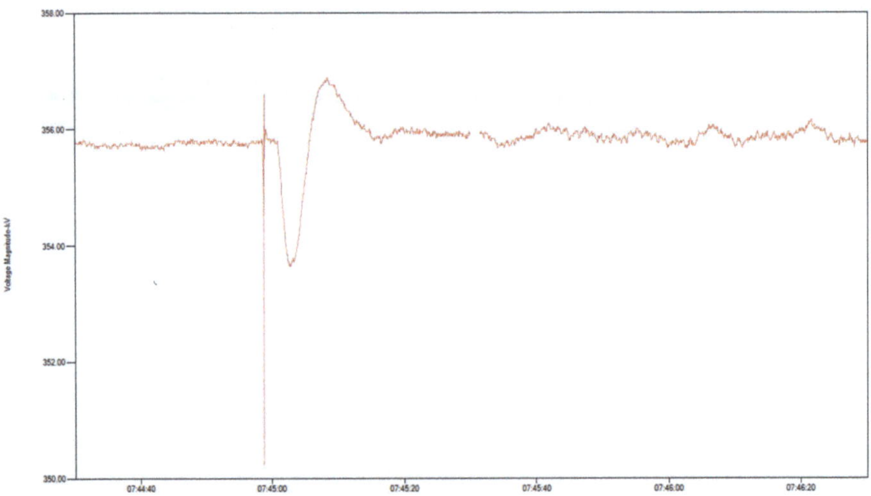

Fig. 15.10 PMU with largest voltage magnitude swing due to loss of generation (345 kV nominal)

15.3.3 Voltage Angle Swings

Voltage angle swings are analysed in PGDA in a similar manner as voltage magnitudes. All voltage angle signals are plotted with reference to a centralised PMU, and the plots are detrended by first value with the start time at Point "A". Since voltage angle is not as susceptible to large swings due to faults as voltage

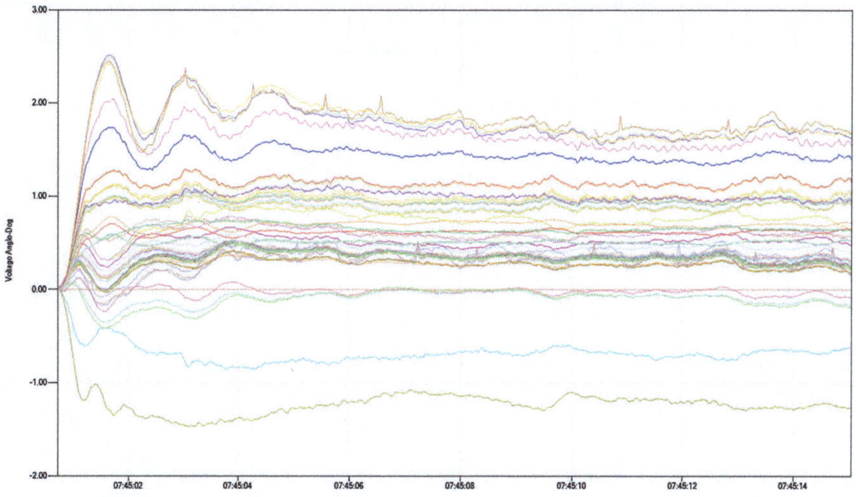

Fig. 15.11 Voltage angle signals detrended by first value (w.r.t. Central PMU)

magnitude, ERCOT engineers focus on the angle swings due to loss of generation. Figure 15.11 shows the resulting chart for this event.

ERCOT engineers use Fig. 15.11 to determine the PMU location with the largest angle swing. For this the event, the largest angle swing was found in the South, near the Rio Grande River Valley. Since location of the unit trip was in the Northeast, the changes in voltage angle reflect the changes in power flow such that more power is being sent to the area where the loss of generation occurred. Looking at the voltage angle signal on its own in Fig. 15.12, we see the pre-event angle was around −18° w.r.t. to the central PMU. When the loss of generation occurs, the initial swing is to −15.4°, a difference of 2.6° from its pre-event angle. Once the resulting oscillation damps down to its normal ambient fluctuations, the post-event angle settles into a range between −16.0° and −16.5°. Since real power flows from higher voltage angle to lower voltage angle, this resulting change in voltage angle post-event confirms that there is less real power flowing from the central area to the southern area after the loss of generation.

15.3.4 Oscillation Modes

To analyse oscillation modes excited during an event, a signal is identified that has a clearly defined oscillatory response to the event as shown in Fig. 15.12. The time window is set from the start of the initial angle swing to when the oscillation is damped down and is indistinguishable from the normal ambient fluctuations found in a typical power system due to constant changes of generation and load balance. ERCOT engineers use PGDA's Ringdown Analysis tool to determine the

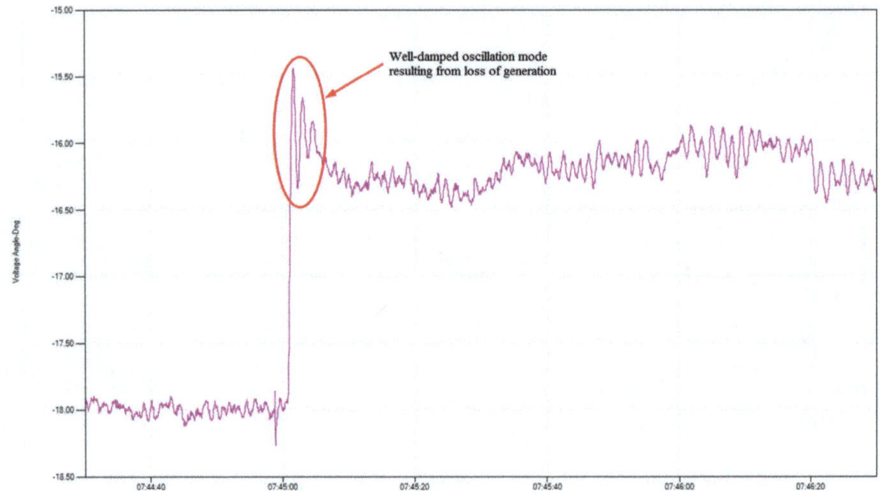

Fig. 15.12 Largest voltage angle swing from Southern PMU (w.r.t. Central PMU)

associated modes found in the oscillatory response. The time frame can be slightly adjusted so that the reconstructed signal may best match the original signal, as shown in Fig. 15.13. From Fig. 15.13, we see that there are three modes associated with the response, they are at 0.13, 0.48 and 0.7 Hz.

To determine the dominant oscillation mode, ERCOT engineers may use the magnitude of the mode found in the mode shape chart (top left), or turn the different

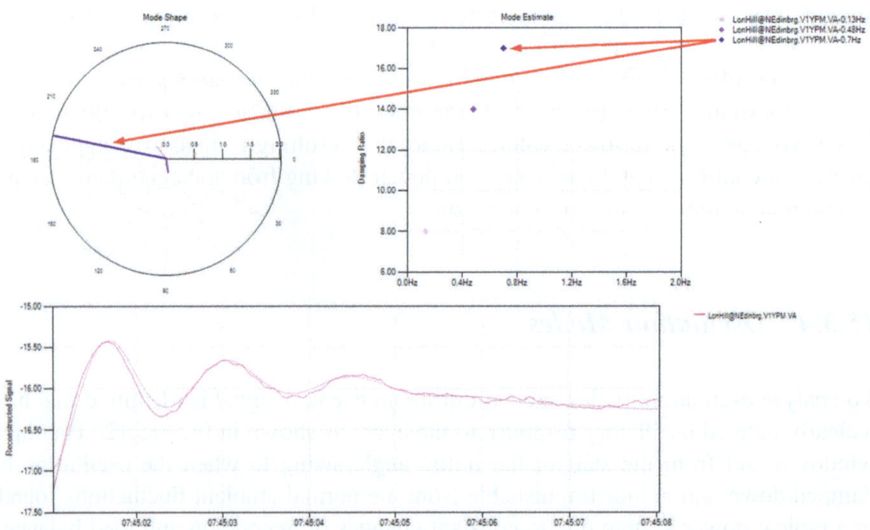

Fig. 15.13 Ringdown analysis of voltage angle from Southern PMU (w.r.t. Central PMU)

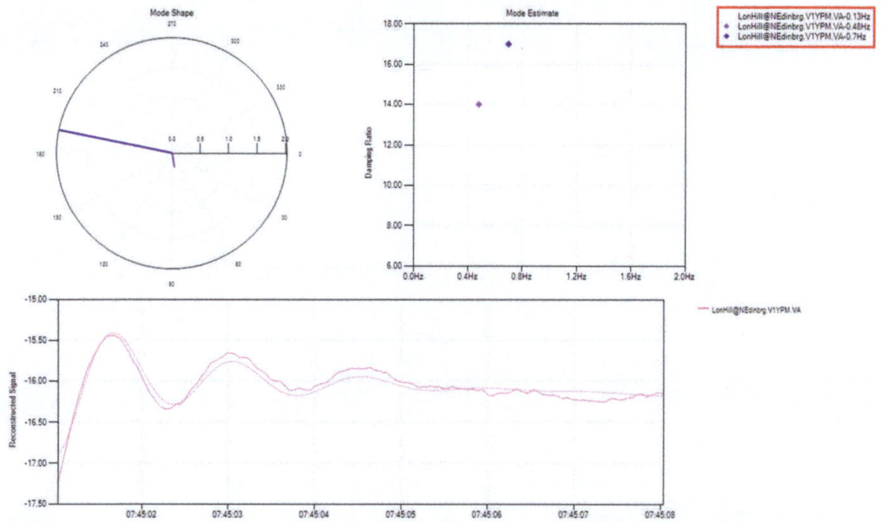

Fig. 15.14 Ringdown analysis of voltage angle w/o 0.13 Hz mode (w.r.t. Central PMU)

modes on and off to determine their impact on the reconstructed signal. For this event, we see that the 0.7 Hz mode has the largest magnitude in the mode shape diagram. Also, if we alternate turning off the three different modes, we see that the 0.7 Hz mode has the greatest impact on the reconstructed signal, as shown in Figs. 15.14, 15.15 and 15.16. From this analysis, it is clear that this oscillatory

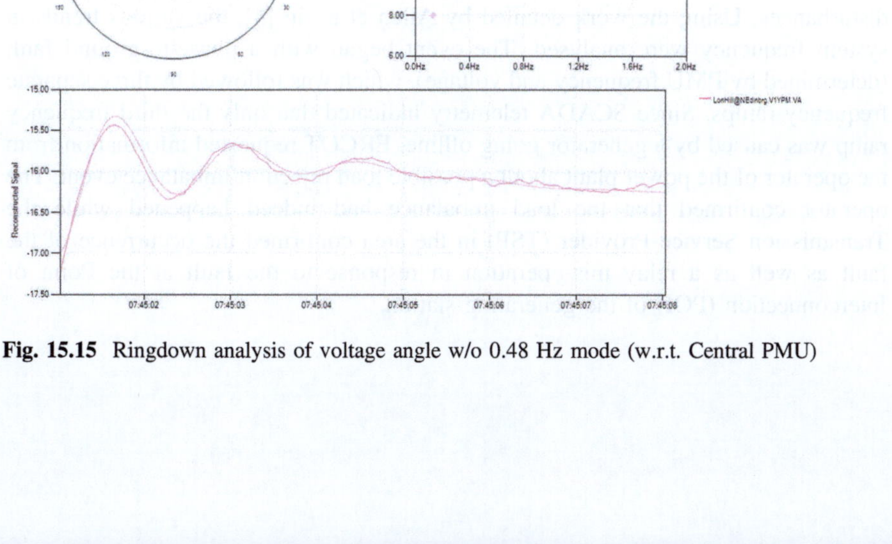

Fig. 15.15 Ringdown analysis of voltage angle w/o 0.48 Hz mode (w.r.t. Central PMU)

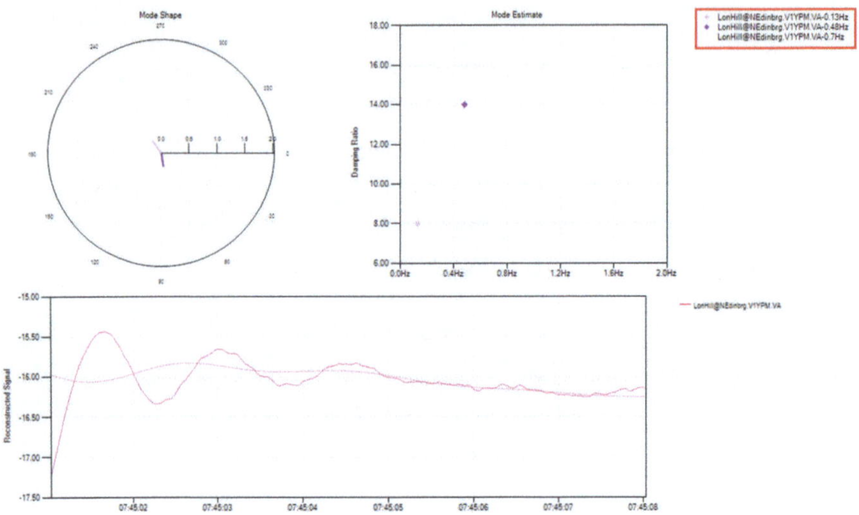

Fig. 15.16 Ringdown analysis of voltage angle w/o 0.13 Hz mode (w.r.t. Central PMU)

response has a single dominant mode of 0.7 Hz, with 17.0% damping (damping for each mode found in top right chart). This is typical of large loss of generation events in ERCOT, with the occasional second dominant mode in the 0.9–1.0 Hz range.

15.4 Case Study 2—Compound Event Inducing a Loss of Generation Event

This section presents the post-event analysis conducted by ERCOT engineers for a compound event on the grid. In the process of analysing a loss of generation event using PMU data, ERCOT engineers discovered other disturbances immediately before the loss of generation. This triggered a more in-depth analysis of these disturbances. Using the work detailed by Allen et al. in [4], the various trends in system frequency were analysed. The event began with a phase-to-ground fault (determined by PMU frequency and voltage), which was followed by three separate frequency ramps. Since SCADA telemetry indicated that only the third frequency ramp was caused by a generator going offline, ERCOT requested information from the operator of the power plant about a possible load rejection/imbalance event. The operator confirmed that the load imbalance had indeed happened while the Transmission Service Provider (TSP) in the area confirmed the occurrence of the fault as well as a relay mis-operation in response to the fault at the Point of Interconnection (POI) of the generating station.

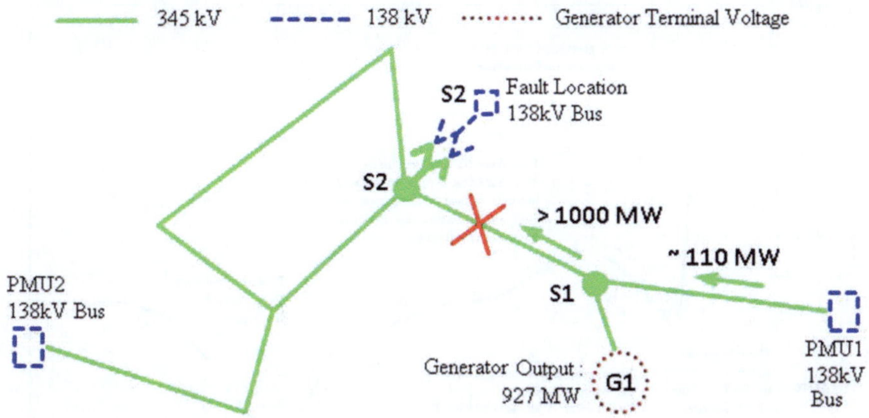

Fig. 15.17 One-line of the 345 kV system near the event and the relative location of the PMUs

Figure 15.17 shows a one-line of the region. A phase-to-ground fault on a 138 kV bus section at the substation marked as S2, one bus away from the POI of generating unit G1, was followed by mis-operation of relays at the POI (substation S1 in the one-line). The clearing of the fault caused the 345 kV circuit connecting the substations S1 and S2 to be tripped, causing a transient imbalance between the loading and mechanical power of unit G1. This finally led to the tripping of the generator and a total frequency swing of 0.217 Hz.

15.4.1 The Power Load Unbalance (PLU) Relay

The PLU relay is designed to rapidly close control/intercept valves under load imbalance conditions in order to reduce generation and protect the turbine from an over-speed event. The relay compares the loading on the generator to the mechanical energy in the turbine [5, 6] and trips if the difference is more than a set percentage, typically 40% or 0.4 per unit, and the load decreases faster than a set rate, typically equivalent to going from rated load to no load in 35 ms [5]. Once the unbalanced condition has cleared and a set time delay has passed, the Power Load Unbalance (PLU) relay is reset and the intercept valves open, allowing load to be restored to the generator. However, the load reference used by the valves does not return to the value before the event, but remains at the value reached when the PLU relay is reset [5].

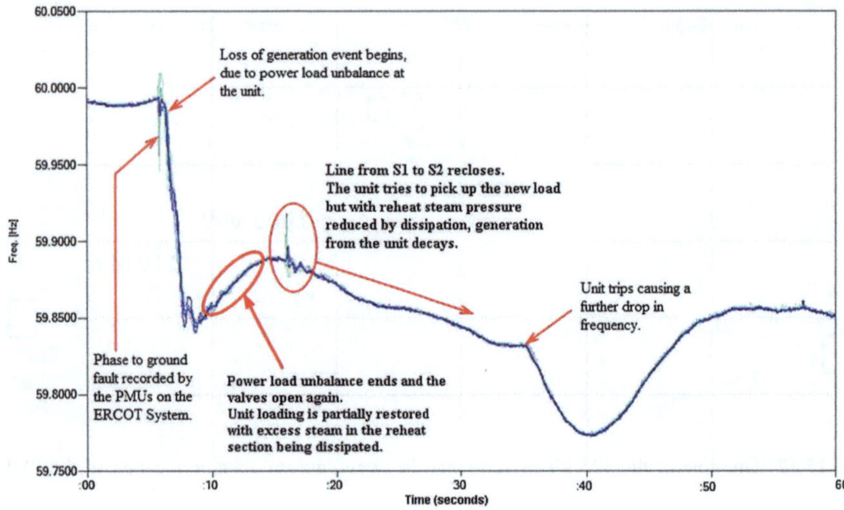

Fig. 15.18 System frequency recorded by the PMUs on the ERCOT Grid during this event

15.4.2 System Frequency

Figure 15.18 shows the frequency trend captured by the PMUs on the ERCOT grid over the course of this event. A. Allen et al. classified frequency trends captured by PMUs during various types of system events into three categories [4]:

1. Frequency Impulse
   a. Half Impulse
   b. Full Impulse
2. Frequency Transient
3. Frequency Ramp

The frequency characteristic (Fig. 15.21) recorded by the PMUs on the ERCOT system shows two of these three distinct characteristics—the Impulse and the Ramp.

15.4.3 System Phase-to-Ground Fault

The phase-to-ground fault on the system occurred approximately 30 cycles before the onset of the loss of generation event. This is seen to be represented by one of the characteristic frequency trends identified in [4]—specifically, the half/full impulse (see Fig. 15.19).

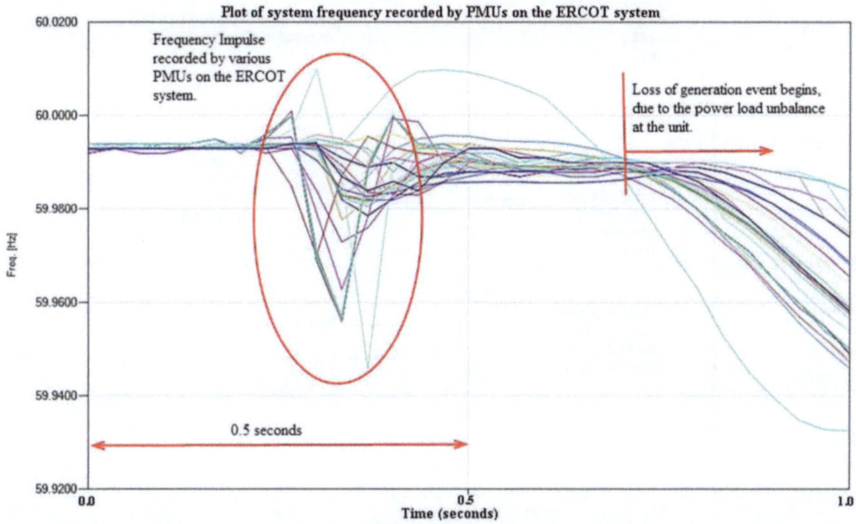

Fig. 15.19 Representation of a system fault in the measured frequency from the PMUs

The authors note here, that such a frequency impulse from PMU data records is not a characteristic of the physical/rotational inertia of the system, but rather a result of the fact that the frequency measured by a PMU is a function (derivative) of the voltage phase angle of the point of measurement [7].

That the event which occurred was indeed a fault was confirmed by examining the voltage magnitudes at the two PMUs (PMU1 and PMU2) identified in the one-line diagram of the region (Fig. 15.17). As expected, there is a sharp (near instantaneous) drop in the voltage recorded by the PMUs, which is the characteristic of voltage near a fault.

15.4.4 Loss of Generation

The loss of generation event can be easily identified from the frequency trend recorded by the PMUs. Following the classifications in [4], the frequency ramp, marked FR1 in Fig. 15.21, corresponds directly with the first loss of generation event.

As Fig 15.18 shows, there were three such ramp events involving the loss of generation in this compound event. They are marked FR1, FR2 and FR3 in Fig. 15.21. FR1 and FR3 were caused by near-instantaneous loss of generation events and FR2 was the slow decay of frequency over several tens of seconds that happened in the interval between FR1 and FR3.

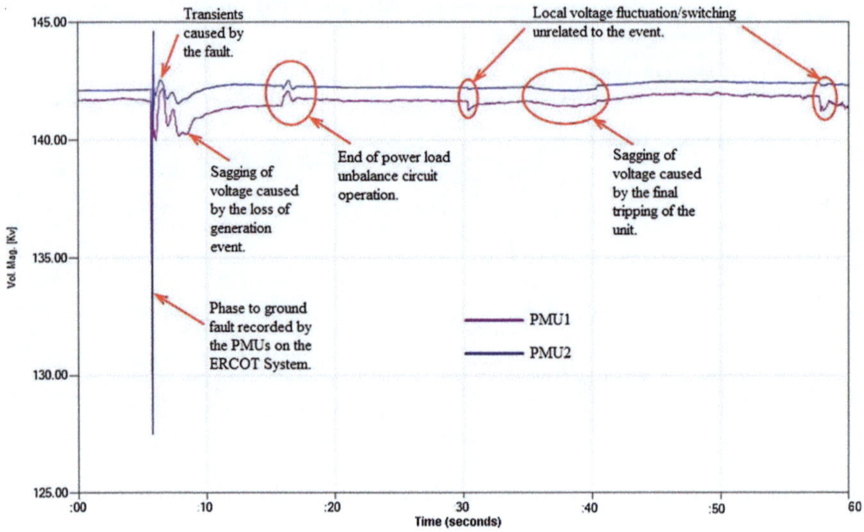

Fig. 15.20 Phase voltages measured by PMUs near the event

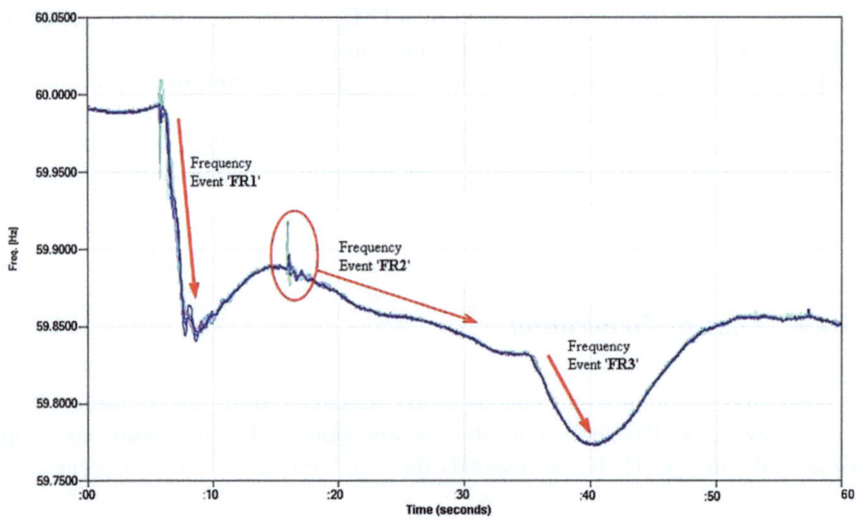

Fig. 15.21 Three frequency sub-events “FR1”, “FR2” and “FR3”

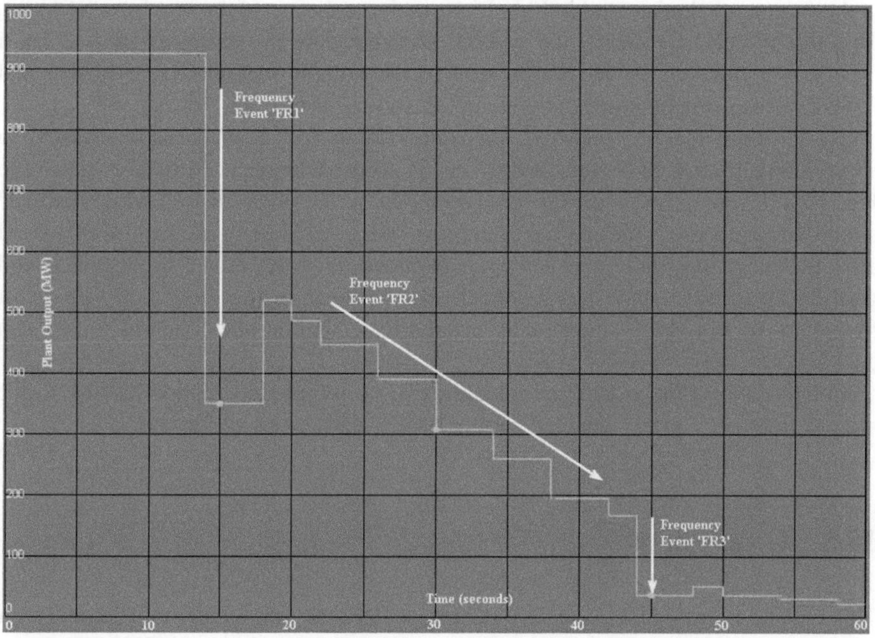

Fig. 15.22 Unit output as telemetered by SCADA (PI chart, times are not exact)

Following the fault on the system and clearance with the tripping of the line between substations S1 and S2 in Fig. 15.17, the system experienced the first loss of generation event. In this first event, the generation from the unit "G1" dropped by approximately 575 MW, causing the frequency to decline from 59.99 to 59.846 Hz, a drop of 0.144 Hz, in around 3 s. With the loading on the unit being drastically reduced due to the loss of one of the lines interconnecting the unit to the ERCOT system, a power load unbalance greater than 40% was detected by the generator-turbine control system. This caused the activation of the PLU circuit, leading to the steam intercept valves being rapidly closed in order to limit the turbine mechanical energy and protect the generator from over-speed. This was the cause of the first frequency event detected by the PMUs on the ERCOT synchrophasor network. However, the unit did not go offline during this event.

Following the arrest of frequency decline by the system's inherent inertia, the governor response of the system set in with AGC acting to restore system frequency to its pre-event state. The PLU condition cleared (difference between mechanical energy and electrical load reduced below 40%) and following the time delay, the PLU control circuit disengaged and the intercept valves were opened allowing the generator to take on load again. Approximately 10 s after inception of event FR1, the line connecting stations S1 and S2 reclosed. This appears as the second impulse in the frequency demarcated in Fig. 15.18 as the reclosing of the line from S1 to S2. That this impulse frequency did indeed correspond to a reclosing event and not a fault was confirmed by an examination of the voltage magnitudes of PMU1 and PMU2 at the corresponding time-stamp as shown in Fig. 15.20.

By the time the line reclosed, the unit loading had been partially restored to approximately 500 MW (according to SCADA telemetry from the unit, see Fig. 15.22). The excess steam that was trapped in the reheat section while the PLU circuit was active was dissipated. Therefore, when the loading on the unit increased after the reclosing, there was not enough pressure in the reheat to sustain it and generation decayed once again, which can be seen as the slow ramp down of frequency in Fig. 15.21 marked as event FR2'. This is also referred to as "run back".

Finally, due to the generation becoming too low, the unit tripped leading to the second frequency event detected by the PMUs designated as FR3.

15.5 Conclusions

This chapter presented the use of synchronised phasor measurement technology for post-event analysis in the Electric Reliability Council of Texas (ERCOT) interconnection. Various aspects of event analysis were discussed. It demonstrates the benefit of synchrophasor data and understanding the system behaviour of different events that happen in the system.

References

1. Final Report on the August 14, 2003 Blackout in the United States and Canada: Causes and Recommendations, Report by the 'Us-Canada' Power System Outage Task Force, April 2004, (Web: https://energy.gov/sites/prod/files/oeprod/DocumentsandMedia/BlackoutFinal-Web.pdf)
2. FERC and the NERC, Report on 'Arizona-Southern California Outages on Sept 8, 2011: Causes and Recommendations', April 2012 (Web: https://www.ferc.gov/legal/staff-reports/04-27-2012-ferc-nerc-report.pdf)
3. ERCOT Protocols (Web: http://www.ercot.com/mktrules/nprotocols/current)
4. Allen A, Santoso S, Muljadi E (2013) Algorithm for screening phasor measurement unit data for power system events and categories and common characteristics for events seen in phasor measurement unit relative phase-angle differences and frequency signals. NREL Technical Report
5. Kundur P (1994) Power system stability and control. McGraw Hill Inc.

6. Kure-Jensen J, Barker W (1996) Speedtronic™ Mark V steam turbine control system. In: GE power generation paper GER-3687C
7. Phadke AG, Thorpe JS (2008) Synchronized phasor measurements and their applications. Springer, New York

Chapter 16
Validation and Tuning of Remedial Action Schemes in Indian Grid Operations Using Synchrophasor Technology

P. K. Agarwal and Chandan Kumar

16.1 Introduction

Generation, transmission, and distribution capacity is being augmented in a planned manner in the Indian Grid. Occasionally, there may be a mismatch in the commissioning of few segments of the network or variation between the forecasted and actual demand. This may result in skewed utilization of the network and cause transmission and generation constraints. Under these scenarios, System Protection Schemes (SPSs) are being deployed in Indian Power System for optimizing the utilization of resources without compromising the system security and reliability. These schemes trigger automatic actions when the predesigned logic is satisfied and provide adequate action to bring back the system to normalcy [1]. Apart from these, with the large size and increased complexity of the Indian power grid, it is also subjected to situations such as congestion, over/under frequency, over/under voltage, system load adjustment, power swings. To detect and take preventive/ protective actions for these conditions, System Protection Schemes are developed to address in order to bring the system back to equilibrium and under Normal State of operation from any Alert or Emergency State. Apart from these, Indian Power System has been designed with N-1 and N-1-1 criteria, and if under N-1-1 criteria system parameters are going above stability limits, System Protection Scheme has to be provided in order to enhance the optimum utilization of available resources along with keeping the grid reliability as paramount importance.

As per CIGRE Task force 38.02.19, SPS is classified into various types based on its input variables, its impact on power system and operating time [2].

P. K. Agarwal (✉) · C. Kumar
Power System Operation Corporation Ltd, New Delhi 11016, India
e-mail: pkagarwal@posoco.in

C. Kumar
e-mail: chandan@posoco.in

S. Nuthalapati (ed.), *Power System Grid Operation Using Synchrophasor Technology*, Power Electronics and Power Systems,
https://doi.org/10.1007/978-3-319-89378-5_16

This SPS definition includes under frequency load shedding (UFLS), under voltage load shedding (UVLS), fault condition that must be isolated, and out of step (OOS) relaying. However, NERC definition of SPS excludes these and includes only the schemes which act while detecting abnormal or predetermined system conditions [3]. In Indian Power System also, the SPS has been discussed in terms of scheme separate from the UFLS, UVLS, and OOS.

A System Protection System is different in concept and implementation from a conventional protection scheme which is being used in substations (transformer load trimming scheme, line loading-based load trimming scheme) as these schemes are intended to provide a safety net for the electrical grid during unplanned/extreme/ N-1-1 contingency conditions or when system or operating constraints could not allow meeting the optimum resource utilization. Deployment of SPS may be a cost-effective measure than building new infrastructure (constriction of new transmission lines may cause non-optimum utilization); however, the risks associated with SPS are as follows:

1. Risk of failure on demand and of inadvertent activation;
2. Risk of interacting with other SPS in unintended ways;
3. Increased management, maintenance, coordination requirements, and analysis complexity [4]; and
4. Improper implementation of envisaged automated action.

The SPS may fail to operate for several reasons, among which prominent ones are:

1. Hardware failure;
2. Faulty design logic;
3. Software failure; and
4. Human error [3].

Recognizing the importance of SPS in any large grid, its mis-operation or failure would have a severe impact. Because of this, SPS design first must address the following performance aspects [2]:

1. Dependability; which is a measure of certainty to operate when required [2].
2. Security; which is a measure of certainty that SPS will not operate when not required [2].
3. Selectivity; which is the ability to affect the least amount of action when performing its intended function [2].
4. Robustness; which is the ability to work correctly over the full range of expected steady state and dynamic system conditions [2].

As on date, 129 SPSs have been deployed in Indian Grid [5]. In line with NERC [6], these 129 SPSs do not include UFLS, UVLS, and OOS relaying. Out of these, 71 SPSs are installed in the Western Regional (WR) Grid in India. These constitute of 34 Wide Impact SPSs, and 37 are Local Impact SPSs [2]. Wide Impact SPSs are those where action envisaged occurs at various locations far away from each other at the same time when its associated SPS acts. While in the Local Impact SPS, the action is localized in nature to provide immediate relief.

Most of these SPSs are designed to operate within milliseconds to few seconds for providing the automatic response during contingencies. Earlier, the SPS operation could not be visualized or analyzed with the help of conventional SCADA system which is having scan rate of more than 10 s. However, Synchrophasor measurements of power system with their capability to provide a synchronized and dynamic view of the system have been found to be better suited for performance evaluation and review of the SPS design due to their high data rate. This chapter discusses utilization of Synchrophasor measurements in the Indian Grid for monitoring, evaluation, and review of the System Protection Scheme.

16.2 SPS Performance Evaluation

In order to ensure the optimum performance of designed and implemented SPS scheme in the grid, there is always a need to evaluate its performance during its actual operation. In order to find the performance level, it has to be classified into various categories. An SPS operation can be classified into basically four categories which are as quoted below:

1. Successful Operation;
2. Failure Operation;
3. Unsuccessful Operation; and
4. Unnecessary Operation.

Based on enumeration of above-mentioned operations, following three indices are defined in [7] to evaluate the performance of SPS.

$$\text{Effectiveness Index (EI)} = \frac{n_1}{n_1 + n_2 + n_3} \tag{16.1}$$

$$\text{Dependability Index (DI)} = \frac{n_1}{n_1 + n_2} \tag{16.2}$$

$$\text{Unnecessary Operation Rate (UOR)} = \frac{n_4}{n_y} \tag{16.3}$$

where

n_1 Number of Successful Operations;
n_2 Number of Failures;
n_3 Number of Unsuccessful Operations;
n_4 Number of Unnecessary Operations;
n_y Number of Scheme—Years of Operation;

Table 16.1 Wide Impact SPS in western region in India

Type of SPS based on control action	Wide Impact SPS in WR grid	
	Number of SPS	Percentage of SPS (%)
Load rejection	20	59
Generation rejection	7	21
Load and generation rejection	6	18
HVDC control	1	3

Table 16.2 Local Impact SPS in western region in India

Type of SPS based on control action	Local Impact SPS in WR grid	
	Number of SPS	Percentage of SPS (%)
Load rejection	34	92
Generation rejection	3	8

Based on the above indices, the performance of any SPS is calculated which provides the feedback regarding the efficiency as well as the design review of the SPS. With the help of Synchrophasor data, the indices of SPS in Western Regional Grid have been calculated and presented in this chapter.

However, before proceeding to calculation, there is a need to further classify the type of SPS whose performance is being studied with these indices based on their control action [6]. Western Regional Grid SPS which amounts to number 71 is further classified based on type of control actions, i.e., load/generation reduction, load and generation rejection, and HVDC control, as given in Tables 16.1 and 16.2.

16.3 WAMS in Indian Grid

In India, Synchrophasor initiative was started in the year 2010 [8]. It was started with five regional pilot projects wherein PMUs were placed at few of the strategic locations in the regional grids and the PMU measurements were integrated at the regional level. Subsequently, the regional projects are integrated at the national level. At the beginning of the year 2015, there were 62 PMUs installed at 62 substations. After recognizing potential benefits of PMUs, two state utilities have also come forward with their own Synchrophasor projects in the last two years. These state-level projects have now been integrated with the National PDC via the regional PDC. By the end of the year 2015, the population of PMUs grew to 154 covering 103 substations. The WAMS network (as in 2016) in Indian Grid is as shown in Fig. 16.1. The next section provides an overview of how the Synchrophasor can be used for SPS monitoring and performance evaluation.

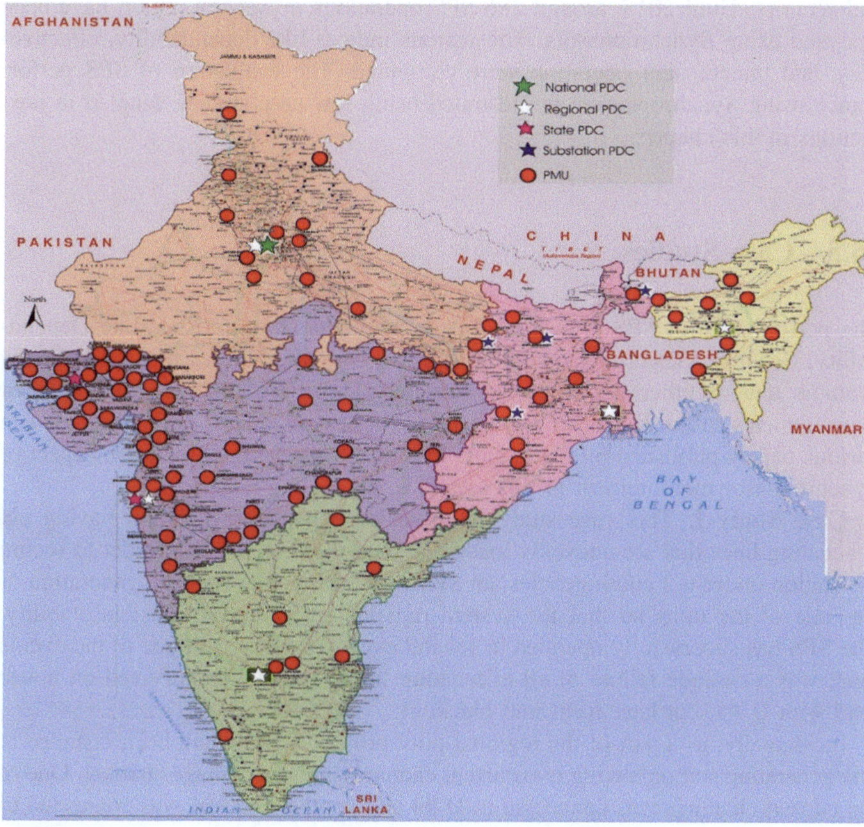

Fig. 16.1 WAMS network in India in 2016

16.4 Monitoring of SPS Using Synchrophasors

SPSs are designed to detect the abnormal conditions in the system and take pre-planned automated corrective actions to ensure system security and reliability [9]. After detection of abnormal system conditions, the SPS actions take place in fraction of sub-seconds in order to ensure that system parameter is not outside the permissible limit for large duration of time and enter transient instability zone. To monitor SPS actions and to evaluate its performance using conventional SCADA which takes long scan time of 2–14 s is of major concern. Synchrophasor measurements can be considered as perfect instrument for monitoring SPS actions and evaluating its performance as they monitor power system states [10] in fraction of a second. Even though number of PMUs is limited, measurements from these PMUs have helped in monitoring the performance of the deployed SPS in the Indian Power System with their suitable installation at selected substations. In many cases, modifications in the SPS were done to improve its performance for better reliability and security. Until 2015, around 286 SPS operations in western region have been analyzed using Synchrophasors. The various indices like dependability, effectiveness, and unnecessary operation were computed. The evaluation of SPS performance using Synchrophasors is illustrated using few case studies detailed in next sections in this chapter.

16.5 Case Studies

The performance of SPS is evaluated in terms of its characteristics like Dependability, Security, Selectivity, and Robustness. The case studies illustrated below evaluate one or other performance characteristics of SPS using Synchrophasor measurements in the Indian Power System. These case studies have been used in various papers published by the authors in order to spread the utilization aspect of Synchrophasor measurement in this area [11, 12].

Case Study I: This case study is about large generating station having six evacuating lines that are unevenly loaded as shown in Fig. 16.2. In order to secure the station under n-2 contingencies, an SPS was deployed to runback generation or trip one of the units so that the system remains stable under transient stability. The SPS has successfully operated in several events. However, in one of the events there was a cascade failure of all evacuating lines. The station survived on house load with U #4, but later went into blackout.

Incidentally, as a part of the regional pilot project, a PMU had been installed at this generating station having two current channels and one voltage channel. One of the current channels was connected to U #4 and another channel was connected to circuit/Line 6. The performance of the SPS during the event was analyzed with the help of data from this PMU. The SPS logic is illustrated in Fig. 16.3.

The event occurred at $t = 0$ resulting in tripping of circuit 3. After $t_1 = 11$ min 13 s 3 ms due to local breaker backup (LBB) action at Bus C, circuit 4 tripped. This resulted in overloading of the existing four lines. Thus at time t_1, the criteria for SPS-2 were satisfied while at time t_2 (t2 = 11 min, 13 s, 724 ms) the criteria for SPS-3 were satisfied. The generation reduction could have prevented cascade failure. The chronology of events is as follows:

1. $t = 0$: Occurrence of an event at Bus C causing tripping of circuit 3.
2. $t_1 = 11$ min 13 s 3 ms: Tripping of circuit 4 satisfying SPS-2 criteria followed by envisaged action; U #4 unloading.
3. $t_2 = 11$ min 13 s 724 ms: Tripping of circuit 1, 2, 5 and 6 on power swing causing the station to island. Generating Units survived on house load. SPS-3 criteria also satisfied for U #4 tripping.

The event was analyzed using Synchrophasors. The Synchrophasors measurements revealed that there was a delay in the unloading of units as well as tripping of U #4 as shown in Fig. 16.4. Simulations indicated that the station islanding could

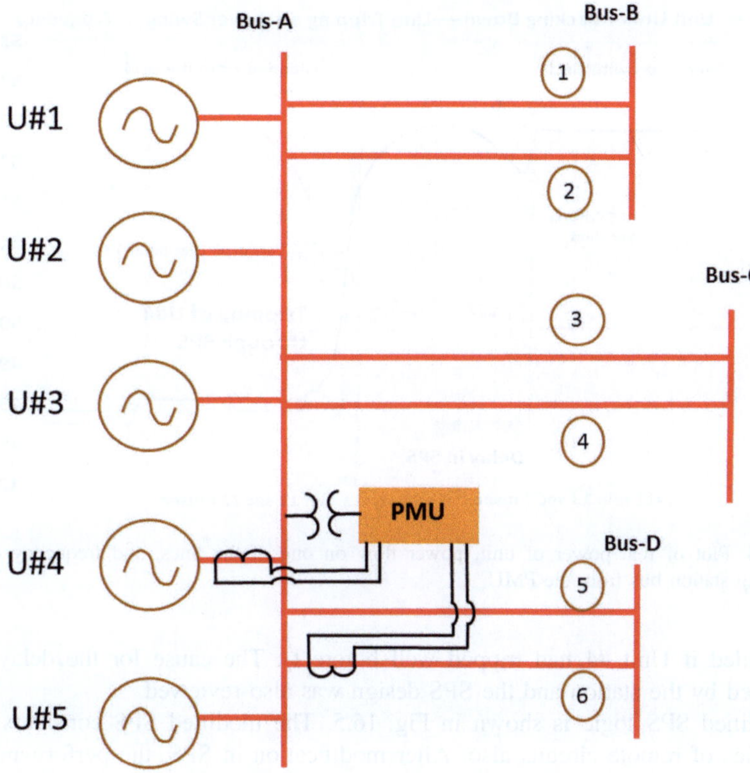

Fig. 16.2 Generation power plant

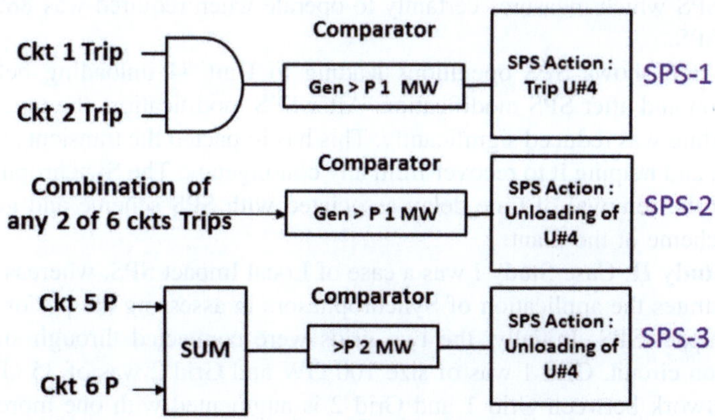

Fig. 16.3 Local Impact SPS scheme implemented at the generation plant

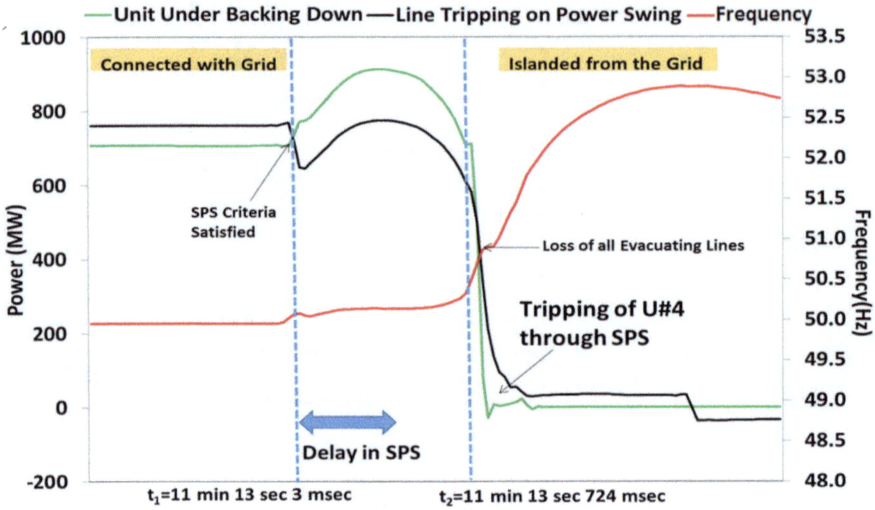

Fig. 16.4 Plot of real power of unit, power flow on one of the lines, and frequency of the generating station bus from the PMU

be avoided if Unit #4 had tripped well before t_2. The cause for the delay was addressed by the station and the SPS design was also reviewed.

Modified SPS logic is shown in Fig. 16.5. The modified SPS considers contingencies of remote circuits also. After modification in SPS, the performance of SPS was improved.

The modified SPS operated on 8 occasions out of which 7 were successful and only 1 failure occurred which was due to human error. The dependability index of modified SPS which measures certainty to operate when required was 88% for the modified SPS.

Figure 16.6 shows SPS operations leading to Unit #4 unloading before SPS modification and after SPS modification. After SPS modification, the time delay in unit unloading was reduced significantly. This has impacted the transient stability of the system and helping it to recover from any contingency. The Synchrophasor data assisted in the removal of time delay associated with SPS scheme and generation runback scheme at the plant.

Case Study II: Case Study I was a case of Local Impact SPS, whereas this case study illustrates the application of Synchrophasors in assessing the performance of a Wide Impact SPS. Initially, the two grids were connected through single 765 transmission circuit. Grid 1 was of size 100 GW and Grid 2 was of 35 GW. Later on, the network between Grid 1 and Grid 2 is augmented with one more 765 kV circuit as shown in Fig. 16.7.

Few SPSs were designed to handle contingencies in either Grid 1 or Grid 2. One of the SPS logic is illustrated in Fig. 16.8. The SPS would trigger if flow on D/C is greater than or equal to 1500 MW for more than 2.5 s or if flow on D/C is greater

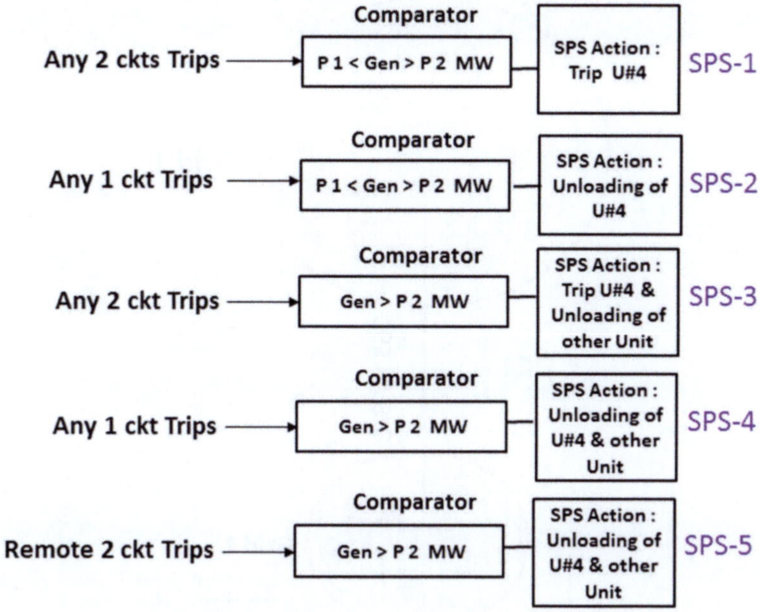

Fig. 16.5 Modified Wide Impact SPS scheme implemented at the power plant

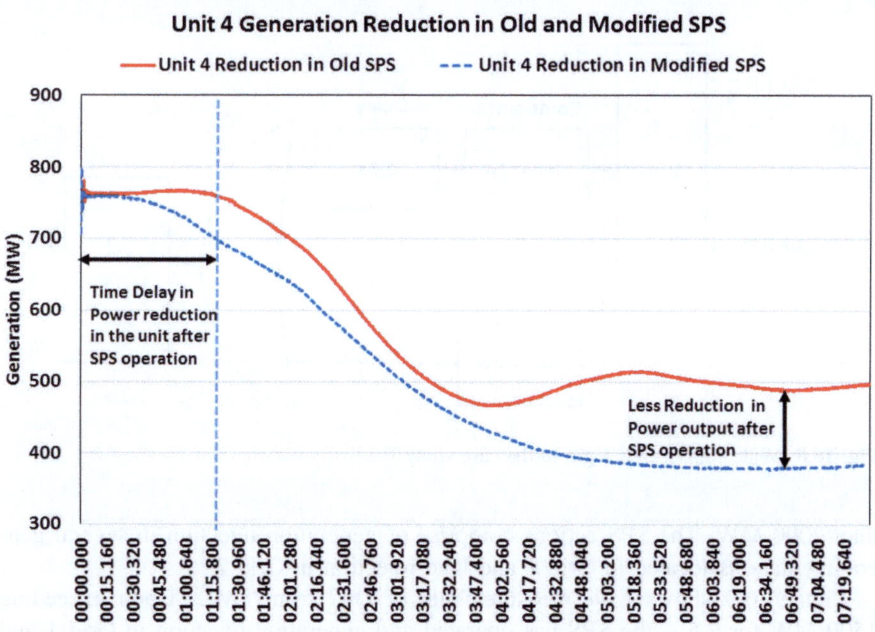

Fig. 16.6 Plot of generation reduction on old and modified SPS operation

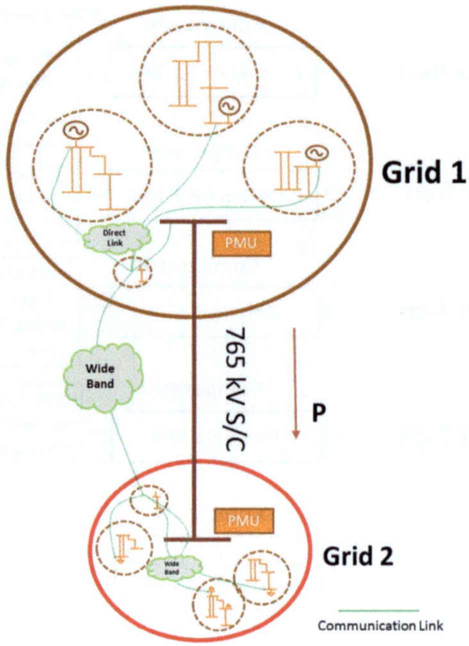

Fig. 16.7 Wide Impact SPS scheme on the inter-regional tie lines

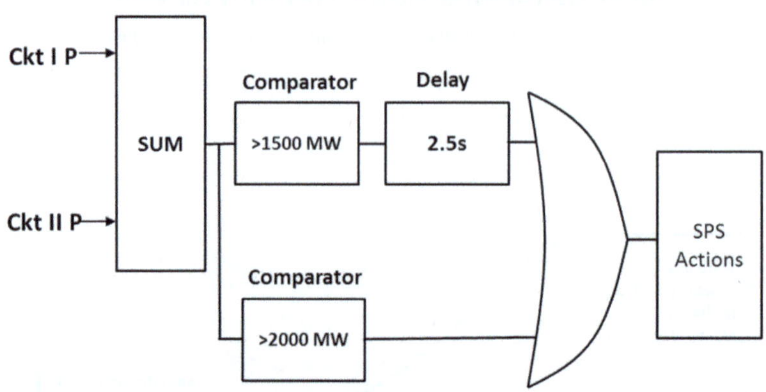

Fig. 16.8 Wide Impact SPS logic in the case study II

than 2000 MW. The SPS actions comprise of generation rejection in several generating stations located in Grid 1 and load rejection in Grid 2.

Figure 16.9 shows MW measurements of D/C from PMU. Upon exceeding 1500 MW for 2.5 s, the SPS has operated and generation rejection in Grid 1 and load rejection Grid 2 happened due to which the power flow on the circuits has come down. The delay of 2.5 s has helped in preventing frequent operation of SPS.

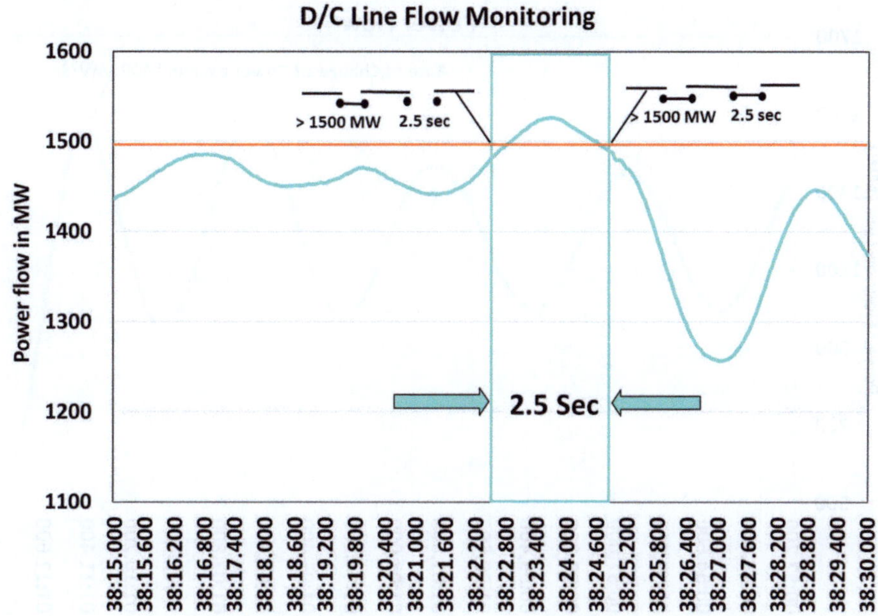

Fig. 16.9 Plot of power flow on the tie line showing the dependability of the SPS scheme

This SPS has operated on 15 occasions with 100% dependability. There has been no unnecessary operation or insecure operation of this SPS which has led to its high effectiveness index, and it has been validated using Synchrophasor measurements.

Case Study III: This case study explains SPS logic when a single tie line is in service between Grid 1 and Grid 2 as shown in Fig. 16.7. This SPS operates when there is huge power flow from Grid 1 to Grid 2 during contingencies happening in either Grid 1 or Grid 2.

The implemented scheme is shown in Fig. 16.10. The SPS would operate when dp/dt exceeds 400 MW/s in Grid 1 to Grid 2 direction. The desirable SPS operation is shown in Fig. 16.11. SPS got triggered correctly and was effective in damping the oscillations that seem to be growing in amplitude.

It was observed that out of 24 operations, the SPS operated 21 times successfully while on two occasions directionality condition was not satisfied and on one occasion the ramp rate condition was not satisfied as shown in Fig. 16.12. Due to

Fig. 16.10 Wide Impact SPS logic in case study III

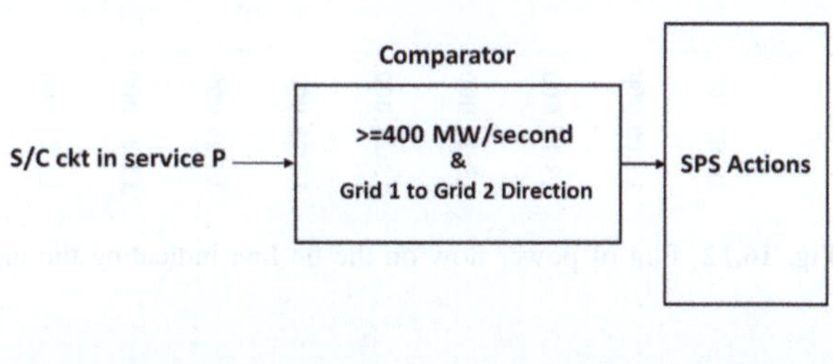

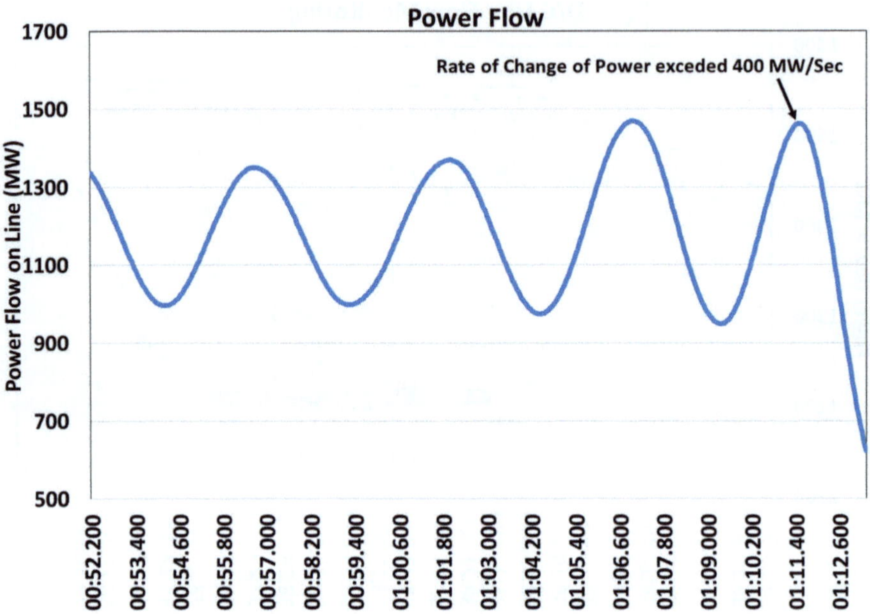

Fig. 16.11 Plot of power flow on the tie line indicating the correct operation of SPS

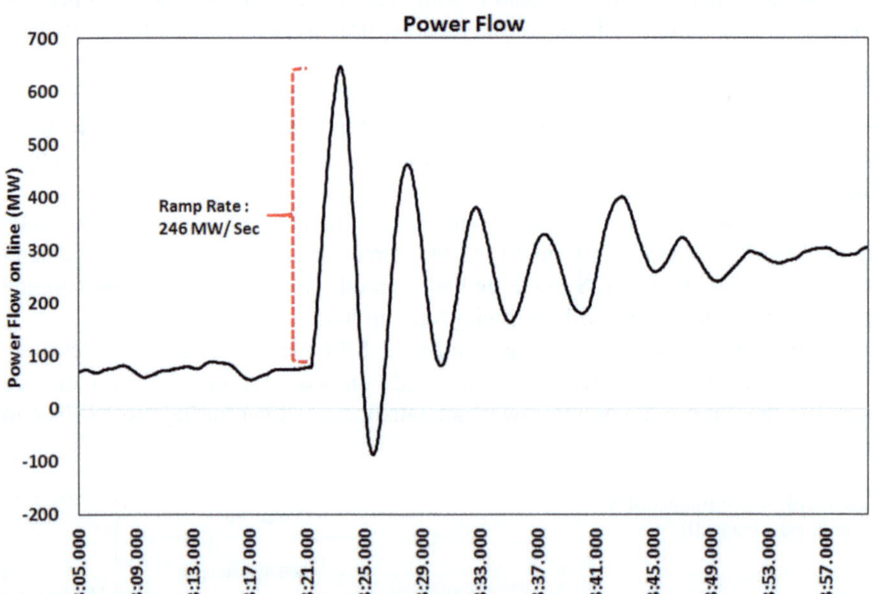

Fig. 16.12 Plot of power flow on the tie line indicating the undesirable operation of SPS

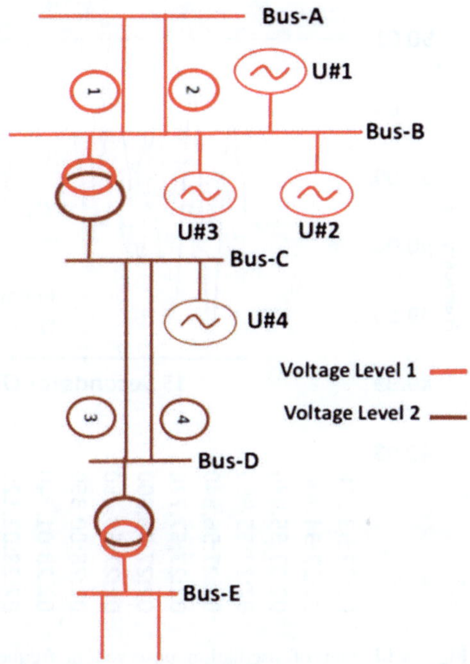

Fig. 16.13 Generators connectivity at two different voltage levels

triggering of SPS irrespective of the direction of power flow at occasions and the undesirable SPS operations, this SPS was disarmed.

Case Study IV: Apart from evaluating performance indices of SPS scheme, Synchrophasor measurements has also provided key input to the design of the SPS scheme. There is a generating station having units and evacuation lines connected at two different voltage levels (Bus B and C). It was observed that tripping of any one of these lines leads to power swing on other line and low-frequency oscillations in the grid. The generating plant connectivity and oscillations observed in the grid using Synchrophasor measurements are shown in Fig. 16.13 and Fig. 16.14, respectively. The oscillations observed were of inter-plant nature and associated with generators and its connectivity with the rest of the network.

The oscillations in the grid are due to tripping of one circuit which further resulted in tripping of other circuit in Zone 1 power swing. This has resulted in the loss of evacuation lines from the power plant. In order to improve system reliability and transient stability, first the PSS tuning of generating station was carried out. After that, an SPS was designed in which detection of power swing was set as base criteria for SPS action. The SPS action either trip or reduce generation as per the logic shown in Fig. 16.15.

After PSS tuning and implementation of SPS based on power swing, reliability of the system has improved. In an event due to the tripping of one circuit, the SPS has avoided tripping of other circuit on power swing and Fig. 16.16 illustrates that oscillations are reduced significantly after implementation of SPS.

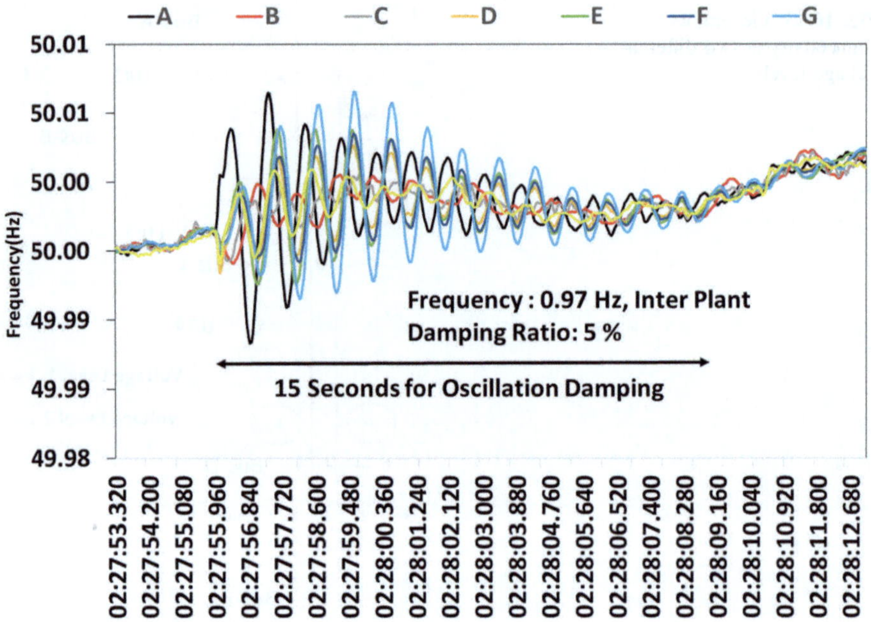

Fig. 16.14 Plot of oscillation observed in frequencies near to substation due to tripping of evacuation line from Bus C and Bus D

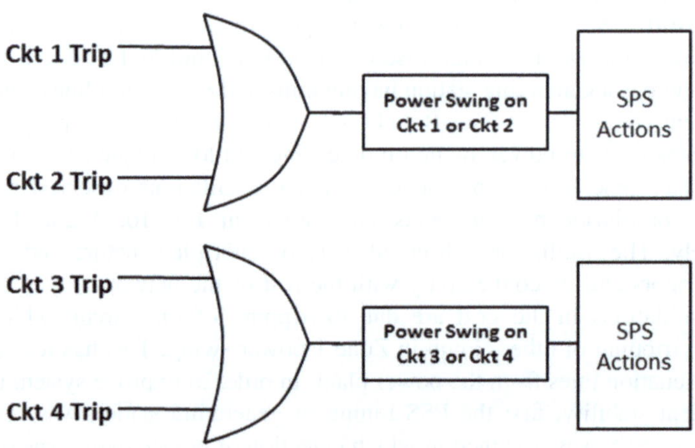

Fig. 16.15 Scheme of SPS implemented at generating station based on detection of power swing and circuit tripping

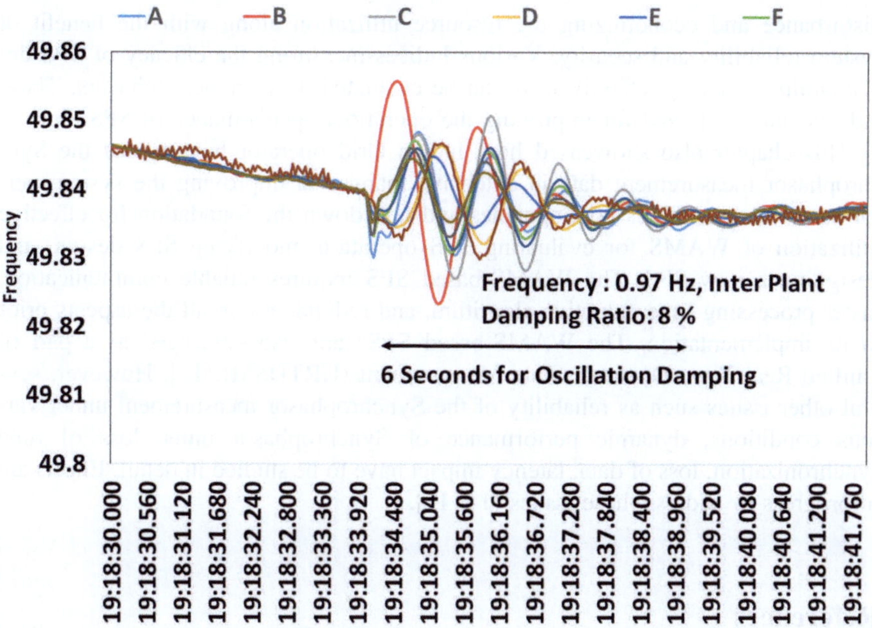

Fig. 16.16 Plot of oscillation observed in frequencies near to substation due to tripping of evacuation line from Bus C and Bus D

Table 16.3 Pefromance indices of SPS discussed in case studies

Sl. no	n_1	n_2	n_3	n_4	n_y	DI (%)	EI (%)	UOR (%)
Case study I	7	1	0	0	71	88	88	0
Case study II	15	0	0	0	71	100	100	0
Case study III	21	0	3	3	71	100	100	4
Case study IV	3	0	0	0	71	100	100	0

From these case studies, it can be seen that how the Synchrophasor data has helped in improving the design of SPS as well as designing a new SPS. Summary of performance indices of four case studies as calculated from Synchrophasor data is shown in Table 16.3.

16.6 Conclusion

This chapter has illustrated through various case studies how Synchrophasor technology can help system operators in performance evaluation of SPS and provide guidance in improving its design. These have helped in adverting major grid disturbance and economizing the resource utilization along with the benefit of system reliability and security. Various indices measuring the efficacy of SPS dependability, security, effectiveness can be evaluated using Synchrophasors. These indices can be utilized for improving the operational performance of SPS.

This chapter also showcased how Indian Grid operator has utilized the Synchrophasor measurement data in extensive manner in improving the system performance under various contingencies and laid down the foundation for effective utilization of WAMS for evaluating SPS operation, modifying SPS design, and designing a new SPS. The WAMS-based SPS requires reliable communication, faster processing time, adaptive algorithm, and redundancy in all the aspects prior to its implementation. The WAMS-based SPSs are also envisaged as a part of Unified Real Time Dynamic State Measurement (URTDSM) [13]. However, several other issues such as reliability of the Synchrophasor measurement under various conditions, dynamic performance of Synchrophasor units, loss of time synchronization, loss of data, latency impact have to be studied in detail. Efforts are in progress to address these issues [14, 15].

References

1. Agrawal VK et al (2011) Deployment of system protection schemes for enhancing reliability of power system: operational experience of wide area SPS in northern regional power system in India. In: International Conference on Power and Energy Systems (ICPS), 22–24 Dec. 2011
2. CIGRE 38.02.19 (2001) System protection schemes in power networks, June 2001
3. NERC, Special Protection Systems (SPS)/Remedial Action Schemes (RAS): assessment of definition, regional practices, and application of related standards
4. PSERC Final Project Report (2010) System protection schemes: limitations, risks, and management, Dec 2010
5. POSOCO, System protection scheme in Indian grid, Internal report
6. Fu W et al (2002) Risk assessment for special protection systems. IEEE Trans Power Syst 17(1)
7. Anderson PM, LeReverend BK (1996) Industry experience with special protection schemes. IEEE Trans Power Syst 11(3):1166–1179
8. POSOCO (2012) Synchrophasors initiative in India, New Delhi, Technical report, July 2012. http://www.erldc.org/Other_Doc/synchrophasor_initiatives_in_india.pdf
9. Anderson PM (1999) Power system protection, chapter 21. IEEE Press, pp 902–909
10. Kundur Prabha et al (2004) Definition and classification of power system stability IEEE/CIGRE joint task force on stability terms and definitions. IEEE Trans Power Syst 19(3):1387–1401
11. Mukhopadhyay P, Pandey V, Kumar C, Seshadri P, Chitturi S (2016) Performance evaluation and review of system protection schemes using synchrophasor measurement. In: 2016 IEEE innovative smart grid technologies—Asia (ISGT-Asia), Melbourne, VIC, 2016, pp 856–861
12. Mukhopadhyay P, Pandey V, Anumasula R, Kumar C, Patil S, Chitturi S, Mahendranath M (2016) Performance evaluation and review of system protection scheme design with the help of synchrophasor measurements. In: NASPI work group meeting and first international synchrophasor symposium, 22–24 Mar 2016. https://www.naspi.org/sites/default/files/2016-10/posoco_mukhopadhyay_performance_evaluation_20160322.pdf

13. POSOCO (2013) Synchrophasors initiative in India, New Delhi, Technical report, Dec 2013. http://www.nldc.in/WebsiteData/Reports/OtherReports/Synchrophasors%20Initiatives%20in%20India%20Decmber%202013%20-%20Web.pdf
14. Agarwal PK, Rathour HK, Anumasula R, Maurya P, Mohan S (2016) Encounter with leap second—experience in Indian WAMS. In: NASPI work group meeting and first international synchrophasor symposium, 22–24 Mar 2016. https://www.naspi.org/sites/default/files/2016-10/posco_agarwal_encounter_with_leap_second_20160323.pdf
15. Sanodiya PK, Pandey V, Anumasula R, Kumar C, Patil S, Chitturi S (2017) Selection of reference node and angular baselining using synchrophasors measurement for real time operation. In: NASPI work group meeting, Mar 2017. https://www.naspi.org/sites/default/files/2017-03/02_Sanodiya_Selection_Reference_Node_Angular_Baselining_20170322.pdf

Chapter 17
Indian Power System Operation Utilizing Multiple HVDCs and WAMS

Rahul Shukla, Rahul Chakrabarti, S. R. Narasimhan and S. K. Soonee

17.1 Introduction

Interconnection of power systems provides significant advantages in optimization of resources to achieve quality in operation, economy as well as minimize impact on the environment. The electricity grid in India evolved from five (5) regional grids in the 1980s to a synchronously interconnected national grid in 2013. As of March 2017, the grid has over 300 GW of generating capacity which caters to a peak load of the order of 160 GW. HVDCs played a key role in integration of the regional grids, facilitating HVDC back-to-back interconnections between different regions initially, followed by long-distance HVDC for bulk inter-regional transfer of power.

The power system operation in India is being coordinated by the State, Regional, and National Load Despatch Centres (SLDCs/RLDCs/NLDC). They are expected to maintain vigil against threats and vulnerabilities in the system and take preventive measures to avoid outages. In the event of outages, it is desirable that the system is restored to its normal state quickly. The challenges in power system operation in India are increasing manifold day by day as a result of enlarged system size, brisk pace of capacity addition, long-distance power flows, multiple players, increasing competition in the electricity market, emphasis on pan India optimization, climate change, large-scale integration of renewable energy sources in certain pockets, and increasing customer expectations. The ability of the system operators to take decisions in real time is dependent on their 'situational awareness' derived from the data/information available with them in real time. Conventionally, the analog and digital information related to the power system, such as circuit breaker status, frequency, voltage, and power flow (MW/MVAR) measured at the substation level is presented in the Load Despatch Centre through the supervisory control and data acquisition/

R. Shukla (✉) · R. Chakrabarti · S. R. Narasimhan · S. K. Soonee
Power System Operation Corporation Ltd., New Delhi, India
e-mail: rahulshukla@posoco.in

S. Nuthalapati (ed.), *Power System Grid Operation Using Synchrophasor Technology*, Power Electronics and Power Systems,
https://doi.org/10.1007/978-3-319-89378-5_17

Energy Management System (SCADA/EMS). In India, there is a hierarchical architecture through which the information is routed and updated (every 10–20 s) at the respective Load Despatch Centres. With such high data updation time, many power system events are difficult to capture and thus limit the system operator action.

The synchrophasor technology along with the high-speed wideband communication infrastructure from substation to control center has now overcome the above limitation. These schemes based on synchrophasor technology are also known as wide area measurement system (WAMS). With the help of WAMS it is now possible to monitor the phase angles at the control center. In addition, this technology enables visualization of magnitude and angle of each phase of the three-phase voltage/current, frequency, rate of change of frequency, and angular separation at every few millisecond intervals (say 40 ms) in the Load Despatch Centre. Thus, the transient/dynamic behavior of the power system can be observed in near real time at the control center which hitherto was possible only in offline mode in the form of substation disturbance records or through offline dynamic simulations performed on network models. Angular separation between coherent groups of generators within a synchronous grid is representative of the grid stress.

The phasor measurement unit (PMU) is the basic building block of wide area measurement system (WAMS). The PMU measures the system state, viz. voltage and angle of a particular location at a rate of multiple samples per second (say 25 samples per second). This data is time-stamped through a common reference and transmitted to the phasor data concentrator (PDC) installed at a nodal point, through high-speed wideband communication medium (such as Optical Fiber). The PDC aligns the time synchronized data and presents it to the User/Historian. The Historian archives the data for retrieval and post-dispatch analysis of any grid event. In control room, the synchrophasor data has helped in improving/enhancing situational awareness through real-time monitoring of frequency, df/dt, angular separation and voltage. It is possible to recognize the occurrence of transmission line tripping/revival within a flow gate by observing the step change in angular separation, step change in voltage magnitude, step change in line current (MW and MVAR). It is also possible to recognize the occurrence of generator tripping by observing the frequency decline, increase in df/dt, change in angular separation, and decrease in voltage magnitude. Occurrence of load crash/load throw off can be observed by sustained high frequency, sustained abnormal phase angle separation, sustained high voltage.

The transmission network of India has 11 functional HVDCs (four HVDC Back-to-Back, three Inter-regional HVDC bipoles, three Intra-regional HVDC bipoles, and one multi-terminal HVDC) ranging from capacities 500–3000 MW as well as more than 33,000 circuit kms of 765 kV AC lines and 164,588 circuit kms of 400 kV AC lines across the length and breadth of the country as on August 2017. There is one cross-border HVDC link for power transfer to Bangladesh. The bipole HVDCs were commissioned at ±500 kV voltage level and the more recent ones commissioned at ±800 kV. The inter-regional HVDC bipoles and HVDC back-to-back stations which initially connected two systems asynchronously have now become embedded in the transmission network with power flow of parallel AC network affected by HVDC set points. The operational HVDC in Indian power network are shown in Fig 17.1. One more 6000 MW HVDC bipole and another bipole based on voltage source converter (VSC) technology (the first in India) is under construction and expected by the year 2020.

With such a highly meshed system with 11 embedded HVDCs, the importance of reliable power system operation also assumes importance. The integrated operation of the pan India grid is a major challenge for the system operator, while maintaining reliability and economy in operations. Synchrophasors have been adopted in Indian power system as a pilot project in 2010 and have been ramped up to around 77 numbers as of 2017. Around 1700 more phasor measurement units (PMUs) are under different stages of implementation in the transmission system. Availability of PMUs has provided situational awareness to the Indian power system operator, thus raising confidence of all stakeholders in power system operation. Availability of synchrophasor data aided in synchronization of southern part of Indian grid with rest of the grid.

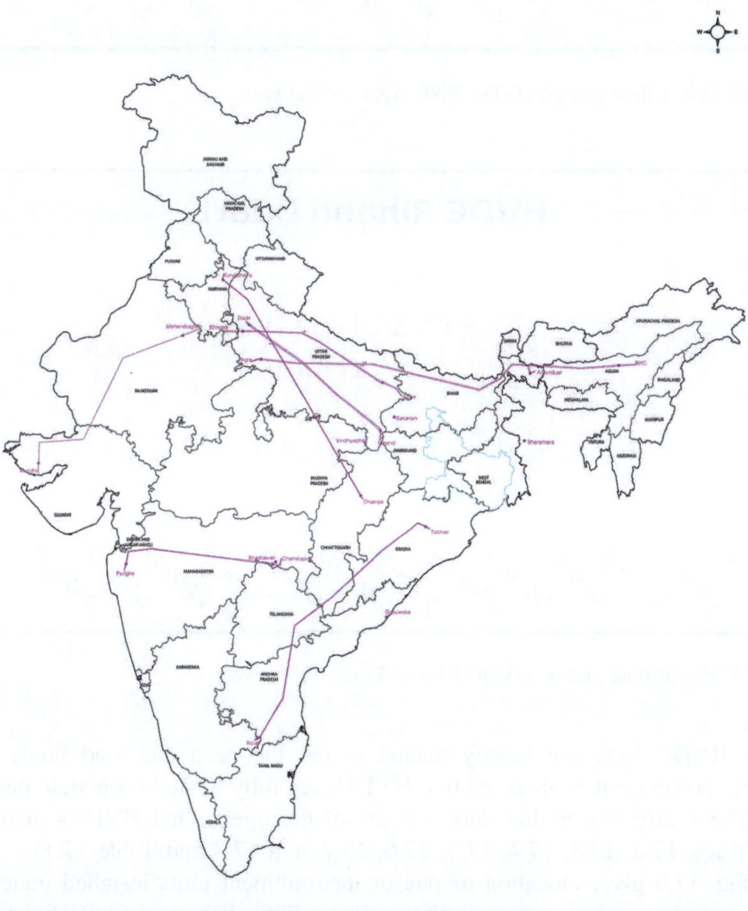

Fig. 17.1 Indian power map highlighting operational HVDC links

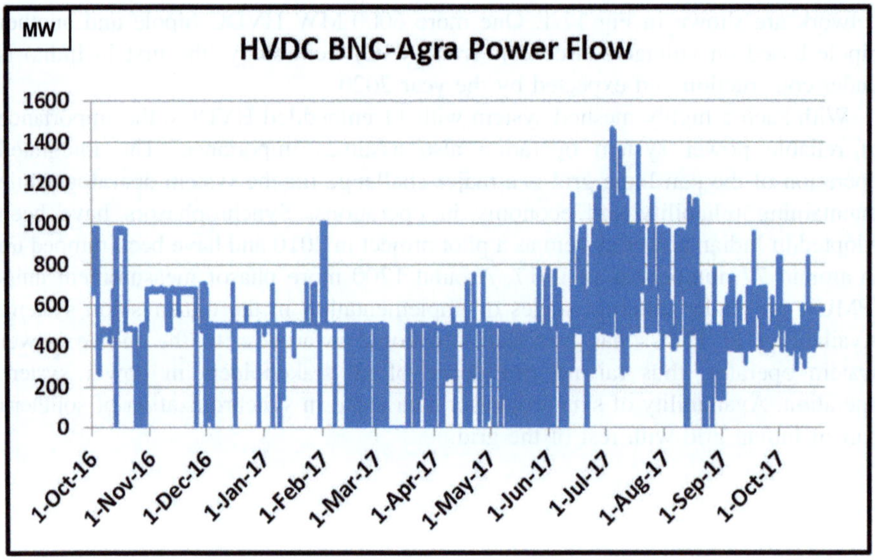

Fig. 17.2 Power flow plot on HVDC BNC-Agra for one year

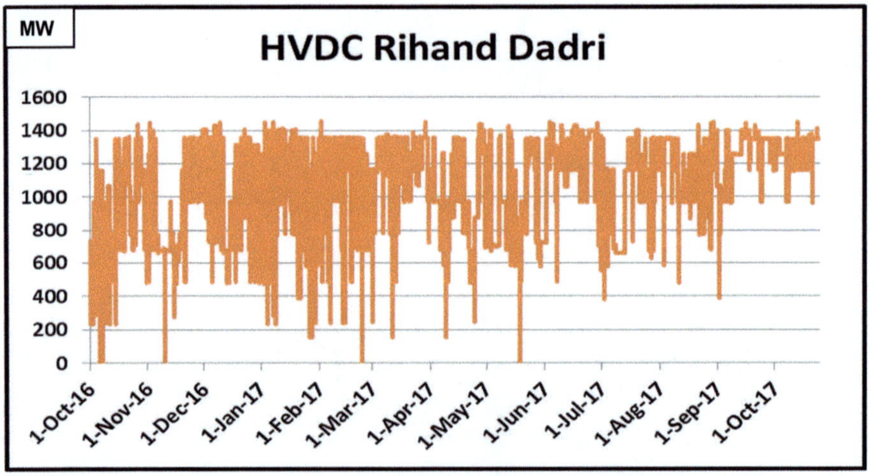

Fig. 17.3 Power flow plot on HVDC Rihand–Dadri for one year

The HVDC links are mostly loaded as per the need and load flows in the network; however, it is observed that HVDCs are fully loaded for certain period of time. The yearly power flow plot for few of the operational HVDCs in India is given (Figs. 17.2, 17.3, 17.4, 17.5, 17.6, 17.7, and 17.8 and Table 17.1).

In Fig. 17.9 given, location of phasor measurement units installed under pilot project in Indian power system has been shown. The pilot project in India, with the

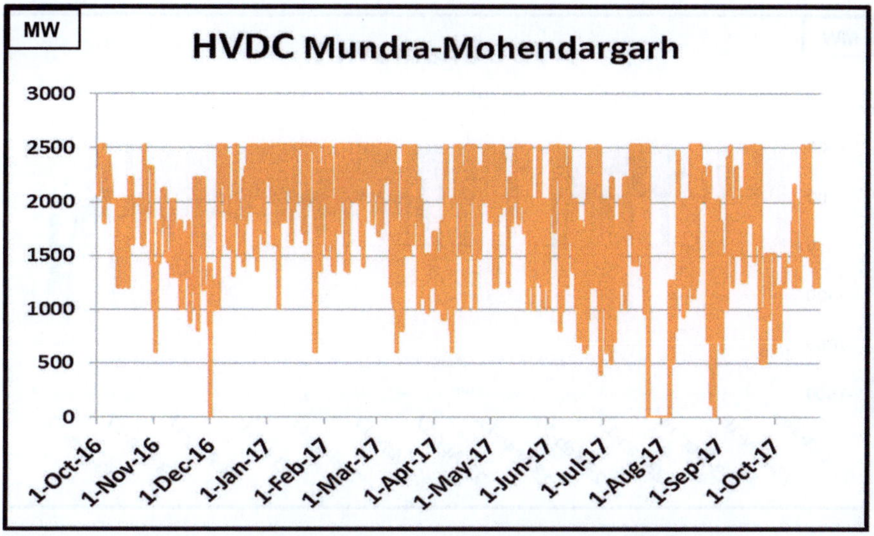

Fig. 17.4 Power flow plot on HVDC Mundra–Mohendargarh for one year

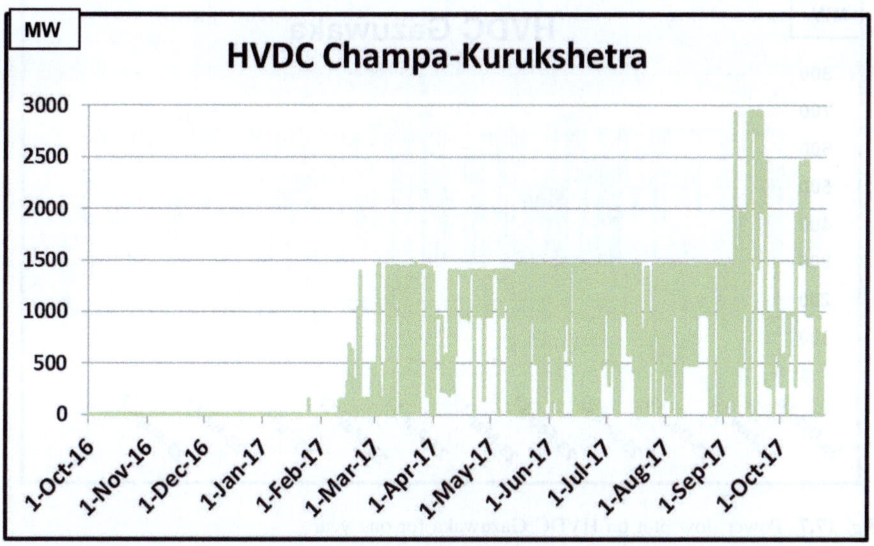

Fig. 17.5 Power flow plot on HVDC Champa–Kurukshetra for one year

initiative taken by the Central Electricity Regulatory Commission (CERC), started with the implementation of a very simple project consisting of four PMUs and one PDC along with data historian and operator console in May 2010 in northern Region. Subsequently, other pilot projects were taken up in different regions. Later

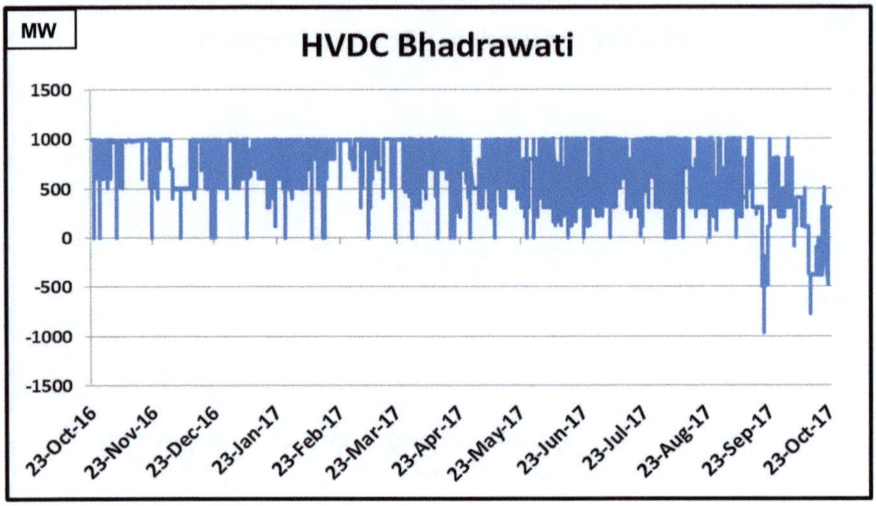

Fig. 17.6 Power flow plot on HVDC Bhadrawati for one year

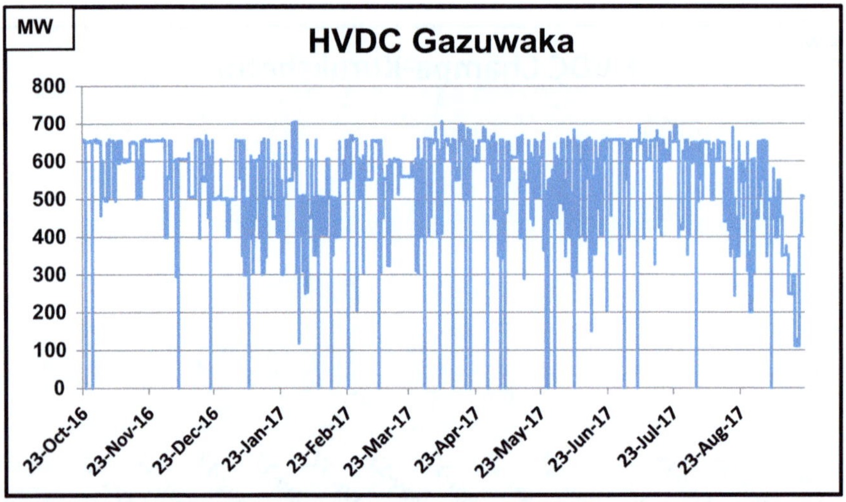

Fig. 17.7 Power flow plot on HVDC Gazuwaka for one year

on these all projects were integrated at national level by providing a PDC at National level in National Load Despatch Center located at Delhi.

PMUs installed under these projects provide time-stamped synchronized measurements to phasor data concentrators (PDCs) installed at control center at a reporting rate of 10, 25, and 50 frames persecond. Most of the PMUs installed at EHV substations are reporting in IEEE C37.118-2005 protocol combination of either one/two sets of voltage or one/two sets of current signals

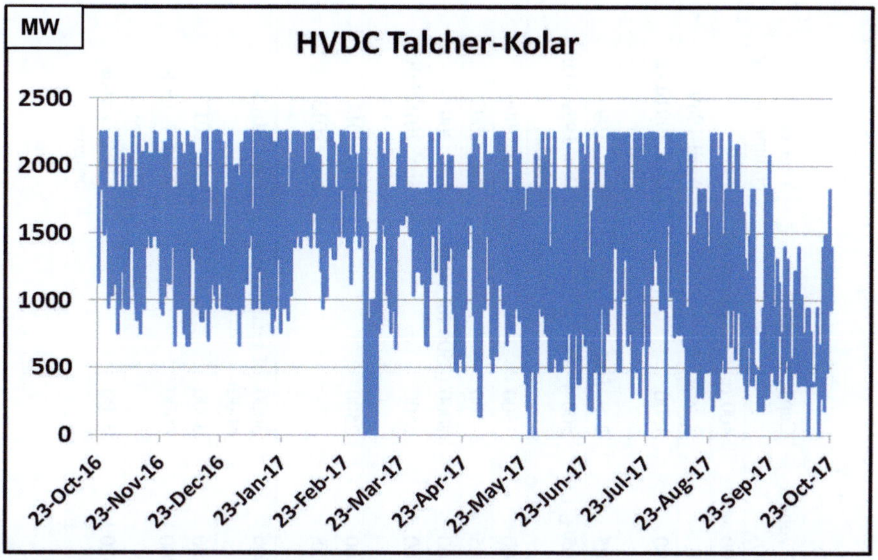

Fig. 17.8 Power flow plot on HVDC Talcher–Kolar for one year

Synchrophasors in operation of HVDC links of India have been used in a variety of ways:

- Monitoring of transients during filter and pole switching in HVDCs
- Detection of commutation failure in HVDCs due to faults close to HVDC terminal stations
- Problem in communication links or station switchyard leading to oscillation of HVDC power order
- Close monitoring of power order setting and HVDC pole outages
- Optimization of angular spread in the grid by changing power order of HVDCs
- Tuning simulation study models using measurement data from PMUs
- Monitoring sub-synchronous resonance in HVDCs.

A number of case studies in Indian grid based on monitoring of HVDCs using synchrophasors are indicated below:

Case 1: HVDC set point optimization using PMU data and offline simulations
The power flow in the highly meshed AC system in India depends on the grid topology, the impedances of the transmission lines, the load as well as the generation location. Embedded HVDCs offer a tool for power flow control over the parallel AC system. As the power flow over a HVDC link can be controlled more easily than AC network, it is possible to estimate the optimal value of power to be transferred through the HVDC link. With the simultaneous operation of 11 HVDC systems in the Indian grid it becomes very important to identify the correct set point for each HVDC. The set point has to satisfy multiple objective functions from line flow control, voltage control, transmission loss, total transfer capability, etc. PMU

Table 17.1 List of HVDCs commissioned/expected in Indian grid

Sl. No.	Name of HVDC	Category	Supplier	DC voltage (kV)	Capacity (MW)	Year of commissioning
1	Sileru–Barsoor	Monopole	BHEL	±200	400 (not in operation)	1989
2	Vindhyachal	Back to back	GEC-Alstom	±70	2 × 250	April 1989
3	Rihand–Dadri	Bipole	ABB/BHEL	±500	1500	June 1991/Dec 1990
4	Bhadrawati	Back to back	Alstom	±205	2 × 500	Dec. 1997
5	Gazuwaka	Back to back	Alstom/ABB	±205/± 177	500 + 500	Feb. 1999/March 2005
6	Chandrapur–Padghe	Bipole	ABB	±500	1500	Nov. 1999
7	Sasaram	Back to back	Alstom	+205	500	Sep. 2002
8	Talcher–Kolar	Bipole	Siemens	±500	2000 (2500 for 10 h)	Feb. 2003
9	Balia–Bhiwadi	Bipole	Siemens	±500	2500	Sep. 2010/July 2012
10	Mundra–Mahendergarh	Bipole	Siemens	±500	2500	July 2012
11	Bheramara (Indo-Bangladesh interconnection)	Back to back	Siemens	±158	500	Sep. 2013
12	Biswanath Chariali—Alipurduar-Agra	Multi-terminal	ABB	±800	3000 (of total capacity 6000)	Nov. 2015
13	Champa–Kurukshetra	Bipole	Alstom	±800	3000	Sep. 2017
14	Raigarh–Pugalur (under commissioning)	Bipole	ABB	±800	6000	Work is under progress
15	Pugalur–Thrissur VSC	Bipole	Siemens/Sumitomo electric	±320 kV	2000	Work is under progress

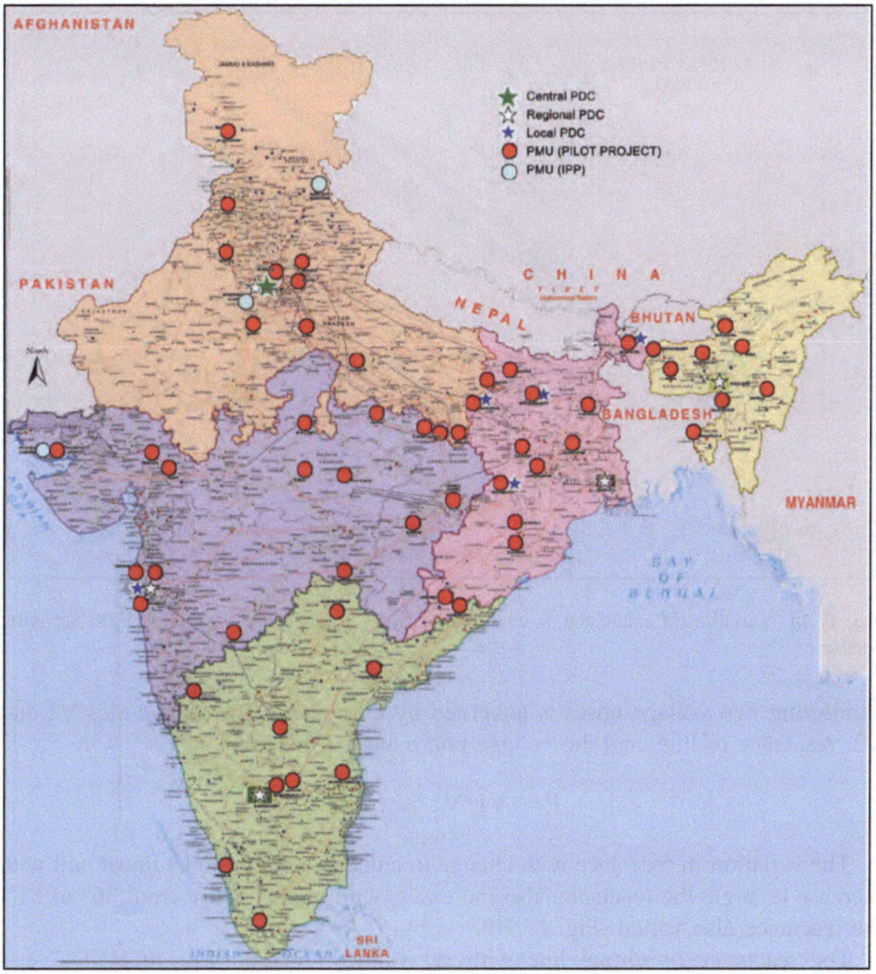

Fig. 17.9 Map showing geographical distribution of PMU installed stations

can provide the data of voltage at both the end terminals of HVDC as well as angular separation between the end terminals. As the voltage angles depend directly on the power flow to a node, the difference between two voltage angles relates to the overall power flow of all transmission paths connecting those two nodes. The calculations were done for identifying the power flow on virtual AC line which could replace the HVDC link (AC line emulation technique [1]).

A study was carried out where for a set power order on one HVDC link, the relationship of reactance with variation in angular difference was found and plotted. In this method, the control would emulate the behavior of an AC line increasing/decreasing the energy transmitted by the HVDC link as a function of the difference of the angles of both HVDC sides. The real power flow P along a transmission line

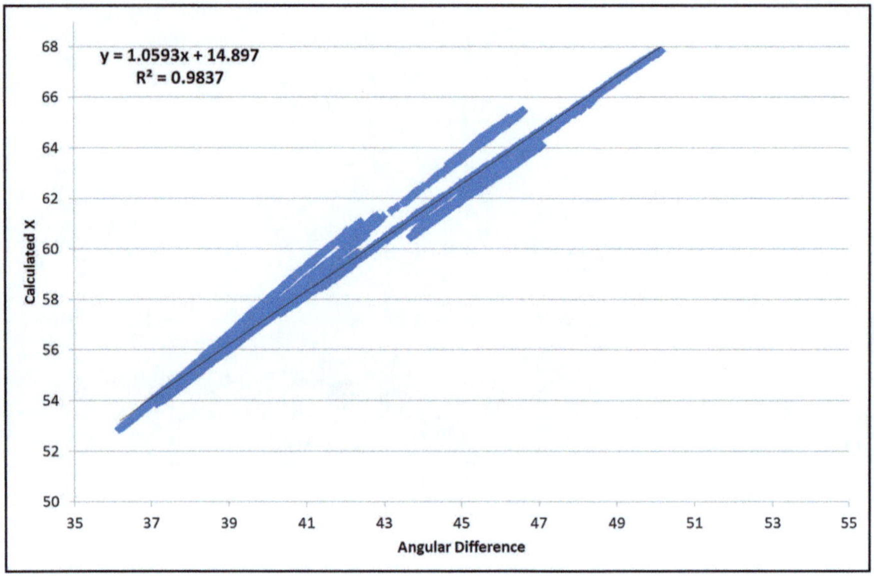

Fig. 17.10 Variation of calculated X with change in angle difference keeping HVDC set point constant

connecting two voltage buses is governed by the two voltage magnitudes V1 and V2, reactance of line, and the voltage phase angle difference, as

$$P = (V1*V2*\sin\delta)/X$$

The variation in reactance with change in angle was found to be linear and with increase in angle the reactance also increases; with angle change from 36° to 51°, the reactance also varied (Fig. 17.10).

The reactance of virtual line with set value of power flow in HVDC was identified over samples of angle difference, and the other parameters are also under study based on similar approach. The procedure is under development, and operator closely monitors the angular separation between end terminals. It helps in avoiding unnecessary wheeling of power over large AC network and overloading of network in some cases, thereby minimizing the transmission losses. While the above exercise is done for one HVDC at a time, optimizing 11 HVDC set points would be a major challenge as it would have to factor the following:

- Transfer capability between two areas
- Transmission losses on a pan India basis
- Angular spread across the country
- Voltage profile and filter bank switching required at different power order on HVD
- Cross-border HVDCs (only one at present but more in the pipeline)

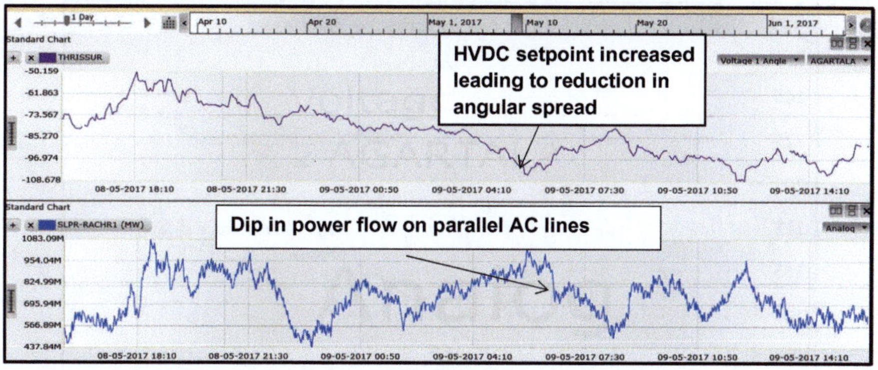

Fig. 17.11 Using angle for HVDC set point control

The behavior can be understood by observing the angular separation plot of one day of two far ends of Indian grid. The sample day chosen is of high demand period when angular difference crosses 100° in the morning peak of loads (Fig. 17.11).

The HVDC set point increase affected the angular separation and difference reduced by approximately 20°. This sudden increase in HVDC set point can be validated by corresponding decrease in power flow on parallel AC lines. The timely action by control room operator in managing the angular separation across the grid helped in safe and reliable power system operation. HVDC set point change is one of the most important tools in the hands of system operator for managing the power transfer in the grid. On an average, it is found that approximate 30 number. of HVDC set point instructions are given on a daily basis by control room operator.

Further exercises are on between the operational planning, SCADA/EMS and real-time teams to develop a real-time tool for the operator to nudge them toward a HVDC set point change factoring all the above factors. This area holds immense promise as far as the Indian power system is concerned.

Case 2: Fault detection in inter-regional HVDC link carrying power from a ultra mega power plant

Ultra-Mega power plants (UMPPs) in Indian grid refer to thermal power stations of capacity 4000 MW or above, or solar parks of capacity 500 MW and above. For reliable power evacuation from UMPPs, HVDC links are installed in Indian grid. One such HVDC caters to power transfer from an UMPP in western region to load centers in northern region. It is a ±500 kV bipole link with capacity of 1250 MW per pole. Under normal conditions, HVDC power order is kept above 2000 MW. Power order on this link is often toggled to control the flow on parallel inter-regional lines or control voltages in the AC system near HVDC terminal stations.

On one incidence, there was a Y phase to ground fault in one of the evacuating AC transmission lines from inverter end as shown in Fig. 17.12. At the same instant, two other circuits tripped in AC system. As observed from PMU, fault

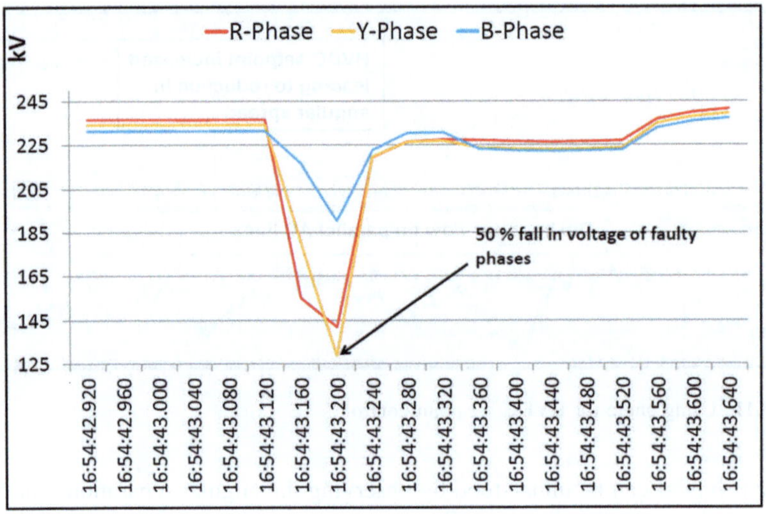

Fig. 17.12 Fault in AC system near inverter terminal

clearance time was 200 ms. Pole-2 of HVDC also tripped due to commutation failure at one end, due to AC system voltages dipping to nearly 50% of pre-fault value in faulty phases. System protection scheme designed for the pole tripping operated, and load loss due to same was captured in PMU plots (See Fig. 17.13).

From the frequency rise of nearly 0.13 Hz, the estimated load loss due to the incidence was around 800 MW, which is in accordance with the planned loss. Observations as above from PMU aid the real-time operator in cases of contingencies on HVDC link.

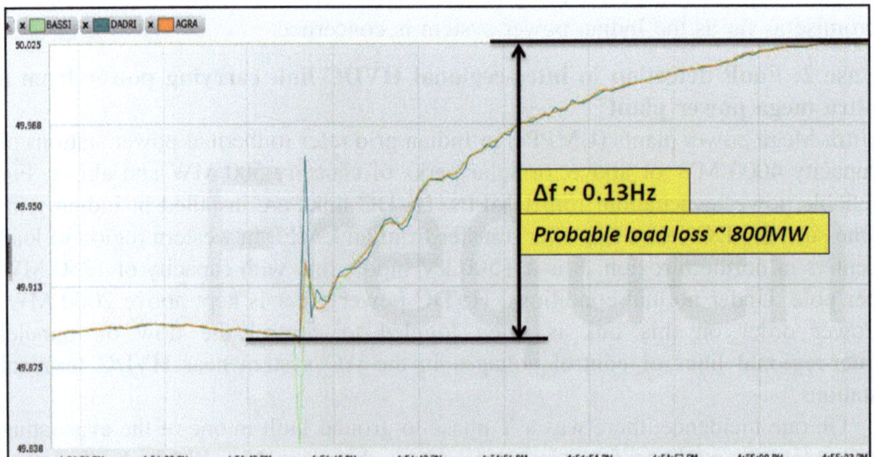

Fig. 17.13 Rise in frequency due to SPS operation

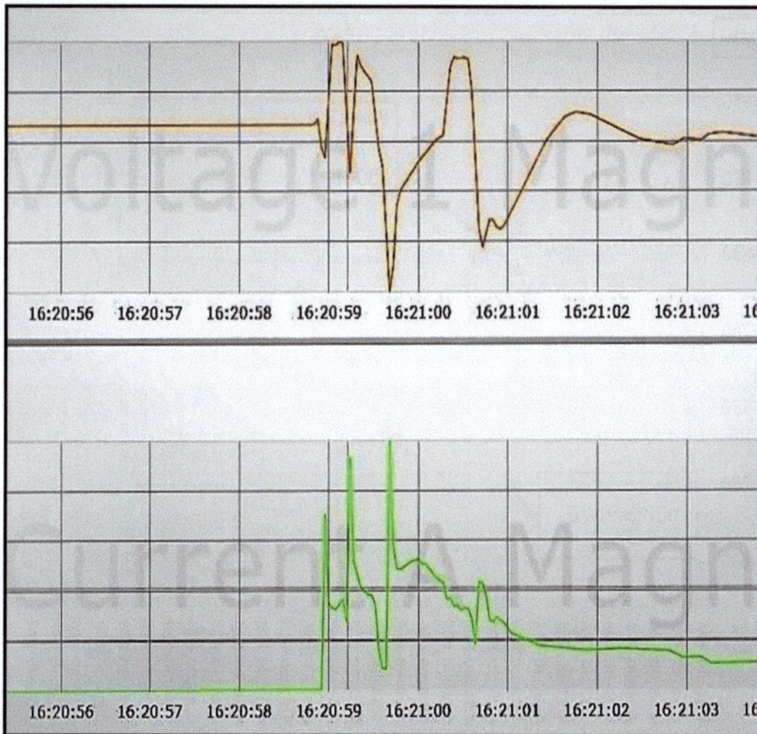

Fig. 17.14 Successive transients in HVDC link captured from PMU voltage plots of nearby AC bus

The link passes through highly polluted regions (salt as well as particulate matter in different stretches) with prevalent foggy and humid conditions and thus prone to repeated transmission line faults on certain days. To maintain power flow, the HVDC is taken in reduced voltage mode of operation (RVO). With reduced voltage mode of operation, the power flow is also less than that of normal mode of operation. On some occasions, when demand in northern region is very high, the power flow on the HVDC link assumes much importance. The transfer capability between western region and northern region is reduced during such periods and restored back with revival to normal operating voltages. During high demand periods in northern region, the HVAC lines operate around their maximum system protection schemes (SPS) threshold value, and any increase in that case would trigger SPS which causes load shedding. The control room operators need to be updated about possible RVO to take immediate actions so that parallel AC line flows remain within limit or SPS satisfying criterion is not met. On some occasions, flashover on insulators causes DC line faults which appear as transients in PMU located near UMPP station. During one such incident, HVDC went through three successive transients as in Fig. 17.14, after which the HVDC pole went to reduced voltage

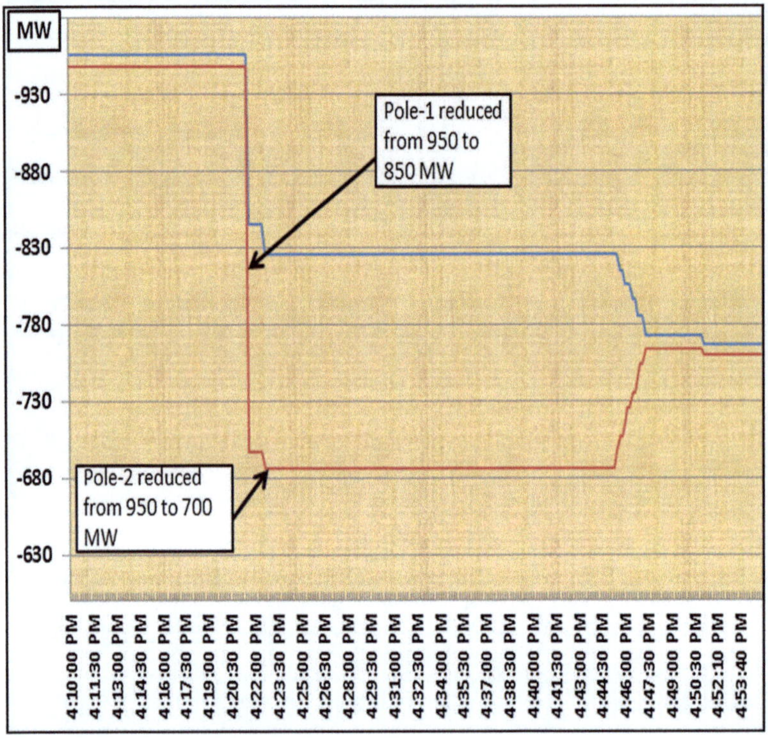

Fig. 17.15 SCADA plots showing power flow reduction with reduced voltage mode of operation

mode of operation. Any transients are captured in PMU located near converter station. From the settings, it is understood that during HVDC line faults, the HVDC will attempt fixed number of autorestarts after which it will operate in RVO mode. The PMU plots (Fig. 17.14) captured the operation and SCADA power flow plots (Fig. 17.15) confirmed the same. The continuous power transmission capability for reduced DC voltage operation with 400 kV DC voltage (80%) is 1800 MW in bipolar operation and 900 MW for monopolar operation as required. In RVO mode of operation, power order is based on a current carrying capability curve which explains the behavior of current with voltage. The power flow reduction in poles is thus proportional to the fall in voltage during RVO, and from Fig. 17.15, the bipole power order settling at different values can be seen.

The operator is thus able to take immediate advance actions after receiving the warning signs of any transient in HVDC (from the nearby PMUs) and interacting with the HVDC terminal personnel. Announcements regarding reduction in transfer capability across two regions are also made after receiving such warnings.

Case-3: Runback in HVDC link due to delayed clearance in nearby AC system fault

One ±500 kV HVDC bipole link is installed between eastern and Southern Regions of India, with rated power transfer capacity of 1000 MW per pole. Considering increased demand of power transfer from generating plants in eastern region to load centers in Southern Region, the capacity of the link was enhanced to operate at 1250 MW per pole continuously for 10 h duration. The reliability of HVDC link is important for meeting the demand of Southern Region. For the reliable operation of link, several system protection scheme criteria have also been designed. There is a scheme of runback in HVDC which is designed to reduce the power flow on link when evacuation lines from inverter end go out of service.

In one incident, there was multiple tripping of lines near one end of HVDC link along with 400/220 kV ICTs. One line tripped on operation of Main-I protection at inverter end on three-phase-ground fault while other line tripped on operation of Main-II protection with single-phase to ground fault in Zone-II (delayed fault clearance). Delayed fault clearance led to multiple tripping of lines near the HVDC link as seen in Fig. 17.16. This in turn led to runback in HVDC and power flow reduction to 75% of initial value, i.e., (2000–1500 MW), as per design.

The reduction in power flow of HVDC link led to the spontaneous increase in power flow of parallel AC lines. However, due to load loss in the incident on account of tripping of 400/220 kV interconnecting transformers (ICT) at the inverter end, the power flow toward Southern Region reduced which is evident in power flow plot of AC line given as Fig. 17.17. Damped Oscillations in power flow

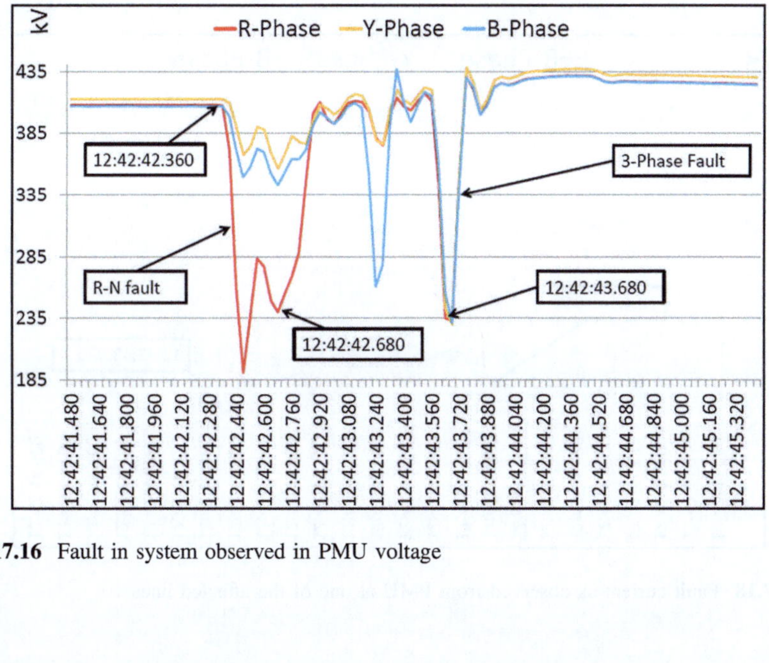

Fig. 17.16 Fault in system observed in PMU voltage

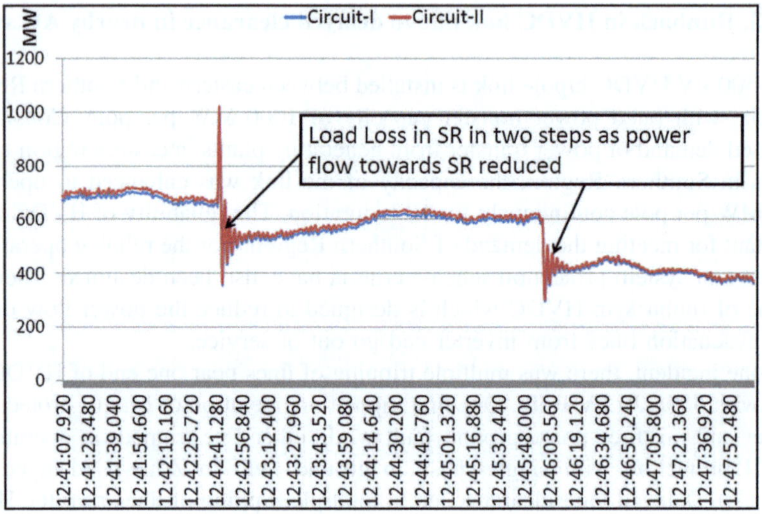

Fig. 17.17 ICT tripping has led to load loss which has led to reduction in power flow toward SR

could also be observed on the AC links, due to step reduction in loads in Southern Regional grid (Figs. 17.18 and 17.19).

Case-4: Periodic spikes in PMU due to damage in Electrode line

One ±500 kV, 1500 MW HVDC bipole link is installed in western region for transfer of power from generating plants in Eastern part of the region to load centers in western part of the region.

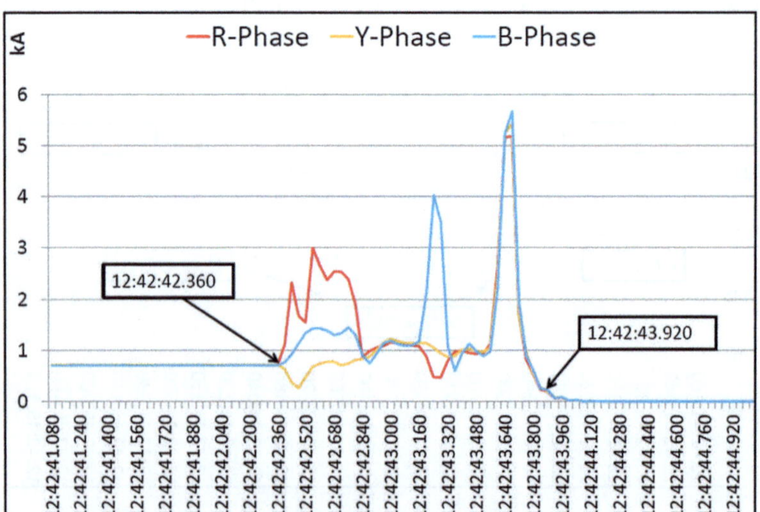

Fig. 17.18 Fault current as observed from PMU at one of the affected lines

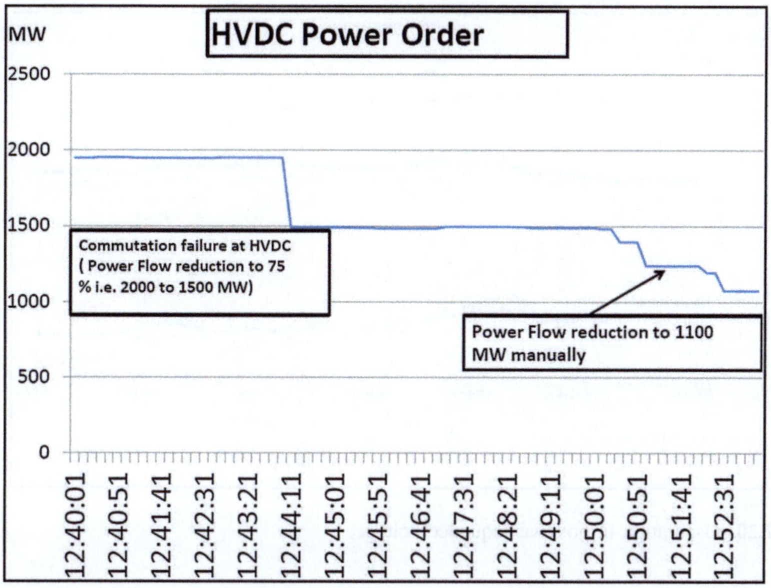

Fig. 17.19 HVDC power order reduction observed from SCADA

In one incident, spikes were observed in voltage, current, power, and frequency parameters in PMUs located at the stations across region when one of the HVDC pole out of bipoles was taken out for annual maintenance works. HVDC power flow on other healthy pole was 750 MW. The long periodicity of repetition of spikes indicated that the source of oscillations would likely be HVDC link instead of generators or generator control systems.

Reduction of power order of healthy pole to 600 MW damped out the sustained periodic oscillations. On further investigation from concerned station, it was informed that one of the electrode line conductor jumpers was damaged. In HVDC, there are basically four protections for function for electrode conductor: electrode line open circuit protection, electrode line impedance supervision, electrode cable longitudinal differential protection, electrode line unbalance supervision. On investigation, it was found that as the ground return was done for other pole after outage of one pole, the loose jumper would have partially snapped (not completely) due to heating resulting from high current flow. This led to unbalance current in both electrode conductor and unbalance in current causing overload of one of the electrode. This resulted in operation of electrode current supervision which basically checks difference between the direct current in each electrode line conductor (IDEL 1 and IDEL 2) and operates when the difference between the two measured currents exceeds a preset level the criterion for a ground fault or an open circuit on one of the conductors is fulfilled. The alarms appeared after every 2 min at one end, and spike was observed in grid during same time.

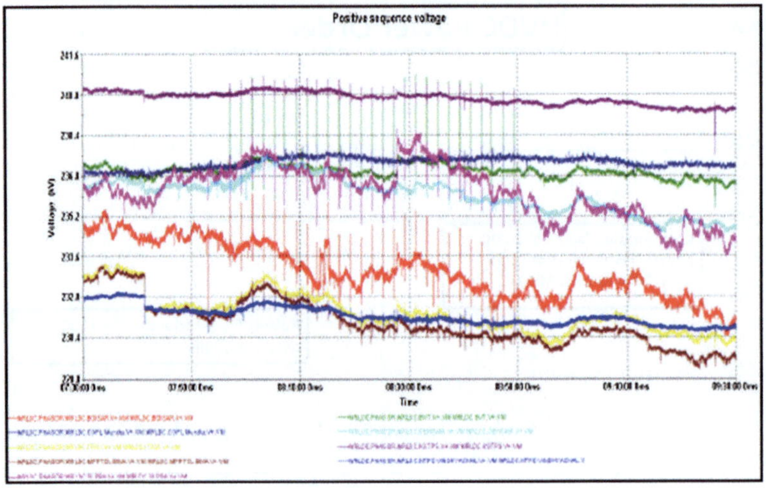

Fig. 17.20 Oscillation in positive sequence voltage

The spikes observed in voltage as shown in Figs. 17.20 and 17.21 resulted from loop which repeated continuously unless the unbalance criteria was reset with power order change instruction from control room. Lesser power order means lesser current in both the different impedance electrode lines (one was healthy and other having few strands burnt down causing unbalance impedance) which reduced the difference between these two current to reduce below 25 A.

Observations from PMUs helped the real-time operator visualize the problem with HVDC link and take appropriate action. Without the availability of synchrophasor data, operator would not have observed the incident in real-time operation.

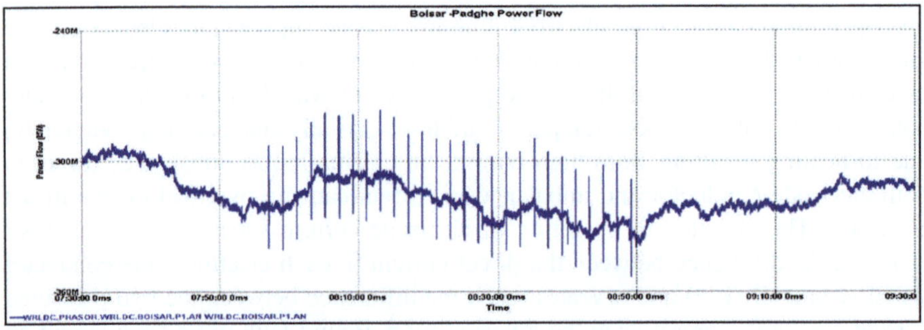

Fig. 17.21 Oscillation in line flow

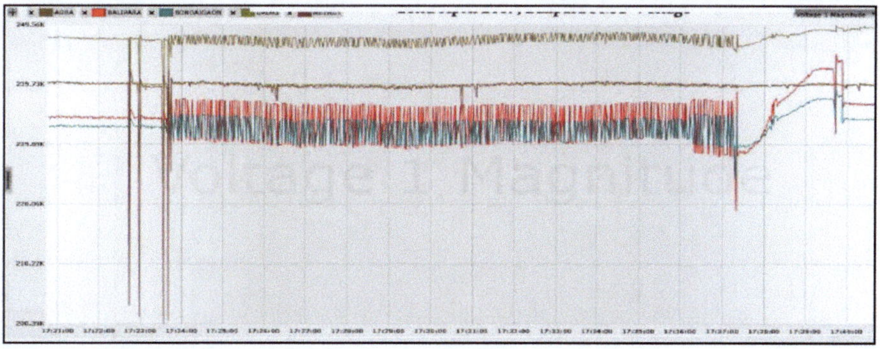

Fig. 17.22 Oscillation in voltages from nearby PMUs

Case-5: Oscillations in ±800 kV HVDC Bipole between Northeastern Region and northern region

For bulk power transfer from generating stations in northeastern region of India to load centers in northern region, a ± 800 kV multi-terminal HVDC link connecting northeastern region–eastern region–northern region with rated capacity of 6000 MW has been installed. The length of HVDC is approximately 1728 kms in total. The availability of this link has helped in regulating the power modulation in parallel AC system and tackling the problem of overvoltage in adjoining areas.

In one incident, sustained oscillations of 5 Hz frequency observed in PMUs installed near HVDC terminal stations as shown in Fig. 17.22. The maximum voltage variation observed in 400 kV nearby voltage was 12 kV (phase-phase). During the incident, the MTDC was operating in bipolar mode from northeastern region to northern region, with a power order of 500 MW and 250 MW, respectively, on Pole-1 and Pole-2.

Low-frequency electromechanical oscillations in the grid generally originate from generators or generator controls usually have frequencies ranging up to 2 Hz. The oscillations observed were of approx. 5 Hz which indicated the likelihood of involvement of HVDC controllers.. The real-time grid operator got alarmed by these system-wide oscillations and in order to prevent the oscillations from causing any threat to grid security, control room gave instructions for blocking of HVDC pole.

By investigating the PMU voltage plot as shown in Fig. 17.22, it was inferred that the voltage dip which corresponds to line fault triggered oscillations in voltage and with blocking of one pole of HVDC the oscillations damped out. From PMU plot, it is clearly visible that oscillations subsided with blocking of HVDC pole.

Case-6: Incident of Runback in ±800 kV HVDC Bipole link

On the ±800 kV HVDC Biswanath Chariali—Agra link, control system initiated runbacks were observed on two consecutive days. The power flow on the link got reduced from 500 to 137 MW itself. The reason for these runbacks was investigated by information from concerned stations and control room data. The first instance of runback was initiated by valve cooling system failure, and it was found that the

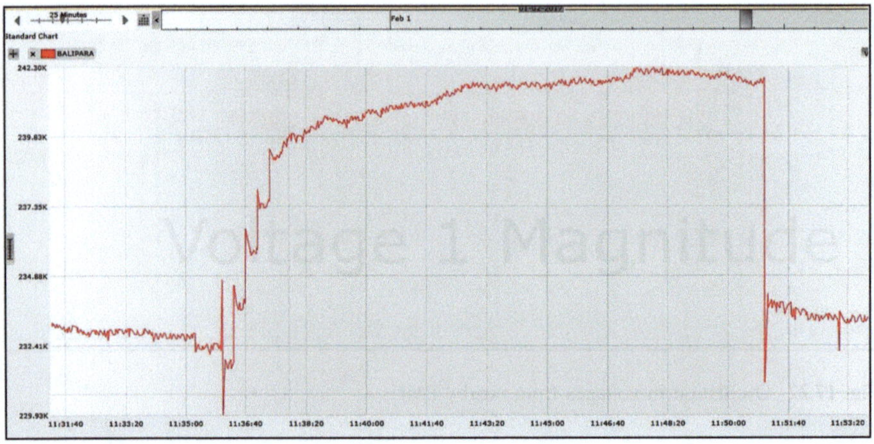

Fig. 17.23 Positive sequence voltage magnitude (first instance)

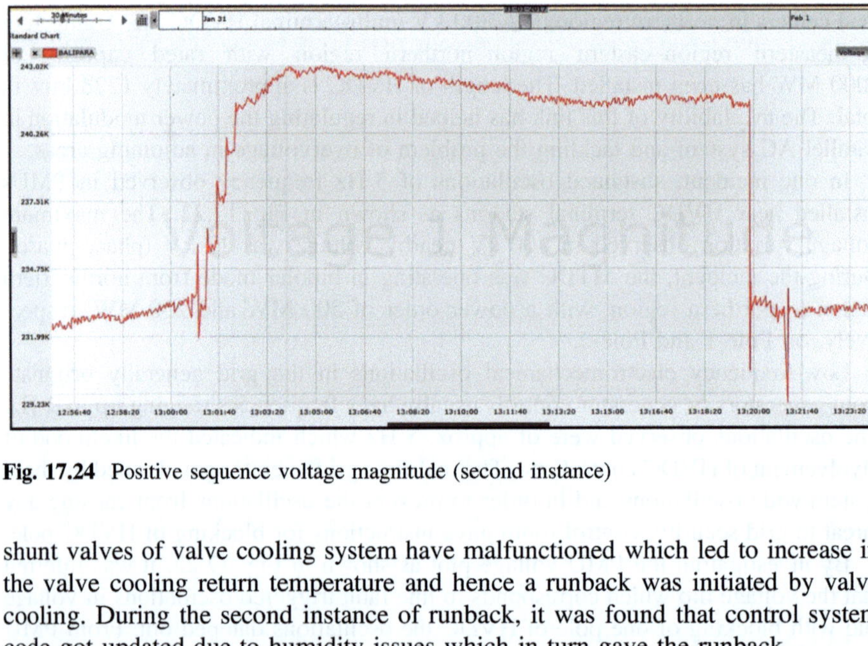

Fig. 17.24 Positive sequence voltage magnitude (second instance)

shunt valves of valve cooling system have malfunctioned which led to increase in the valve cooling return temperature and hence a runback was initiated by valve cooling. During the second instance of runback, it was found that control system code got updated due to humidity issues which in turn gave the runback.

The runbacks were visualized by the stepwise voltage rise at PMU located near one of the terminal stations as given in Figs. 17.23 and 17.24. The presence of PMU at nearby station of HVDC helped in enhanced visualization and proper feedback.

Case-7: Incident of tripping in HVDC back-to-back link
Between eastern and southern grid, for transfer of power to Southern Region, a 2 × 500 MW Back-Back HVDC link is installed. The power from eastern region is fed to southern region via two 400 kV AC lines, i.e., 400 kV Jeypore–Gazuwaka-1 and 2.

During one incident, there was major disturbance in eastern region due to tripping of several lines. The tripping of lines in different parts of adjacent network caused severe voltage issues and unstable island formation. The disturbance led to the blocking of HVDC back-to-back station, the oscillations, islanding, and voltage variations can be observed from the PMU three-phase voltage plot near HVDC station Fig. 17.25.

The fault MVA of stations near this HVDC Station is very low, and further outage of any element causes further reduction in fault MVA. The incident started with tripping of one line and further aggravated due to multiple tripping of lines in lower voltage network. Due to the above trippings, an island was formed comprising of hydro generators and associated loads in system with HVDC (under reduced power order). Heavy fluctuations in voltage and frequency were observed at adjacent bus subsequent to the island formation. The fall in voltage due to these oscillations can be observed in plot, and the voltage fell to as low as 50 kV.

In HVDC, both poles were carrying 300 MW each just prior to the island formation. However on island formation, a severe low voltage occurred and power order ramped down to counter this low voltage. HVDC Pole-2 tripped on 'DC voltage low' with other HVDC pole carrying around 300 MW power. As per event logger output from HVDC station, fast ramp down of HVDC Pole-1 occurred from

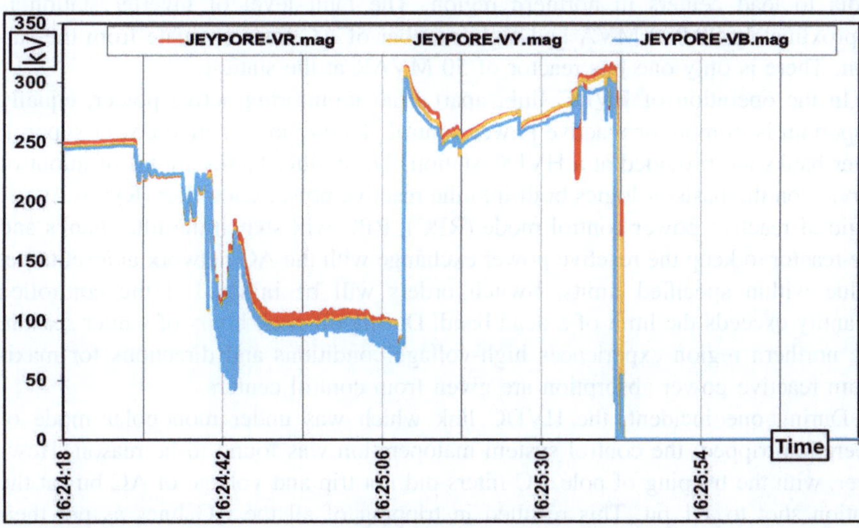

Fig. 17.25 Three-phase voltage of Jeypore station

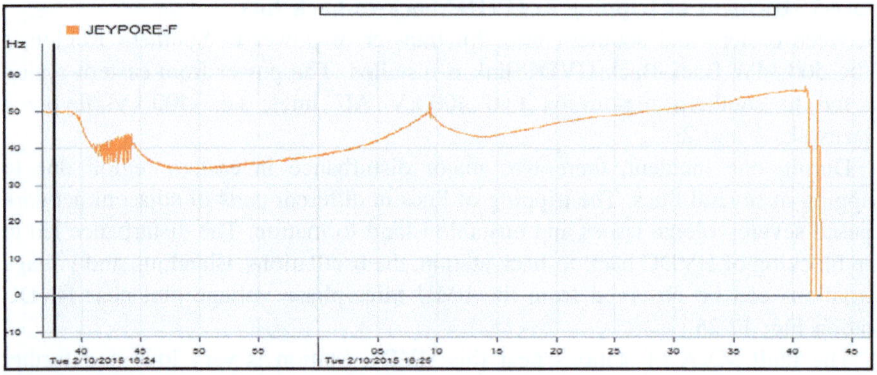

Fig. 17.26 Frequency of Jeypore station which observed wide variations during the sequence of tripping

300 to 80 MW or below. The island survived for 16 s with frequency varying from 50 to 35 Hz as observed from PMU plot of Frequency as given in Fig. 17.26.

Availability of PMUs helped pinpoint the reason for tripping and sequential islanding and tripping of loads as well as generation. It was understood that when the complete network is in place, there are no sudden overvoltages while if one leg of adjacent AC network is open, the system is prone to overvoltages on account of reduced fault levels. The matter was further investigated, and power demand override settings for the HVDC were reviewed.

Case-8: Filter bank switching operation monitoring

There is a HVDC link which is for transfer of power from generation rich Central India to load centers in northern region. The fault level of inverter station is approximately 20,000 MVA and eight number of AC lines emanate from the station. There is only one bus reactor of 50 MVAR at the station.

In the operation of HVDC link, apart from monitoring active power, equally important is to monitor reactive power control. To provide reactive power support, filter banks are provided at a HVDC station. These filter banks are taken in/out of service on the basis of logics built into the reactive power controller (RPC). As per logic of reactive power control mode (RPC), RPC will switch the filter banks and the reactor to keep the reactive power exchange with the AC network at a set target value within specified limits. Switch orders will be initiated if the controlled quantity exceeds the limit of a dead band. During off peak hours of winter season, the northern region experiences high-voltage conditions and directions for maximum reactive power absorption are given from control centers.

During one incident, the HVDC link which was under monopolar mode of operation tripped; the control system maloperation was found to be reason. However, with the tripping of pole AC filters did not trip and voltage of AC bus at the station shot to 1.1 pu. This resulted in tripping of all the AC lines as per their overvoltage settings. The PMU plots captured the sudden voltage rise as well as

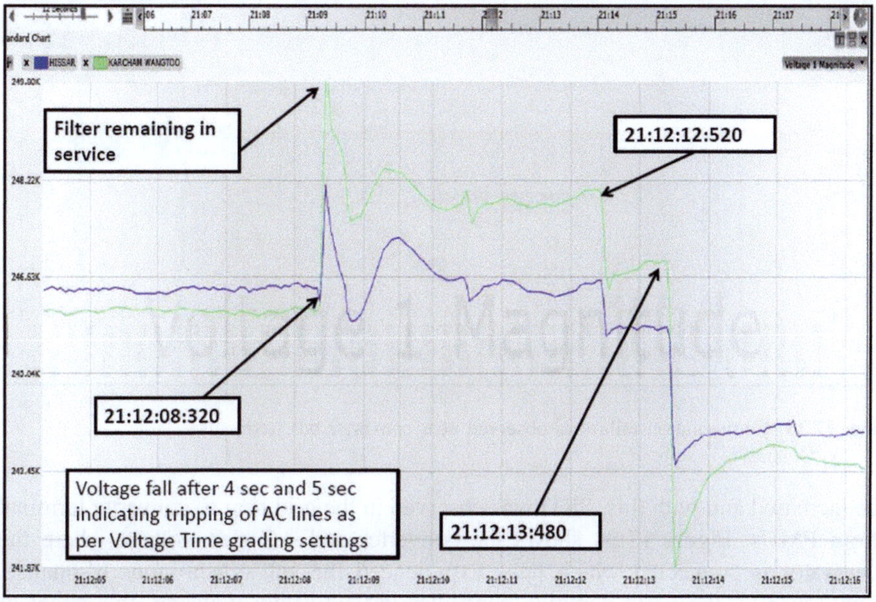

Fig. 17.27 PMU voltage plot of nearby AC bus

voltage relief when lines got tripped. PMU plot in Fig. 17.27 represents the AC bus voltage of nearby station where sudden rise can be observed which continues for 4–5 s which conforms with overvoltage time grading settings for the emanating lines.

Synchrophasor data helped operator in analyzing the situation and prepare for further actions. The monitoring of PMU helped operator in managing situations during charging/discharging of the pole. Based on the feedback from the incident, magnitude grading of overvoltage tripping was done on a few lines to ensure that a situation of complete blackout on the inverter side AC substation was avoided.

Case-9: Sub-Synchronous torsional interaction with generator near HVDC converter station

The sub-synchronous torsional interaction (SSTI) is observed when a HVDC converter terminal is located near to a generating complex. SSTI is observed on two different operating conditions, i.e., low generation at the plant and multiple commutation failure in the HVDC due to fault on the line. Two oscillation cases on sub-synchronous torsional interaction were observed in a generating plant connected at the converter terminal of HVDC system under two different scenarios. The first case was when multiple commutation failure occurred in HVDC during faults near to the inverter substation under inclement weather. These faults were reflected on the converter end of the HVDC as dip in the voltage and rise in the current. With these commutation failures, HVDC power flow was also varying and current controller was trying to keep the power constant. During these faults, voltage dip is observed and in order to keep the current same, firing angle α has to

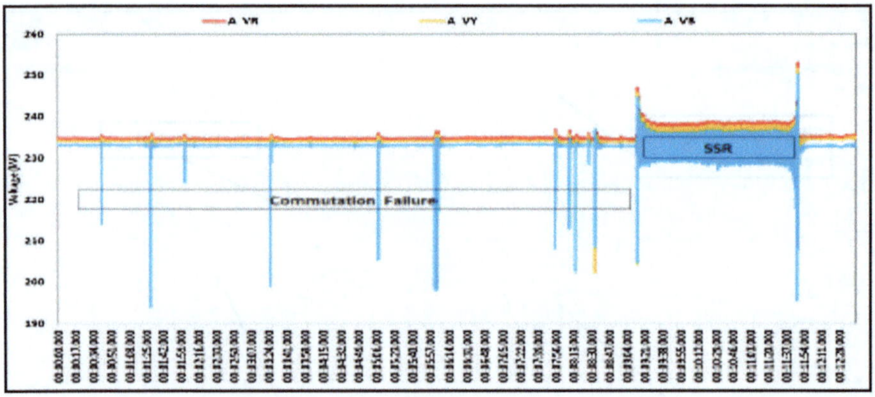

Fig. 17.28 Commutation failure as observed near converter bus terminals

be increased and with this, SSTI was observed in the grid near to converter terminal from PMUs. Figure 17.28 shows the converter end AC side voltage where the generator is connected which have experienced the sub-synchronous resonance. Subsequently, the SSTI damping controller got activated; however, the extent of SSTI was high even after damping controller activation resulting in tripping of both poles and subsequent SPS operation leading to the tripping of some of the units.

Whether PMU can measure SSTI or SSR oscillation is a matter of discussion. PMUs in India, which are reporting at 25 Samples per second, can measure the oscillation up to 12.5 Hz as per the Nyquist criteria. Any higher frequency beyond 12.5 Hz will be reported as 25-Fo in the Synchrophasor due to aliasing. So a SSR having frequency of 15 Hz will be observed in PMU as 10 Hz.

Case-10: Observing impact of intermittency in communication between two ends of HVDC link

There is one HVDC link of 1500 MW within northern region for power transfer from generating stations to load centers over bipole. The blocking of one pole will lead to part overloading of remaining healthy pole as per design. For successful shifting of power post blocking of one pole, it is important that communication between both the terminal stations is healthy. During an incident, one pole was manually tripped due to fire alarm in valve hall. Immediately before the tripping, the bipole was having a power order of 1225 MW. Immediately after the tripping of Pole-1, the loading in HVDC Pole-2 got increased to 1000 MW (5 s overload capacity of 33%) and then got settled to 10% overload capacity, i.e., 825 MW. The blocking of pole and initial shift of power are given as Fig. 17.29. Immediately after that, an anomalous power flow was observed in HVDC power flow, and a sample of the same is as shown in Fig. 17.30. Initially, it was suspected that there is some kind of oscillation in the system but when the modal analysis of the frequency plot was checked, there was no oscillation found. Moreover, none of the machines

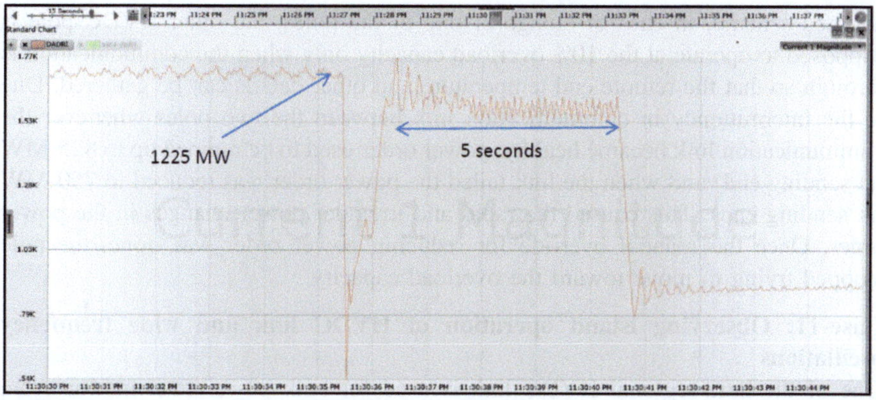

Fig. 17.29 PMU plot of AC current flow in one of the two bus sectionalizers at Dadri end

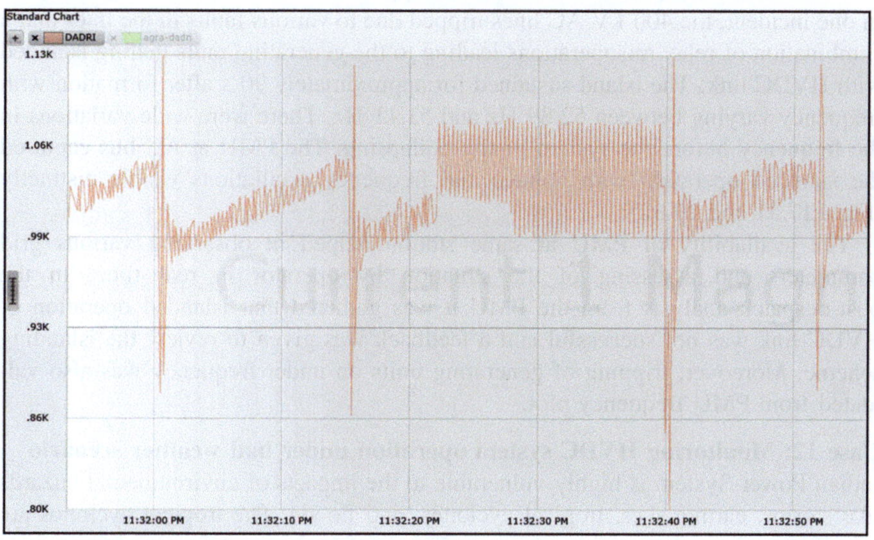

Fig. 17.30 Anomalous pattern observed in current flow

near terminal stations too had reported any oscillations. Also, the pattern of dips and rises was also not cyclic so as to warrant being called an oscillation.

We can clearly make out that the variation is between two set of values (as measured at converter end). The variation is between 1.1 kA at one end and 0.92 kA at the other end. It seems that the current flow is trying to increase to around 1.1 kA value but is brought down in a sudden manner by some other factor. The fluctuations were finally stopped by reducing the power order manually to 750 MW. As per telephonic discussion with HVDC site, it was gathered that there was intermittency in telecommunication between the two ends of the HVDC link. As per the adopted scheme, in case of outage of one of the poles, the other HVDC pole is supposed to operate at the 10% overload capacity only when the communication is through so that the remote end temperature and other details can be gathered. Due to the intermittency in communication link between the two poles whenever the communication link became healthy, power order used to be ramped up to 825 MW (at sending end) and when the link failed the power order was reduced to 750 MW (at sending end). This caused the rapid and irregular power changes in the power order. Once the manual override for reducing power order was done, the pole stopped trying to move toward the overload capacity.

Case-11: Observing island operation of HVDC link and wide frequency oscillations

One of the intra-regional HVDC link is used for bulk power transmission from coal-based thermal power station in southeast part of northern region to load centers in central part of northern India. There are four numbers of 400 kV AC lines and a 1500 MW HVDC bipole for evacuation of approximately 2500 MW of generation. In one incident, the 400 kV AC lines tripped due to various faults in the lines and a combination of relay misoperations leading to the generating units getting islanded with HVDC link. The island sustained for approximately 90 s after formation with frequency varying between 52.89 Hz and 51.11 Hz. There were wide variations in the frequency before the system finally collapsing. The PMU at AC bus captured the islanded operation with voltage and frequency oscillations visible distinctly (Figs. 17.31 and 17.32).

The availability of PMU at same station helped in observing various grid parameters and assessing of the situation by operator in real time. In the post-despatch analysis from the PMU it was gathered that islanded operation of HVDC link was not successful and a feedback was given to review the islanding scheme. Moreover, tripping of generating units on under frequency was also validated from PMU frequency plot.

Case 12: Monitoring HVDC system operation under bad weather scenario

Indian Power System is highly vulnerable to the impacts of environmental hazards like storms, earthquakes, tropical cyclones, and floods. The tropical cyclones are classified as those that originate between Tropic of Cancer and Tropic of Capricorn, and the wind speeds usually exceed 62 kmph. Cyclones occur mainly in the months of May–June and October–November, with primary peak in November and secondary peak in May. The Indian subcontinent mainland with a coastline of 5422.6 kms faces nearly 10% of the world's tropical cyclones. Although there are nine coastal States and two Union Territories in India that are affected by cyclones, the States most prone to cyclone are Andhra Pradesh, Odisha, Tamil Nadu, West Bengal, and Puducherry on East Coast and Gujarat on West Coast. Out of the cyclones that develop in the Bay of Bengal, roughly 58% approach and cross the East Coast in October and November. Only 25% of the cyclones that develop over the Arabian Sea approach the West Coast. In the pre-monsoon season, corresponding figures are 25% over Arabian Sea and 30% over Bay of Bengal.

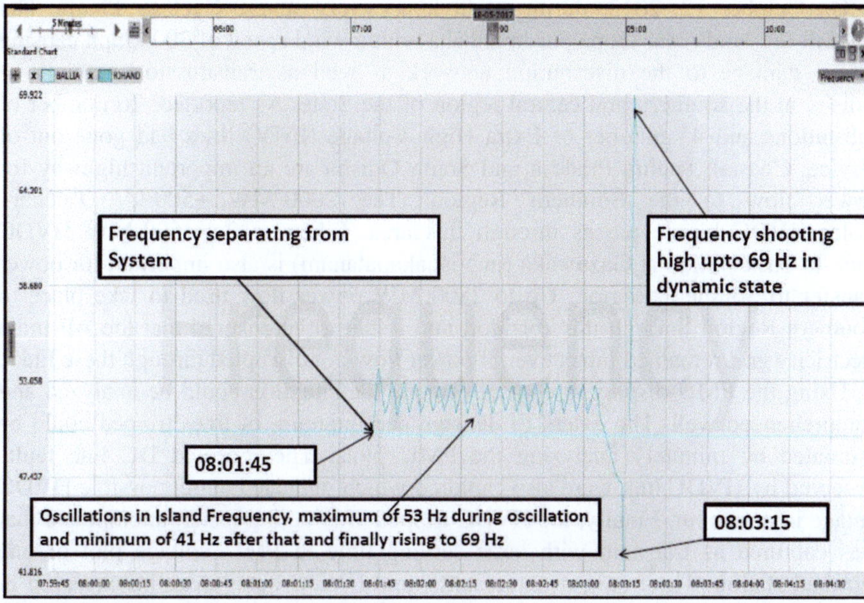

Fig. 17.31 Islanded operation of HVDC with frequency of AC bus separating and oscillating

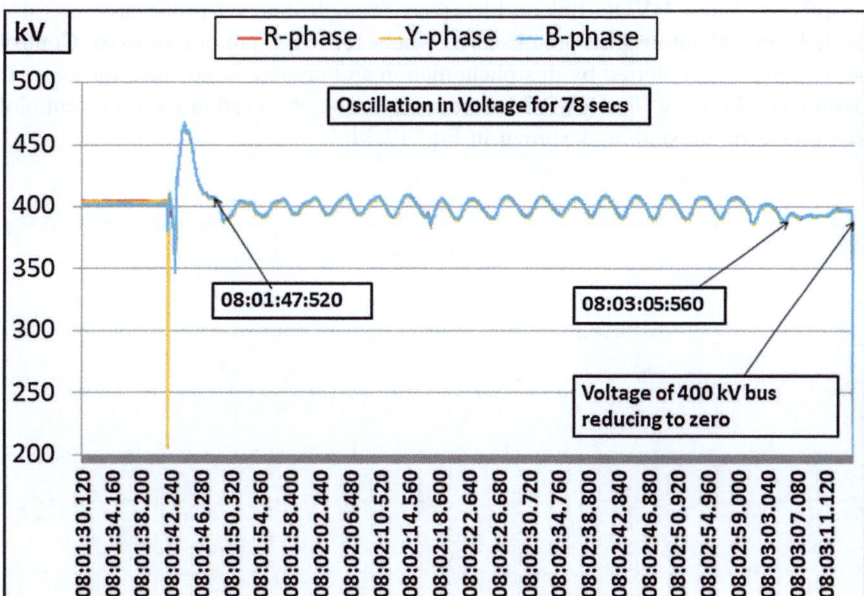

Fig. 17.32 Islanded operation with AC bus voltage oscillation and collapse

On October 12, 2013, during night hours, very severe Cyclone Phailin had landfall (hit land) near Gopalpur in Odisha with a wind speed of 200 kmph causing severe damage to the distribution network as well as transmission network of Odisha in the southern and central region of the State. As reported, 26 number of substations and 45 number of Extra High Voltage (EHV) lines had gone out of service. Coastal Andhra Pradesh and South Odisha are an important highway for power flow to the Southern Region. The 2000 MW, ±500 kV Talcher–Kolar HVDC bipole passes through this area. Likewise 2 × 500 MW HVDC back-to-back station at Gazuwaka (in Vishakapatanam) is also important for power transfer to Southern Region. Up to 2800 MW power flow used to take place to Southern Region through this corridor, and it had to be ensured that the All India electricity grid remained intact even if power flow got disrupted through these links.

Using the PMU display at control centers, the situation could be analyzed and comprehended well. The extent of damage and quantum of lines tripped could be estimated by minutely analyzing the PMU plots. The repeated DC line faults observed by HVDC link were also captured which indicated about possible HVDC outage in next hour. Finally, the HVDC tripped after taking restart attempt and that was captured as transient with large settling time in PMU voltage plot of one terminal station. The frequent autorestarts, and HVDC outage can be observed in Fig. 17.33. As the system operators had already reduced flow on the HVDC to minimum levels, its complete outage did not result in any impact on the system.

During winter season, northern region of India experiences severe smoggy conditions. Due to this, insulator flashover takes place which causes tripping of transmission lines. HVDC links which pass through this fog-prone area are also affected; several autorestarts are observed whenever foggy conditions exist. Control room operators get alerted by this phenomenon and prepare in advance for the next possible contingency. The HVDC autorestarts can be observed in PMU current plot from one terminal station as shown in Fig. 17.34.

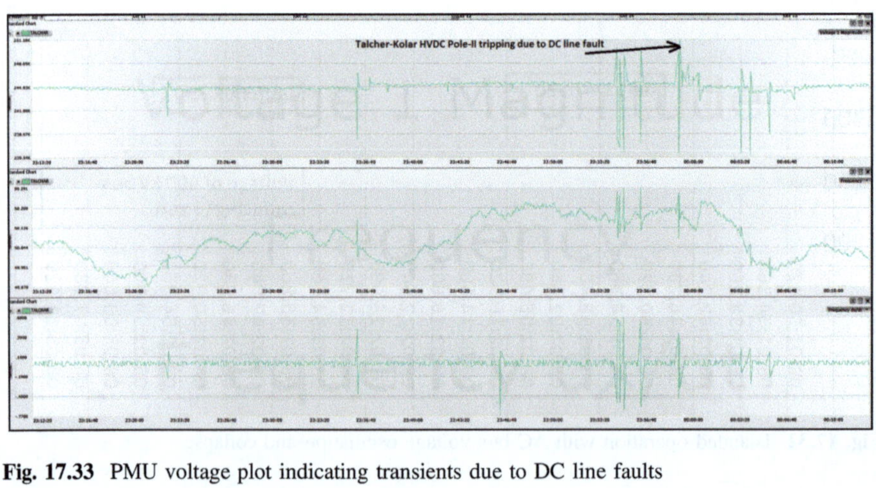

Fig. 17.33 PMU voltage plot indicating transients due to DC line faults

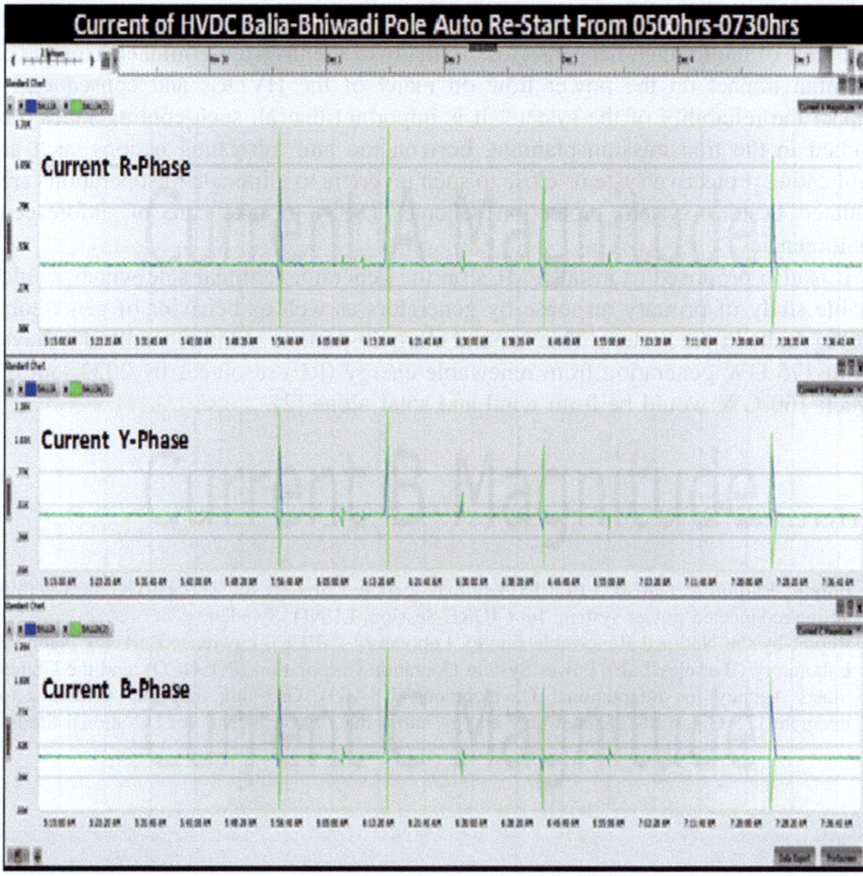

Fig. 17.34 HVDC autorestarts during foggy weather captured by PMU current plot

17.2 Conclusions/Way Forward

As illustrated in the previous sections, availability of PMUs has greatly enhanced the visualization at the control centers and significantly helped monitoring of HVDC links which are in parallel with AC links. It has helped the operator in taking steps in time for secure system operation. Apart from improving real-time monitoring, the synchrophasor technology would also help in optimizing the 11 HVDC set points in the system as well as the cross-border HVDC links. This is an area which holds immense promise. Various controllers available in HVDCs such as frequency controllers, power oscillation damping (POD), runback features, and SSTI become important, and their performance under various system conditions can be tracked through synchrophasors and feedback given to the HVDC terminals for any change in settings.

One of the concerns with 11 HVDC systems is the possibility of a delayed clearance of fault which happens close to many of the inverter terminals. This has a potential impact on the power flow on many of the HVDCs and consequently impact the reliability of the system. It is important that all such contingencies are studied in the transmission planning horizon too and mitigating actions such as duplicating protective systems close to such inverters to eliminate misoperations are initiated besides system protection schemes (SPS) to take care of unforeseen contingencies.

It is also proposed to install PMUs on the generator terminal side which would enable study of primary response by generators as well as behavior of generators during faults in the system. This is particularly important as India is poised to have about 175 GW generation from renewable energy (RE) resources by 2022, out of which 160 GW would be from wind and solar alone [2].

References

1. Frey K, Rudion K, Christian J (2015) Optimal operation strategy for VSC HVDC links within an interconnected power system. In: CIGRE Session, LUND, Sweden
2. Report by the National Renewable Energy Laboratory (NREL), Lawrence Berkeley National Laboratory (Berkeley Lab), Power System Operation Corporation (POSOCO), and the United States Agency for International Development (USAID). Greening the grid: pathways to integrate 175 Gigawatts of renewable energy into India's electric grid, vol I, National Study

Chapter 18
Model Validation Using Synchrophasor Technology

Brian Thomas and Slaven Kincic

18.1 Introduction

Modeling power systems assets accurately in online and off-line tools and applications are very important to reliably operate and plan power systems. The accuracy of the power system models used in power systems planning and operations can only be ascertained by performing frequent model validation studies. Power systems planning studies are performed using power system models to determine new infrastructure and capital investments based upon economic and reliability benefits. In operation arena, power system studies are performed using power system models to determine System Operation Limits (SOLs) and thereby calculate Available Transmission Capacity (ATC). Once SOLs are determined, power system is constrained to operate within these limits for N-1 and credible N-2 contingencies. In this regard, SOLs account for reliability and transmission capability of the power system. More transmission capacity is available if the SOLs are higher.

Accuracy of power system studies and, consequently reliability of the power system and system limits, depends on the quality of the power system models (steady state and dynamic) as well as on the quality of study assumptions. In operations, SOLs are typically determined based on seasonal studies using assumed forecasted conditions. These seasonal studies are only performed few times in a year. The SOLs based upon seasonal studies are then fine-tuned based on operating conditions. There are large number of factors used in study assumptions for forecasted future conditions. Some of these factors are load forecast, cost of the fuel, forced outages, weather conditions which have uncertainty associated with it.

B. Thomas (✉)
GE Energy Consulting, Schenectady, NY, USA
e-mail: brian.thomas.iit@gmail.com

S. Kincic
Peak Reliability, Vancouver, WA, USA

S. Nuthalapati (ed.), *Power System Grid Operation Using Synchrophasor Technology*, Power Electronics and Power Systems,
https://doi.org/10.1007/978-3-319-89378-5_18

The uncertainty associated with some of these factors increases for long-term forecasts. Large penetration of the renewables, especially wind, further increases these uncertainties. For this reason, it is beneficial to calculate SOLs in real time using real-time (online) power system model. Similarly planning studies are performed using base cases for the entire interconnection. The base case consists of power-flow and dynamics models of power system equipment in the interconnection. These base cases are assembled by various utilities in the interconnection and reflect near-term and long-term forecasted conditions including system upgrades. In Western Interconnection of United States of America (USA), Western Electricity Coordinating Council (WECC) assembles and publishes dozens of planning cases each year for all their members in the Western Interconnection. The planning studies are conducted by utilities to make sure that power system can be operated reliability without any violations for forecasted conditions. The power system models used in these planning studies reflect the physical power system asset in the field.

In this chapter, authors discuss different ways of validating power system models used in planning or operations using field data such as PMU measurements. The process of validating power system models would be referred to as model validation in the rest of this chapter. Power system model would include power-flow and dynamics model of equipment used in traditional transient and voltage stability studies. There are two primary methods of model validation:

- Validating power systems models of individual or select power plants (Power plant model validation);
- Validating all the power system models used in the entire interconnection wide case (System wide model validation).

This chapter provides brief overview of both the above approaches.

18.2 Plant Model Validation

A given power plant might have single or multiple generators based upon the configuration of the power plant. Dynamic models of generators in the power plant are intended to mathematically characterize the behavior of the power plant under different system conditions. Each generator in a power plant might have dynamic model of generator, excitation system, turbine-governor system and PSS. To ensure that the dynamic models of generators in power plants in planning and operation studies represent the real system accurately, they need to be periodically benchmarked against measured quantities and operational practices of the power system.

In North America, reliability entities require the dynamic models of generators, generator excitation system or plant volt/var control functions and turbine/governor and load control or active power/frequency control systems to be validated/verified once every 5 or 10 years depending upon the size of the generators in the three major interconnections [1, 2]. This validation/verification helps reflect any updates

Table 18.1 Criteria to determine the generators required to be tested in the major interconnections in United States

Interconnection	Individual generating unit (gross nameplate rating) (MVA)	Individual generating plant consisting of multiple generating units that are directly connected at a common BES bus with total generation (MVA)
Eastern or Quebec	>100	>100
Western	>75	>75
ERCOT	>50	>75

in the power plant such as generator upgrades or use of more accurate dynamic models. Verification/validation is done by conducting series of stage tests to determine different set of parameters [3]. Table 18.1 shows the criteria used to determine the generators required to be tested in the three major interconnections in USA.

Some recent reliability standards such as NERC MOD-033 [4] in North America require dynamic models to be validated against measured field quantities such as PMUs once every two years or even more frequently to verify the accuracy of the planning models. Disturbances of various types occur very frequently in the power grid. The installed base of PMUs has been growing in the past few years. Some interconnections require that generator owners add PMUs at the point of interconnection or at the generator terminals for the new power plants. Regular disturbances along with availability of PMUs provide the opportunity to use these disturbance events along with the PMUs to periodically validate power plant models used in transient stability studies and complement the stage testing.

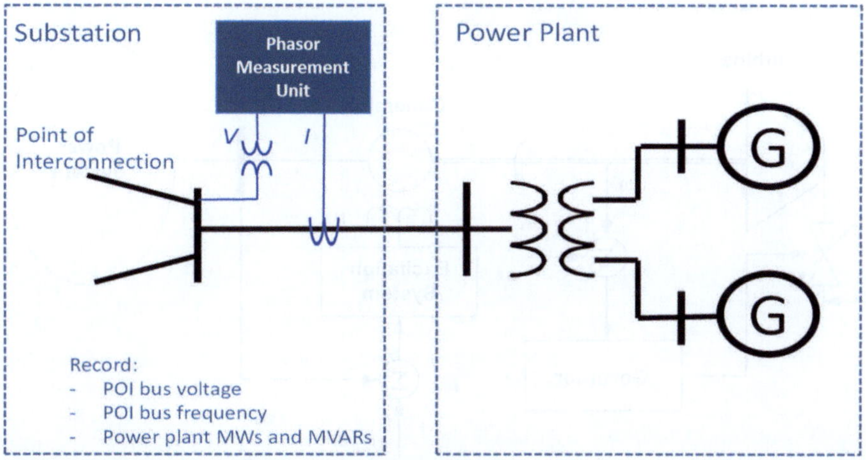

Fig. 18.1 Positioning of PMU for plant model validation

PMUs used for generator validation are generally located at the Point of Interconnection of the power plant as shown in Fig. 18.1 or after the high side of the generator step-up transformer.

Power plant model validation is traditionally done using commercial software packages such as GE PSLF or Siemens PSS/E by injecting (playing in) voltage, frequency, active and reactive power from PMU. This process of playing-in field measurements to drive the simulation is known as playback feature. GE PSLF is one of the most widely used commercial software in the Western Interconnection of US. GE PSLF provides advanced playback capability such as allowing users to play-in field voltage (E_{fd}), turbine mechanical power (P_m), voltage reference for excitation system (V_{ref}), and governor reference for turbine controls (ω_{ref}) as shown in Fig. 18.2 [5–7].

The typical procedure for power plant model validation is described below.

1. Start with any interconnection wide planning case or operations EMS case.
2. Extract the power plant subsystem data from the point of interconnection (POI) or wherever the PMU is located up to the power plant.
3. Load the PMU data.
4. Adjust the various power systems setpoint data and solve the case till the boundary conditions (flow and voltage conditions at the point of interconnection) are met. This is an iterative process.
5. Create the playback simulation setup in PSLF to play in the PMU data and drive the simulation.
6. Compare the output of the simulation with PMU data to check the accuracy of the models.

All the above steps for power plant model validation are completely automated by General Electric in their PSLF software as part of a collaboration effort between

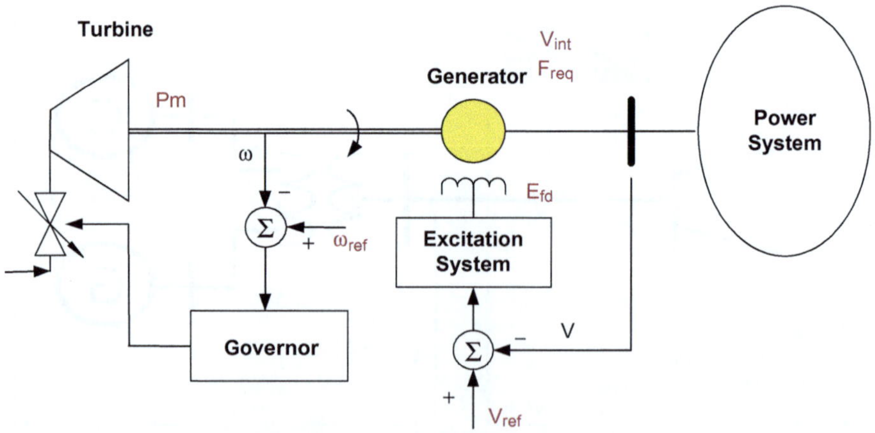

Fig. 18.2 Power plant model validation using playback feature

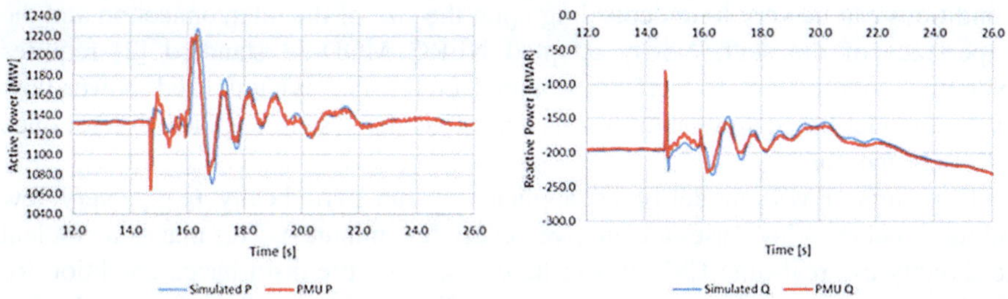

Fig. 18.3 Active and reactive power (PMU recording vs. simulation) for play-in frequency and voltage

Bonneville Power Authority, Peak Reliability, and General Electric for power plant model validation [7].

Figure 18.3 shows the power plant model validation summary for a large nuclear power plant steam generator. In this case, PMU recording of frequency and voltage on POI is played into the simulation while calculated active and reactive power-flow (P and Q) from the simulation are compared to the measured power-flow from the PMU recording.

In this case, the power plant model validation indicates that the simulation response from model does not match well with measured PMU data, and the parameters of the dynamic models can be calibrated using commercial software tools such as GE PSLF or EPRI's PPPD. The commercial software tools use different optimization algorithm such as curve fitting techniques, particle swarm optimization techniques, Kalman-based optimization techniques, or proprietary optimization algorithm and techniques [8–12].

18.2.1 System Model Validation

In system model validation, all the power systems models in the entire interconnection wide case are used and important quantities such as major path flows, voltage and frequency in the region of interest are validated. This is different than power plant model validation where only dynamic models in the power plants are used and validated. System model validation poses its own challenges due to large number of components and their interaction. Although PMU-based plant model validation tools have been successfully developed and used [8–12], system model validation remains challenging, mainly due to time required to adjust entire power-flow system model to match pre-disturbance operating conditions. Traditionally, in system model validation, a planning (off-line) power-flow model is adjusted based upon pre-event operations snapshot case, obtained from Energy Management System (EMS), to reflect pre-disturbance conditions. This is done to correctly initialize dynamic simulation from pre-disturbance conditions. The time required to adjust off-line steady-state (power-flow) model to pre-disturbance conditions can be very long depending upon the size of the interconnection and the experience of the staff. Newly adopted NERC MOD-33 standard [2] requires planning coordinators to perform comparison of the dynamic and power-flow system model-based simulation results against actual system response at least once every 24 months.

Real-time (EMS) model-based applications run periodically (e.g., every few minutes and real-time cases are archived every few minutes). This makes it efficient to directly use real-time EMS power-flow case from pre-disturbance condition for model validation, as the pre-disturbance conditions would reflect the actual operating conditions just before the disturbance. This eliminates the need to manually adjust the power-flow case.

The real-time model representing Western Interconnection is named West-wide System Model (WSM). This is node-breaker model that is adjusted from real-time measurements automatically every 10 s. Model is adjusted using more than 120,000 SCADA measurement representing breaker statuses, voltages, active and reactive power-flows. Because of this, WSM represents real-time operating conditions. Such adjusted model is used by Reliability Coordinators such as Peak Reliability for real-time and near real-time assessment. Continuous validation of real-time model leads to higher confidence in simulation results allowing real-time calculation of system limits. Consequently, use of real-time model allows to minimize uncertainties and to unlock additional transmission capacity and enhance reliability. For this reason, it is crucial to validate real-time model for various operating conditions. Currently, Western Reliability Coordinators (Peak Reliability) receive inbound 4000 **PMU** signals of nearly 400 PMU from **18** WECC utilities. The data obtained from these PMUs is used for model validation.

This section explains how synchrophasor data is used to validate the system model and help to enhance the dynamic transfer capability of the grid in WECC.

There are two primary objectives of System Model Validation in Western Interconnection:

- To make sure that there are no significant modeling errors that affects system study results (for both steady-state and dynamic model);
- To come up with the study error margin for WECC major transmission corridors under various levels of system stress.

System Model Validation is performed by comparing results of simulation to PMUs measurements for different types of system disturbances. It can include frequency and non-frequency events. At Peak Reliability (Reliability Coordinator), the state estimator case is archived every 5 min. This makes it possible to retrieve pre-disturbance archived EMS case for any disturbance. This case is taken "as it is" and no adjustments are performed within the power-flow unless any switching has happened between the moment the case was archived and the moment when the event occurred. The EMS case is in full topology node-breaker format.

The framework of system model validation is established by directly linking operations and planning models using EMS labels instead of bus numbers [13].

The reason for matching two models is due to the fact that dynamic model database in WECC is created for planning models and not for real-time model. EMS labels, which uniquely identify power system assets in full topology model, are stored as long identifiers (long ids) in the planning models. The EMS labels are standardized by Peak RC and WECC and are guaranteed to be unique in the western interconnection as they are formed by combination of different identifiers like substation name, base voltage, and device label. These long ids can be used as unique power system asset identifier in planning case in addition to the traditional bus identifiers. The presence of EMS labels for power system assets in the two different models makes it possible to link any planning model directly to WSM and vice versa. This in-turn makes it possible to use any WECC planning dynamics database with any operations WSM case for model validation or to perform disturbance analysis.

A one-time mapping effort is required to map existing power system assets like generators between planning and operations models. The mapping effort is focused on mapping each generator in the WSM model (based upon the substation, node, device labels and EMS labels) to the planning case (based upon the bus identifier and generator id) as shown in Fig. 18.4. The result of the mapping includes EMS labels and is stored in a centralized data base or file.

A mapping automation tool has been developed which makes it easy for WECC and Reliability Coordinators to seamlessly update the mapping file for any change in operations model and/or planning base cases. The tool facilitates incremental updates or complete update of mapping file as well as validation of existing mapping based upon WSM model and planning base case. The tool exports the mapping data which can be imported in PSLF to populate the long ids. This process makes it efficient and easy to automatically populate EMS labels (long ids) of all generators in planning case from operations model with minimum support from system planners and WECC members.

New tools, methods, and program features have been developed in PSLF which enable WECC members to read any WSM model and link it to any WECC planning master dynamics database using long ids from the mapping file [14]. This architecture makes it possible to validate WECC planning dynamics model using WSM disturbance cases by running transient stability simulations in PSLF as shown in Fig. 18.5. The results of simulation are then compared to PMU measurements as shown in Fig. 18.6 to validate system response based upon dynamic models used in WECC. This eliminates the need to adjust any power-flow case and decreases the time for perform system model validation from man-months to less than a day.

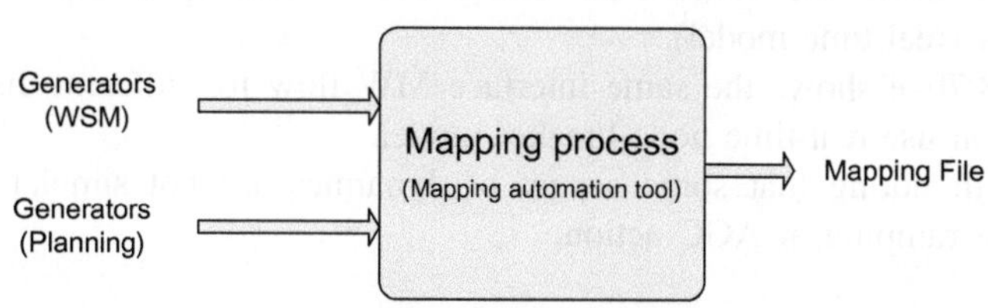

Fig. 18.4 Mapping process

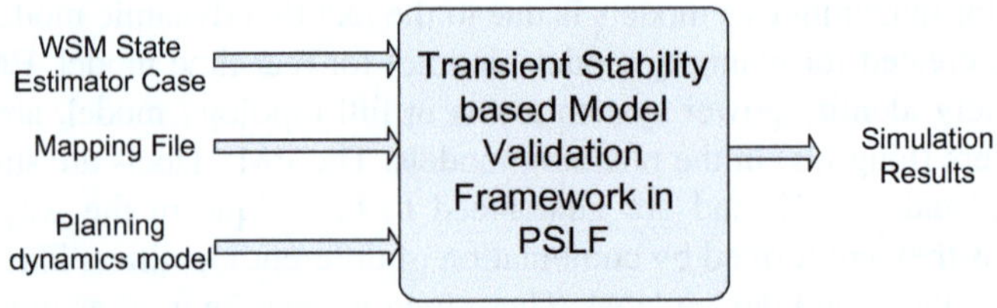

Fig. 18.5 Transient stability based model validation framework in PSLF

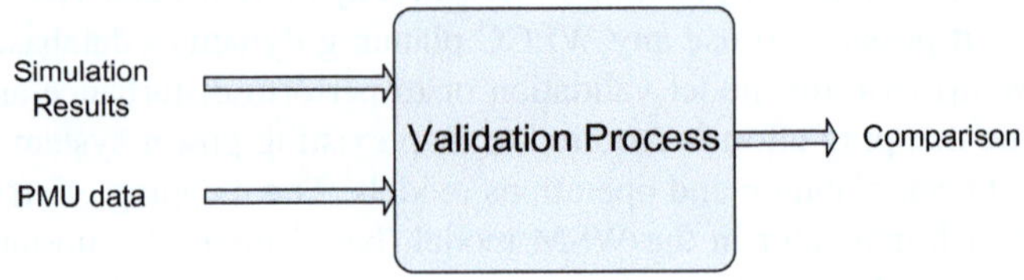

Fig. 18.6 Model validation (simulation vs. measured data)

The success of this approach for model validation depends upon properly aligning the operations and planning power system models and reconciling the modeling differences in addition to accurately mapping the power system assets.

The next section below shows some of the System Model Validation results.

18.2.2 Benchmarking System Interface

In calculating system limits, it is important to be aware of expected simulation errors. If error is not known, then that limits are set very conservatively blocking available transmission capacity. Performing frequent system model validation for given transmission interface for different system operating condition will enable to set limits more accurately. Figure 18.7a–e illustrates results of one such exercise for different operation patterns using different system disturbances for the same interface in Western Interconnection. Figure 18.7a shows MW flow over interface for disturbance in Pacific Northwest and consequent RAS generation trip. Generation is tripped to decrease flow over interface. Red trace shows PMU signal while blue and black traces show the same flow using bus branch (planning model) and node-breaker (real-time model).

Figure 18.7b–e shows the same interface MW flow for different disturbances. All simulation use real-time node-breaker model.

It is worth noting that some aspect of dynamics are not simulated such as eventual unit ramping or AGC action.

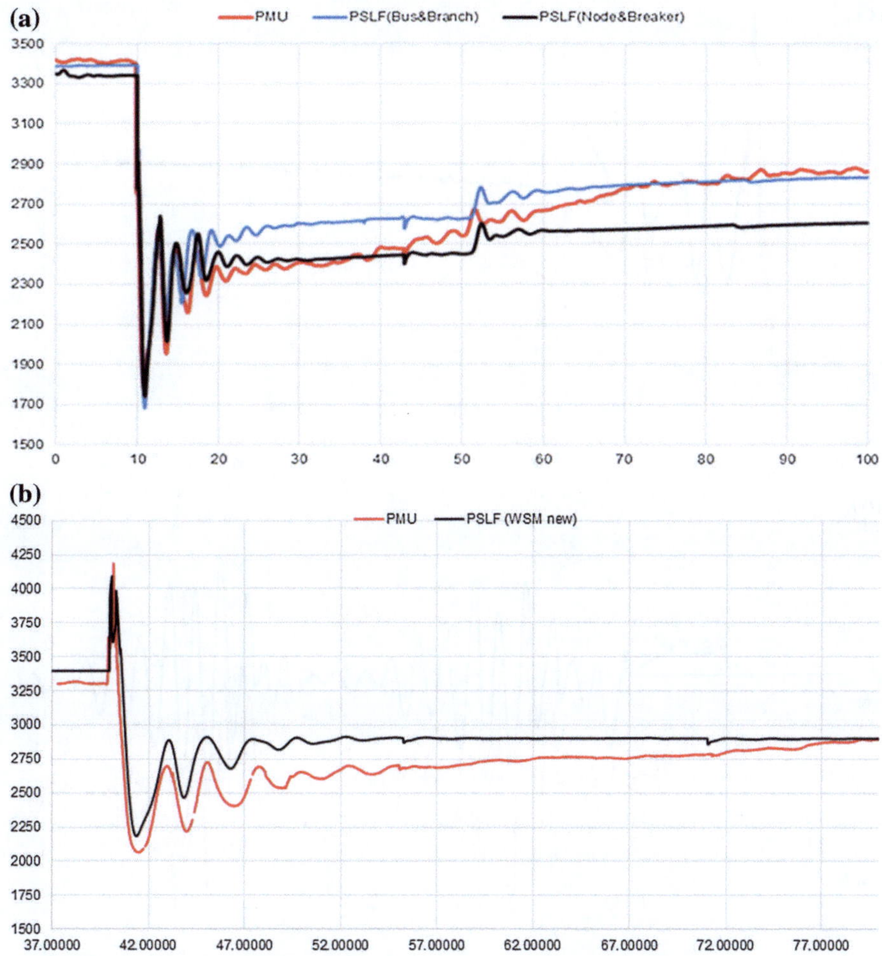

Fig. 18.7 **a** Interface path flow (measurement vs. planning basecase versus real-time node-breaker WSM model for RAS event resulting in 2.563 MW generation trip), **b** Interface path flow for simulated event (measurement vs. real-time node-breaker WSM model for 2.826 MW generation trip), **c** Interface path flow for simulated event (measurement vs. real-time node-breaker WSM model for RAS event resulting 1,708 MW generation trip), **d** Interface path flow (measurement vs. real-time node-breaker WSM model for double reclosing and tripping of Garrison-Taft line), **e** Interface path flow (measurement vs. real-time node-breaker WSM model for brake insertion into the system), **f** Interface path flow (measurement vs. real-time node-breaker WSM model for loss of large generation unit), **g** Interface path flow (measurement vs. real-time node-breaker WSM model for loss of two large generation unit)

(c)

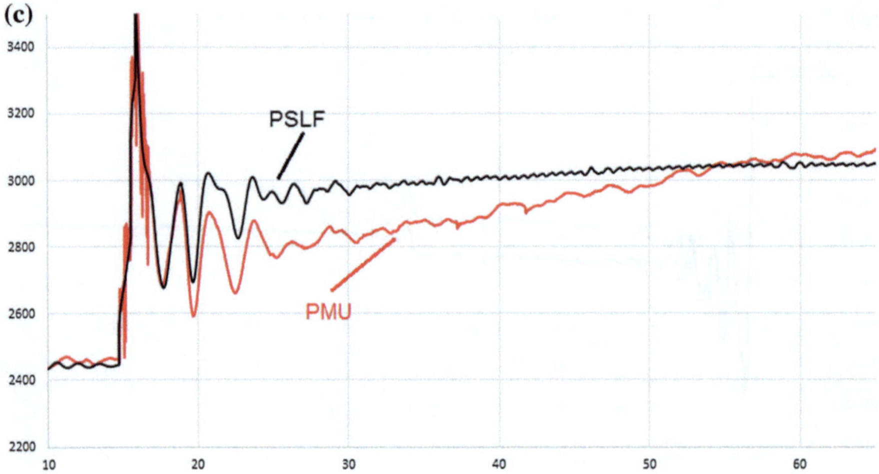

(d)

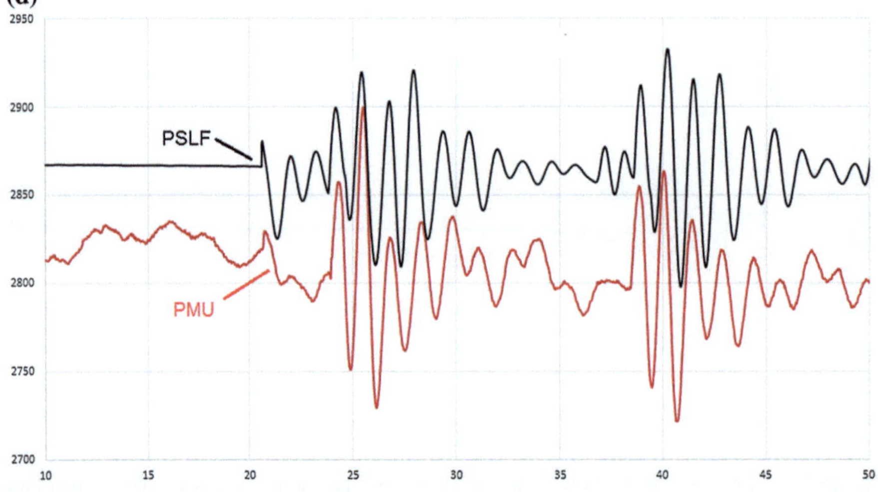

Fig. 18.7 (continued)

18.2.3 Benchmarking Frequency Response of Interconnection

Frequency response of the system is important characteristic of the system that changes with time of the day, season, and generator mix. In general, it is calculated after the fact, after generation tripping disturbance. Large increase in renewable resources can lead to possible deterioration in frequency response and it is important to be able to estimate frequency response in real time. For that reason, it is important to validate frequency response of the system for past event and compare simulated response to actual response and to correct eventual modeling

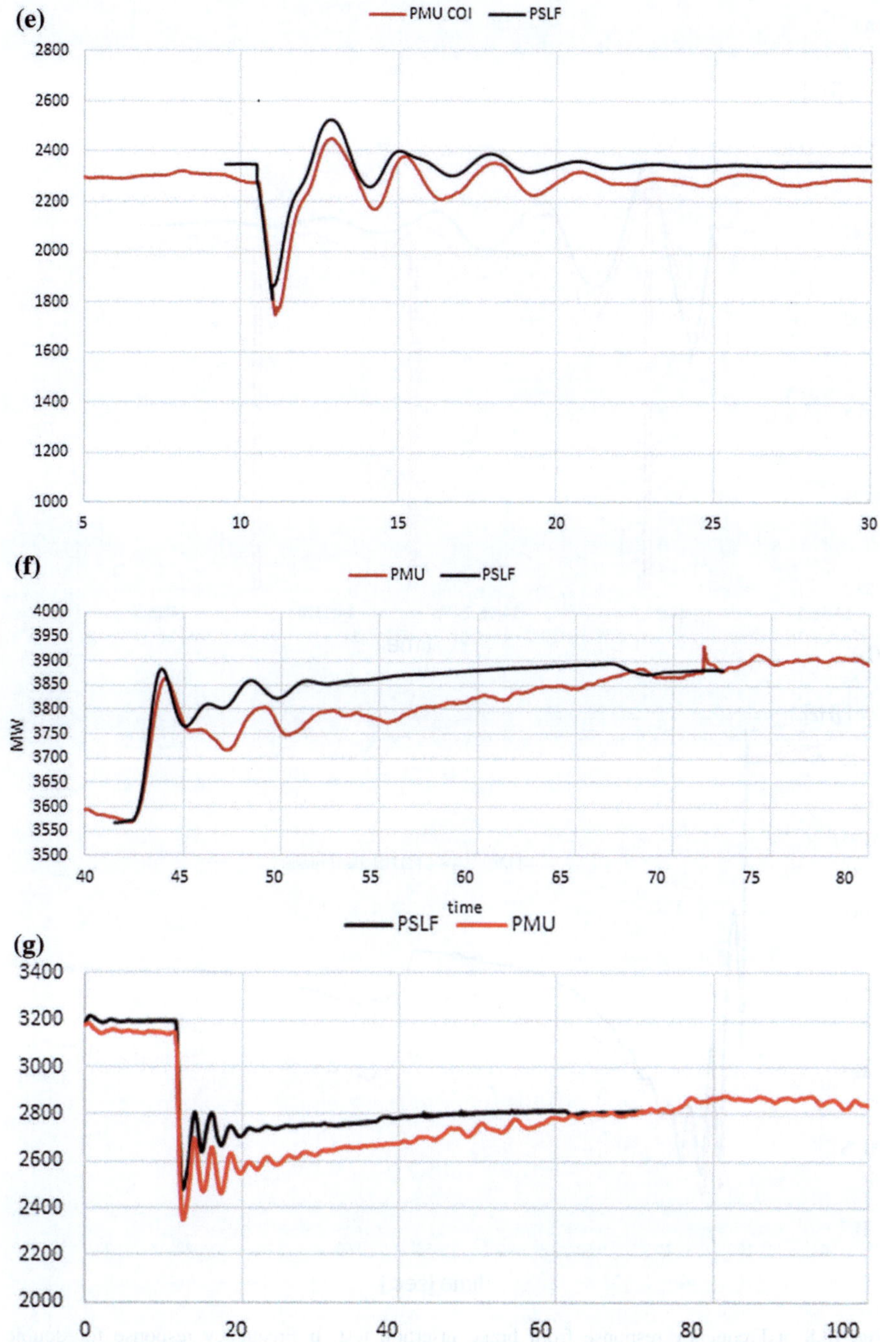

Fig. 18.7 (continued)

(a)

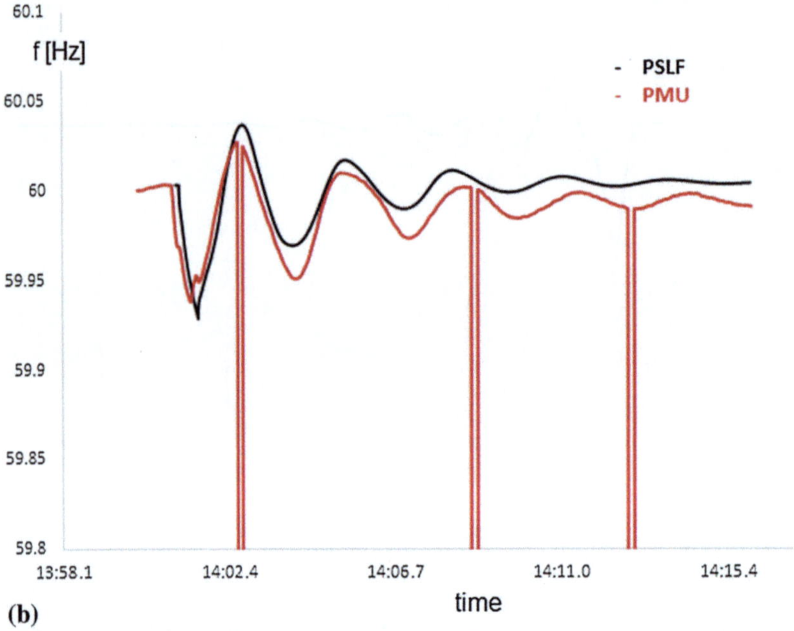

(b)

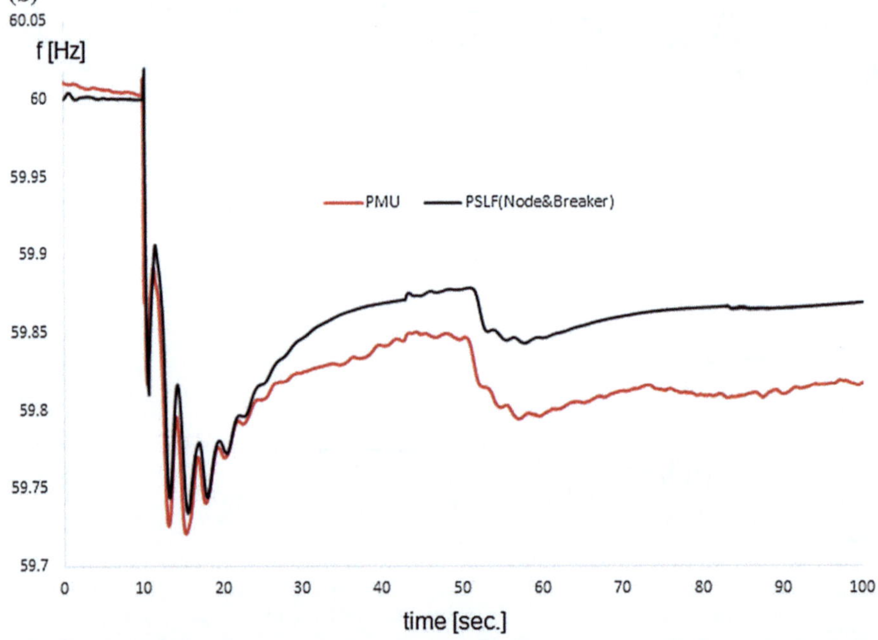

Fig. 18.8 **a** Frequency response from brake insertion test. **b** Frequency response for double generation trip event. **c** Another brake insertion test. **d** Frequency response for event HVDC trip and consequent RAS generation trip of 2800 MW. It can be seen that simulation results (black) does not mach frequency measurement (red)

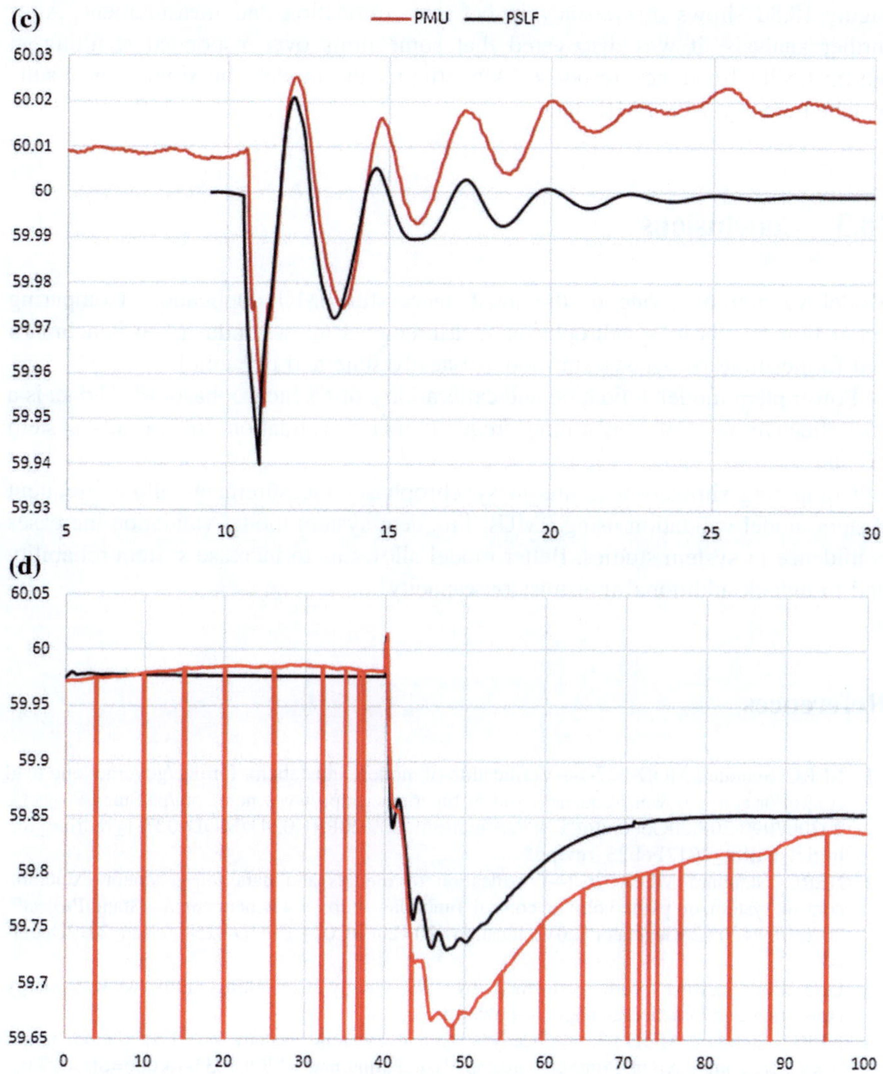

Fig. 18.8 (continued)

issues. A number of recent disturbances in WECC were successfully used to validate frequency response.

Figure 18.8a–d shows the comparison of simulation results for frequency response for different system events. Figure 18.8a shows brake insertion test. Brake test is performed periodically in WECC for the assessment of system damping. Brake is a large resistor that is inserted on 230 kV level for 30 cycles.

Figure 18.8b shows results for Pacific Northwest event that resulted in RAS generation tripping in 10 seconds and additional generation tripping in 50 seconds.

Figure 18.8d shows discrepancy in between simulation and measurement. After further analysis, it was discovered that some units over-responded resulting in optimistically frequency response. After tuning the model, the simulation results matched closely with PMU data.

18.3 Conclusions

Model validation is one of the most successful PMU application. Comparing simulation results to synchrophasor measurements allows industry to benchmark and further tune power system models (steady state and dynamic).

Power plant model validation and calibration using synchrophasor PMU data is a cost-effective way of performing frequent model validation for various system events.

Comparing simulation results to synchrophasor measurements allows frequent system model validation using PMUs. Frequent system model validation increases confidence in system studies. Better model allows us to increase system reliability and to unlock additional transmission capacity.

References

1. NERC, Standard MOD-027-1—Verification of models and data for turbine/governor and load control or active power/frequency control functions. http://www.nerc.com/pa/Stand/Project%20200709%20%20Generator%20Verification%20%20PRC0241/MOD-027-1_redline_to_initial_ballot_2012Feb23_rev.pdf
2. NERC, Standard MOD-026-1—Verification of models and data for generator excitation control system or plant volt/var control functions. http://www.nerc.com/pa/Stand/Project%20200709%20%20Generator%20Verification%20%20PRC0241/MOD-026-1_clean_2012Dec11.pdf
3. Grid Code testing work performed by GE Energy Consulting. http://www.geenergyconsulting.com/practice-area/global-power-projects
4. NERC, Standard MOD-033-1, Steady state and dynamic system model validation. http://www.nerc.com/_layouts/PrintStandard.aspx?standardnumber=MOD-033-1&title=Steady%E2%80%90State%20and%20Dynamic%20System%20Model%20Validation&jurisdiction=null
5. GE PSLF program user's manual
6. http://www.nerc.com/pa/RAPA/rg/ReliabilityGuidelines/Reliability%20Guideline%20-%20Power%20Plant%20Model%20Verification%20using%20PMUs%20-%20Resp.pdf
7. Application note 171 in GE PSLF program user's manual
8. Huang R, Diao R, Li Y, Sanchez-Gasca J, Huang Z, Thomas B, Etingov P, Kincic S, Wang S, Fan R, Matthews G, Kosterev D, Yang S (2017) Calibrating parameters of power system stability models using advanced ensemble Kalman filter, Accepted for Transaction on Power Systems
9. Huang Z, Du P, Kosterev D, Yang S (2013) Generator dynamic model validation and parameter calibration using phasor measurements at the point of connection. IEEE Trans Power Syst 28(2):1939–1949

10. Pacific Northwest National Laboratory, Power Plant Model Validation Tool (PPMV). https://svn.pnl.gov/PPMV
11. Li Y, Diao R, Huang R, Sanchez-Gasca J et al An innovative software tool suite for power plant model validation and parameter calibration using PMU measurements. In: IEEE PES general meeting 2017. Chicago, IL
12. Quint RD (2016) Power plant model verification guideline. https://www.wecc.biz/Administrative/Power%20Plant%20Model%20Verification%20-%20Ryan%20D.%20Quint.pdf
13. Ramanathan R, Tuck B, Kincic S, Zhang H, Davies D (2017) Equipment naming convention methodology for node/breaker model, UPEC
14. Thomas B, Kincic S, Davies D, Zhang H, Sanchez-Gasca J (2016) A new framework to facilitate the use of node-breaker operations model for system model validation in WECC, PES-GM, Boston, MA

Chapter 19
A Software Suite for Power System Stability Monitoring Based on Synchrophasor Measurements

Jian Ma, "Mani" Vaithianathan Venkatasubramanian, Scott Feuerborn, Clifton Black, Mark Halpin and Shih-Min Hsu

19.1 Introduction

Synchrophasor technologies use phasor measurement units (PMUs) to collect time-synchronized data from hundreds of installed devices and to create a detailed, comprehensive wide-area assessment of system conditions. Incorporating synchrophasor technologies into real-time operations dramatically increases system operators' situational awareness of the grid activity. Synchrophasor technologies provide system operators with unprecedented data on situations that could radically affect operational reliability. Several synchrophasor-based applications have been

This material is based upon work supported by the U.S. Department of Energy under Award Number DE-OE0000700.

J. Ma (✉)
Burns & McDonnell, Kansas City, MO 64114, USA
e-mail: jma@burnsmcd.com

M. V. Venkatasubramanian
Washington State University, Pullman, WA 99164, USA
e-mail: mani@eecs.wsu.edu

M. Halpin
Auburn University, Auburn, AL 36849, USA
e-mail: halpism@auburn.edu

C. Black · S.-M. Hsu
Southern Company Services, Birmingham, AL 35291, USA
e-mail: crblack@southernco.com

S.-M. Hsu
e-mail: smhsu@southernco.com

S. Feuerborn
Burns & McDonnell, Maitland, FL 32751, USA
e-mail: sfeuerb@burnsmcd.com

S. Nuthalapati (ed.), *Power System Grid Operation Using Synchrophasor Technology*, Power Electronics and Power Systems,
https://doi.org/10.1007/978-3-319-89378-5_19

available in the current market, such as PhasorPoint from General Electric, Real-Time Dynamics Monitoring System (RTDMS) from Electric Power Group, and Synchrowave from Schweitzer Engineering Laboratories. However, there is still a need for more production-level software tools to help system operators to monitor and analyze power grid stability based on real-time synchrophasor measurements.

In this chapter, a comprehensive software suite for real-time power system stability monitoring and analysis based on synchrophasor measurements is presented. The software suite (named Grid Stability Awareness System—GSAS) consists of five integrated analytical and monitoring tools, namely an Oscillation Monitoring Tool, a Voltage Stability Monitoring Tool, a Transient Instability Monitoring Tool, an Angle Difference Monitoring Tool, and an Event Detection Tool. Working together, these tools are targeted at addressing the three key stability problems in power system operations, i.e., small signal stability, voltage stability, and transient stability, while providing operators with wide-area situational awareness in terms of system stress. These tools have been integrated into one framework to provide power grid operators with both real-time stability status and historical information about system stability status of a power grid. The software provides capabilities for both real-time monitoring and off-line event playback analysis for power grid operators and planning engineers.

In addition, the results of performance evaluation based on off-line simulations are provided and discussed. It is necessary for operators to have confidence in the results provided by the tools related to sustained or lightly damped oscillations, potential voltage instabilities, and angular separation of portions of the system. To begin to develop this confidence, a series of off-line simulations were developed to test, evaluate, and validate the tools. The initial base cases and the simulation methodology are described first, followed by the specific cases developed and the rationale for them. Expected and actual results are summarized as appropriate for each case. Conclusions and recommendations are provided at the end of this chapter.

19.2 GSAS Architecture

19.2.1 Design Consideration

GSAS is designed for power grid operators to use in a control room environment to monitor real-time power system stability status. Meanwhile, GSAS provides functionalities for off-line historical data playback and analysis to facilitate real-time monitoring. GSAS is implemented in an open platform, in which interoperability and integration are the primary design objectives. An open platform is an open innovation framework that leverages open standards to enable third parties to add functionality. An open platform enables GSAS's data to be both accessed directly by users and published in open formats and enables GSAS to add more features or functionalities in the future

Furthermore, the advanced analytics are coupled with GSAS design considerations, such as open platform, scalable and pluggable architecture, and modular design, in the software suite development and deployment. These design considerations make GSAS scalable for different sizes of control rooms, pluggable for different types of synchrophasor applications, and extendable for different functionalities in an application. Thus, the GSAS software suite will provide the foundation for many advanced applications in the future.

19.3 System Architecture

The innovation of the GSAS software suite is also reflected from the functionality of its key components. A short description of the key components is provided as follows. Figure 19.1 shows the GSAS system architecture.

- *Analytical Engines*—Perform analytical calculation for oscillation analysis, voltage stability analysis, and transient instability analysis, event detection, and wide-area angle difference monitoring.
- *Visualization*—Visualize real-time synchrophasor data and analytical outputs, including both static information and time series data. Provide human–machine interaction, such as changing analytical engine parameters, showing warning messages, performing historical event playback.
- *Event/Alarm Trigger and Archival*—Detect event, trigger alarms, and archive alarm data based on real-time data and analytical results.
- *Geographic Information Display*—Demonstrate real-time geographic information about events, disturbances, system stability status, and trends. Advanced display techniques based on Geographic Information System (GIS) provide grid operators with an efficient way to enhance wide-area situational awareness.
- *Database*—Archive selected time sequence synchrophasor data and event/alarm data for a certain time around an event/alarm. Archive outputs of analytical engines for real-time operation and off-line analysis.

Southern Company Services (SCS) serves as the reliability coordinator for Southern Balancing Authority Area (SBAA). SBAA geographically consists of most of the state of Georgia, three-quarters of the state of Alabama, the western portion of the Florida Panhandle, and the southeast corner of Mississippi.

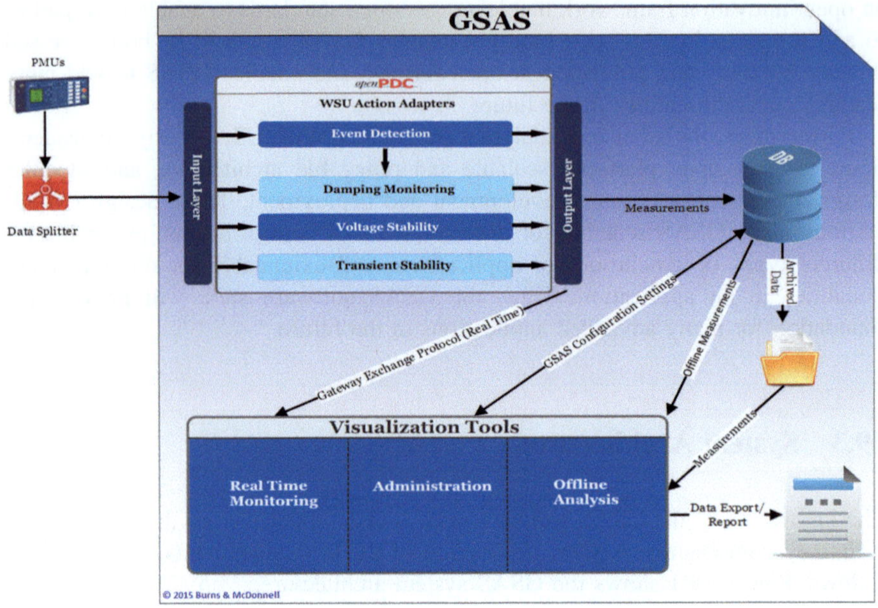

Fig. 19.1 GSAS architecture

GSAS is installed on SCS's synchrophasor infrastructure. openPDC is used to publish real-time synchrophasor measurement data to applications. Stability analysis engines developed previously by Washington State University (WSU) are implemented in the openPDC platform as action adaptors to perform the stability analysis and to produce results to the GSAS application. GSAS communicates with openPDC and analytical engines through Gateway Exchange Protocol (GEP). GSAS visualizes the real-time synchrophasor measurement data and calculated results from analytical engines using visualization tools, which consists of real-time monitoring, off-line analysis, and administration modules. GSAS has its own database to archive event data, messages, configuration settings, and relevant measurements.

Figure 19.2 shows the installation and configuration of GSAS at SCS. Synchrophasor stream splitter is used to split PMU data to different applications. openPDC is used for providing PMU data to GSAS. GSAS Data Service provides data flow management for GSAS. GSAS Database is used to archive relevant real-time PMU data and analytical result data. GSAS Client is the major graphical user interface (GUI) of the GSAS for visualization and situational awareness.

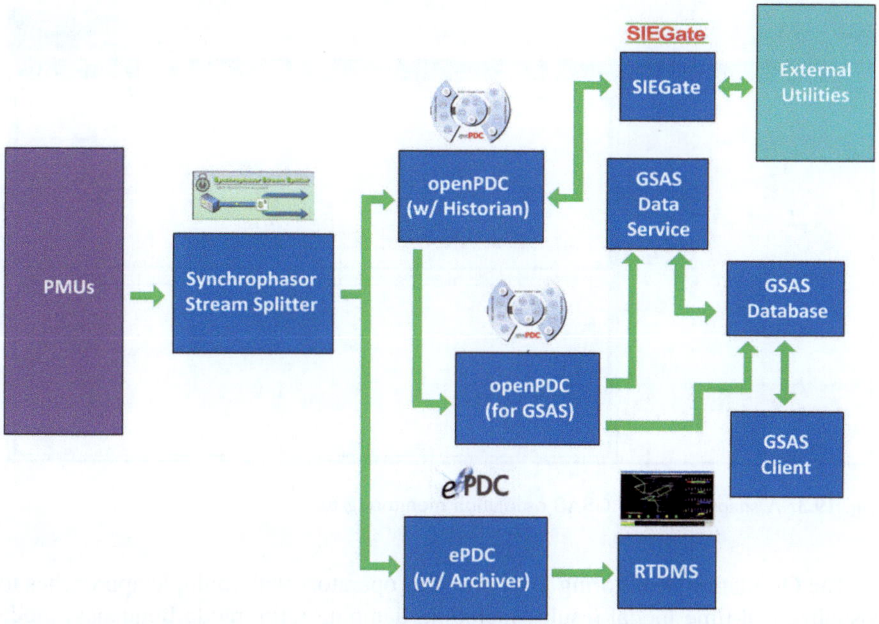

Fig. 19.2 GSAS installation at Southern Company Services (SCS)

19.4 GSAS Stability Monitoring Modules

In this section, a brief description on GSAS's key modules for real-time stability monitoring are provided.

19.4.1 Oscillation Monitoring Tool

The Oscillation Monitoring Tool monitors synchrophasor measurement data in real time to detect growing or poorly damped oscillations in early stage of oscillatory instability by performing extended frequency domain decomposition (FDD) analysis of ambient synchrophasor measurement data. The main idea of FDD is to apply singular value decomposition (SVD) to the power density spectrum matrix. The resulting singular values correspond to individual modes under specific conditions. The damping ratio, modal frequency, and the mode shape of poorly damped oscillatory modes can be directly determined from ambient synchrophasor measurements [1, 2].

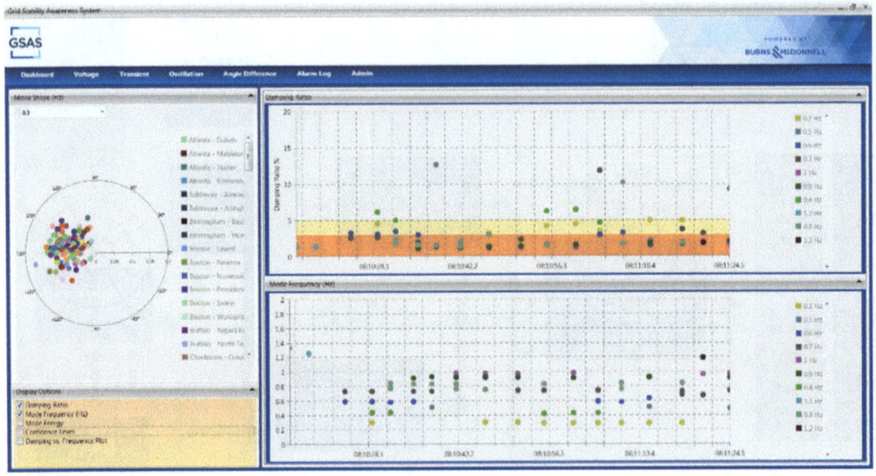

Fig. 19.3 A snapshot of the GSAS oscillation monitoring tool

The Oscillation Monitoring Tool provides operators with multiple approaches to visualize real-time modal results, including damping ratio, mode frequency, mode energy, confidence levels, and mode shape. Operators can highlight modal data for a mode. Figure 19.3 shows a snapshot of the GSAS visualization of the Oscillation Monitoring Tool.

Operators can set thresholds for warning and critical alarms limits for damping ratio going outside the limit violations, and damping ratio duration exceeding limit violations. These thresholds are shown as yellow and orange bands in damping ratio real-time plots in Fig. 19.3. Operators can enable and disable a certain alarm trigger function and define how many minutes of data to store around an alarm.

19.4.2 *Voltage Stability Monitoring Tool*

The Voltage Stability Monitoring Tool indicates voltage stability stress for a large area of the system using wide-area synchrophasor data by calculating local and remote voltage stability indexes and their associated confidence levels on each monitored line (see Sect. 19.6.3.1 for technical details). The dQ/dV type of sensitivity analysis is performed on real time. Figure 19.4 shows a snapshot of the GSAS visualization of the Voltage Stability Monitoring Tool.

The Voltage Stability Monitoring Tool provides operators with different ways to visualize real-time voltage stability index time sequence data and their associated confidence levels. Operators are also able to choose different plots for different measurements, like line active power flow, reactive power flow, and voltage magnitude time sequence measurement data.

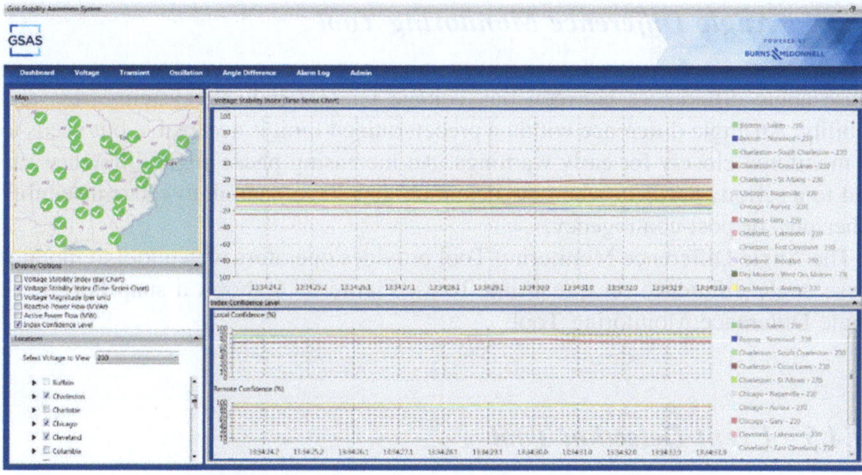

Fig. 19.4 A snapshot of GSAS voltage stability monitoring tool

19.4.3 Transient Instability Monitoring Tool

The Transient Stability Monitoring Tool aims at monitoring dynamic behaviors, identifying system disturbances, and providing early warning of system transient instability trend based on analysis of phasor angle separation and accumulated phase angle separation among the critical areas after a significant event occurs. The novelty of the approach is that all the decisions are made in real time purely based on wide-area synchrophasor measurements without any knowledge of the details of relay actions that may have resulted in the angle instability phenomenon [3].

Operators can set thresholds for warning and critical alarms limits for accumulated angle separation exceeding limit violations. These thresholds are shown as yellow and orange band in real-time plots in Fig. 19.5.

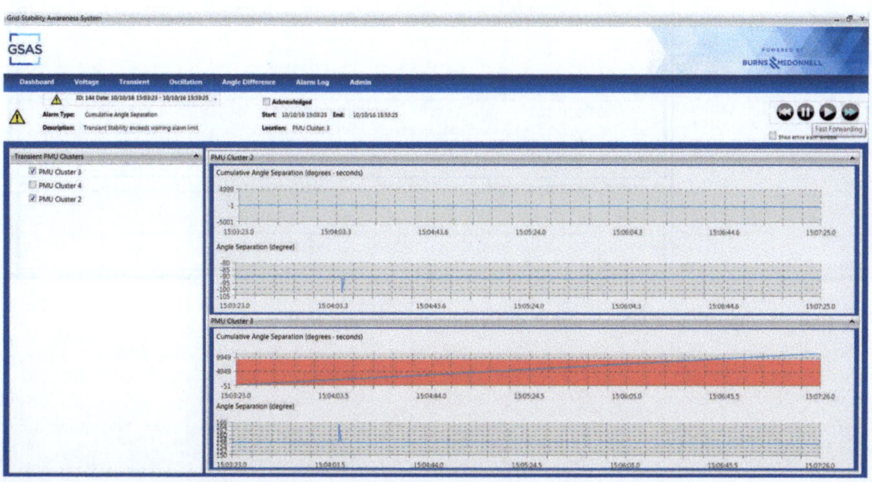

Fig. 19.5 A snapshot of GSAS transient stability monitoring tool

19.4.4 Angle Difference Monitoring Tool

Phase angle differences between bus voltage phasors indicate system stress and stability. An angle difference within a predetermined limit is acceptable but needs to be monitored closely for early warnings. An increasing phase angle difference can lead to a serious problem when the deviation gets large enough to cause instability either pre- or post-contingency.

The Angle Difference Monitoring Tool provides operators with a tool to monitor phase angle difference on a real-time basis. Figure 19.6 shows a snapshot of the Angle Difference Monitoring Tool.

19.4.5 Event Detection Tool

The Event Detection Tool detects occurrence of large disturbances based on monitoring of real-time synchrophasor measurement data and determines onset of an event if certain user-defined criteria are met after a large disturbance is detected. And then the tool uses three methods, namely the Prony method [4, 5], the Matrix Pencil method [6] and the Hankel Total Least Square (HTLS) method [7, 8], for automated oscillatory ring-down analysis [12]. Modal results for the event data are also provided by the Event Detection Tool. The GSAS visualizes the modal results after an event is detected.

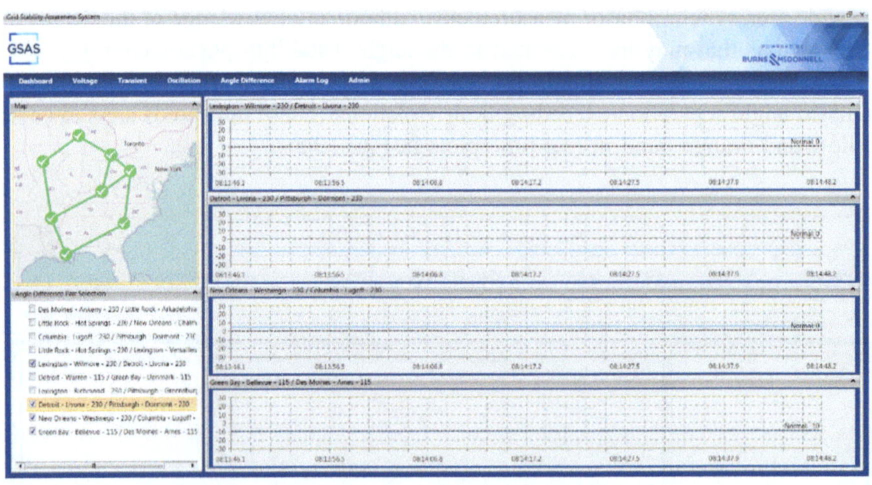

Fig. 19.6 A snapshot of GSAS angle difference monitoring tool

19.5 GSAS Alarming

19.5.1 GSAS Alarming Mechanisms

The alarms of GSAS are triggered by monitoring real-time outputs of analytical engines. After an alarm is triggered, GSAS will create a message with sufficient details about type, time, location, and other properties of the alarm. Two types of messages indicate two levels of severity: "Warning" refers to threshold violations that may cause problem to the system and need to bring operators' attention, and "Critical Alarm" referrers to more severe threshold violations than "Warning".

GSAS monitors different signals to create alarms for different tools. Table 19.1 shows the monitored signals for the five tools in GSAS.

19.5.2 GSAS Alarming Dashboard

GSAS dashboard is an easy to read, single page, real-time user interface, showing a graphical presentation of the status and historical trends of power grid stability (see Fig. 19.7). The dashboard provides at-a-glance views of power grid stability index status and messages relevant to the stability status. Meanwhile, the dashboard also provides a way to show historical warning/alarm messages for severe system status based on the measurements and stability indexes.

On the dashboard, the four indicators show real-time stability status of a power grid, in terms of voltage stability, transient stability, oscillatory stability, and angle difference. Different statuses are indicated by different colors and shapes of the indicators. For each of the indicators, there are four statuses, namely critical, warning, stable, and idle. A red circle with an exclamatory mark indicates critical alarm status. A yellow triangle with an exclamatory mark indicates warning status. A green circle with a true mark indicates stable status. A gray hexagon with a word "OFF" indicates there is no information for the status.

Table 19.1 Monitored attributes for GSAS alarm triggers

Tools	Monitored Signals	Number of Output Signals
Oscillation	Damping ratio	Number of modes
Voltage stability	Voltage stability index (local and remote)	Number of monitored lines
Transient instability	Angle separation and accumulated angle separation	Number of user-defined generator clusters
Angle difference	Angle difference pairs	Number of user-defined angle difference pairs
Event detection	Frequency, voltage magnitude, or voltage phase angle	Output "Yes" if there is an event detected

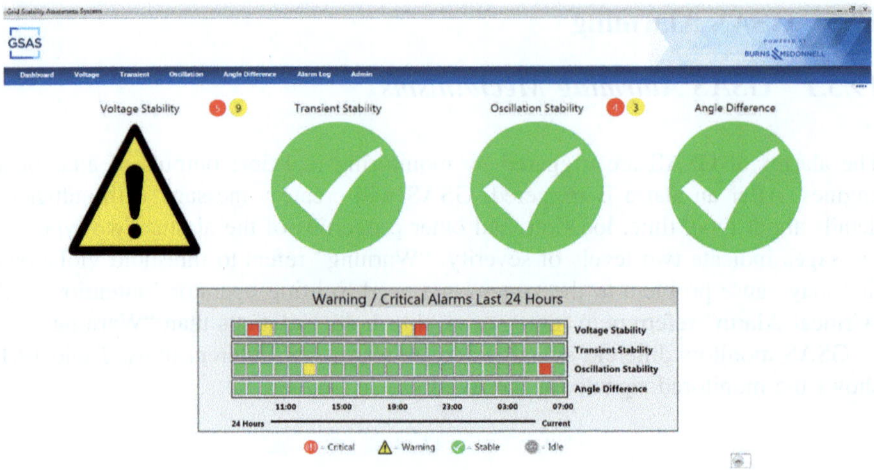

Fig. 19.7 A snapshot of GSAS alarming dashboard

On the dashboard, the stability status of last 24 h is shown on the horizontal bars for each stability category. Each bar shows the worst stability status in term of stable, critical alarm, or warning of one hour. A green bar indicates there is no warning and critical alarm in the hour. A red bar indicates the worst status of that hour is a critical alarm or there is at least one critical alarm in the hour. A yellow bar indicates the worst status of that hour is a warning alarm or there is at least one warning alarm in that hour.

Below the historical status bar, the timeline shows the current time and the historical time (last 24 h based on current hour). "Current" means the bars at the right side indicate the current time. "24 h" means the bars at the left side indicate historical hours. The hours on the timeline will be updated based on real-time timestamp.

Red or yellow circles with a number inside indicate the number of un-acknowledged critical alarms or warning alarms for a specific stability status. These red or yellow circles are clickable and linked to the details of un-acknowledged alarms. Operators can click the red or yellow circles to open a list of un-acknowledged alarm message for event/alarm data playback.

Three different layout schemes of dashboard are provided in GSAS to show information on different level of details.

Layout 1: Dashboard shows both real-time status indicators and historical status bars for real-time and historical stability status. This is the default layout for dashboard.
Layout 2: The dashboard shows only real-time status indicators indicating real-time stability status.
Layout 3: The dashboard shows only historical status bar indicating historical stability status.

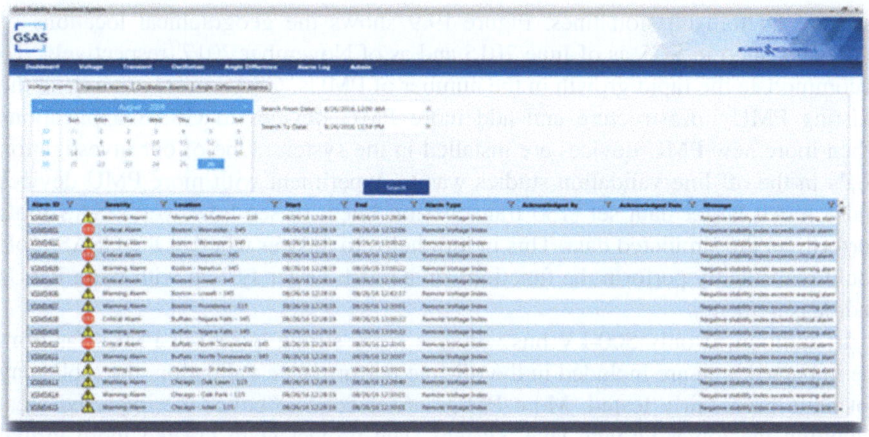

Fig. 19.8 A snapshot of GSAS alarming logs

19.5.3 GSAS Alarming Logs

GSAS's alarming logs (see Fig. 19.8) provide operators with an efficient approach to retrieve historical event/alarm data based on date and time to facility off-line analysis. All the alarms are grouped by types, namely voltage alarms, transient alarms, oscillation alarms, and angle difference alarms. Operators can search and filter alarms based on different criteria, such as date, hour, period of time. These alarms are those that have already been acknowledged by operators.

19.6 Off-line Validation of GSAS Performance

19.6.1 Overview

The objectives of the off-line validation are (1) to evaluate the immediate suitability of the tools, in their present state of development, for use in an operational environment, (2) to identify and support potential improvements on the tools, and (3) to identify potential limitations of the tools due to existing constraints, particularly the existing and planned deployment of PMUs, within SBAA. In the off-line performance validation, the modules validated include the voltage stability monitoring tool, the oscillation monitoring tool, and the transient stability monitoring tool. Simulated cases were used for the off-line performance validation.

The deployment of GSAS used in SCS was based on existing PMU devices when the GSAS validation was performing, which include 17 PMU devices and measure 54 transmission lines. Figure 19.9 shows the geographical locations of PMUs installed in SCS as of June 2015 and as of November 2017, respectively. To accommodate the rapid growth in the number of PMUs, GSAS users can extend the existing PMU infrastructure and add more PMU devices to GSAS applications when more new PMU devices are installed in the system. One of our investigation goals in the off-line validation studies was to experiment with more PMU devices and a much larger data set (130 transmission lines) to see how well GSAS tools perform on the simulated data. This in turn helps to answer whether the GSAS tools could realistically perform the functions on a much larger system with more PMUs andmeasurements.

In addition, if only 500-kV bus voltages (very strong buses) or PMUs far from the impacted area are included in the off-line performance validation, the tools may not have been fairly tested. More PMUs are needed to cover the impacted area, including the lower voltage level system. The project team created more virtual PMU devices and assigned to selected locations in SCS's territory. These additional PMUs, with several more in the impacted area of the simulated cases, should provide a lot more information for the GSAS tools to detect targeted oscillation and voltage stability problems. One of the goals was to determine whether GSAS tools could include more PMUs, and what hardware or software improvements might be required if more PMU devices are installed in the system.

General procedure for the off-line performance validation on the extended PMU structure is described as follows. First, PMU devices and lines monitored are defined. Next, PSS/E simulations are performed based on the extended PMU structure as described. Then, GSAS is configured to reflect the extended PMU structure. Finally, the simulated data is fed into GSAS and the system behavior can be observed.

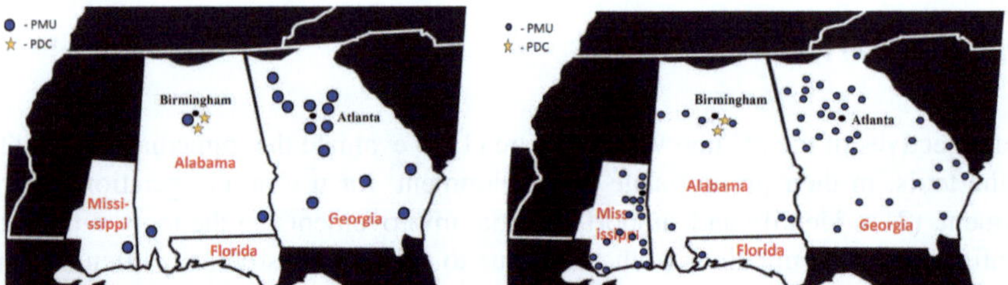

(a) Installed PMUs as of June 2015 (b) Installed PMUs as of November 2017

Fig. 19.9 Locations of installed PMUs in SBAA

19.6.2 *Procedure and Methodology for Off-line Performance Validation*

19.6.2.1 Validation Procedure

Two transmission planning base cases (summer peak load level and 50% of summer peak load level cases) prepared by SCS team were used as the starting cases in the GSAS off-line performance validation. The validation cases for various scenarios were created in consultation with the SCS team and the overall project team. Power system simulation software PSS/E was used for preparation of validation cases. Excel files emulating PMU data were exported from PSS/E simulation outputs. Burns and McDonnell team ran the emulating PMU data through openPDC and GSAS and archived the results coming out of GSAS and the analytical engines. WSU team analyzed the GSAS results and the output data to interpret results and provided recommendations. Generally, the off-line performance validation is a type of blind tests for all simulated cases. WSU team had no access to power flow and stability models. The details of simulated cases were not known initially to WSU team.

The percentage of random variation can be easily varied in the simulated cases. In the off-line simulation, we settled on 1% random load variation. The intermediate output needs to be managed in some way to reduce the runtime. Writing the intermediates to a log file creates 50 Mbps of simulation time, so this is not an acceptable solution.

19.6.2.2 Validation Cases

SCS provided two base cases and developed scenarios based on the base case. Summer peak load level and 50% of summer peak load level power flow base cases were used to create various power flow cases for dynamic simulations. Validation scenarios were created for voltage stress/stability, oscillation, and transient stability/angular separation.

Total simulation time for each scenario was 10 min. 1% random load variation was simulated for each scenario to mimic the random changes in system load in real operating conditions. Contingencies were applied at $t = 5$ min. The outputs of PSS/E simulations were selected based on the actual PMU locations and measurements. The output measurements include bus frequency, bus voltage magnitude, bus voltage phase angle, line current magnitude, and line current phase angle. Active power and reactive power are calculated based on the voltage and current output measurements. Simulation outputs were provided as Excel files for playback via openPDC and GSAS for testing purpose.

Eighteen cases in total were created for off-line validation purposes. Among these eighteen cases, six cases were used to verify general setup process, playback capabilities, and ambient performance of tools. One case was created to verify voltage stress/stability identification of known areas of significant voltage stress. Four cases were created to verify oscillation identification based on turning off generator power system stabilizer (PSS) and applying disturbance. Three cases were created to verify voltage stress/stability identification of known areas of voltage sensitivities within other part of SBAA. Three cases were created to verify angular stability/separation for known transient instability contingencies. One case was created to verify forced oscillation identification based on artificial oscillatory stimulus of generator governor limits. In addition, six general cases were designed to establish and verify overall scheme for future testing and ambient performance of all tools.

A significant issue was the determination of the size of ambient load variation required to excite "normal" system behavior characteristics. Key features of tools are based on changes relative to ambient conditions. 1, 2 and 4% load variations were evaluated. 1% load variation was determined to be sufficient. There should be no alarm or abnormal conditions reported by the tools.

19.6.2.3 Testing Approach

The two cases, namely peak and off-peak (50%) cases, were modified as necessary to simulate various scenarios/contingencies which were known or believed to be associated with oscillatory, voltage stability, or angular stability problems. Standard planning models and datasets were used for all equipment in the PSS/E simulation software. In general, each simulation was 10 min in length and a time step of half cycle was used. An output rate of 30 Hz, corresponding to 30 incoming PMU measurements per device per second, was implemented.

A critical functional aspect of the tools is that they operate based on distinguishing new or emerging system conditions with respect to some normal background condition. Given that the tools are evaluating dynamic phenomena, it is necessary to have dynamic background conditions similar to those present in actual power systems due to normal load variation. For this reason, a 1% random load variation was introduced into the simulation based on a uniform distribution, 0–1%. This random variation was applied every two cycles throughout the 10-min simulation.

For each 10-min simulation, regardless of the contingency implemented or the tool being tested, the first 5 min of the simulation contained only the random background load variation. At $t = 5$ min, the contingency was applied. The simulation was then allowed to run for a total time of 10 min. Note that the random load variation remained throughout the entire 10-min simulation period.

The 18 total cases developed consist of six cases for the purposes of general setup testing, data exchange, and parameter determination, four cases for the purposes of evaluating the voltage stability tool, five cases for the purposes of evaluating the oscillation tool, and three cases for the purposes of evaluating the angular stability tool. These cases and the tests performed are described further in the following sections.

19.6.2.4 Six General Cases

The first six cases developed were intended to (1) determine the percentage of background load variation necessary for proper tool performance, (2) develop and implement the overall simulation procedures and methods for transferring results (very large files) between project team members, and (3) develop and implement methods for converting simulation results into equivalent PMU data streams.

There were no contingencies applied for these six cases. Background load variations of 1, 2, and 4% were evaluated on the peak and 50% base cases. The 1% variation was selected for all future simulation studies based on the feedback from the tool developers. Because there were no contingencies applied, the tools were not expected to produce output with meaningful results. However, the voltage stability tool was able to correctly identify areas of the system that are known to be more sensitive to voltage stability problems based only on the background load variations.

During these initial evaluations, the project team concluded that it is not realistic to compare results from the peak and 50% cases because there are too many differences between the two base cases. Of particular importance is the obviously different generation dispatch in the two cases, and it is well known that load level significantly impacts the dynamic phenomena which are evaluated by all the tools. For this reason, all future simulation studies utilized the peak load level case even if this might not normally be the case used in the planning environment for a particular dynamic simulation.

19.6.3 Performance of Voltage Stability Monitoring Tool

The objective of the voltage stability monitoring tool is to estimate the proximity of the system operating condition to static voltage instability limit in the sense of the classical QV margin at any bus by using a few PMU measurements. The method is based on power flow to attain line sensitivity by simulating the load change at some bus by inserting a reasonable amount of shunt device to capture V and Q changes. Although this method needs power flow calculation, the process is simple and not time-consuming. The off-line performance of voltage stability monitoring tool is evaluated using historical synchrophasor measurement data.

19.6.3.1 Overview of Voltage Stability Analysis Method

The voltage stability analysis engine calculates voltage stability indexes for each measured line, and estimates the QV slope from real-time PMU data by individually estimating QV line sensitivity on each of the lines connected to bus *i* and by taking the sum [9]:

$$\Gamma_i = \frac{\Delta Q_i}{\Delta V_i} = \sum_j \frac{\Delta Q_{ij}}{\Delta V_i} \tag{1}$$

QV slope Γ_i from the line sensitivities is directly extracted by the above summation. This method is a direct estimation of line sensitivities $\Delta Q_{ij}/\Delta V_i$ from PMU data by exploiting fundamental nature of power system causality. This method is purely based on PMU measurement data and does not need system model information for real-time voltage stability analysis. Line sensitivities $\Delta Q_{ij}/\Delta V_i$ can be calculated for each of the lines connected to bus i giving us the net sum Γ_i in Eq. (1) which is then the voltage stability index for bus i. If the value Γ_i goes below a predefined threshold the system near the monitored bus is approaching static voltage stability limit.

The voltage stability index of each bus consists of the sum of all line sensitivities connected at that bus. Theoretically, the line sensitivity is the total differential of line reactive power versus the corresponding bus voltage. In a small region, $\Delta Q/\Delta V$ is used to replace that total differential without unacceptable error. Since all power transfers occur through transmission lines or equivalent lines and this voltage stability index has capacity of gathering all information from line sensitivities, it can represent that bus voltage state of stability and security. This index is similar to the slope of certain point on V-Q curve or the degree of stability in static analysis about voltage stability study. By checking the minimum index in one system at some operation condition, the weakest bus can be easily found. Meanwhile, the level of system voltage stability can be estimated, such as whether the system is close to certainly critical limit. It can be used to mark the impact which a contingency or load change gives as well. The following rules can be used to assess the system voltage security at any bus [2]:

- A "high" value of Γ_i indicates a "strong" bus in terms of being distant from static voltage instability limit.
- If there is any bus with Γ_i value near "0", the system is close to static voltage stability limit related to saddle-node bifurcation, and the bus with the lowest Γ_i is likely in the critically voltage stressed part of the system.
- If one system has the bus whose Γ_i is less than some critical value, and then the system may be vulnerable toward voltage instability caused by either of saddle-node or limit induced bifurcations. We need to pay attention to this system voltage security.

GSAS software uses voltage stability indexes (local index—LI, and remote index—RI) calculated by the voltage stability monitoring engine to indicate the system's voltage stability status. The voltage stability monitoring engine calculates continuously LI and RI indexes for each measured line based on the synchrophasor measurements in a moving window. Users can specify the size of the window (say 20 s continuous data) and the length of interval between two windows (say 3 s between the updates of LI and RI indexes). LI and RI indexes provide operators with a quantified indicator to evaluate system's (in terms of measured lines) voltage stability status on a near-real-time manner. Based on LI and RI indexes, operators are also able to know which lines are worse compared with other lines in term of voltage stability index.

19.6.3.2 Results of Validations Based on PSS/E Simulations

(1) Simulation Cases

For the first voltage stress test, a single case was created based on known conditions leading to excessive voltage stress (limiting) conditions within SBAA. The disturbance is a specific unit outage and a 500-kV line outage. The voltage stress/stability tool should indicate locations of increased voltage stress/sensitivity. For the second voltage stress test, three cases were created with increasing voltage stress levels in another part of SBAA. The conditions considered in the simulations include: (1) two lines out of service; (2) one group of capacitors out of service; and (3) additional group of capacitors out of service.

The project team discussed a reasonable voltage stress case and have developed what the team thinks is a reasonable scenario to evaluate. The simulation is 10 min total with the stress applied at $t = 5$ min. Hopefully, this will ensure that there is sufficient time for the tool to properly initialize and ultimately to properly characterize the situation. The team saw no use in performing this scenario simulation on anything other than a peak load case because there are no known voltage issues under light load conditions.

Cases 1 and 2 are pilot testing simulation cases for testing the voltage stability monitoring tool. These files are 5-min simulations with random 1% load variations. For these two pilot testing cases, the team is looking for is differences in the indexes for the two different cases. For example, what is the index at a certain place (or in a certain region) in Case 1 v.s. what is the value of the same index, at the same place or region, in Case 2. Cases 3, 4, 5, 6, and 7 are simulation cases for testing the voltage stability monitoring tool. Cases 12, 13, and 14 are the voltage stress cases with various combinations of capacitor banks out based on the existing PMU structure. Cases 12-1, 13-1, and 14-1 are the voltage stress cases with various combinations of capacitor banks out based on the extended PMU structure. In these three cases, there are many more PMU channels included in the results. Hopefully, this will improve the ability of the GSAS tool to pick up the sensitive areas of the system.

(2) Voltage Stress Cases

Four cases were created to test the voltage stability tool. All four cases represented problematic voltage conditions in the SBAA. These four cases were created in two groups. The first of the four cases were created based on a known problem area following a specific unit and one 500-kV line out of service. The remaining three cases were created based on a different known problem area associated with various combinations of line and capacitor bank outages. Specifically, these three cases were created to simulate (1) two 500-kV lines out of service, (2) two 500-kV lines out of service and a group of capacitors out of service, and (3) two 500-kV lines out of service and a second (in addition to the first) group of capacitors out of service.

For the first test case, the voltage stability tool properly identified the known problem area in the SBAA following the contingency. Tool test results were inconclusive for the remaining three test cases. One likely reason for the inconclusive results for the three sets of cases is that there are not many PMUs located in the known problematic area. It is not realistic to expect the tool to be able to reveal potential system problems when there is no incoming data from PMUs that contain the information that the tool requires to perform its function. In these cases, for example, lack of PMUs in the impacted 115-kV network may be a significant contributor to the inconclusive results obtained.

This lack of PMU data is a potentially significant limitation in the performance of the developed tools. It must be recognized that none of the tools can be fairly expected to recognize or predict problems when they do not have sufficient information. A determination of the extent to which the tools can "see" beyond potentially limited input data is an important objective for future work.

(3) Outputs of Analytical Engines Based on Simulated Data

We use average of LI and RI indexes to verify that the tools are in fact doing what they are supposed to do. The occurrence of any contingency in simulation or actuality will result in some type of temporary change in the indexes which will need to be ignored by the operator. It appears that the event detection procedures would, or could, be used to block the potentially misleading indexes during any period of surrounding an event. We would suggest that this "blanking time" window be related to the actual response time available to an operator. If any operator cannot realistically take any action in 10 s (for example), there is no reason to report values for 10 s after an event has been detected.

From the perspective of validating the tools, we use averages to indicate that the tools are performing correctly. We all know that transient events are occurring and that they will slightly impact a 5-min average. We perform averaging calculation for before and after the contingency. To avoid the impact of the transient period on the LI and RI indexes, we use $0 < t < 5$ and $6 < t < 10$ min for averaging. There should not be any initial transients. The base case drift runs are perfectly flat and no disturbance, other than the requested random load variation, is applied until $t = 5$ min.

Table 19.2 shows average LI and RI indexes, active power and reactive power of Case 12 for existing PMUs. Average local and remote indexes, active power and reactive power for other test cases were also analyzed and compared.

Table 19.2 Average local and remote indexes, active power and reactive power for Case 12

PMU Names	LI (1–5 min)	LI (6–10 min)	RI (1–5 min)	RI (6–10 min)	P (1–5 min)	P (6–10 min)	Q (1–5 min)	Q (6–10 min)
Sub1-Line1	14.75	15.07	−18.69	−19.60	217.85	219.05	−2.68	−2.53
Sub1-Line2	14.69	12.40	−17.94	−14.65	378.80	378.30	109.85	109.87
Sub1-Line3	8.49	8.52	−13.01	−13.26	12.69	11.64	−0.37	−0.37
Sub2-Line1	34.45	36.32	−38.23	−34.76	1382.45	1377.27	147.55	147.78
Sub2-Line2	34.43	36.30	−38.24	−34.76	1382.45	1377.27	147.47	147.70
Sub2-Line3	20.18	19.38	−25.94	−25.87	−407.10	−398.81	91.66	89.51
Sub2-Line4	16.98	16.27	−22.18	−22.28	112.92	112.42	25.68	27.03
Sub4-Line1	13.33	N/A	−15.69	N/A	−110.71	−111.70	19.26	19.30
Sub4-Line2	19.19	19.47	−24.89	−23.82	−212.03	−211.56	−125.33	−125.38
Sub4-Line3	7.11	7.26	−8.75	−8.04	−12.16	−11.32	1.83	1.75
Sub4-Line4	7.70	7.39	−10.54	−10.08	64.08	64.04	−6.83	−6.79
Sub5-Line1	29.35	29.17	−36.66	−36.81	613.00	618.73	−17.05	−0.52
Sub5-Line2	14.64	14.30	−18.31	−17.41	381.88	385.76	32.30	34.20
Sub5-Line3	17.83	17.11	−23.63	−23.87	−492.81	−504.87	−30.64	−31.67
Sub5-Line4	11.48	10.91	−14.90	−13.67	313.56	318.33	33.06	36.11
Sub6-Line1	9.63	9.74	−13.30	−12.55	3.57	3.74	77.39	76.97
Sub6-Line2	7.58	7.61	−9.26	−9.41	27.57	25.74	15.98	16.57
Sub6-Line3	7.44	6.01	−7.54	−6.23	145.96	145.50	−2.75	−2.67
Sub6-Line4	N/A	N/A	N/A	N/A	−181.40	−180.44	20.48	20.42
Sub6-Line5	35.11	34.71	−40.65	−42.11	701.30	711.05	6.71	7.57
Sub7-Line1	14.43	12.15	−20.59	−19.10	−676.73	−676.67	−45.35	−45.67
Sub7-Line2	29.59	N/A	−37.64	N/A	−222.45	0.00	−19.78	0.00
Sub7-Line3	N/A	N/A	N/A	N/A	−59.81	−59.25	54.88	54.70
Sub8-Line1	20.83	31.91	−27.91	−40.67	638.82	917.80	−88.60	−24.64
Sub8-Line2	27.37	31.92	−34.88	−38.51	737.75	592.61	−91.29	−70.64
Sub8-Line3	16.49	11.67	−22.30	−18.25	−879.75	−879.69	1.82	−27.00
Sub8-Line4	14.49	11.48	−19.41	−15.52	−534.30	−533.76	−8.16	−7.78
Sub8-Line5	33.28	32.88	−40.79	−37.42	1164.77	1165.26	−26.42	−29.07
Sub9-Line1	22.89	20.59	−29.00	−27.00	−935.69	−925.39	−4.27	−6.02
Sub9-Line2	12.68	12.18	−14.93	−15.63	205.16	205.13	16.70	16.77
Sub9-Line3	28.16	24.76	−33.15	−31.92	−1046.21	−1056.54	−87.43	−82.38
Sub9-Line4	24.73	24.81	−31.30	−29.68	110.69	102.81	−94.92	−94.56
Sub10-Line1	30.29	30.45	−34.48	−37.35	193.72	192.56	−22.29	−22.27
Sub10-Line2	20.72	18.16	−27.71	−24.12	−972.88	−961.69	−62.69	−65.91
Sub10-Line3	25.08	19.89	−31.37	−26.81	−1159.64	−1160.12	43.88	46.64
Sub11-Line1	19.24	19.89	−25.79	−25.00	84.75	73.78	−73.24	−72.10
Sub11-Line2	15.44	12.96	−20.56	−16.99	−886.18	−908.77	11.03	9.03
Sub11-Line3	25.40	23.38	−31.04	−30.00	−38.78	−18.33	−136.93	−129.98
Sub12-Line1	22.76	21.59	−29.82	−29.04	−19.98	−22.86	−41.94	−54.07
Sub12-Line2	56.88	56.81	−60.38	−59.39	1052.42	1062.87	91.08	88.24
Sub12-Line3	20.44	18.28	−26.97	−26.06	−899.90	−899.81	13.60	8.90
Sub12-Line4	20.22	17.96	−26.96	−25.14	−928.90	−928.81	17.97	13.29
Sub13-Line1	35.77	39.43	−42.69	−43.63	418.87	343.66	−96.08	−96.55
Sub13-Line2	20.06	19.65	−25.43	−26.46	−84.71	−73.75	5.30	4.05
Sub14-Line1	27.02	21.79	−35.02	−28.84	−1007.89	−1019.19	−176.64	−175.07
Sub14-Line2	31.39	N/A	−40.09	N/A	433.48	0.00	−23.74	0.00

(continued)

Table 19.2 (continued)

PMU Names	LI (1–5 min)	LI (6–10 min)	RI (1–5 min)	RI (6–10 min)	P (1–5 min)	P (6–10 min)	Q (1–5 min)	Q (6–10 min)
Sub14-Line3	20.22	18.51	−25.91	−26.22	−734.33	−590.38	20.06	−22.43
Sub15-Line1	24.33	N/A	−28.26	−0.34	223.02	0.00	−161.87	−17.49
Sub15-Line2	20.72	19.65	−26.13	−26.00	−1195.22	−072.52	54.52	−17.56

19.6.4 Performance of Oscillation Monitoring Tool

19.6.4.1 Overview of Oscillation Analysis Method

The Oscillation Monitoring Tool monitors both ambient synchrophasor measurement data and event data. The tool uses FDD to perform real-time analysis on ambient PMU measurements [10]. The main idea of FDD is to apply a SVD to the power spectrum matrix from multiple measurements [11]. The resulting singular values correspond to individual modes under certain conditions and can be analyzed separately. The oscillation monitoring tool extends the traditional FDD method for reasonably short time-interval monitoring of poorly damped oscillatory modes (say about 3–4 min) to enable its applicability for power system operation [1].

Oscillation monitoring tool uses three modal analysis methods, namely, Prony's Method, Matrix Pencil method and HTLS method, for oscillation analysis based on event data [12]. All these methods try to fit a sum of exponential terms to the uniformly sampled data. Prony analysis has become one of complimentary standard approaches in power system small signal stability problem. The idea of Matrix Pencil method comes from the pencil-of-function approach, and now is used in system identification and spectrum estimation. The SVD step in Matrix Pencil method acts as a filtering process, thus this method can extract dynamic information from noisy data. The HTLS method is able to extract system eigenvalues from multi-channel noisy measurements.

The Prony analysis of real-time data or event recordings in power systems is especially challenging because the power system responses contain both linear and nonlinear phenomena. Moreover, presence of noise in the measurements can upset the accuracy of results. Typically, there are also many discrete switching events that occur during a routine disturbance event, and these damping of the modes can change after each of these switching events. In the oscillation monitoring engine developed by WSU, rules have been developed to crosscheck results from (1) moving time-window analysis, (2) multiple signal groups that contain modal responses, and (3) different types of signal processing engines, in other to ensure consistency of the modal analysis [12].

19.6.4.2 Results of Off-line Validation Based on PSS/E Simulations

(1) Oscillation Cases

Five cases were developed in two groups for testing the oscillation tool. The first group of four cases was based on turning off power system stabilizers from specific units at a generating plant and performing line out and fault contingencies. The second group, consisting of a single case, was based on artificially applying a sinusoidally-varying maximum power limit to a specific unit governor so as to artificially create a sustained oscillation.

For the first group of cases, the four cases consisted of (1) a 500-kV line-out contingency and the PSSs in service, (2) a 500-kV line-out contingency and the PSSs out of service, (3) a 500-kV line tripping to clear a 3-cycle fault contingency with the PSSs in service, and (4) a 500-kV line tripping to clear a 3-cycle fault contingency with the PSSs out of service.

As was the case with the voltage stress tool, the oscillation tool should not be expected to provide conclusive results when the incoming PMU data does not include sufficient data related to the disturbance or event. The extent to which the tool can "see" beyond the incoming data stream to identify potential problem situations should be considered an important future work item.

(2) Simulation Cases

Cases 9-1, 9-2 and 9-4 are for the oscillation tool and the only difference between them is the amount (percentage) of random load variation. Case 9-1 has 1% random load variation, Case 9-2 has 2%, and Case 9-4 has 4%. The purpose of these variations is to determine if the level of load variation has any impact on the ability of the oscillation tool to predict oscillatory problems/issues/concerns.

For the first oscillation test, four cases were created to test the oscillation monitoring tool. In the simulation, outage of PSSs at two generators and one line were used to promote sustained oscillations. The conditions considered in the simulations include: (1) line out, PSSs in service; (2) line out, PSSs out of service; (3) fault applied then line out, PSSs in service; and (4) fault applied then line out, PSSs out of service. The oscillation tool should identify location of oscillation and increasing severity of situation.

For the second oscillation scenario, one case was created with a forced sustained oscillation. A single governor (upper) power limit was varied sinusoidally in the simulation to create an approximately 1-Hz sustained oscillation. The location chosen where PMU was present on generator GSU high-voltage side. The oscillation tool should identify location of oscillation and severity of situation.

(3) Outputs of Analytical Engines and GSAS Tools

For the first set of cases with the outage of PSS units at two generators, the event detection tool showed that the local oscillatory mode of one of the generators was strongly excited by the disturbance even though the damping ratio of the local mode was still strong at about 8%. Ambient tool, namely the FDD estimates, did not identify this mode because the energy of this local mode was weak compared to the main inter-area oscillatory modes in the PMU data.

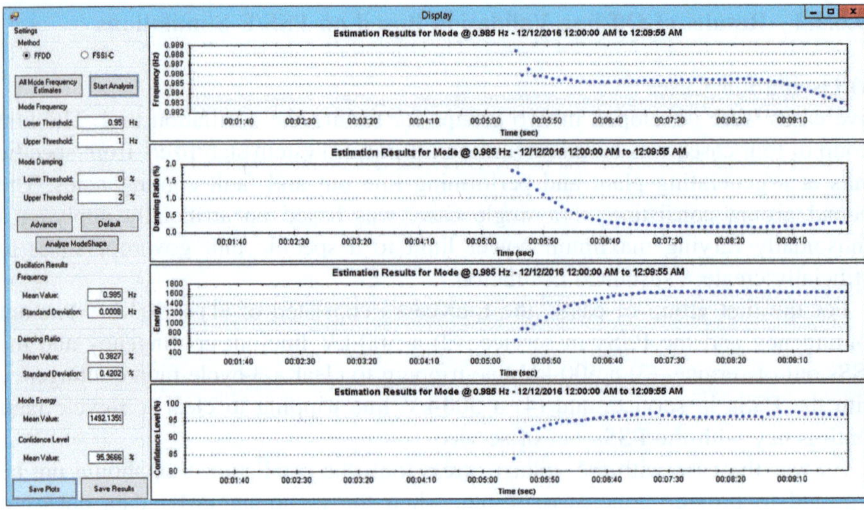

Fig. 19.10 FDD estimation results for the forced oscillation test case

For the forced oscillation simulation case, the modal estimates from FDD when using the current magnitudes as analysis signals are presented in Fig. 19.10. In the simulation, forced oscillation is applied at a specific generator from $t = 5$ min. Figure 19.10 shows that FDD estimates show the presence of forced oscillation shortly after $t = 5$ min with the presence of a 1-Hz estimate that has very low damping values. When current magnitudes are used as oscillation analysis signals, the corresponding mode shape is shown in Fig. 19.11. The most dominant signal in the mode shape correctly points to the source location where the forced oscillation was applied in the simulation. The current magnitude from the PMU close to the source location has the largest mode shape magnitude among all the signals and clearly stands out as the strongest contributor to the oscillation. When voltage phase angles are used, the corresponding mode shape shown in Fig. 19.12 is also able to identify the source location well.

19.6.5 Performance of Transient Stability Monitoring Tool

19.6.5.1 Overview of Transient Stability Monitoring Method

The transient (angular) instability monitoring method is based on real-time tracking of the deviation of angle measurements of different control areas with respect to a weighted system reference angle that is computed as a real-time approximation of the system center of inertia reference frame. The algorithm analyzes the phase

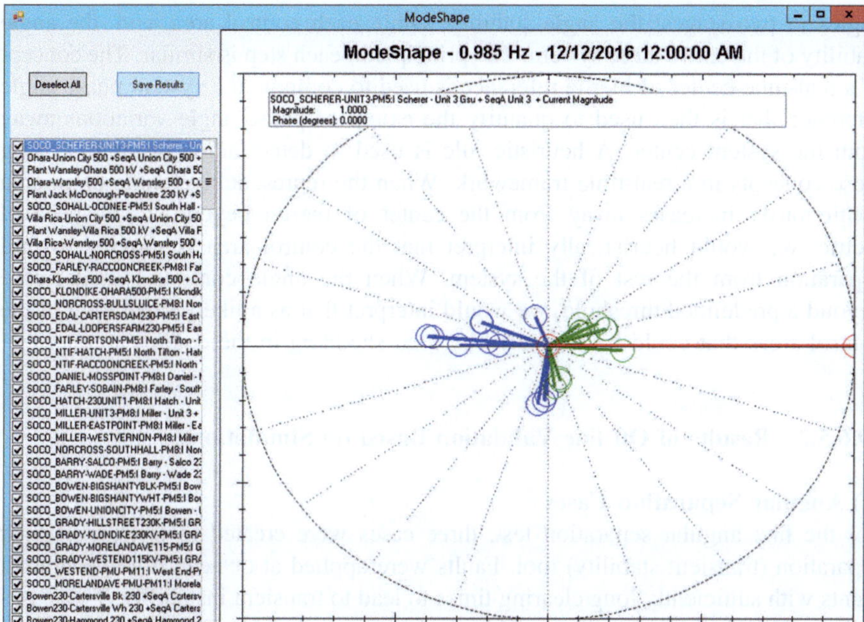

Fig. 19.11 Mode shape of the low damping mode estimate from FDD based on current magnitudes

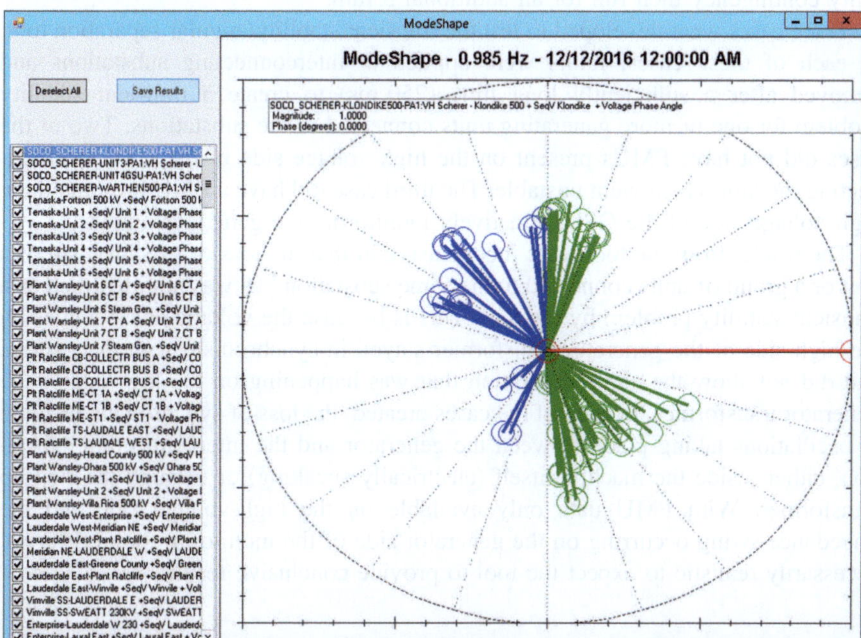

Fig. 19.12 Mode shape of the 1 Hz mode from FDD using voltage phase angles

angles in two stages: the angle stability within each control area, and, the angle stability of the entire large system. The principle in each step is similar. The concept of a real-time center of inertia reference is used to compute the system phase angle reference that is then used to quantify the extent of phase angle variations away from the system center. A heuristic rule is used to detect angle instability using these concepts in a real-time framework. When the representative angle of an area continuously increases away from the center of inertia beyond a pre-specified metric, we would heuristically interpret that the control area is moving toward separation from the rest of the system. When the angle continues to decrease beyond a predefined threshold, we would interpret that as a likely separation of the control areas that could be countered by load shedding in the control areas.

19.6.5.2 Results of Off-line Validation Based on Simulations

(1) Angular Separation Cases

For the first angular separation test, three cases were created to test the angular separation (transient stability) tool. Faults were applied at or near large generating plants with sufficiently long clearing times to lead to transient instability. Two cases without PMUs present on the unstable unit(s). One case with PMUs present on the unstable unit(s). The angular separation tool should identify units losing synchronism with the system to demonstrate the effectiveness of tool without adequate PMU data was unknown. For the transient stability cases, simulation cases have the 1% random load variation and run the simulation for 5 min then apply the instability contingency then run for an additional 5 min.

Three cases were developed to test the transient stability/angular separation tool. In each of these cases, faults were applied at interconnecting substations and removed after a sufficiently long time (250 ms) to create a transient stability problem for one or more generating units connected to the substations. Two of the cases did not have PMUs present on the high-voltage side of the GSU interconnecting the unit which went unstable. The third case did have a PMU present on the high-voltage side of the GSU, effectively monitoring the generator output.

The results from the tool were inconclusive in that the loss of synchronism of a unit or a group of units connected to the same substation bus was not identified as a transient stability problem by the tool. This is because the voltage phase angles on the high side of the generator transformer stayed in synchronism with the system, and did not show the angle separation that was happening on the low side of the generator transformer. In each of the cases created, the loss of synchronism resulted in oscillations taking place between the generator and the interconnecting substation, either inside the machine itself (electrically speaking) or through the step-up transformer. With PMU data only available on the high-voltage side and the impedance swing occurring on the generator side of the monitoring point, it is not necessarily realistic to expect the tool to provide conclusive results.

In addition to the issue of insufficient PMU data, a limitation of the simulation software was also encountered. One to two seconds after the fault contingency was cleared, and long after the unit had gone unstable, the simulation software apparently reset the machine rotor angle to a stable and essentially constant value. This would appear to the tool as a normal condition and no predictive (undesirable) result would be expected. Whether or not the tool should have predicted the instability event based on only 1–2 s of limited data is a subject for possible future work.

It is important to note that the tool is designed primarily to detect islanding events where large portions of the system lose synchronism with other large portions. Each large portion would naturally be assumed to have multiple PMUs present, and all the angles within one large portion would most certainly pull away from the angles in another large portion. No test case for this functionality was created due to the unrealistic nature of this type of event in the SBAA; such a test is a subject for possible future work.

(2) Outputs of Analytical Engines and GSAS Tools

Figures 19.13 and 19.14 show all frequency and voltage phase angles of Case 15-1.

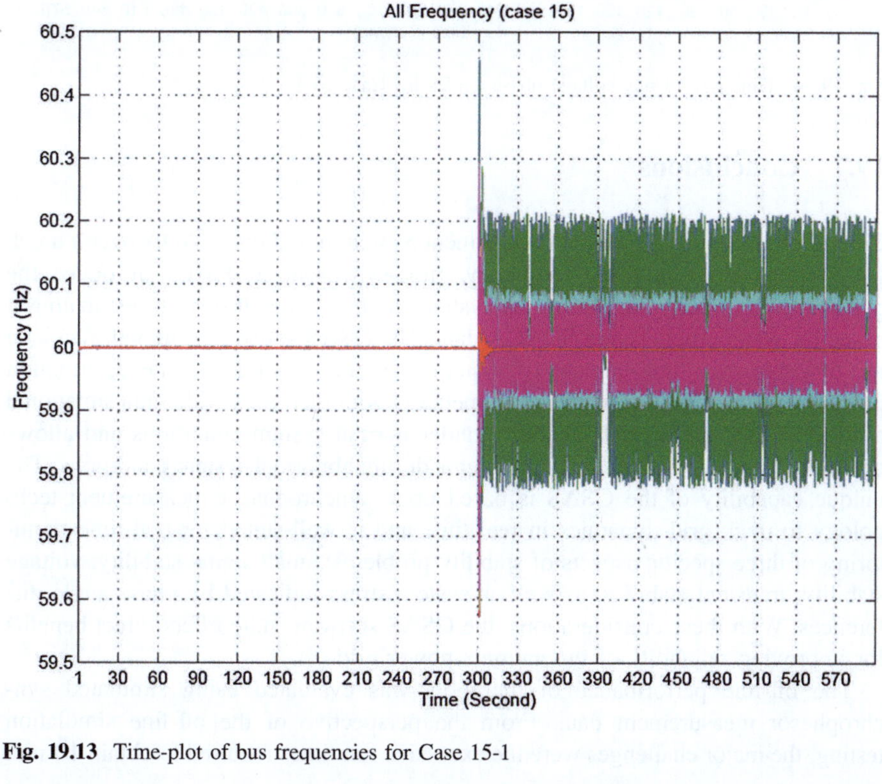

Fig. 19.13 Time-plot of bus frequencies for Case 15-1

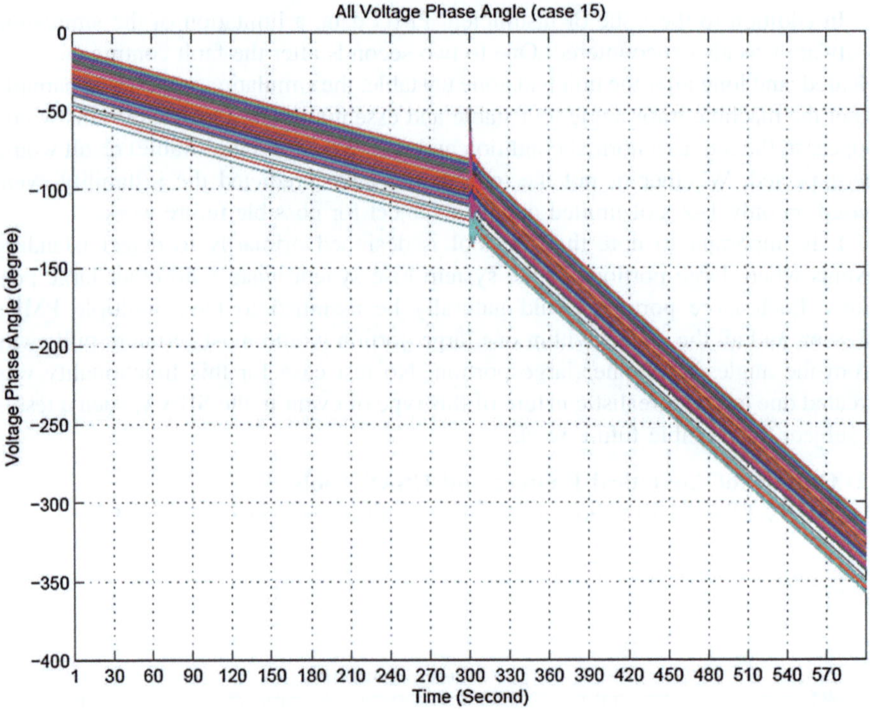

Fig. 19.14 Time-plot of bus voltage phase angles for Case 15-1

19.7 Conclusions

Compared to a current Energy Management System based on SCADA over a much lower monitoring sampling frequency, time-synchronized PMUs introduce the possibility of directly measuring the system stability status instead of estimating it based on system models and SCADA data. The GSAS software suite will facilitate the dynamic, real-time capture of system operating conditions. The information provided by the GSAS provides grid operators with increased wide-area situational awareness and operational reliability under normal system conditions and allows operators to anticipate and detect problems during abnormal system conditions. The unique capability of the GSAS is based upon synchrophasor measurement technology to track grid dynamics in real time and is well-suited for real-time monitoring of three specific aspects of stability problems: small signal stability, voltage stability, transient stability, as well as system stress indicated by phase angle differences. With these considerations, the GSAS software suite offers direct benefits for improving reliability of the nation's power grid.

The off-line performance of the tools was evaluated using simulated synchrophasor measurement data. From the perspective of the off-line simulation testing, the major challenges were the excessive run times required to obtain 10 min of simulated PMU data and the large files associated with the amount of output involved. While these challenges are not debilitating, they do serve to significantly complicate the overall testing process and make it very difficult to develop the large number of off-line tests that may ultimately be required to fully evaluate the capabilities of the tools.

In general, the off-line testing showed that the tools accurately predicted the events for which they were designed provided that they were given sufficient incoming PMU data. In every case in which insufficient incoming data was present, the tools produced inconclusive results. In every case in which sufficient incoming data was present, the tools produced conclusive and accurate results.

The major conclusion is that it is critically important to consider the applications for which the data will be used when installing PMUs. Information from PMUs which might be important for one application could be less valuable for another application. For the tools to function properly, they must receive PMU input which adequately reflects the disturbance or system condition to be detected. Examples of this include installing PMUs in lower voltage portions of the transmission system (the 115-kV network in the particular portion in SBAA) and installing PMUs on the low-voltage side of generator step-up transformers. To date, the very vast majority of the installed PMUs in SBAA are on the higher-voltage portions of the transmission system and the number of monitored generators is very limited. This practical reality has a direct impact on the ability of all the tools to perform their intended functions.

Based on the testing conducted and the existing deployment of PMUs within SBAA, it appears that limited use of the tools is possible in the operations environment provided that proper training is provided and realistic expectations exist. Only through additional experience and testing, including in the operational environment with realistic constraints, can the tools be further improved, while possibly prioritizing PMU deployment across the SBBA so as to maximize the potential of the tools.

References

1. Liu G, Venkatasubramanian V (2008) Oscillation monitoring from ambient PMU measurements by frequency domain decomposition. In: 2008 IEEE international symposium on circuits and systems, Seattle, WA, 18–21 May 2008
2. Venkatasubramanian V, Liu X, Liu G, Zhang Q, Sherwood M (2011) Overview of wide-area stability monitoring algorithms in power systems using synchrophasors. In: 2011 American control conference, San Francisco, CA, 29 June–01 July 2011
3. Sherwood M, Hu D, Venkatasubramanian V (2007) Real-time detection of angle instability using synchrophasors and action principle. In: 2007 IREP symposium—bulk power system dynamics and control—VII, revitalizing operational reliability, Charleston, SC, 19–24 Aug 2007
4. Hauer JF, Demeure CJ, Scharf LL (1990) Initial results in Prony analysis of power system response signals. IEEE Trans Power Syst 5(1):80–89

5. Trudnowski DJ (1999) Making Prony analysis more accurate using multiple signals. IEEE Trans Power Syst 14(1):226–231
6. Sarkar TK, Pereira O (1995) Using the matrix pencil method to estimate the parameters of a sum of complex exponentials. IEEE Antennas Propag Mag 37(1):48–55
7. Crow ML, Singh A (2005) The matrix pencil for power system modal extraction. IEEE Trans Power Syst 20(1):501–502
8. Papy JM, Lathauwer LD, Huffel SV (2006) Common pole estimation in multi-channel exponential data modeling. Sig Process 86:846–858
9. Liu X (2012) Real time voltage stability monitoring and control using synchrophasors. PhD thesis, Washington State University, Pullman, WA, Aug 2012
10. Wang T, Zhang L (2006) Frequency and spatial domain decomposition for operational modal analysis and its application. Acta Aeronautica et Astronautica Sinica 27(1):62–66
11. Brincker R, Ventura CE, Andersen P (2001) Damping estimation by frequency domain decomposition. In: Proceedings of international modal analysis conference (IMAC), Kissimmee, Florida, pp 698–703
12. Liu G, Quintero J, "Mani" Venkatasubramanian V (2007) Oscillation monitoring system based on wide area synchrophasors in power systems. In: 2007 IREP symposium—bulk power system dynamics and control—VII, revitalizing operational reliability, Charleston, SC, 19–24 Aug 2007

Chapter 20
A Cloud-Hosted Synchrophasor Data Sharing Platform

Eugene Litvinov, Xiaochuan Luo, Qiang Zhang, Ken Birman, Theodoros Gkountouvas, Dave Anderson, Carl Hauser and Anjan Bose

20.1 Introduction

During the August 14, 2003 blackout, automatic protection systems worked well at the interface between New York and New England, limiting sustained outages in New England to approximately 2,500 MW of load in Connecticut, Massachusetts, and Vermont. Nonetheless, the outage left 50 million people without power in nine states and one Canadian province. A major factor contributing to the blackout was inadequate situational awareness: Transmission operators were uncertain of the state of the regional network and lacked ways to obtain trustworthy "ground truth" information or to coordinate their remedial actions. As a result, the US-Canada Power System Outage Task Force recommended development, evaluation, and adoption of a new generation real-time tools for operations, using time-synchronized data for wide-area situational awareness [1, 2].

One of the post-2003 blackout recommendations was that synchrophasor technology be deployed within the power transmission network, and an effort to achieve this goal has been underway for some time. The pace of rollout accelerated with national investment through the U.S. Department of Energy's Smart Grid Investment Grant (SGIG) program, funded by the 2009 stimulus bill [3]. As a result, synchrophasor state tracking is now a viable option, with the potential to provide substantial value for power system operations.

E. Litvinov · X. Luo (✉) · Q. Zhang
ISO New England Inc., Holyoke, USA
e-mail: xluo@iso-ne.com

K. Birman · T. Gkountouvas
Cornell University, Ithaca, USA

D. Anderson · C. Hauser · A. Bose
Washington State University, Pullman, USA

S. Nuthalapati (ed.), *Power System Grid Operation Using Synchrophasor Technology*, Power Electronics and Power Systems,
https://doi.org/10.1007/978-3-319-89378-5_20

In this chapter, the authors discuss the ISO New England (ISO-NE) synchrophasor deployment and key issues with today's PMU data exchange among different regions, and then present a new cloud-hosted synchrophasor data sharing platform, named GridCloud, to capture PMU data into a cost-effective data storage and analysis platform for wide area monitoring and situational awareness. GridCloud is a free, open-source platform that includes a PMU-based state estimation application, and can easily be customized to run additional applications. The authors envision that over time, the system will scale to capture more and more data, and to host an ever-growing collection of analytic tools for system operations.

20.1.1 Synchrophasor System at ISO New England

With the support from the SGIG, ISO-NE, and the region's seven major transmission owners (TOs) have invested $14.9 million on a Synchrophasor Infrastructure and Data Utilization (SIDU) project. The project started on July 1, 2010, and its implementation phase was completed on June 30, 2013. This was followed by a two-year observation phase, which completed on June 30, 2015. The core objective of the SIDU effort was to create a synchrophasor infrastructure and technology platform, upon which advanced analysis and wide area visualization tools could be developed and operated to enhance situational awareness.

At the time of this writing, the major components of the New England synchrophasor system include:

- PMU installations at 44 substations in New England, with adequate coverage to achieve full visibility of the 345 kV network, and redundancy to maintain visibility under various PMU failure scenarios.
- Eight Phasor Data Concentrators (PDCs) at the ISO and seven transmission owners.
- Communication infrastructure to support streaming PMU data from substations to the transmission owners and then to the ISO.
- GE's PhasorPoint application for enhanced wide-area monitoring and situational awareness.
- V&R Energy System Research's Region of Stability Existence (ROSE) application to compute the operational stability boundary and margin.
- In-house developed online PMU Data Quality Monitoring System (DQMS) to monitor and alert on PMU data validity and health.

Figure 20.1 shows the overall architecture of the New England synchrophasor system completed through the SIDU project in June 2015. The synchrophasor data network is shown in what is referred to as a "bottom-up tree convention." Starting at the bottom of the figure, each PMU sends data to Transmission Owner's Phasor Data Concentrator (PDC) through the corporate WAN network. The PDC receives synchrophasor data from different PMUs, time-aligns the records, and then relays

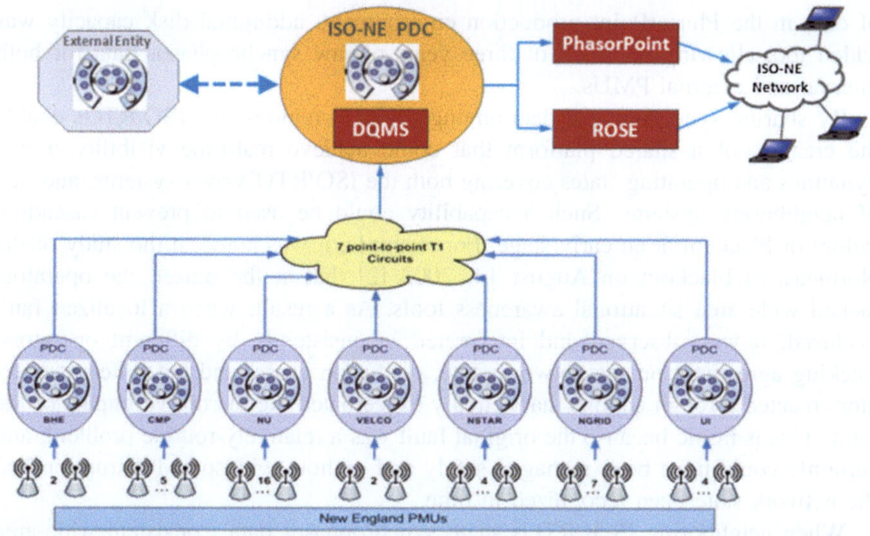

Fig. 20.1 New England SIDU project implementation in June 2015

the resulting processed data "up" through point-to-point T1 circuits to the ISO-NE's PDC. ISO-NE's PDC performs a second stage of time alignment as it receives data from different transmission owners, and then feeds advanced applications with one concentrated real-time data stream.

20.1.2 PMU Data Exchange with External Regions

Although the SGIG program makes it possible for each of the major regional operators to instrument its own network, data sharing is a prerequisite for wide-area regional situational awareness. The New England SIDU project initially lacked an infrastructure to acquire external regions' PMU data. Accordingly, in 2016 ISO-NE initiated an effort aimed at an exchange of PMU data with NYISO, PJM, and MISO.

Each external region's PMU data is acquired over the existing Eastern Interconnect Data Sharing Network (EIDSN), thereby avoiding any additional costs for communication infrastructure, routers, and firewalls. The EIDSN is designed to allow authorized operating entities to securely share operational reliability data, including both SCADA and synchrophasor data, and is intended to promote reliable and efficient operation of the Eastern and Quebec Interconnections, a goal consistent with the new effort.

New PDC servers were installed at the ISO-NE to receive and concentrate external PMU data, stream the resulting data to the existing PDCs, and then send the full external dataset to the PhasorPoint application together with internal ISO-NE PMU data for wide area monitoring. In order to support the added volume of data in the PhasorPoint production environment, additional disk capacity was added too, allowing retention of three years of raw synchrophasor data for both internal and external PMUs.

By sharing synchrophasor data among different regions, the ISO/RTOs enable the creation of a shared platform that could achieve real-time visibility of the dynamics and operating states covering both the ISO/RTO's own systems, and that of neighboring systems. Such a capability could be used to prevent cascading failure or blackout at an early stage. For example, it was noted in the study of the Northeastern blackout on August 14, 2003 [2] that at the outset, the operators lacked wide area situational awareness tools. As a result, when a localized fault occurred, it was observed and interpreted inconsistently by different operators. Lacking agreement on the network state, confusion arose, and the different operators reacted through actions that actually exacerbated the disruptive impact of the fault. This is ironic because the original fault was a relatively routine problem, and certainly could have been managed safely and without widespread disruption had the network state been recognized in time.

When neighboring ISOs/RTOs share synchrophasor data, consistent wide-area situational awareness is feasible: Even distinct grid operators working in different control centers will still see a consistent picture of the evolving network state, in real-time, because they could work from the direct state measurements across regions based on the shared underlying synchrophasor input data streams. As a result, operators are able to coordinate actions both implicitly (because they react on the basis of the identical situational information) and explicitly (when a contingency requiring collaborative planning or coordinated reaction arises, the individuals who need to take action can base that coordinated response upon an agreed-upon network state). But if each ISO/RTO *separately* collects data and then runs its own private state estimation and analytics, then during a crisis they might still disagree on system state. The key to successful sharing is that they should see the *same* data, should agree on the wide-area network state, and then base actions on a consistent, agreed-upon, joint perspective.

Today's most pressing opportunity centers on wide-area regional state estimation and coordination, but tomorrow, the same concept could be taken even further. Sharing on an even larger scale would permit cooperative reasoning to be taken even further, enabling a new kind of "over the horizon" capability, with great potential for further benefits. Not only could the national grid be better managed in this way, it could also be better protected against disruptions such as large-scale weather events or earthquakes, deliberate attempts to destabilize the grid.

20.1.3 Problem Statement of the Existing Implementation of PMU Data Exchange

Similar to the ISO-NE, most synchrophasor data networks in North American were implemented in a bottom-up tree structure that facilitates unidirectional data flow. Data generated from PMUs are sent through communication networks and several

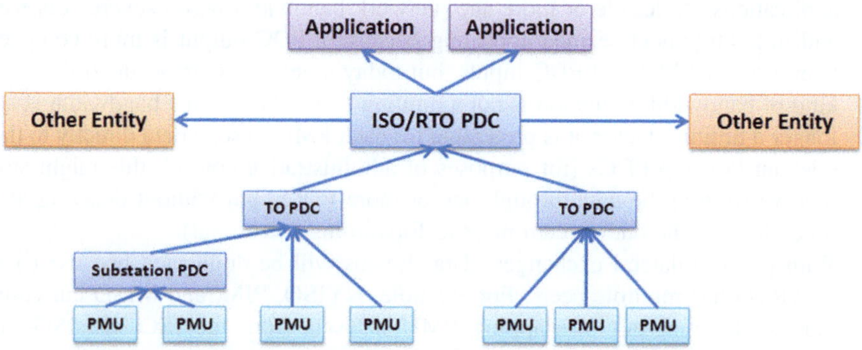

Fig. 20.2 Typical PMU/PDC structure

layers of PDCs using point-point real-time network flows (IP Unicast) to ISO/RTO PDCs, from which data are passed onto various applications. As seen in Fig. 20.2, PDCs are designed to time-align measurements and combine different PMU/PDC streams into one stream, the intent of which is to save communication bandwidth and simplify configuration. Another reason for time-aligning data is that many of today's applications operate on vectors of PMU values obtained at a single timestamp. Several data protocols are in use today including IEEE C37.118-2005, IEEE C37.118-2011, C37.118.2, and IEC 61850-90-5.

Although widely used, today's implementation of PMU data exchange has a number of drawbacks:

- This chained PDC network accumulates and increases the total time delays. Small communication network delays arise as a side effect of the way that modern computer networks handle overload when traffic from multiple sources traverses a shared network link or router. A PDC needs to wait if no data has been received from some PMUs because transient congestion on a stream can easily delay it. The PMU or communication failures will result in a maximum latency (the configured PDC waiting time). When data reaches the next level, the receiving PDC inherits the previous latencies and may experience further delay. This delay is exacerbated because PDCs must wait for multiple streams, and thus the worst delay on any single incoming stream tends to be the performance-limiting factor for the combined output stream.
- There is substantial complexity to managing a PDC hierarchy. For example, the PDC operator must configure the PDC to coordinate waiting time across the hierarchy: not an easy task.
- The time alignment at each PDC hierarchy might be unnecessary. As a recommended practice, most entities have each PDC perform time alignment and only send the concentrated data stream to PDCs at the next level or advanced applications. However, from an application's point of view, time alignment only needs to occur once, at the last step. Therefore, the relaying PDCs do not need to time-align data (additional delays), except for the end ones that feed

applications. A decade or more ago, network bandwidth was a severe concern, and time alignment seemed appealing because a PDC output is more compact than a set of PMU and PDC inputs, but today's networks are so rapid that this kind of bandwidth reduction is not a limiting factor. Given that bandwidth is no longer a limiting factor, it is preferable for each PMU to send data directly to the relevant last-hop PDCs (for purposes of administrative control, this might still involve *routing* the data through one or more PDCs, but without delaying and time-aligning the data stream prior to forwarding the output).

- With purely bilateral exchanges, data streams will be duplicated because each ISO/RTO has multiple peers. For example, NYISO, PJM, and MISO currently request different but overlapping PMU datasets from the ISO-NE. ISO-NE therefore is required to configure three different output streams, sending each over the EIDSN. Similar situations arise at NYISO, PJM, and MISO.
- Each entity has to model and maintain its own and other regions' PMU data. For example, ISO-NE has to model PMU data from NYISO, PJM, and MISO in the PhasorPoint application for wide area situational awareness. Similarly, NYISO has to model PMU data from PJM, MISO, and ISO-NE in their advanced visualization applications.
- Today most entities focus on sharing PMU data, instead of sharing common displays among regional grid operators. Since each ISO/RTO uses different vendors and applications for wide area monitoring and situational awareness, the grid operators may not be able to communicate straightforwardly when they call each other after identifying a wide area problem.

With the continued growth of synchrophasor technology by US utilities and the inefficiency of the current implementation of PMU data exchange for wide area monitoring, there is a need to look at new solutions which are building on mature and tested technologies, and use them in a small development process that focuses on leveraging the best of breed solutions in ways that have been successful in the past.

In the next section, the authors present a solution to this problem. Researchers from the ISO-NE, Cornell University and Washington State University have built a platform based on the existing cloud computing infrastructure, "hardening" it for the purpose of PMU data exchange and wide area monitoring. The resulting platform offers a shared infrastructure, is highly robust, and represents a response to the various concerns elaborated above.

20.2 Why the Cloud?

20.2.1 Overview of Cloud Computing Technology

The term cloud computing itself is extremely "flexible," and is used in conjunction with multiple styles of computing and communication. According to U.S. National

Institute of Standards and Technology (NIST), cloud computing can be defined as "*a model for enabling ubiquitous, convenient, on-demand network access to a shared pool of configurable computing resources (e.g., networks, servers, storages, applications, and services) that can be rapidly provisioned and released with minimal management effort or service provider interaction*" [4]. The cloud computing paradigm centers on easy access to large-scale computing resources via a network (i.e., Internet) at relatively low cost, and with the capability of quick reconfiguration to add resources or release them, using a pay-as-needed cost model. Cloud computing can be a basis for a robust and efficient solution that might briefly use computing resources far beyond the economic reach of most organizations, and that has outstanding operational efficiency, resilience, and sustainability.

While cloud failures do occur, it is very rare for an outage to impact more than one data center at a time. Thus, much as for other power grid reliability goals, one can use redundancy to anticipate and overcome the types of failures seen in operational cloud settings. Further, cloud security has become stronger and stronger: One can already protect cloud systems against a wide range of attacks, and with new hardware security options emerging at a fast pace, it will not be long before cloud systems can even be protected against intrusion by the data center operator itself.

Various drivers are contributing to the surge of cloud computing technology, including the dramatic decrease in costs for hardware optimized to support cloud computing, the simultaneous increase in computing power and storage capacity, the advent of multiple-core and multi-thread computing architectures, customer demand for stronger and stronger security, and the exponential growth in computing and data storage demands on the part of scientific and commercial applications.

To accommodate the needs of different customers, cloud computing services can be provided in different ways. According to the levels of the capacity and service provided, cloud computing services can generally be divided into at least the following three categories [5, 6]:

- Infrastructure as a Service (IaaS): A cluster of virtualized resources (e.g., CPU capacity, data storage, network bandwidth, and basic operating system) is provisioned. Once being provisioned, customers run software stacks on the top of the virtualized resources and are able to resize the scale of resources as needed.
- Platform as a Service (PaaS): An environment on which customers can easily develop and deploy applications on a large scale of computing infrastructure to meet changing demands.
- Software as a Service (SaaS): Online applications are delivered as Web services to users which allows for alleviation of the burden of software maintenance and upgrade on the customers' side.

According to the physical location and distribution of cloud computing services, a cloud can be further classified as [5, 6]:

- Public cloud: A third-party provider offers and maintains a multi-tenant cloud infrastructure and/or services to customers on a subscription basis that allows for the optimum utilization of computing resources. Here the physical server might be simultaneously used by other clients.
- Private cloud: Cloud computing infrastructure that uses cloud software but runs within a company's own data center and is dedicated to internal users and/or partner users.
- Hybrid cloud: Mixed usage of the client's own computing resources and resources hosted on a public cloud. Such an approach allows the customer to employ public cloud services for specific purposes that the customer's own capacity is inadequate to address, securely combining the cloud data and output with the customer's own data and systems.

Commonly recognized benefits of cloud computing technology include:

- Operational efficiency and cost savings
- Operational resilience through state-of-the-art backup services and security schemes as well as fully redundant hardware infrastructure and services
- Scalable and Flexible
- Almost unlimited storage
- Shortened development life cycle
- Ease of interoperation
- Shared data and applications for easy collaboration among different entities.

Increasing attention has been drawn to potential security and privacy issues with cloud computing implementation. The risk is evident: In a cloud setting, a third-party provider hosts data that the end users traditionally have stored on their internal computers behind many layers of firewalls. However, customer demand for ironclad cloud security has fueled a wave of creative work, resulting in powerful new security mechanism with better and better data and computation protection, and remarkably good options for secure interoperability, and for integration of cloud and private systems.

Today's top-tier cloud service providers such as Amazon Web Service (AWS) deploy state-of-the-art security schemes for enhanced data security and privacy protection [7]; while the out-of-box configuration is often not an exact match to the needs of the power sector, past experience shows that the underlying security building blocks are powerful and flexible, and that the needed solutions can be created by configuring them properly. These features center on the AWS virtual private cloud (which uses software encryption to protect data and implement firewalls). If one trusts the software used to implement them, the resulting security is extremely good.

Meanwhile, new and even more exciting options are emerging, in which one would need to trust the hardware but not the software or the data center owner. For example, with Intel's SGX architecture [8, 9], a cloud hosts computing "enclaves" that even the data center owner itself cannot penetrate: The data center hosts the system and provides computing cycles and storage, yet the computation and stored data is encrypted at all times, using keys that reside only at the customer's site. Thus, even an intruder who has fully compromised the cloud data center would have no ability to observe the data streams, network state, or other analytics carried out in a system like GridCloud: That attacker's power would be limited to crashing or overloading computers. The project team plans to do experiments with SGX, but in fact SGX is just one example among an array of proposed hardware technologies for security.

The conclusion is that for a properly configured system, Internet and cloud security is already surprisingly good today, and far better than one might imagine without looking closely at the choices. Moreover, the authors see strong reasons for optimism that security will continue to improve in the future.

20.2.2 GridCloud Platform

The GridCloud platform was created by Cornell University and Washington State University, as a research effort sponsored by the DOE's ARPA-E GENI program. GridCloud aims to achieve a highly scalable, reliable, and secure computing infrastructure strong enough to operate a nationally critical infrastructure such as the power grid. It extends a standard commercial cloud computing infrastructure, configuring it for maximum security and enhancing real-time and fault-tolerant features through use of redundancy and replication. GridCloud thus seeks a balance between the cloud's widely cited cost-effectiveness and flexibility and the special requirements of PMU data sharing and wide area monitoring.

The core technologies used in GridCloud draw upon decades of experience and systems work by the project team on issues of reliability, consistency, fault-tolerant, and real-time availability. A book-length treatment of the issues that arise, the underlying theory, and the protocols employed to achieve correct behavior is available [10]. Here the authors will discuss higher level solutions created using those techniques, but without delving deeply into the theory or the structure of the software tools.

The resulting solution is quite flexible. In particular, GridCloud can be configured to run on any desired number of cloud data centers by using data and code replication for both fault tolerance and performance. The system defines a standard data collection infrastructure, which can include standard databases and solutions such as OpenPDC. By hosting shared applications on machines that are rented only as needed, GridCloud can offer the ISOs and TOs authorized access to the platform —a highly effective collaborative infrastructure.

GridCloud proves that mission-critical real-time solution could be hosted on today's cloud by careful attention to security policy together with additional technology aimed at high-availability and real-time performance. The system supports a mix-and-match approach to deployment and has following key components:

20.2.2.1 The GridCloud Security Architecture

The GridCloud system obtains security by careful configuration of Amazon's virtual private cloud and network link encryption options. No new mechanisms were required. The security solution trusts the core security features offered by Amazon in support of its virtual private cloud and network link-layer security, including implementation of the software, and Amazon's correct operation of their infrastructure. Further, it trusts Amazon's system operators not to attack the active system, for example by freezing a copy of the system and then inspecting it offline using forensic analysis software tools. Such actions are not possible for other cloud customers, and they are intrusive enough that they would trigger audit and other monitoring alarms, but are certainly attacks that data center owner intent on breaking into a virtually private cloud could conduct. Thus, GridCloud is secure against a large class of possible attacks, but its security model could be defeated if the cloud-hosting company were using flawed software, or had been severely compromised in one of the ways the authors have outlined.

In the future, if GridCloud begins to take more advantage of hardware security, the "attack surface" of the system can be reduced in many ways. On the other hand, like any technology, it will still ultimately be dependent upon the hardware on which it runs, and core elements of the software used to operate that hardware.

20.2.2.2 The GridCloud Data Collection Layer

The GridCloud data collection layer captures incoming PMU data over one or more cryptographically secured network links, writes all received data into files to create a historical record, and also relays it to the state estimation subsystem. This system automatically establishes secure, replicated connections and will automatically reestablish connectivity if disruption occurs. Redundant data collection is employed to overcome network timing issues and ensure that critical data will still flow into the system even if some network links fail.

20.2.2.3 The GridCloud Archival Data Storage Subsystem

The GridCloud data collectors operate by writing the received data into files and simultaneously forwarding the data for state estimation processing. GridCloud stores data in a variety of formats, favoring standards. For example, streams of PMU data are stored into append-only logs in the IEEE C37.118 format. Network state estimates are stored as a table, in which each row describes an element of the network model for the grid and gives that element's estimated state. The storage system also records both the original timestamp of the PMU data, and the platform time at which the data was actually archived. By so doing, the project team can sense situations in which a GPS time receiver malfunctions and streams containing data with incorrect timestamps, and can also support time-based retrieval (snapshots of the exact system state at a designated instant in time, accurate to the millisecond).

Before deciding to implement any solution, the authors evaluated a number of candidate file systems relative to the requirements posed for this archival storage role. Thorough review and experiments revealed limitations that precluded use of existing solutions: Prior systems lacked ways to exploit embedded timestamps, and the prior approaches to retrieving snapshots embodied temporal inaccuracies, performance issues, and inconsistencies when files are being updated rapidly [11, 12]. Accordingly, the authors developed a new solution: the Freeze Frame File System (FFFS) [13] which offers a quick way to access data at any desired temporal resolution, from any point in the past or present history of the system, and employs a memory-mapped access structure and hardware assisted data transfers (RDMA) to achieve high performance in support of smart grid computational tasks. FFFS also provides logical consistency in the sense of Chandy and Lamport [14].

20.2.2.4 Cloud Manager

Management of the GridCloud infrastructure posed challenges: Existing cloud management tools focus on end-users who are accessing Web pages, a scenario remote from wide area monitoring. Accordingly, the GridCloud required a new robust system management tool. Cloud Manager (CM) is an extension of the UNIX "Makefile" infrastructure. The tool supports the standard dependency-triggered model used in Make (and the standard syntax, too), but extends the model to also react to system events such as nodes failing or joining, events being sensed in the incoming data stream. CM encodes such events as XML files and when something occurs, the file update triggers reactive behavior as specified in the CM script, which can be customized for each use-case. Actions are similarly initiated: CM outputs a file, and it is used to drive the desired response. An off-the-shelf constraint solver is used to carry out the optimizations needed for purposes such as job placement on the available cloud nodes, deciding which data collector will handle which data stream.

CM also manages the desired suite of applications. At present, CM application support is focused on a state estimator and visualization tool, but new applications can easily be added to the system and either configured to run continuously, 24 × 7, or on demand. CM can also automatically configure applications with redundancy. The authors have worked with either one or two Amazon data centers, and in one case, went further and also triple-replicated the entire cloud infrastructure within one Amazon AWS data center, to understand the extent to which doing so could conceal scheduling and other latency factors.

20.3 Proof-of-Concept Cloud-Hosted Wide Area Monitoring System

In 2015, ISO-NE established a collaboration with Washington State University and Cornell University to employ GridCloud as a proof-of-concept infrastructure, collecting synchrophasor data in a wide-area monitoring deployment and then carrying out cloud-hosted state estimation in real-time. This is the first such application of cloud computing in the electricity industry, which has been cautiously turning to the cloud, migrating various roles as the needed reliability, security and privacy concerns have been addressed. The project seeks to demonstrate a reliable, secure, and fault-tolerant cloud-hosted synchrophasor platform and wide area monitoring system and to carefully evaluate performance, scalability, and cost, but not to actually use the system in production.

20.3.1 Conceptual Overview of the Cloud-Hosted Synchrophasor Platform

Figures 20.3 and 20.4 give a conceptual overview of the resulting cloud-hosted synchrophasor platform. The synchrophasor data are streamed from PMUs in the substation directly to the cloud or from PDCs at the TO or ISO/RTO to the cloud-hosted data collectors. The cloud data collectors then route the PMU data to various cloud-hosted applications, such as Regional Data Repository and Real-Time Phasor State Estimator (RTP-SE). The RTP-SE performs an

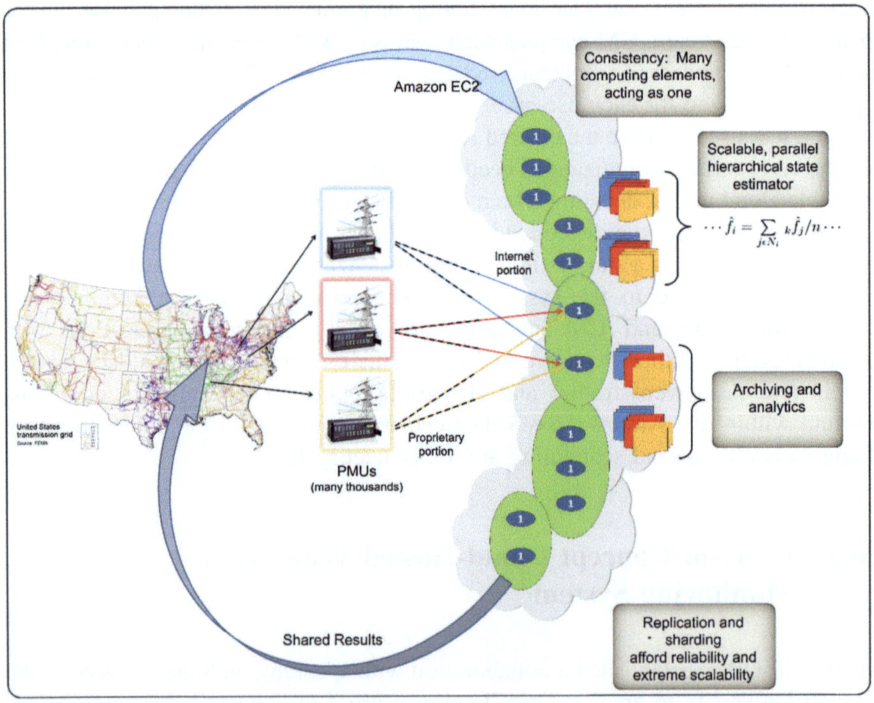

Fig. 20.3 GridCloud infrastructure diagram

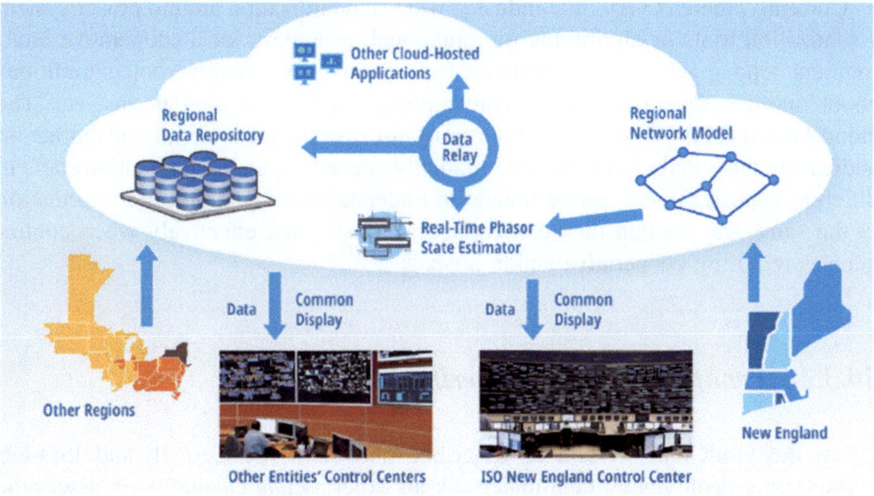

Fig. 20.4 Overview of the cloud-hosted synchrophasor platform

interconnection-wide, time-synchronized assessment of the voltage magnitudes, phase angles, and transmission line flows and provides common visualization displays for real-time collaboration among grid operators. The regional entities can also subscribe to each other's PMU data or retrieve historical data from the central repository. Because data feeds, the data archive, and the state estimator are shared, distinct entities can see consistent information at all times.

The cloud-based RTP-SE is a core component which performs state estimation at one-second intervals or faster—a much higher update rate than the current SCADA-based EMS state estimation which occurs at 3-min intervals in the ISO-NE. With sufficient computational resources, the RTP-SE could be updated even at the PMU data rate so that the power system dynamics are captured.

The RTP-SE offers the capability to:

- Provide time-synchronized state estimation that leverages PMU data and accurate GPS time information to eliminate state estimation errors caused by time-skew;
- Calculate "pseudo" PMU measurements at unmeasured buses and lines and fill in missing PMU data;
- Cross-validate the PMU data at different substations, allowing data quality enhancement;
- Employ cutting-edge linear state estimation methods, which are fast, have no convergence issues, and work properly even under islanding conditions;
- Use a high sampling rate to achieve dynamic visibility of disturbances, disturbance propagation, frequency response, and system oscillation;
- Establish real-time models to determine stability margin.

Currently, there is very little industry-wide standardization around power system visualization tools or algorithms for situational awareness for a cooperative environment among entities. Different control centers have deeply rooted traditions about one-line schematics, color conventions, overloading alert/alarms, etc. The cloud-hosted RTP-SE offers an opportunity for operating entities to collaborate on addressing this vitally important and mutually needed capability. Operation staffs in different control centers can be trained to understand and interpret the information in the same way so that they can collaborate easily and effectively when contingencies requiring cooperative action arise.

20.3.2 Proof-of-Concept Experiment Setup

Using the GridCloud system as described above, WSU, Cornell, and ISO-NE carried out a deployment, then undertook an experimental evaluation to assess the performance of the system, specifically focusing on both internally and externally observable latencies, on the time to recover from failures, and on whether the system consistently delivered the same data and results when running multiple, parallel instances. In addition, the demonstration system was instrumented to evaluate any additional latency introduce by the use of encryption of communication channels.

The experiment uses real PMU data captured by ISO-NE. Twenty-one seconds of recorded historical data were replayed at 30 measurements/second with their timestamps altered to reflect the current time. The 21-s of data were "looped" to provide continuous data streams. The results below are from a 25-min run of the system.

There were a total of 73 PMUs. The data were split across two data source machines, with 31 PMUs streaming from ISO-NE and 42 PMUs streaming from Cornell to the cloud, to mimic a situation in which two or more ISOs might collaborate to share data via the cloud platform. Each data source was emulated, and the data was transmitted via a PMU data replay and streaming software, designed to transmit each measurement at the proper instant in time.

The overall experimental design is seen below in Fig. 20.5. PMU data was sent from Cornell and ISO-NE to the AWS cloud via SSH tunnels in IEEE C37.118 format. These data are captured by a set of data collectors in the cloud, then archived in the cloud, as well as processed by the RTP-SE application software in the middle and a synchrophasor data visualizer on the right. The RTP-SE results were also relayed back to the Cornell in the form of additional C37.118 feeds, representing the reconstructed power system states after the state estimation solutions. The raw PMU streaming from ISO-NE was also relayed back to the Cornell in the form of C37.118 format, representing raw PMU data exchange between different entities.

To evaluate the costs of redundancy for fault tolerance, all data were mirrored, with one copy sent to a GridCloud system running on an AWS data center in Virginia, and the other, in Oregon.

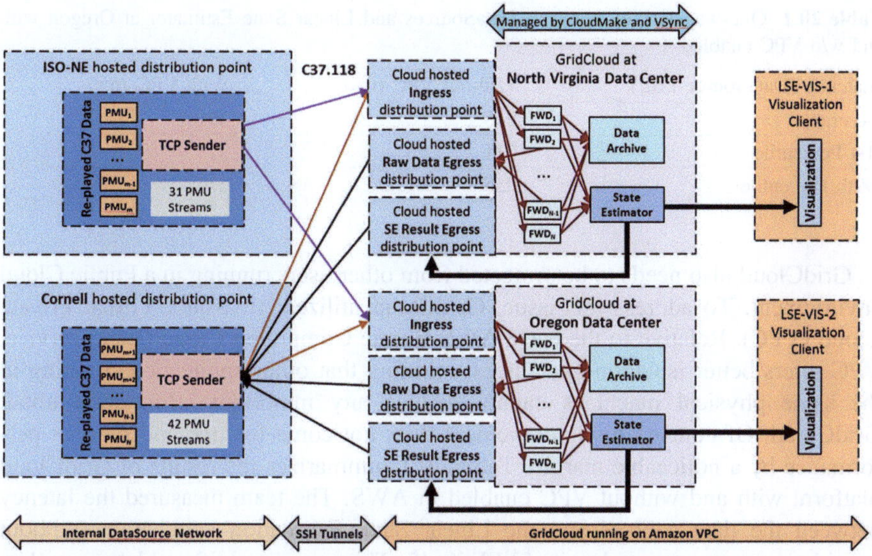

Fig. 20.5 Overall design of the Proof-of-concept project

The RTP-SE used in the experiment was the Linear State Estimator (LSE) created at Washington State University by Anjan Bose and his colleagues [15]. This is a complex application that includes an internal communication bus and OpenPDC, as a front-end for ensuring that only time-aligned data are passed to the actual application, and of course an implementation of the linear state estimator algorithm. Cloud Make is used to track the health of the LSE's components and, if anything fails, can cleanly shut down and then reboot the entire LSE application.

The power system network model used by the LSE was based on a PSS/E planning model that matched the operational topology at the time when the PMU data was captured by the ISO-NE.

20.3.3 *Experiment Findings*

20.3.3.1 Security

Three defenses are utilized at different parts of GridCloud in order to protect the system against a variety of security threats. First, we employ SSH tunnels as secure connections between data sources (where PMU data is replayed) and datacenters (where PMU data are processed). The SSH tunnels add less than 2 ms latency (less than 1%) to Round-Trip Time (RTT) on average, relative to unencrypted TCP. Based on these measurements, the authors consider SSH tunnels to be a cost-effective solution for securing the external (to the datacenters) communications.

Table 20.1 One-way latency between DataSources and Linear State Estimator at Oregon with and w/o VPC enabled

Latency (DataSource-LSE)	Without VPC (ms)	With VPC (ms)
Average	245	261
1st Percentile	211	228
99th Percentile	255	270

GridCloud also needs to be protected from other users running in a Public Cloud environment. To address this issue, GridCloud utilizes Amazon's Virtual Private Cloud (VPC). Relative to the basic AWS Elastic Computing Cloud (EC2) service, VPC offers better isolation guarantees, meaning that other applications running in the same physical machines cannot deduce any important information about GridCloud. Of course, this improvement does not come for free and reduces performance by a noticeable margin. Table 20.1 summarizes the results of GridCloud platform with and without VPC enabled in AWS. The team measured the latency between the data sources and the Linear State Estimation application (without considering computational cost of LSE itself). The average additional latency, when VPC is used, is 16 ms (approximately 6.5%). This overhead is considered to be acceptable by the team, since the VPC configuration greatly improves the protection of sensitive data against malicious behavior by other cloud users.

A third security concern arises when maintaining data confidentiality for information stored into persistent storage (SSD or Hard Disk). GridCloud uses AWS S3 storage with AES 256 encryption to protect against attacks of this type, and AWS, for its part, takes steps to ensure that no person with physical access to a storage device would also have a way to obtain the storage encryption keys. Experiments showed that there is no measureable performance impact when GridCloud enables the S3 encryption option.

20.3.3.2 Latency

Figure 20.6 shows three different types of latencies measured in the experiments. L1 is the one-way latency from data sources up to the point where data are available for application (Linear State Estimation in the measurements). L1 measurements are summarized at Table 20.1. In less than 300 ms, more than 99% of the data produced by PMUs is available to the LSE application.

It is useful to understand how much time is spent in the network between data sources and datacenters and inside the datacenter. L2 is the round-trip latency between the point when data enters the datacenter and when the corresponding processed data exits the datacenter. Figure 20.7 shows that average value of L2 is around 300 ms. From previous work, it is known that most of that is due to the LSE computation itself. Since L2 is measured entirely internal to each datacenter, as expected there are no major differences between the two datacenters.

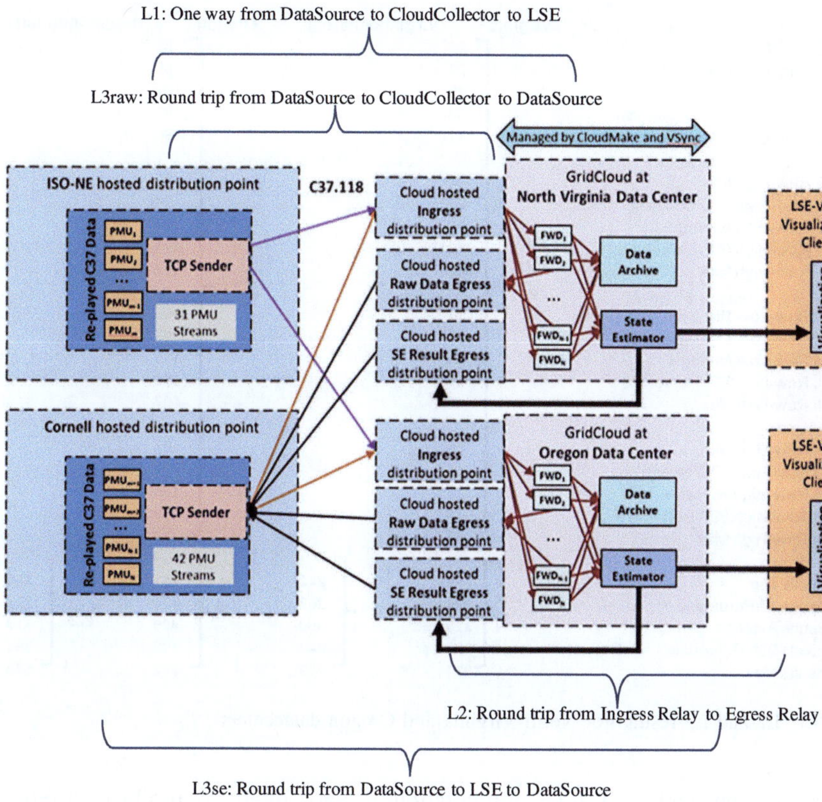

Fig. 20.6 Latencies measured in GridCloud platform

L3 is the full-path latency between the data sources and the data consumers located outside the cloud. L3 Raw is the latency for raw unprocessed data while L3 SE is the latency for LSE results. As can be seen in Fig. 20.7, Oregon latencies are approximately 60 ms larger than the Virginia ones, since Oregon datacenter is much further away from data sources (network latency is higher). Moreover, it is evident that the Linear State Estimator contributes around 270 ms of latency (comparison between L3 Raw and L3 SE). Figures 20.8 and 20.9 provide additional details on the latency distribution of L3 Raw and L3 SE.

20.3.3.3 Data Consistency and Fault Tolerance

When using standard data mirroring approaches with two different datacenters, inconsistencies can arise. GridCloud is designed to minimize this problem, but cannot fully mask it: If a failure occurs, some raw data streams might not reach one or both datacenters. Furthermore, LSE calculations are based on data that arrive *on*

		Virginia		Virginia-Internal		Oregon		Oregon-Internal
ISONE Raw-Low Min		20				88		
ISONE Raw-Low 1st Percentile		22				89		
ISONE Raw-Low Average		**25**				**102**		
ISONE Raw-Low 99th Percentile		58				152		
ISONE Raw-Low Max		611				696		
ISONE Raw-High Min		22				90		
ISONE Raw-High 1st Percentile		25				99		
ISONE Raw-High Average		**46**				**127**		
ISONE Raw-High 99th Percentile	L3 Raw	82			L3 Raw	179		
ISONE Raw-High Max		612				697		
Cornell Raw-Low Min		17				90		
Cornell Raw-Low 1st Percentile		17				91		
Cornell Raw-Low Average		**18**				**115**		
Cornell Raw-Low 99th Percentile		20				191		
Cornell Raw-Low Max		49				407		
Cornell Raw-High Min		18				91		
Cornell Raw-High 1st Percentile		18				92		
Cornell Raw-High Average		**19**				**120**		
Cornell Raw-High 99th Percentile		20				199		
Cornell Raw-High Max		49				413		
SE Results Min		279	L2	242		351		240
SE Results 1st Percentile	L3 SE	294		267	L3 SE	370		273
SE Results Average		**325**		**300**		**409**	L2	**317**
SE Results 99th Percentile		384		348		490		393
SE Results Max		911		469		962		642

Fig. 20.7 L2 and L3 results for North Virginia and Oregon datacenters

time at the datacenter. A LSE computation is very likely to produce different outputs when there are missing inputs. In this experiment, if LSEs do not have the full array of data, they do not do any computations. Thus, the authors expect to have missing LSE result points rather than incorrect results.

The results of the 25 min experiments can be seen in Table 20.2. All the raw data were delivered in both datacenters (N. Virginia, Oregon). There are some missing LSE results in both datacenters (mostly in Oregon), but the rate of data missing is very small. The team also witnessed a couple of different LSE results (between the two data centers) for the same stream and same timestamp. This should not normally happen, but seems to arise because of a long delay each time the 21-s data loop repeats: In all cases, these occurrences were seen on the first data point of a looped 21-s epoch. The team also confirmed that all data received by each data center was stored (indeed, it was confirmed that both data centers had identical data at the end of the run). This seems to suggest that the rare inconsistencies that arose in the experiments were due to some form of long delay when data sources reset for a new 21-s epoch: The same data ultimately reached both data centers, but in one, a data point arrived too late and was excluded from the LSE computation, while in the other, it was included.

Although there was no node crashes during the study, the team did experiment with complete data center shutdowns. Restart required approximately 175 s, during

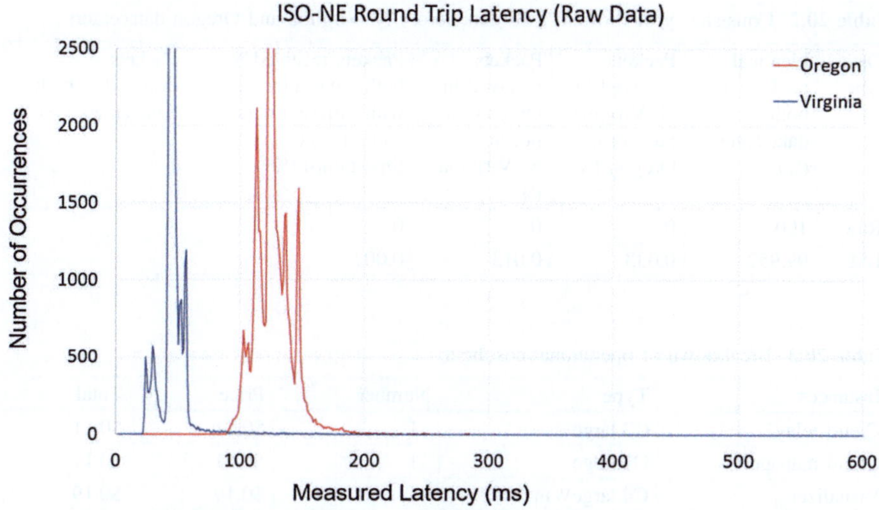

Fig. 20.8 L3 Raw latency to Oregon and Virginia datacenters

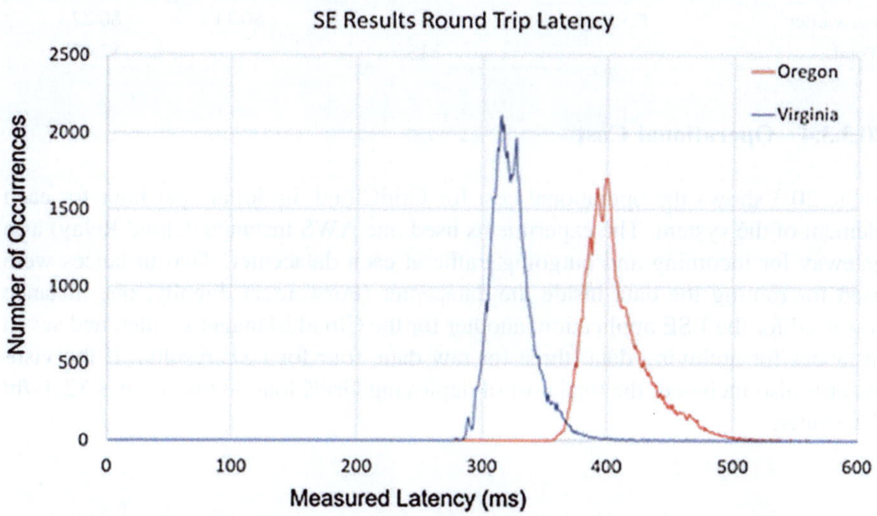

Fig. 20.9 L3 SE latency to Oregon and Virginia datacenters

which no data was lost because the Oregon data center continued to operate while the Virginia one was recovering. The vast majority of the 175 s was consumed by EC2 rebooting the Windows LSE instance. Then, the delays of reconfiguration and initial setup of the LSE instance are non-significant compared to 175 s. At the end of the 175 s period, the GridCloud system was back to full function (state estimation included) and full redundancy.

Table 20.2 Consistency of raw and LSE data across N. Virginia and Oregon datacenters

Data type	Identical packets in both datacenters (%)	Packets received in N. Virginia but not in Oregon (%)	Packets received in Oregon but not in N. Virginia (%)	Packets received at both datacenters but with different data value (same timestamp) (%)	Packets dropped at both datacenters (%)
Raw	100	0	0	0	0
LSE	99.952	0.033	0.013	0.002	0

Table 20.3 Breakdown of operational cost/hour

Instances	Type	Number	Price	Total
Cloud relay	C3.large	1	$0.11	$0.11
Cloud manager	C3.large	1	$0.11	$0.11
Visualizer	C4.largeWin	1	$0.19	$0.19
LSE	C4.xlargeWin	1	$0.39	$0.39
Raw archiver	C3.xlarge	3	$0.21	$0.63
LSE archiver	C3.xlarge	4	$0.21	$0.84
Forwarder	C3.large	2	$0.11	$0.22
Total		13		$2.47

20.3.3.4 Operational Cost

Table 20.3 shows the operational cost for GridCloud, in dollars per hour for each element of the system. The experiments used one AWS instance (Cloud Relay) as a gateway for incoming and outgoing traffic at each datacenter. Two instances were used for routing the data inside the datacenter (Forwarder). Finally, one instance was used for the LSE application, another for the Cloud Manager Leader, and seven instances for archiving data (three for raw data, four for LSE results). If the visualizer is also included, the total cost of deploying GridCloud at this scale is $2.47/h/datacenter.

20.4 Challenges and Future Research Directions

The preliminary work with GridCloud confirms that the system is a robust basis for capturing PMU data, archiving it, and for carrying out state estimation. Additional applications will be added in the future, as well tools for carrying out exploratory data analysis on the archive of past system states.

The authors see a number of challenges ahead. First, as more ISO/RTO partners are added to the system, a wide range of regulatory and security-policy questions will inevitably arise. The system is believed to be adequately secure and robust to respond to these requirements, but until such steps are carried out, it is impossible to be certain. In particular, the primary policy document for the area of work, namely the NERC CIP v5, was not designed with cloud-hosted PMU data monitoring in mind.

A second question concerns leveraging advanced security technologies, notably the Intel SGX architecture, which can protect an application even against a malicious or intrusive cloud host. With SGX, the authors believe that GridCloud could operate so securely that Amazon itself would be unable to see any of the data captured or archived within it, even by freezing and examining the states of compute nodes, disks, or carrying out other forms of attack on the system. However, to be fully convinced that this is the case, the authors would need to carefully evaluate the exact manner in which SGX was employed, to confirm that the configuration itself is correct and trustworthy. Further, to be practical, the performance and cost impact would need to be measured and determined to be within an acceptable range. SGX is just one of several options that might be considered.

A third question involves scalability of the platform with growing numbers of users and growing numbers of PMU data streams. Although the preliminary experiments suggest that GridCloud could scale to a deployment that would instrument and monitor the entire North American continent, careful monitoring of performance and costs will be required as the system expands over time.

Finally, the project team is interested in the concept of GridCloud as a "big data" platform for the bulk power community, offering tools aimed at better understanding power management, improving operational efficiencies, and tracking down the sources of disruptions or instabilities. Such tools could leverage some of the existing technologies available in today's cloud, but would need to also go beyond today's most common solutions simply because power grid analytics are unique in many ways: the real-time nature of the problem, the specific problem formulations that are of interest, and the kinds of analysis required, all depart from anything seen in common web hosting platforms or big data systems used to understand human behavior, advertising placement, or other questions of interest to the primary cloud community today. With further work, considerable progress can certainly be made on this question.

References

1. U.S.-Canada power system outage task force final report on the August 14, 2003 blackout in the United States and Canada: causes and recommendations. https://www.npcc.org/Library/Blackout%20Recommendations/BIT_Conclusions.pdf
2. Conclusions of the NPCC 2003 blackout investigative team assessing the power system collapse of August 14, 2003. https://www.npcc.org/Library/Blackout%20Recommendations/BIT_Conclusions.pdf
3. Smart Grid Investment Grant (SGIG) program final report. https://www.smartgrid.gov/document/us_doe_office_electricity_delivery_and_energy_reliability_sgig_final_report.html

4. The NIST definition of cloud computing. https://www.nist.gov/sites/default/files/documents/itl/cloud/cloud-def-v15.pdf
5. Cloud computing. https://en.wikipedia.org/wiki/Cloud_computing
6. Zhang Q, Cheng L, Boutaba R (2010) Cloud computing: state-of-the-art and research challenges. J Internet Serv Appl 1:7–18. https://doi.org/10.1007/s13174-010-0007-6
7. www.amazon.com
8. Anati I et al (2013) Innovative technology for CPU based attestation and sealing. In Proceedings of the 2nd international workshop on hardware and architectural support for security and privacy, vol 13
9. Costan V, Devadas S (2016) Intel SGX Explained. IACR Cryptol ePrint Arch 2016:86
10. Birman KP (2012) Guide to reliable distributed systems. Building high-assurance applications and cloud-hosted services. Texts in computer science. Springer
11. Shvachko K et al (2010) The hadoop distributed file system. In: 2010 IEEE 26th symposium on mass storage systems and technologies (MSST)
12. Weil SA et al (2006) Ceph: a scalable, high-performance distributed file system. In: Proceedings of the 7th symposium on operating systems design and implementation. USENIX Association
13. Song W et al (2016) The freeze-frame file system. In: Proceedings of the ACM symposium on cloud computing (SoCC'16)
14. Chandy KM, Lamport L (1985) Distributed snapshots: determining global states of distributed systems. ACM Trans Comput Syst (TOCS) 3(1):63–75
15. Tao Y, Sun H, Bose A (2009) Two-level PMU-based linear state estimator. In: 2009 IEEE/PES power systems conference and exposition, Seattle, WA, pp 1–6

Index

S. Nuthalapati (ed.), *Power System Grid Operation Using Synchrophasor Technology*, Power Electronics and Power Systems,
https://doi.org/10.1007/978-3-319-89378-5

MIX
Papier aus verantwortungsvollen Quellen
Paper from responsible sources
FSC® C105338

If you have any concerns about our products,
you can contact us on
ProductSafety@springernature.com

In case Publisher is established outside the EU,
the EU authorized representative is:
Springer Nature Customer Service Center GmbH
Europaplatz 3, 69115 Heidelberg, Germany

Printed by Libri Plureos GmbH
in Hamburg, Germany